Weak Interactions as Probes of Unification
(Virginia Polytechnic Institute – 1980)

AIP Conference Proceedings
Series Editor: Hugh C. Wolfe
Number 72
Particles and Fields Subseries No. 23

Weak Interactions as Probes of Unification
(Virginia Polytechnic Institute – 1980)

Editors
G. B. Collins, L. N. Chang and J. R. Ficenec
Virginia Polytechnic Institute

American Institute of Physics
New York   1981

Copying fees: The code at the bottom of the first page of each article in this volume gives the fee for each copy of the article made beyond the free copying permitted under the 1978 US Copyright Law. (See also the statement following "Copyright" below). This fee can be paid to the American Institute of Physics through the Copyright Clearance Center, Inc., Box 765, Schenectady, N.Y. 12301.

Copyright © 1981 American Institute of Physics

Individual readers of this volume and non-profit libraries, acting for them, are permitted to make fair use of the material in it, such as copying an article for use in teaching or research. Permission is granted to quote from this volume in scientific work with the customary acknowledgment of the source. To reprint a figure, table or other excerpt requires the consent of one of the original authors and notification to AIP. Republication or systematic or multiple reproduction of any material in this volume is permitted only under license from AIP. Address inquiries to Series Editor, AIP Conference Proceedings, AIP.

L.C. Catalog Card No. 81-67184
ISBN 0-88318-171-1
DOE CONF- 801244

Foreword

A three-day workshop on "Weak Interactions as Probes of Unification" was arranged for December 4-6, 1980 in connection with a visit to VPI by Professor Chou Kuang-chao of the Institute for Theoretical Physics in Peking. The Local Organizing Committee (listed below) was augmented by Professors J. D. Bjorken, T. D. Lee and C. N. Yang. An attempt was made to recapture the relaxed atmosphere of the early "Rochester" conferences by limiting the number of participants and by making ample provision in the program for informal comments and discussion. Indeed, the last session was devoted exclusively to a presentation of experimental and theoretical reactions to and speculations on the general theme of the workshop.

The original plans did not call for any proceedings. However, the exchange of views proved to be so stimulating that the participants voted for the preparation of Workshop Proceedings in the A.I.P. book format. The Editors (listed on the title page) with the assistance of David Scherer and Gordon Pusch are to be congratulated for putting together such a useful volume in so short a time. The management of the workshop and the production of these proceedings were in the hands of Janet Manning and other members of the VPI secretarial staff, to all of whom thanks are due.

The financial support of the Department of Energy and VPI is gratefully acknowledged.

R. E. Marshak
(for the Local Organizing Committee: A. Abashian, M. Blecher, L. N. Chang, K. C. Chou, G. B. Collins, J. Ficenec, K. Gotow and L. Mo)

TABLE OF CONTENTS

I. First Generation Problems – C. N. Yang, Chairman

   A. Leptonic Tests of Electroweak Models

   1. Neutrino Mass
      C. S. Wu 1

   2. Dirac vs Majorana Neutrinos: Low Energy Tests
      Riazuddin 21

   3. Proposed Ultrasensitive Investigation of
      Neutrinoless and Two-Neutrino Double Beta
      Decay of ^{76}Ge: A Progress Report
      F. T. Avignone, III et al. 34

      Discussion

   4. Possible Deviations From (V-A) Charged Currents:
      Precise Measurement of Muon Decay Parameters
      M. Strovink 46

      Commentator: J. D. Bowman 71
      Discussion

   5. A Planned Experiment on Parity Violation for
      Atomic Hydrogen
      V. Hughes 78

   B. Baryon and Lepton Number Non-Conservation

   1. Probing the Design of Grand Unification
      Through Conservation Laws
      J. Pati 84

      Discussion

   2. SU(5) Predictions: τ_p, $\sin^2\theta_W^{exp.}$, m_W and m_Z
      W. J. Marciano et al. 107

      Discussion

   3. Nucleon Decay Experiments
      K. Lande et al. 128

      Discussion

4. On Neutron Oscillations
   N. P. Chang 142

   Discussion

5. A Proposed Experiment for a Sensitive Search
   for $\Delta B = 2$ Transitions Via n-$\bar{n}$ Mixing
   G. R. Young et al..................................159

   Commentators: H. L. Anderson......................176
                  S. Ratti...........................183
                  T. Kamae 189
   Discussion

6. CP Violation and the Development of Cosmological Baryon Asymmetry
   G. Senjanovic 192

   Commentator: M. Turner...........................224
   Discussion

7. Lepton Number Non-Conservation
   R. D. Peccei......................................244

   Discussion

C. <u>Weak Scalar Boson and Vector Boson Phenomenology</u>

   1. Current Phenomenological Status of Higgs
      Physics and Technicolor
      G. L. Kane.....................................257

      Commentator: L. L. Wang 264
      Discussion

   2. Axion
      S.-H. H. Tye...................................270

      Discussion

   3. Search for "Axion-Like" Particles
      L. W. Mo 277

   4. W's and Z's
      L. Trueman.....................................285

      Discussion

II. Multi-Generation Problems - L. N. Chang, Chairman

   A. Leptons

      1. Neutrino Oscillations
         V. Barger..309

            Commentators: T. Goldman.........................319
                           P. Rosen..........................320
         Discussion

      2. Solar Neutrino Experiments and Neutrino
         Oscillations
         R. Davis, Jr. et al.................................322

      3. Neutrinos and Cosmology
         M. Turner...335

         Discussion

      4. Rare Muon Processes
         R. Mischke..355

         Discussion

      5. Implications of Neutrino Masses and Mixing
         for Weak Processes
         R. E. Shrock 368

         Discussion

      6. Rare Kaon Processes
         P. Rosen 407

         Discussion

   B. Quarks

      1. CP Violation
         L. Wolfenstein......................................413

         Discussion

      2. Phenomenology of CP Violation from the
         Kobayashi-Maskawa Model
         L. L. Wang..419

         Discussion

3. B Meson Physics at CESR
   W. A. Loomis 432

4. Heavy Quark Decays
   N. Cabibbo 445

   Discussion

C. <u>Multi-Generational Models of Quarks and Leptons</u>

1. Horizontal Symmetries
   K. C. Wali 455

2. Calculable Masses in GUTS: An E_6 Example
   P. Ramond 467

   Discussion

3. Extended Unification Models
   A. Zee 475

   Discussion

4. Beyond Grand Unification
   E. C. G. Sudarshan et al. 481

   Discussion

5. Gauge Hierarchy and Decoupling
   Y. P. Yao 494

6. Physical Implications of Dynamical Symmetry Breaking
   M. A. B. Bég 505

7. Some Ideas on Composite Quarks and Leptons
   N. Srednicki 515

   Discussion

   Commentators: H. S. Tsao 523
                 K. Huang 528
   Discussion

8. Composite Models
   S. Adler 534

   Discussion

   Commentators: L. C. Biedenharn et al. 553
                 O. W. Greenberg et al. 564
                 M. Yasue 574
                 F. Mansouri 594

III. Comments and Speculations, T. D. Lee, Chairman

    A. Presentations

        1. Review of New Experimental Upper Limits on Forbidden Decay Modes of the Tau Lepton
    M. L. Perl et al. 602

        2. Near-Term Prospects for Information on b Decay
    E. Thorndike 614

        3. Axial U(1) Anomaly and Chiral Symmetry-Breaking in QCD
    K. C. Chou 621

        4. Mass Spectrum Above 40 GeV
    M. Veltman 627

        5. Strings and Vortices
    Y. Nambu 633

        6. Exceptional Groups for GUTs and Remarks on Octonions
    F. Gursey 635

        7. Left-Right Symmetry, Composite Models and Baryon Number of Universe
    R. N. Mohapatra 647

        8. γ_5 Invariance, B-L Symmetry and Naturalness
    R. E. Marshak 665

    B. Panel Discussion 669

    C. Audience Discussion 679

IV. Participants 685

FIRST GENERATION PROBLEMS
Leptonic Tests of Electroweak Models

Neutrino Mass

C. S. Wu
Columbia University, New York, NY

Introduction:

The rest mass of the neutrino was considered to be vanishingly small even when the hypothesis of the neutrino in β-decay was first proposed. When Pauli presented his view of the neutrino[1] for the first time at the Solvay Congress (1933), he said: 'With regard to the properties of these neutral particles, we first learn from atomic weights that their mass cannot be much larger than that of the electron'. F. Perrin was also present at the discussion. Immediately following the Solvay Congress, he examined the shape of the continuous β-spectrum and concluded, even at that early stage, that the neutrino had a zero intrinsic mass like the photon[2].

Since then, all the experimental evidence indicates that the rest mass of the neutrino is very small and may, in fact, be zero. Hence, for the sake of simplicity it is justifiable to neglect it in calculations in classical β-decay

However, in the present theory of β-decay and other weak interactions, the question about the rest mass of the neutrino assumes a rather important position:

(I) It is known that according to the two-component neutrino theory,[3] if lepton number is conserved, then the rest mass of the neutrino is expected to be identically <u>zero</u>.

(II) Secondly, it has been experimentally established that the neutrino involved in β-decay and that in μ-decay are two different particles[4]. The newly discovered Tau neutrino ν_τ,[5] is also distinguishly

ISSN:0094-243X/81/720001-20$1.50 Copyright 1981 American Institute of Physics

different. Naturally, one is anxious to find out what properties and characteristics differentiate these light neutral particles? Could be their masses?

(III) A non-vanishing mass for one or more of the neutrinos has many indirect consequences because it leads to the possibility of a Cabibbo-like mixing for the neutrinos and to the possibility of neutrino decays or oscillations. The hypothesis[6] of mixing of two neutrinos (Majorana as well as Dirac) with masses m_1 and m_2 is not in contradiction with the existing data if the mass difference $|m_1^2 - m_2^2| \lesssim 1(ev)^2$.

(IV) The present limits on the three neutrino masses from direct experiments reported in published journals are:

$m(\nu_e)$ [7]	$m(\nu_\mu)$ [8]	$m(\nu_\tau)$ [9]
14ev< <46ev	<0.52Mev	<250Mev

<u>The Mass of Electron Neutrino, $m(\nu_e)$</u>:

The determination of the mass difference $|m_1^2 - m_2^2|$ depends on the assumed mixing angle θ and the analyzed results of neutrino oscillatron experiments under various constraint conditions. In this talk, we will have time to discuss only the determination of the mass of the electron neutrino $m(\nu_e)$.

There are two indirect methods of measuring the neutrino masses:
(I) One is based on the measurement of nuclear recoils[10] in β-decays or K-electron captures; for $m(\nu_\mu)$, in pion decay in rest or in flight. The principle of the method is to determine the relation between the missing energy and momentum, which are presumably carried away by the neutrino. Because of the difficulties inherent in this method, i.e., preparing a thin recoil surface and detecting and measuring the spectrum of recoil

nuclei, it is not easy to attain high precision. Nevertheless, the relations between the missing energy and momentum of the neutrino in β-decay such as in ^{32}P or ^{37}A, etc... illustrate the simple kinematic relatrons which can be related by the Equation: $E = pc$ as that in the case of photon.

(II) The other method is the investigation of the detailed shape of the β-spectrum in the vicinity of the upper limit.[7][11] This method promises to give a more precise measurement of the neutrino rest mass.

The precise shape of the allowed β-spectrum can be expressed by

$$N(p_i) = A \sum_K S(p_k, m_\nu, E_0) \{1 + \alpha(p_0 - p_k)\} R_{ik} + B \tag{1}$$

where A - Normalization

m_ν - Neutrino mass

E,p - Electron Energy, Momentum.

α - Correction for detection efficiency.

B - Background

$R_{i,k}$ - Resolution function.

S - Spectrum Shape

For an allowed β-spectrum, the shape "S" is given by the statistical factor[12]. However, in this study, the following expression was used.[11][7]

$$S(p_k, m_\nu, E_0) = F(E)|M(m_\nu)|^2 p^2 \{W_1(E_0 - E)\sqrt{(E_0 - E)^2 - m_\nu^2} + W_2(E_0 - E^* - E)\sqrt{(E_0 - E^* - E)^2 - m_\nu^2}\} \tag{2}$$

where

$$|M(m_\nu)|^2 \propto 1 + a\frac{m_e m_\nu}{E_\beta E_\nu}$$

"a" is known as the relativistic spinor term; it depends on the β-interactions assumed. After the discovery of parity violation, "a" became identically zero. In general, there are atomic effects expressed in the Coulomb Factor F(Z,E). It is very small in this special case of ^{3}H decay. However, for such precision work, as determination of m_ν, one has to take into account another atomic effect which had been heretofore completely neglected until in Bergkvist's study of Tritium β-spectrum[11]. It is known that in the β-decay of a radioactive atom, the sudden change in the atomic Hamiltonian leads to a distribution in energy among the daughter atomic states. By virtue of over-all energy conservation in each individual decay event, there will be a corresponding distribution in the decay energy for the β-process. In the tritium β-decay, if the ^{3}He is a free atom, 70% of the decay will go to the 1s ground state of the ^{3}He$^+$ daughter ion. The remaining 30% will go to higher s-states of the He$^+$ ion, with a dominating contribution of 25% to the 2s-state. The energy difference between the ^{3}He$^+$ 1s and 2s - states is 40.5 ev. The tritium β-spectrum will hence contain one main branch of 70% of intensity, one weaker one at 25% of intensity and with its end-point Energy E_o at 40.5eV less. The remaining 5% of the β will go to slightly smaller end-pt energies with gradually decreasing intensities. No de-excitations to the continuum region were taken into account in these analyses.

The effect of the spread in end-point energy on the shape of the measured β-spectrum will be equivalent with that of measuring a well-defined spectrum with a modified line shape resulting from the combined effect of finite experimental resolution and distribution in end-point energy due to atomic effects in the tritium β-decay (Fig. 1). In both Lyubimov[7] and Bergkvist[11] cases, for simplicity, they have assumed $W_1 + W_2 = 1$; $W_1 = 0.7$; $W_2 = 0.3$; $E^* = 43$ev.

In order to see how the rest mass of neutrino can be determined from the detailed shape of the β-spectrum in the vicinity of the upper limit, Fig. (2) illustrates the influence of a finite neutrino mass on the Kurie plot of an allowed β-spectrum. The sensitivity of the influence of $m(\nu_e)$ on the spectrum shape depends on several factors, the energy resolution ΔE, statistics $N^{-1/2}$, particularly on the Background B^2. The modification of the measured spectrum shape will be sufficiently pronounced only when the energy resolution width ΔE is of roughly the same magnitude of the $m(\nu_e)$. See the comparison between Figs. 3 and 4. On the other hand, the counting rate is proportional to $N \propto (\Delta E)^{5.5}$. Therefore the basic intensity conditions must be improved in a dramatic way by designing magnetic spectrometer with high luminosity in order to have sufficient counting rates and adequate energy resolution. It is also very important to reduce the background "B" in the upper energy region. The uncertainty introduced by the large fluctuations in the background will statistically affect the conclusion on the rest mass of neutrino inclusive(see Fig. 5).

There are generally two different types of instrumentations used for the measurements of β-spectra. One is the magnetic spectrometer and the other is the implantation in Si(Li) detector. The characteristics and merits are listed in Table 1. The first type gives very high resolution but the second type gives nearly 100% efficiency and records the whole spectrum at once. The particular desirability of the Si(Li) detector is that no final atomic structure correction required. However, the energy resolution is limited by the statistic factor of the ion-pair in Si(Li) $\left(\frac{18.6 \times 10^3}{3}\right)^{-1/2}$ which gives a value $\simeq 250$ev at best.

The study of the rest mass of $m(\nu_e)$ by investigating the shape of the tritium spectrum initiated from fifties. In the period from fifties to seventies, the results on $m(\nu_e)$ were limited to < 250ev - 380ev. The most well-known case was Langer and Moffat's value <250 ev.(See Table 2). Nevertheless, Bergkrist devoted several years to improve the luminosity of an electromagnetic spectrometer and finally in 1972[11] obtained a value of $m(\nu_e)$ < 55ev with 90% Confidential level. This value is nearly an order of magnitude smaller than what was published heretofore. Furthermore, Bergkrist called attention to take into account of the final atomic structure correction in ^{3}He.

The Determination of Electron Neutrino mass $m(\nu_e)$ at ITEP:

In 1976, the experimental β-group at ITEP reported their precision measurements of $m(\nu_e)$ at the International Neutrino Conference, Aachen and gave a mass limit $m(\nu_e) \leq$ 35ev. with 90% confidential level. In July, 1980, another experimental paper from ITEP was published on the estimate of the $m(\nu_e)$[7] of 14ev $\leq m(\nu_e) \leq$ 46ev with 99% C.L. This recent result, their C.L. is greatly raised and they gave a non-vanishing rest mass of neutrino between two limits.

Experimental Arrangements

The β-spectrometer used at ITEP is an ironless thorodial β-spectrometer with rectangular thorodial coil as shown in Fig.(6). The total magnetic deflection angle is of 4 x 180° = 720°. It permits to reach a high resolution ≃ 45ev at FWHM at the end of the tritium β-spectrum(18.6keV) as well as a low background (0.03 to 0.1 counts/s). The tritium source was prepared from a standard ^{3}H tagged valine ($C_5H_{11}NO_2$) containing 18% tritium. The thickness of the source was ≃ 2μg/cm^2 of valine. The calibration of the spectrometer was made using conversion lines of ^{169}Yb (L, M, N series of 20.734 keV and 63.117keV) in the energy range 10-60keV.

Three of the lines were in the energy range studied (18.427, 18.644, 18.749 kev).

Results

The Kurie Plot of the measured spectrum of the last eight tritium samples is shown in Fig.(7). It illustrates the statistical significance of the data and as well as the differences between the expected spectra of different values of the neutrino mass.

The detailed analysis of the β-spectrum and the applications of atomic structure corrections described in ITEP papers$^{(7,\ )}$ were very close to that discussed in Bergquist's pioneer paper$^{(11)}$ on the precision measurements of $m(\nu_e)$ from the β-spectrum of ^{3}H. In the most latest work on tritium β-spectrum$^{(7)}$ by ITEP, the χ^2 least fittings (see Fig. 8) were done for each measured data of the 16 samples they used in the last several years. The free parameters were m_ν, E_0, A, α and β. The most probable values of m_ν are also shown in Fig.(8). One clearly sees a peak with a mean value $m(\nu_e)$ = 34.3±4ev.

This measurement is undeniably a meritoriously planned and executed experiment. When such a delicate experiment is dedicated to conclude a crucial and ultimate qestion whether the mass of the elementary particle is identically zero or small but non-vanishing; many small and inconspicuous effects which may affect the results must be going over with detailed scrutinizing. The greatest uncertainties encountered in ITEP analysis were the unknown atomic structure of the ^{3}He$^+$ implanted inside of the valine. The authors pointed out that it could be very different from that of free ^{3}He atom but they had no reliable knowledge of it. For this reason, they made minimum χ^2 fittings of the experimental results by assuming either <u>two</u> level structure of ^{3}He$^+$ (W_2 = 0.3; E* = 43ev) or <u>one</u> level structure of $^3\text{He}^+$; (W_1(or W_2) = 1) to cover the whole range of uncertainties arising from the chemical composition of the source. They also treated the data shown in histogram in Fig.8 either with the shaded area or without (see Fig.9). The mass limit thus obtained was

$$14\text{ev} \leq m_\nu \leq 46\text{ev}$$

with a confidence level equal to 99%. Because such large variation in the theoretical assumptions on the $^3\text{He}^+$ atomic structure still retain the neutrino mass within the range of the non-zero mass; the authors considered this was an indication of a non-zero neutrino mass. It was also seriously questioned by Thompson[14] about the published paper[7] whether the excitation processes to the continuum region which had been neglected in ITEP analysis should be taken into account. It turns out the excitation to the continuum region plays a negligible importance to this question. On the other hand, fabricating of the tritium source is also a very difficult problem. Although tritium could be impregnated into organic material like Valine and many others. A very small amount of ^3H near the surface keeps on emanating into the evacuated spectrometer vessel and contaminating the background continuously. In the ITEP experiment, the background increased by a factor of four in one month which may raise some questions concerning the proper ways to correct for the background. What was the slope of the background? Did they investigate it as the contamination was building up? Furthermore, Valine is an insulating material if it is mounted in an evacuated container but is not well electrically grounded; then, under an extended long period of β-emission from the β-source, the source may produce a small electric potential which could affect the energy distribution of the β-spectrum. Such effects in a plastic detector have been reported[15].

It is truly a great challenge to undertake such an investigation and such a precision determination. The ITEP group and Dr. Bergquist have dedicated many long years of hard work to pioneer in this important study and have shown that the result could turn out to be most exciting and rewarding. However, in order to make the measurements absolutely certain and decisive, the questions raised above must all be satisfactorily answered soon. Experimental and theoretical investigators who have the interests, experience, resources and perseverance should join in the investigation and determination as the stakes are so tantalizing and so rewarding!

REFERENCES

1. W. Pauli, Noyaux Atomiques, in Proceedings of the Solvay Congress, Brussels, 1933(Gauthier-Villars, Paris, 1934), p.324.

2. F. Perrin, Compt. Rend. 197(1933) 1625.

3. T.D. Lee and C.N. Yang, Phys. Rev. 105, (1957), 1671.
    L. Landau, Nucl. Phys. 3 (1957), 127.
    A. Salam, Nuovo Cimento 5 (1957) 299.

4. Danby, G., et.al. Phys. Rev. Lett. 9 (1962), 36.

5. M.L. Perl et.al. Phys. Lett. 63B(1976)466; G.J. Feldman et.al. Phys. Rev. Lett. 38 117 (1977).

6. S.M. Bilenky and B. Pontecorvo, Phys. Rep. "Lepton Mixing and Neutrino Oscillations". May, 1978. North-Holland Publishing Company.

7. V.A. Lubimov et.al. Phys. Lett. 94B (1980) 266.

8. D.C. Lu et.al. Phys. Rev. Lett. 45 (1980), 1066.

9. from recent reports from SLAC or the Mini-Conf. on neutrino mass at Telemark, Wisconsin, Oct. '80.

10. O. Kofoed-Hansen, Alpha-Beta and Gamma-Ray Spectroscopy, Edited by Siegbahn., p. 1400.

11. K. Bergkvist, Nuclear Physics, B39 (1972), 317; B39, (1972) 371.

12. C.S. Wu, Alpha, Beta and Gamma-Ray Spectroscopy, Edited by Siegbahn, p. 1393.

13. E.T. Tretyakov et.al. Proceedings of the Int'l Neutrino Conf. Aachen 1976, p. 663.

14. W.J. Thompson, Univ. of North Carolina. Preprint "Atomic Effects in Estimating the Neutrino Mass from the β-spectrum of Tritium". Recent estimates of the electron anti-neutrino mass from the beta-decay spectrum of tritium are shown to be inconclusive because excitation of continuum states of ^{3}He was ignored. However, the effect due to excitation of continuum state of ^{3}He should be negligible.

15. S.T. Hung et.al. Phys. Rev. C 14 (1976) 1162.

Table 1
Two Different Instrumentations Used

for the Measurements of the

β-Spectrum

Magnetic Spectrometer	Implantation in Si(Li) Detector
Good resolution* (50eV)	Worse (≃ 300eV)
Efficiency (very low)	Efficiency (100%)
Data accumulation (Step wise operation)	Whose spectrum recorded at once. This is desirable.
The final atomic structure correction.	All de-excite to the ground state of ^{3}He.
Present result: I.T.E.P. $14eV < m_{\nu_e} > 46eV$ 99% C.L. in Valine molecule	Present report: Univ. of Guelph. J. Simpson. $m_{\nu_e} < 60eV$ 90% C.L.

Table 2

Historical Records on the Measurements

of m_{ν_e}

Year	m_{ν_e}	Methods	Reference
1952	< 250eV	Magnetic Spectrometer	Langer and Moffat P.R. 88-689
1969	< 300-500eV	"	Davis and St-Pierre N.P. A138-545
1969	≈320eV	Electrostatic Integral Spectrometer	Salgo and Staub N.P. A138-417
1970	380eV	Si(Li) Detector	Lewis, N.P. A151-120
1972	< 55eV 90% C.L.	Magnetic	Bergkvist, K., N.P. B39,317
1976	< 35eV 90% C.L.	"	Tretyekov et.al. P.I.N.C. Aachen 1976 p663
1980	14 < > 46 99% C.L.	"	Lyubimov et.al. Phy. Lett. 94B,266
1980	< 60eV	Si(Li) Detector	J. Simpson. Wisconsin Mini-Conf. on Neutrino Mass

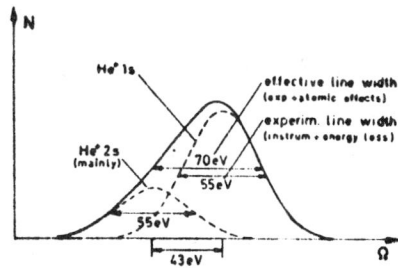

Fig. 1 Effective line shape resulting from the combined effect of experimental resolution and distribution in end-point energy due to atomic effects in the tritium β-decay. Ref 11

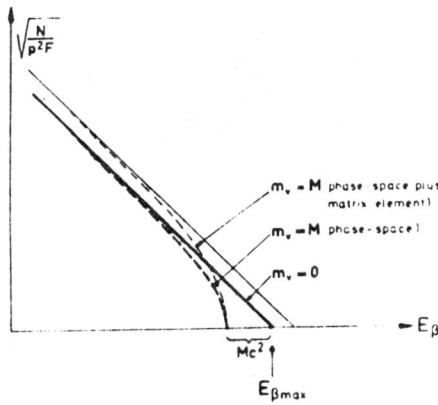

Fig. 2 To illustrate the influence of a finite $m_{\nu e}$ on the Kurie plot of an allowed β-spectrum. Ref 11

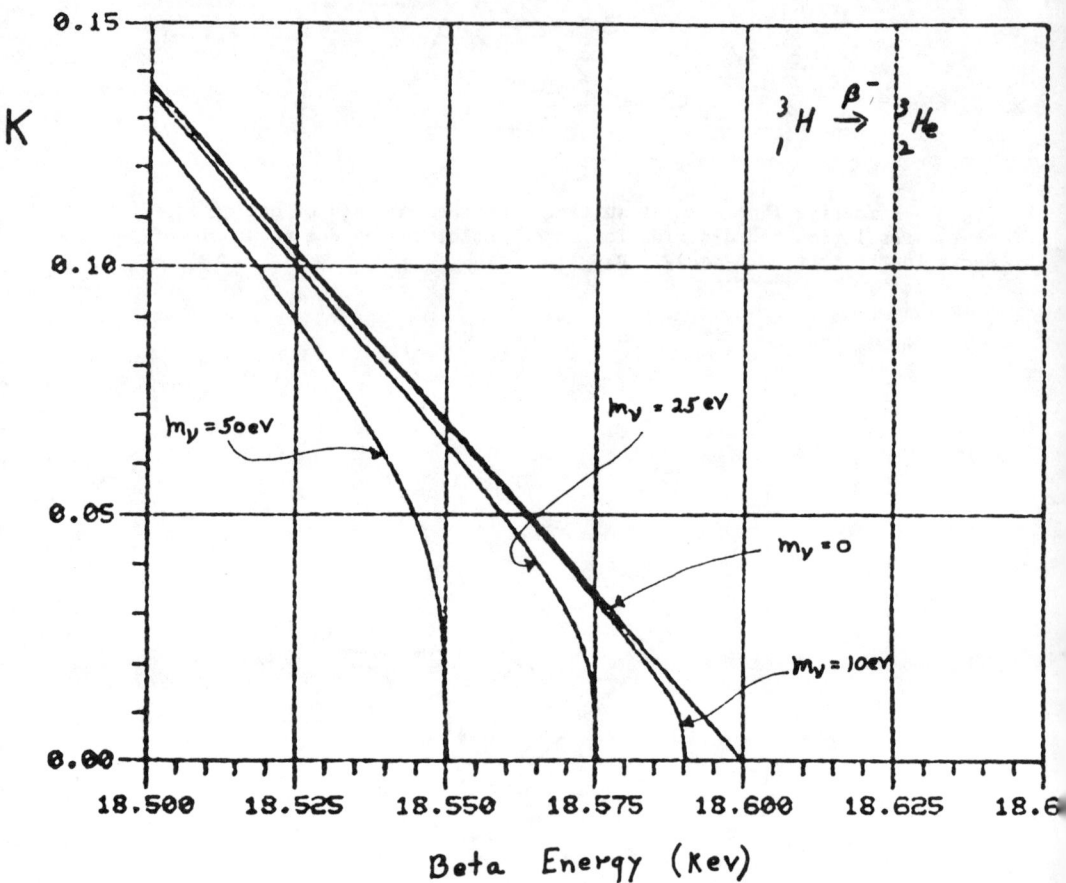

Fig.3 Kurie plots for the beta spectrum of tritium. Plots are shown for different neutrino masses as indicated. The resolution of the beta detector is assumed to be $\Delta E(FWHM) = $ zero; that is, there are no effects due to resolution.

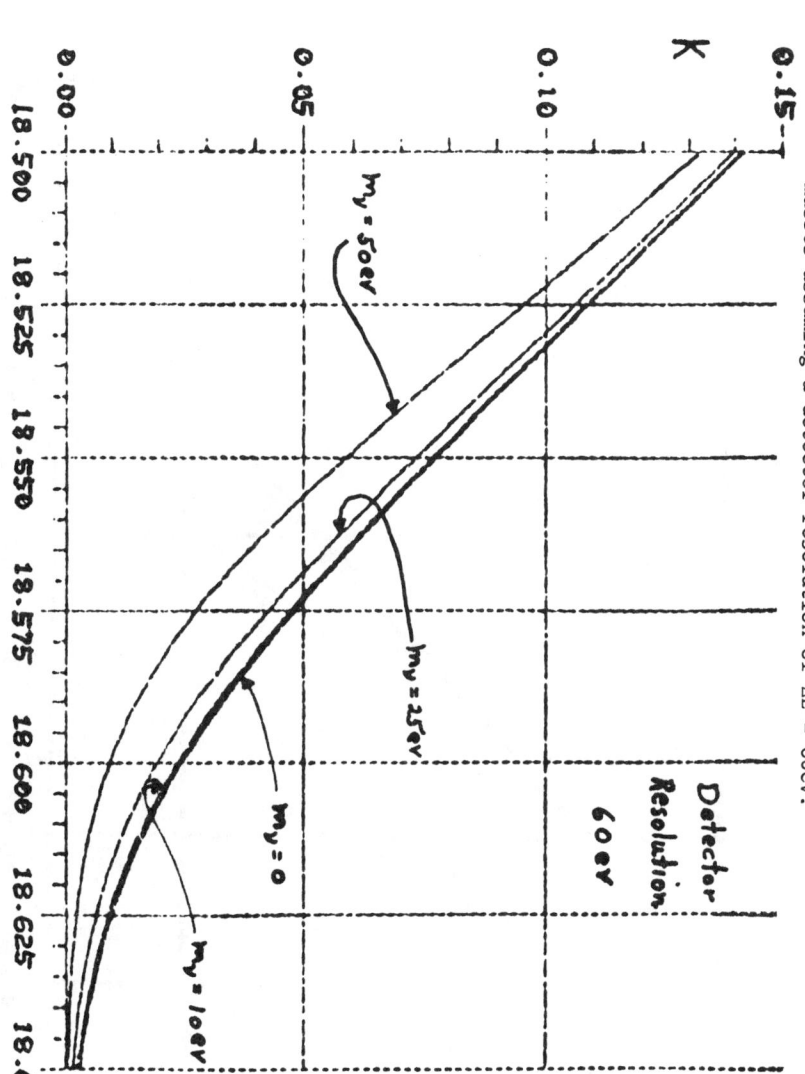

Fig. 4 Kurie plots of the beta spectra of tritium of different neutrino masses assuming a detector resolution of ΔE = 60ev.

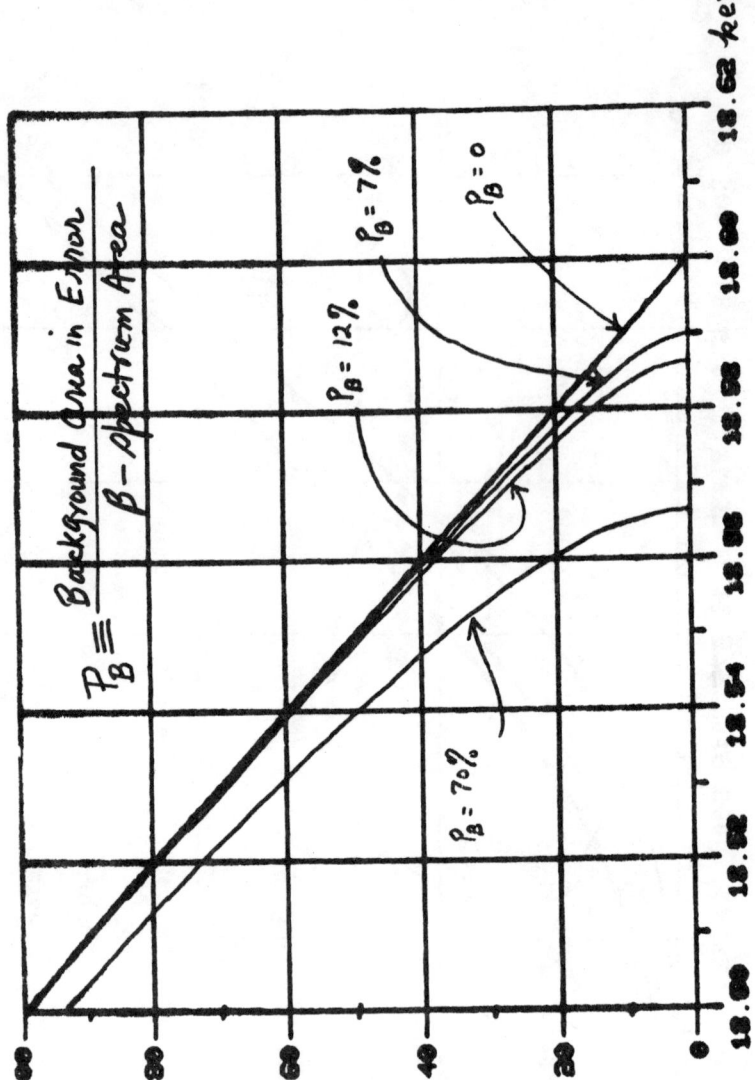

Fig. 5 To illustrate how sensitive that the shape of the β-spectrum near the upper-end is influenced by errors introduced in the corrections of the background area.

Fig.6 A rough sketch of the magnetic spectrometer used for the tritium β-spectrum experiment at ITEP.

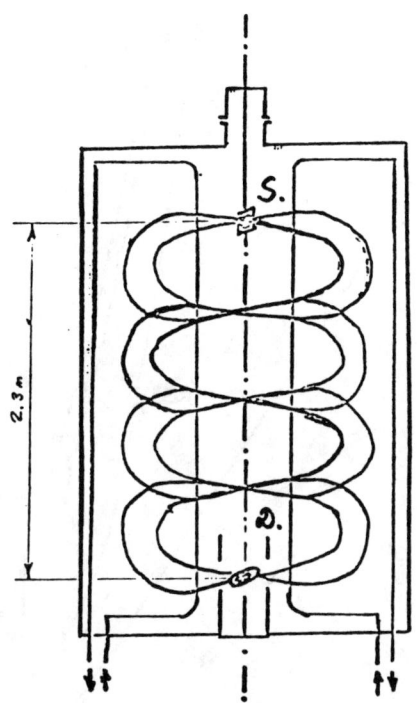

ITEP β-spectrometer

Proceedings of Wisconsin Workshop on
Neutrino Mass, Oct. 2-4, 1980
Telemark, Wisconsin.

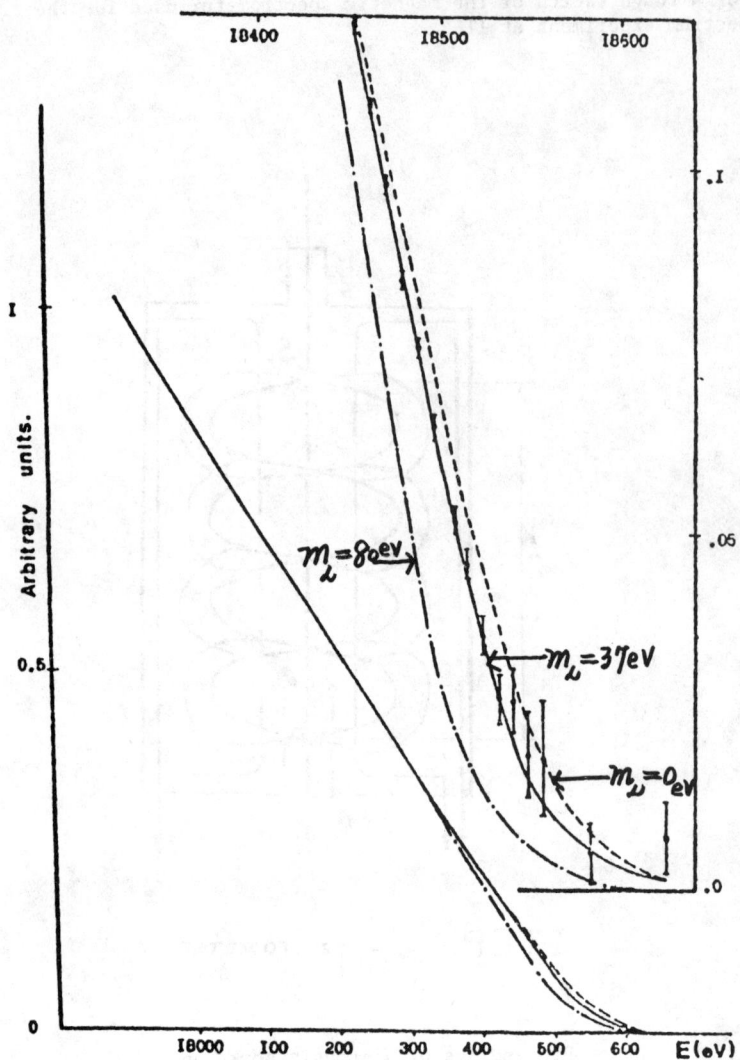

Fig. 7 The Kurie plots of β-spectra of tritium shown are:
1) The solid line represents the x^2 fit for m_ν = 37ev and E_0 = 18.578 kev;
2) The dashed line for m_ν = 0ev and E_0 = 18.574 kev;
3) The dash-dotted line for m_ν = 80ev and E_0 = 18.586 kev.   (Ref 7)

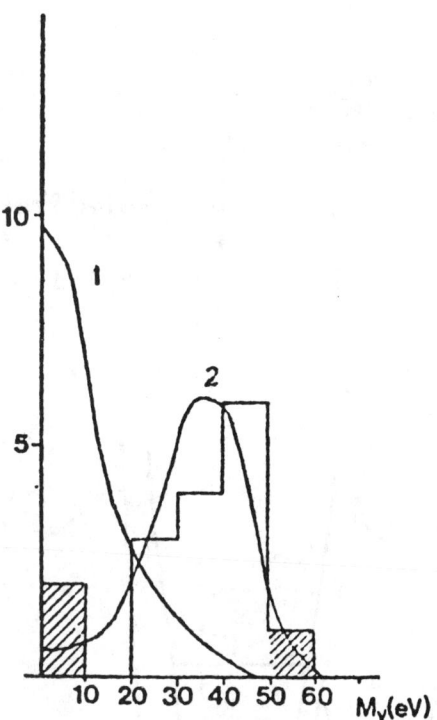

Fig. 8 Experimental histogram of m_ν values and the function $P(m_\nu/m_\nu^*)$:
curve 1 for $m_\nu^* = 0$;
curve 2 for $m_\nu^* = 35$ev.
(Ref 7)
The shaded areas are discarded.
The $P(m_\nu/m_\nu^*)$ are normarized to the experimental histogram.

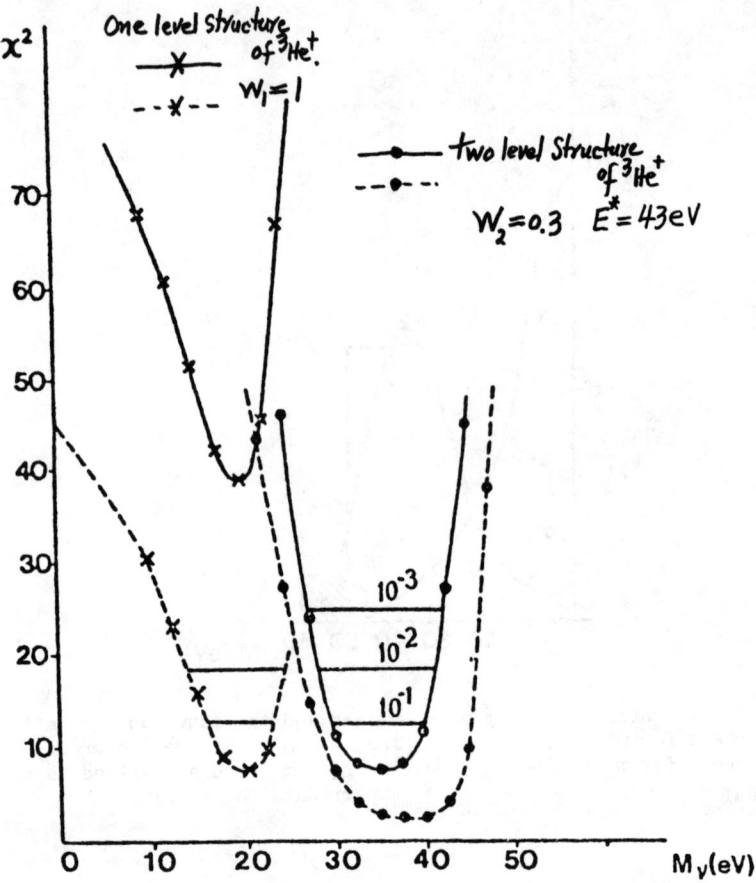

Fig. 9 The χ^2 values as functions of the parameter M_ν for the different theoretical assumptions.

Curves marked with dots are for the two-level final state structure of $^3He^+$. Curves marked with crosses are for the one-level structure of $^3He^+$.

The dashed curves correspond to the histogram of Fig. 8 when the shaded areas are discarded. The horizontal lines indicate different confidence levels. (Ref 7)

Dirac vs Majorana Neutrinos: Low Energy Tests

Riazuddin*
Virginia Polytechnic Institute, Blacksburg, VA

The masslessness of the neutrino was considered[1] by Salam, Lee and Yang, and Landau as a consequence of global γ_5- invariance. There is, however, no corresponding local gauge symmetry to guarantee the masslessness of the neutrino in contrast to the photon where both the masslessness of the photon and charge conservation are consequences of local invariance of Maxwell's equations. Similarly, the conservation of lepton and baryon numbers is not supported by any local group symmetry. One may thus expect both a finite mass of the neutrino and lepton number non-conservation at some level. The main problem is then to understand a finite but small mass of the neutrino, i.e. much smaller than the mass of the associated charged lepton. Allowing both a finite mass for the neutrino and lepton number non-conservation, we can write for an electrically neutral lepton the Lagrangian:

$$\mathcal{L} = -\bar{\psi}(\gamma_\mu \partial_\mu + m_D)\psi + \frac{m_M}{2}(\psi^T C \psi - \bar{\psi} C^{-1} \bar{\psi}^T), \qquad (1)$$

where ψ is a Dirac spinor:

$$\psi = \begin{pmatrix} \xi \\ \eta \end{pmatrix}$$

and the second term in (1) is the Majorana mass term and violates lepton number conservation.

Using a representation in which γ_5 is diagonal,

$$\xi, \eta = \frac{1 \pm \gamma_5}{2} \psi$$

by a unitary change of variables[2]:

$$\phi_1 = \frac{1}{\sqrt{2}} (\xi - i\sigma_2 \eta^*)$$

$$\phi_2 = -\frac{i}{\sqrt{2}} (\xi + i\sigma_2 \eta^*)$$

so that:

$$\phi_{1,2} \xrightarrow{C} \pm \phi_{1,2} ,$$

we can rewrite (1) as a sum of two terms:

$$\mathcal{L} = -[i\phi_1^+ \sigma_\mu \partial_\mu \phi_1 + (\frac{m_D + m_M}{2}) \phi_1^T(-i\sigma_2)\phi_1 + \text{h.c.}$$
$$+ i\phi_2^T \sigma_\mu \partial_\mu \phi_2 + (\frac{m_D - m_M}{2}) \phi_2^T(-i\sigma_2)\phi_2 + \text{h.c.})] \quad (2)$$

with $\sigma_\mu = (\underline{\sigma}, i)$. Thus, if we start with ν_L and ν_R, we can have two Majorana particles of masses $(m_D \pm m_M)$ in the presence of a Majorana mass term. If we start with ν_L only, $m_D = 0$ and we have one Majorana neutrino of mass m_M:

$$\mathcal{L} = -[i\nu_L^+ \sigma_\mu \partial_\mu \nu_L + \frac{m_M}{2}(\nu_L^T(-i\sigma_2)\nu_L + \text{h.c.})] \quad (3)$$

where $\nu_L = \frac{\phi_1 + i\phi_2}{\sqrt{2}}$.

Let us now apply the above considerations to various gauge models. We start with the minimal version[3] of $SU_L(2) \times U(1)$ at the electroweak level and $SU(5)$ at the grand unification level[4] (GUT), where in each case there is no place for ν_R. The table below summarizes the situation[5]:

Electroweak	GUT
$SU_L(2) \times U(1)$	$SU(5)$
$\begin{pmatrix} \nu \\ e \\ \nu \end{pmatrix}_L ; e_R$	Fermions belong to $\{\bar{5}\}$ & $\{10\}$
Higgs:	
$\phi(\frac{1}{2},1) \to \begin{pmatrix} 0 \\ \lambda \end{pmatrix}$	$\{5\}$, $\{24\}$
$\Delta(1,2) \to \begin{pmatrix} 0 & 0 \\ v & 0 \end{pmatrix}$	$S_{ij} = \{15\}$

gives Majorana mass term gives Majorana mass term

<u>One gets Majorana Neutrino of Mass</u>:

$m_\nu \approx (\tfrac{1}{4}|1-\rho|)^{1/2} m_e$ $m_\nu \sim h_5 \langle S_{55}\rangle$

$\rho = \dfrac{m_W^2}{m_Z^2 \cos^2\theta_W}$ $\langle S_{55}\rangle \sim \dfrac{m_W^2}{M_X} \sim 10^{-10}\text{-}10^{-11}$ GeV

$m_\nu < 35$ KeV, not a useful upper limit $m_\nu \lesssim 10^{-1}$ eV

The situation is more interesting in left-right symmetric models[6] and in SO(10)[7] at the GUT level, which admit both left-handed and right-handed neutrinos. Thus, in general, we can write the mass matrix for neutrinos as:

$$\begin{array}{c} \\ \nu_L \\ N_R \end{array} \begin{array}{cc} \nu_L & N_R \\ \begin{pmatrix} m & m_D \\ m_D & M \end{pmatrix} & \end{array} \quad (4)$$

The diagonalization gives:

$$\begin{aligned} m_\nu &\simeq m - \dfrac{m_D^2}{M} \\ m_N &\simeq M \end{aligned} \quad (5)$$

It is a general feature of these models that $m_D \approx m_\ell$ or m_q, $M \simeq m_{W_R}$, m is either much smaller than m_D^2/M or is of order $m_{W_L}^2/m_{W_R}$. The situation is summarized in the table below:

ELECTROWEAK	GUT
$SU_L(2) \times SU_R(2) \times U(1)_{B-L}$	$SO(10)$

ELECTROWEAK

$$\begin{pmatrix} \nu_L \\ e_L \end{pmatrix}, \begin{pmatrix} N_R \\ e_R \end{pmatrix}$$

$Q = I_{3L} + I_{3R} + \frac{1}{2}(B-L)$

Higgs: with associated mass scales

$\Delta_R(0,1\,2) \to \begin{pmatrix} 0 & 0 \\ v_R & 0 \end{pmatrix}$ m_{W_R}

$\Delta_L(1,0,2) \to \begin{pmatrix} 0 & 0 \\ v_L & 0 \end{pmatrix}$ $\simeq 0$

$\Phi(1/2,1/2,0) \to \begin{pmatrix} K & 0 \\ 0 & K' \end{pmatrix}$ m_{W_L}

$SU_L(2) \times SU_R(2) \times U(1)$

$\langle\Delta_R\rangle \neq 0, \langle\Delta_L\rangle = 0 \Rightarrow SU_L(2) \times U(1)$

⎡ $-\Delta I_{3R} = \frac{1}{2}\Delta(B-L)$

⎣ $\langle\Phi\rangle \neq 0$ breaks $SU_L(2) \times U(1)$ to $U(1)_{e.m.}$

↳ implies that breakdown of parity and breaking of local (B-L) symmetry are related[8] and occur at the mass scale of m_{W_R}

This scale also determines the mass of the heavy Majorana neutrino since Δ_R gives Majorana mass term leading to[11]

GUT

Fermions belong to {16}

$SU^c(4) \times SU_L(2) \times SU_R(2)$

[54] → (6   1/2   1/2); M_X is $\simeq 10^{15}$
(from analysis of[9] $\sin^2\theta_W$)

[126] → ($\overline{10}$ 0   1); $m_{W_R} \simeq 10^{13}$ GeV

(10   1   0); $\simeq \dfrac{m_{W_L}^2}{m_{W_R}} \simeq 10$

[10] → (1   1/2   1/2); m_{W_L}

$SO(10) \xrightarrow{54} SU^c(4) \times SU_L(2) \times SU_R(2)$
$\xrightarrow{126} SU^c(3) \times SU_L(2) \times U(1)$
$\xrightarrow{10} SU^c(3) \times U(1)_{e.m.}$

[126] gives Majorana mass term[10] leading to mass matrix of Eq. (4) and one obtains[12]

$m_N \simeq 10^{13}$ GeV

$m_\nu \simeq m \simeq \dfrac{m_{W_L}^2}{m_{W_R}}$

$\simeq 1$ eV

$$m_N \sim \frac{1}{a} m_{W_R}$$

$$m_\nu \simeq a\, m_e^2 / m_{W_R}$$

where a is a parameter involving the gauge coupling g and the relevant Yukawa couplings and is of order 1.

<u>Scale of m_{W_R} is not fixed by Model</u>

However, phenomenology of "low energy" neutral current weak interaction puts:

$$(m_{W_R}/m_{W_L}) \geq 3$$

so that

$$m_\nu \leq 1 \text{ eV}$$

$$m_N \geq 100 \text{ GeV}$$

One may look for "low energy" tests for Majorana neutrinos

However, even without [126] Higgs so that there are no M and m at tree level, the two loop diagrams are of form:

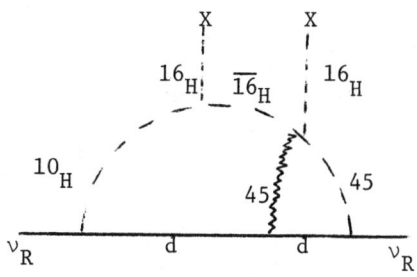

give[13]:

$m(\nu_R) \simeq 10^5$ to 10^6 GeV, although still very large but much smaller than 10^{13} GeV.

$$m(\nu_L) \sim \frac{m_q^2}{m(\nu_R)} \simeq 1 \text{ eV}$$

Hopeless to find any effect of m_{W_R} and/or m_N at "low energies".

Thus, if we adopt a dynamical or composite approach, it may be justified to stay at the electro-weak level, and it may be possible to test some of the above ideas. Below we discuss some of such tests:

(i) ν Dirac: If $N_R \equiv \nu_R$, ν could be a Dirac particle in the absence of a Majorana mass term, although it would be difficult to understand its small mass in a natural manner. This is testable in, for example, measurement of deviations of the ξ parameter in the μ decay spectrum from its (V-A) value[14]:

$$\xi \simeq (1 - 2\,\frac{m_{W_L}^4}{m_{W_R}^4})$$

If the deviations are detected, ν_R may belong to a right-handed doublet making ν a Dirac particle rather than having ν and N as two Majorana particles since ν is expected to have extremely small right-handed coupling (see below). Moreover, such deviations would put a limit on m_{W_R}. If no deviations are detected, they would not necessarily imply that m_{W_R} is very large as we shall see, but that ν (and N) may be Majorana.

(ii) ν and N Majorana particles (ν light and N heavy). In this case, the right-handed coupling of ν is of order γ where γ is the mixing angle between ν_L and N. One expects[15]:

$$\gamma \sim \frac{m_e}{m_{N_e}} \sim 10^{-5}$$

Thus it is very difficult to find any deviations from V-A from charged current phenomenology. We are thus left with either neutral current phenomenology, which may put a limit on Z_R and hence on W_R, or to look for those effects which are sensitive to the Majorana nature of neutrinos, i.e. those which do not conserve lepton number.

(iii) Effect of heavy Majorana Neutrino N_e in rare weak processes.

a) <u>Neutrinoless double β decay</u> $(\beta\beta)_o$.

The intermediate neutrino must be Majorana. The right-handed coupling of ν with W_R gives the lepton number non-conserving parameter[15] (β being the mixing angle between W_L and W_R):

$$\eta \sim 0 \; (\beta\gamma) \sim 10^{-7}$$

too small to be detectable. Similarly[16], $m_\nu \lesssim 30$ eV implies:

$$\eta \lesssim 1.5 \times 10^{-5}, \; T_{\frac{1}{2}} > 4 \times 10^{23 \pm 2} \text{ years}$$

Thus, for $m_\nu \lesssim 1$ eV, $(\beta\beta)_o$ would get most of its contribution from N_R. The heavy Majorana neutrino would simulate an "effective" η to be[5]:

$$\eta = \frac{m_{W_L}^4}{m_{W_R}^4} \frac{1}{m_N} (0.35) \text{ GeV} \tag{6}$$

$$< 3.5 \times 10^{-5}$$

$$T_{\frac{1}{2}} > 8 \; (10^{22 \pm 2}) \text{ years}$$

for $m_N \gtrsim 100$ GeV and $(m_{W_L}^2 / m_{W_R}^2) \lesssim \frac{1}{10}$. Thus, a measurement of a half-life for $(\beta\beta)_o$ decay in the range 10^{21} to 10^{25} years would have an important bearing on the above considerations.

(b) <u>Rare muon processes.</u>

(A) $\mu \rightarrow e\gamma$ $\Delta L = 0$

      $\mu^- + A(Z) \rightarrow e^- + A(Z)$ $|\Delta L_\mu| = 1$

(B) $\mu^- + A(Z) \rightarrow e^+ + A(Z-2)$ $|\Delta L| = 2$

                                                    $|\Delta L_\mu| = 1$

The former (A) can occur independently of whether the neutrinos in the "intermediate state" are Dirac or Majorana while the later (B) occurs only if the neutrinos are Majorana. In each case, there is a mixing between electron-type and muon-type neutrinos and we denote the mixing angle by θ'. The branching ratios for the various processes are summarized below:

$\underline{\mu \to e\gamma}$

$$B_R^{17} = \frac{3\alpha}{32\pi} \left[\frac{m_{W_L}^2}{m_{W_R}^2} \right]^4 \left[\sin\theta'\cos\theta' \frac{m_{N_2}^2 - m_{N_1}^2}{m_{W_L}^2} \right]^2$$

$$< 5 \times 10^{-8} (\sin\theta'\cos\theta')^2 \text{, for } \frac{m_{W_L}^2}{m_{W_R}^2} \lesssim \frac{1}{10} \text{, } \frac{m_{N_2}^2 - m_{N_1}^2}{m_{W_L}^2} \lesssim 1$$

$$\lesssim 2.5 \times 10^{-10} \tag{8}$$

for $\theta' \lesssim \sqrt{m_e/m_\mu} \simeq 0.07$., Present limit[18] on B_R is 1.9×10^{-10}

$\underline{\mu^- + A(Z) \to e^- + A(Z)}$: diagrams which contribute are shown in Fig. 2,

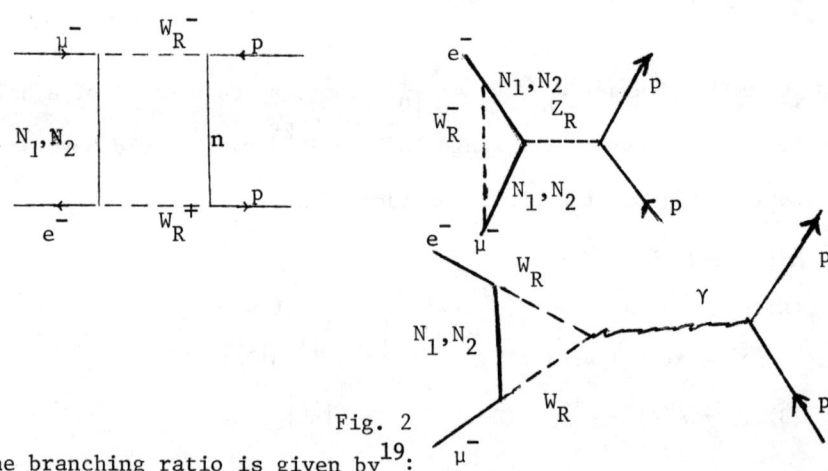

Fig. 2

The branching ratio is given by[19]:

$$B \cdot R(1) = \frac{\Gamma_{\mu^- \to e^-}}{\Gamma_{\mu^- \to \nu}} \simeq 5 \times 10^{-8} \varepsilon^2 \sin^2\theta' \cos^2\theta' \tag{9A}$$

where:

$$\varepsilon = \frac{(m_{N_2}^2 - m_{N_1}^2)}{\sin^2\theta_W m_{W_L}^2} \left(\frac{m_{W_L}^2}{m_{W_R}^2}\right)^2 \left[\ln\frac{4 m_{W_R}^2}{(m_{N_2}+m_{N_1})^2} - 1\right] \quad (9B)$$

Thus:

$$B \cdot R \cdot (1) \lesssim 1 \cdot 2 \times 10^{-10} \sin^2\theta' \cos^2\theta' \quad (9C)$$

for $(m_{W_L}^2/m_{W_R}^2) \lesssim \frac{1}{10}$, $\frac{m_{N_2}^2 - m_{N_1}^2}{m_{W_L}^2} \lesssim 1$. For $\theta' \simeq \sqrt{m_e/m_\mu}$, we find[19]:

$$B \cdot R \cdot (1) \lesssim 6 \times 10^{-13} \quad (9D)$$

A VPI Experiment is now under way at TRIUMF to measure this conversion process at the 10^{-12} level[20].

$\mu^- + A(Z) \to e^+ + A(Z-2)$:

We treat the nucleus as an "elementary particle" and assume that it and the nucleus in the intermediate state have spin 0. We have here only the box diagram shown in Fig. 3:

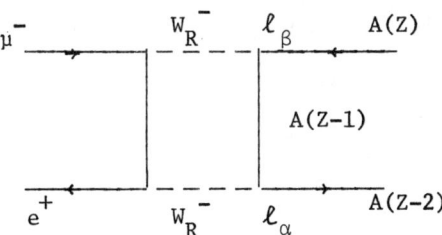

Here ℓ denotes the internal momentum. We have[19]:

$$B \cdot R \cdot (2) = \frac{1}{32} \sin^2\theta' \cos^2\theta' (\alpha/16\pi)^2 \frac{1}{\sin^4\theta_W}$$

$$\times \left(\frac{m_{W_L}^2}{m_{W_R}^2}\right)^2 \left(\frac{m_{N_2} - m_{N_1}}{m_A}\right)^2$$

$$\lesssim 10^{-10} \sin^2\theta' \cos^2\theta' \left(\frac{m_{N_2} - m_{N_1}}{m_A}\right)^2 \quad (10)$$

where m_A is the mass of the target nucleus. Now $\left(\frac{m_{N_2}-m_{N_1}}{m_A}\right)^2$ could easily be greater than 10 giving:

$$\frac{B \cdot R \cdot (2)}{B \cdot R \cdot (1)} \gtrsim 1$$

This result is very sensitive to the spin 0 assumption. For spin $\frac{1}{2}$ nuclei, B·R·(2) is suppressed by $\left(\frac{m_p + m_\mu}{m_{W_R}}\right)^2$.

To summarize, the detection of the processes considered would provide strong evidence for a heavy neutrino in the mass range $\sim$ 100 GeV. Detection of $(\beta\beta)_o$ or $\mu^- + A(Z) \to e^+ + A(Z-2)$ would settle the Majorana character of the heavy neutrino. The estimates of the transition probabilitities for the processes treated were based on values of $M_N \sim$ 100 GeV, $m_{W_R} \sim$ 300 Gev, $\theta' \sim \frac{1}{15}$, all compatible with present experiments. The left-right symmetric electroweak group does not by itself fix the scale of m_{W_R} but only states that breaking of parity is related with that of local (B-L) and that these breakings occur at the energy scale m_{W_R}. Moreover, the theory has two Majorana neutrinos per generation [one light (ν) and one heavy (N)] with a definite relationship of their masses with m_{W_R}:

$$m_\nu \sim \frac{m_e^2}{m_{W_R}} \;,\; m_{N_e} \sim m_{W_R}$$

Since $m_N \sim m_{W_R}$, we can summarize the dependence of various low energy processes on m_{W_R} as follows:

$(\beta\beta)_o$: $(T_{\frac{1}{2}})^{-1} \sim m_{W_R}^{-10}$

$\mu \to e\gamma$ $B \cdot R \cdot \sim m_{W_R}^{-4}$

$$\mu^- + A(Z) \to e^- + A(Z) \qquad B \circ R \bullet \qquad \sim \qquad m_{W_R}^{-4}$$

$$\mu^- + A(Z) \to e^+ + A(Z-2) \qquad B \circ R \bullet \qquad \sim \qquad m_{W_R}^{-2}$$
(spin 0 nucleus)

We cannot judge the refinements possible in the above experiments, but it is evident that reduction due to increased m_{W_R} is by all odds minimized for the last process.

This work was done at VPI, in collaboration with R. E. Marshak of VPI and R. N. Mohapatra of City College. It was supported by the Research Corp. and the Department of Energy.

References

1. A. Salam, Nuovo Cimento, 5, 299 (1957); L. Landau, Nucl. Phys. 3, 127 (1957); T. D. Lee and C. N. Yang, Phys. Rev. 105, 167 (1957).

2. See for example, J. Schechter and J.W.F. Valle, Phys. Rev. D22, 2227 (1980); S. M. Bilenky and B. Pontecorvo, Phys. Rev. 41 C, 225 (1978); K. M. Case, Phys. Rev. 107, 307 (1957); C. Ryan and S. Okubo, Suppl. Nuovo Cimento, Series 1, 2, 234 (1964).

3. S. Weinberg, Phys. Rev Lett. 19, 1264 (1967); A. Salam, "Elementary Particle Theory" ed., N. Swartholm, Stockholm (1968), S. L. Glashow, Nuc. Phys. 22, 579 (1961).

4. H. Georgi and S. L. Glashow, Phys. Rev. Letters 32, 438 (1974).

5. See, for example, R. E. Marshak, R. N. Mohapatra and Riazuddin, "Proc. of Muon Physics Workshop (TRIUMF 1980); Riazuddin, R. E. Marshak and R. N. Mohapatra, submitted to Phys. Rev.

6. J. C. Pati and A. Salam, Phys. Rev. D10, 275 (1974); R. N. Mohapatra and J. C. Pati, Phys. Rev. D11, 566; 2559 (1975); G. Senjanovic and R. N. Mohapatra, Phys. Rev. D12, 1502 (1975).

7. H. Fritzsch and P. Minkowski, Ann. of Phys $\underline{93}$, 193 (1975); H. Georgi, Particles & Fields, 1974 (APS/DPF Williamsburg), Ed. C. E. Carlson (AIP Press, N.Y. 1975).

8. R. E. Marshak and R. N. Mohapatra, Phys Letters $\underline{91B}$, 222 (1980), R. N. Mohapatra and R. E. Marshak, Phys Rev. Letters $\underline{44}$, 1316 (1980).

9. See, for example, H. Georgi and D. V. Nanopoulos, Nucl. Phys. $\underline{B159}$, 16 (1979); ibid $\underline{B155}$, 52 (1979).
G. Lazarides, Q. Shafi and C. Welterich - Univ. of Freiburg Preprint 80/2 (1980).

10. M. Gell-Mann, P. Ramond and R. Slansky, Supergravity, eds. P. Van Nieuwenhuizen and D. Z. Freedman, (North-Holland Publishing Co., 1979, p. 315.

11. R. N. Mohapatra and G. Senjanovic, Phys. Rev. Lett. $\underline{44}$, 912 (1980) and Fermi Lab Preprint Pub-80/61, 1980 (to be published in Phys. Rev. D).

12. See, for example, G. Lanzarides, Q. Shafi and C. Welterich, ref. 9, M. Magg and C. Wetterich, CERN Preprint Th. 2829.

13. E. Witten, Phys. Lett. $\underline{91B}$, 81 (1980).

14. M. A. B. Beg, R. V. Budny, R. N. Mohapatra and A. Sirlin, Phys. Rev. Letters $\underline{30}$, 1252 (1977).

15. R. E. Marshak, R. N. Mohapatra and Riazuddin, Ref. 5; R. N. Mohapatra and G. Senjanovic, Fermi Lab Preprint, Pub-80/61, 1980.

16. A. Halprin, P. Minkowski, H. Primakoff and S. P. Rosen, Phys. Rev. $\underline{D13}$, 2567 (1976).

17. See, for example, S. M. Bilenky and B. Pontecorvo, Phys. Reports 41C, 276 (1978).

18. J. D. Bowman, et al. Phys. Rev. Lett. 42, 556 (1979).

19. Riazuddin, R. E. Marshak, and R. N. Mohapatra, VPI Preprint (1981), submitted to Phys. Rev.

20. K. Gotow and M. Blecher, private communication.

*On leave from Quaid-i-Azam University, Islamabad.

PROPOSED ULTRASENSITIVE INVESTIGATION OF NEUTRINOLESS AND TWO-NEUTRINO
DOUBLE BETA DECAY OF ^{76}Ge: A PROGRESS REPORT.

F. T. Avignone, III
University of South Carolina, Columbia, S.C. 29208

and

R. L. Brodzinski and N. A. Wogman
Battelle, Pacific Northwest Laboratories
Richland, Washington 99352

I. INTRODUCTION

There have been several exciting developments in weak interactions, both experimental and theoretical, which give increased importance to any new fundamental measurements. Reines[1,3] has presented experimental evidence for neutrino oscillations while a new analysis of the beta spectrum from ^{3}H decay seems to give a definite non-zero mass for the electron neutrino.[4] The standard Weinberg-Salam theory has been very successful in explaining phenomena involving weak neutral currents which has led to wide acceptance of a pure left-handed $SU(2)_L \times U(1)$ symmetry with massless neutrinos. An alternative has been proposed based on $SU(2)_L \times SU(2)_R \times U(1)_{L+R}$ symmetry.[5] The spontaneous breaking of the left-right symmetry, for which $M(W)_R \gg M(W)_R$ gives the same results at low energies as the standard model; however, the U(1) generator in this model is identified with (baryon-lepton) number or (B-L) symmetry and a massive Majorana neutrino which could, for example mediate double beta decay with no neutrinos in the final state.[3] A high resolution, ultra-low-background investigation of both $2\tilde{\nu}_e$ and $0\tilde{\nu}_e$ double beta decay of ^{76}Ge is under development and is the subject of this paper.

Until very recently, neither two neutrino nor zero neutrino double beta decay had been observed in a direct experiment. Recently, however, the direct observation of the double beta decay of ^{82}Se was reported by Moe and Lowenthal,[6] and the total half-life reported was $(1.0 \pm 0.4) \times 10^{19}$ years. Previously there was indirect evidence of double beta decay from geological

ISSN:0094-243X/81/720034-12$1.50 Copyright 1981 American Institute of Physics

experiments (see ref. 7 for example). Neither geological experiments nor poor energy resolution direct measurements give strong evidence concerning neutrinoless double beta decay nor can they place sensitive limits on its half-life. The two major goals of the experiments discussed here are: an ultra high sensitivity search for neutrinoless double beta decay and a direct observation of and measurement of the half-life of the two neutrino double beta decay of ^{76}Ge. The general method was published by one of us (FTA)[8], and is an extension of the experiment performed earlier by Fiorini et. al.[9]

If the electron-neutrino ν_e emitted in ordinary single-beta decay is a γ_5-non invariant Majorana particle, then neutrinoless double beta decay will occur as a second order weak process. The Majorana nature allows the exchange of a virtual neutrino between two neutrons and the γ_5 non-invariance ensures that the amplitude will not vanish for helicity reasons alone. The γ_5-invariance of the Hamiltonian breaks down under transformations of the neutrino field when the weak leptonic current contains a small $(1 - \gamma_5)$ term as given in Eq. 1 or when the neutrino rest mass does not vanish, or both. The leptonic current for the first case can be written,

$$L_\gamma = \tilde{\psi}_e \gamma_\lambda [(1 + \gamma_5) + \eta (1 - \gamma_5)] \psi_e, \qquad (1)$$

where η is a parameter to be determined experimentally. The recent direct measurement by Moe and Lowenthal[6] resulted in the value $\eta \sim 3 \times 10^{-4}$. Another mechanism to drive neutrinoless double beta decay could be a, heavy, Majorana neutrino N_e coupled to the electron with the same strength as ν_e and with definite helicity. The leptonic current would then be written[3]

$$L_\lambda = \tilde{\psi}_e \gamma_\lambda (1 + \gamma_5) \psi_{\nu_e} + \tilde{\psi}_e \gamma_\lambda (1 \pm \gamma_5) \psi_{N_e}. \qquad (2)$$

A recent shell model calculation by Haxton and Stevenson[10] indicates that the case of ^{76}Ge should be highly favorable to the investigation of both two neutrion and neutrinoless double decay.

Most attempts to observe double beta decay have been made using detectors of poor energy resolution; however, one high resolution attempt was reported in 1973 by Fiorini and his co-workers[9], and was an attempt to observe the full decay-energy peak in the hypothesized doubled beta decay of ^{76}Ge, which is calculated from the measured mass excess to lie at (2.045 ± 0.003) MeV. The source was the 7.7% abundant, ^{76}Ge in a Ge(Li), detector with an active volume of 68.5 cm^3. The detector was placed in a graded shield and distilled mercury, nylon, copper, and lead surrounded with paraffin and was located in a tunnel under Mount Blanc. The result was that the half-life for the neutrinoless double beta decay is greater than 5 x 10^{21}y for which they reported a value of the lepton-non-conserving parameter $\eta=10^{-3\pm1}$.

A recent value of the double beta decay half-life ratio $T_{\frac{1}{2}}$ (^{128}Te)/$T_{\frac{1}{2}}$ (^{130}Te), reported by Hennecke et. al.,[7] implies a value $\eta = (4.3 \pm 0.1$ x $10^{-5})$. This ratio was determined by observation of the excess isotopic abundant of ^{130}Xe and ^{128}Xe in geological ores of tellurium. These data were interpreted to imply that the double beta decays of ^{128}Te and ^{130}Te are predominantly two neutrino decays with a small but possibly observable mixture of lepton-non-conserving term. Even if the data were far more accurate, one could not state that lepton-number-conservation is violated, becaue the analyses are model dependent. Such a conclusion can only be drawn by observing the total decay energy in the electrons with some degree of confidence.

Extensive Monte-Carlo calculations described earlier,[8] have been performed to determine the effectiveness of various size NaI(Tl) anticoincidence annuli in removing the interference with the 2.045 MeV region of the spectrum of a large germanium detector due to the main components of the background. One of the most serious interferences was found to be that of the 2.615 MeV γ-ray in ^{208}Pb which is found at the end of the thorium chain.

II. SUPPORTING NUCLEAR STRUCTURE CALCULATIONS

Nothing about lepton number conservation can be determined from double beta decay experiments if the nuclear matrix elements are extremely small. For that reason, detailed shell model calculations were performed by Haxton and Stevenson,[10] specially to support this experiment. The LASL version of the Glaskow code, developed by Haxton and Dubach,[11] was used with a subsidiary density matrix code which calculates one and two body density matrix elements between states of good J and T. Neutron and proton states were treated separately and then combined in a weak coupling calculation. The effects of spurious isospin mixing affects the Fermi matrix element, which is small, hence the correction for this effect is probably less than 10%. It is somewhat reassuring that the results of these calculations give a predicted total ββ half-life of ^{82}Se of 1.5×10^{19} years which is in general agreement with the recent measurement of $(1.0 \pm 0.4) \times 10^{19}$ years reported by Moe and Lowenthal.

We note that the active model space, matrix elements and general nuclear structure properties of ^{82}Se are very similar to those of ^{76}Ge. The results obtained by the LASL group for ^{76}Ge are $T_{1/2}$ (2ν-ββ) = 2.4×10^{20} yr. and $T_{1/2}$ (0ν-ββ) = $1.2 \times 10^{13} \eta^{-2}$ yr. The calculations clearly show that ^{76}Ge is very favorable case for investigating both modes of ββ decay.

III. SENSITIVITY LIMITS ON THE DETERMINATION OF η

The Monte-Carlo calculations were an extension of those given in ref. 8. The absolute background γ ray flux was deduced from the experimental spectra given by Fiorini et. al.[9] Similar spectra were also obtained in our own preliminary measurements. This flux was then allowed to impinge on a 245 cm³ Ge detector inside of a 20 x 24 inch cylindrical NaI(Tℓ) anticoincidence annulus. All of the background was assumed to originate inside of the cryostat and recent experimental studies support this hypothesis.[12] The opening which accepts the Ge detector makes very little difference, because most of the γ rays are in coincidence with others, hence if one scatters out of the Ge detector and escapes through the opening, there is still a large probability that the event will be cancelled by a coincident γ ray. The code was run to simulate 1 year of counting assuming three half-lives greater than the upperlimit 5×10^{21} years.[9] These calculations then assume no better radiopurity than that achieved in ref. 9, hence the only improvement is the size and resolution of the detector and the addition of the annulus. For this reason our results are conservative. Vast improvements are expected with lower levels or radioactive background.

The graphical results of our Monte Carlo calculations are sown in Fig. 1, and clearly demonstrate that the present upper limit of 5×10^{21} years can easily be extended by two orders of magnitude to a value of about 5×10^{23} years for 0ν-ββ decay. This prediction is simply based on the fact that after one year of counting we would easily observe the 0ν-ββ peak if its half-life was 5×10^{22} years. If no more counts appeared in the peak than in ten lower channels, an increase of another order of magnitude in the limit results. This conclusion is based on the worst case assumption that we could not achieve radiopurity of the cryostat materials beyond that achieved by Fiorini and his co-workers. The present technique should allow the observation of the process at $T_{1/2} = 5 \times 10^{22}$ years and could place a new lower limit of 5×10^{23} years with reasonable probability.

The result of the nuclear structure (shell-model) calculation[10] gives,

$$T_{1/2} = \frac{1}{\eta^2} \times 1.2 \times 10^{13} \text{ yr},$$

which would allow a determination of a new limit on η of $\cong 5 \times 10^{-6}$. In fact, if η is as large as 1.6×10^{-5}, the (0ν-$\beta\beta$) decay of ^{76}Ge should be directly observable.

It should be mentioned that using a small NaI(Tℓ) anticoincidence shield, a 28% efficient detector, and no preliminary precautions on radiopurity we obtained only 102 counts in 40 days in the 11 keV region about 2.045 MeV. This experiment was performed above ground and would correspond to Fiorini's "dirty" spectrum because of the aluminum end cap. This corresponds to a limit $T_{1/2}(0\nu) > 2.3 \times 10^{21}$ yrs which is not much lower than the present limit 5×10^{21} yrs. All indications are that the sensitivity claims made here are conservative, and that the annulus will play the anticipated role.

A more complete examination of the results of ref. 9 clearly show that all of the background observed in that experiment can be accounted for by contamination of the cryostat materials themselves. The end cap is only one source of thorium and uranium contamination. We have experimentally determined that the cryosorption materials used in all Ge(li) detectors are highly contaminated with radioactive thorium and uranium.[12,13] This can account for a significant portion of the background because of the usually short, straight-line path between this material and the detector. In addition, we have experimentally determined that the materials customarily used in field effect transistors and connections to the detector are radioactive. These sources can be eliminated to a much higher degree than in earlier experiments. The advantage of having the background directly identifiable in high resolution

peaks is unique to this double beta decay experiment. After driving the
background contamination to its absolue minimum, using the series of tests
discussed elsewhere[12,13], and which are routine in our laboratory, we can
correct the two neutrino double beta decay spectrum by subtracting the
continuum of γ rays predicted for each peak observed in the spectrum. The
prediction of the continuua associated with a given background can be accomplished using our Monte Carlo code which has proven to be very accurate.[14]

IV. EXPERIMENTAL DESIGN

The main features of the technique to be used in these measurements have
been discussed elsewhere; however, as stated above since the publication of
ref. 8, we have learned that almost all of the background observed in Ge
detectors comes from within the cryostat, but that virtually none comes from
the diode itself. We have conducted a completely new set of Monte Carlo
calculations which account for the coincidence relationships between the
observed γ rays in the background. Basically the system consists of a large,
high resolution intrinsic Ge diode inside of a large NaI(Tℓ) anti-coincidence
annulus as shown in Fig. 2.

All materials are being precounted in existing high sensitivity multi-dimensional gamma-ray spectrometers to insure radiopurity. The pre-amplifier will be located external to the lead cave because it is known
source of radioactivity. Similarly, the liquid nitrogen cryostat is removed from the line-of-sight of the detector, and the selection and location
of molecular sieve material and being chosen to minimize radiation incident
on the germanium diode. All connections and wires will be fabricated from
99.9999% pure silver. The NaI(Tl) anticoincidence shield will also be
fabricated from precounted "clean" materials. Only four photomultiplier
tubes will be used on the annulus since these tubes are the single largest
source of radioactivity in the system. Pure NaI, quartz, and lead will be
used to shield the system from the phototubes. The internal reflector around
the NaI(Tl) crystal will be $CaCO_3$ instead of MgO, because of higher radio-purity. All metals will be preselected pre-World War II, battleship steel or
stainless steel to avoid the radioactivities ubiquitously present in modern
steels. The cave will consist of 12" of clean pure virgin lead and 2" of
battleship steel. A cadmium sheet will surround the lead cave to absorb

thermal neutrons generated by the spontaneous fission of natural uranium in the surrounding strata. A Tedlar sheet will seal the entire cave and detector system to prevent the intrusion of radon gas.

It is also necessary to eliminate the cosmic rays with a large earthen overburden. It is planned to locate the system in an Idaho silver mine, which typically have depths in excess of ~4200 feet. To provide the necessary environment for the electronics, the counting system will be housed in a small room constructed from low background battleship steel and fitted with an air conditioner. The steel room will be further shielded from primordial radioactivities in the mine walls by a barrier of low background dunite. (See Fig. 3). Since the completion of these calculations, it has been learned that a new technical breakthrough[15] in the growing of large single Ge crystals will make cylindrical detectors of 3 inches in diameter by 3 inches in length readily available. This increases the volume by a factor of 1.4 over the detector used in the present calculations and by a factor of 5 over that used by Fiorini et al.[9]. In addition, Compton continuua from background γ rays will be further reduced which will increase the sensitivity. Calculations based on these new larger detectors are in progress.

The authors are indebted to Professors G. J. Stephenson and W. C. Haxton for their continuous advice and theoretical support and to Dr. T. P. Lang and the Georgia Tech Neutrino Group, for their support during the early phases of Monte-Carlo calculations.

REFERENCES

1. F. Reines, Announcement at the Spring APS meeting session HA4, Bull. Am. Phys. Soc. 25 (1980).

2. F. Reines, H. Sobel, E. Pasierb, Private communication during Minineutrino Conference, Univ. Cal. Irvine, March 1980.

3. F. Reines, H. W. Sobel and E. Pasierb, Phys. Rev. Lett. 16, 1307, (1980).

4. V. A. Lyubimov, E. G. Novikov, V. Z. Nozik, E. F. Tretyakov and V. S. Kosik, ITEP preprint No. 62, Moskow, 1980.

5. A. Halprin, P. Minkowski, H. Primakoff, and S. P. Rosen, Phys. Rev. D13, 2567 (1976), R. N. Mohapatra and R. E. Marshak, Phys. Rev. Lett. 44, 1316 (1980).

6. "An Experimental Investigation of the Double Beta Decay of ^{82}Se," M. K. Moe and D. D. Lowenthal, University of California, Irvine Rept. UCI-10P19-143, Dec. 1979. (To be published)

7. E. W. Hennecke, Phys. Rev. C17, 1168 (1978). (Also see references given in the review by D. Bryman and C. Picciotto, Rev. Mod. Phys. 50, 11 (1978).

8. F. T. Avignone, III and Z. D. Greenwood, Nucl. Instr. Meth. 160, 493 (1979).

9. E. Fiorini, A. Pullia, G. Bertolini, F. Cappellani and G. Restelli, II Nuovo Cimento 13 (1973), 747.

10. W. C. Haxton and G. J. Stephenson, Jr., LASL, (Private Communication).

11. Code developed by J. Dubach and W. C. Haxton at LASL.

12. N. A. Wogman and J. C. Laul, "Natural Contamination in Radionuclide Detection Systems," 1980 Nuclear Science Symposium, held November 5-7, 1980, Orlando, Florida, and in the Symposium on Natural Radiation Environment, January 19-23, 1981, Bombay, India.

13. N. A. Wogman, "Current Methods for Laboratory Analysis of Environmental Levels of Radioactivity," IEEE Trans. Nucl. Sci. NS-23, (1976) 1214.

14. F. T. Avignone, III, Nucl. Instr. and Meth. 174 (1980)555.

15. Engineering staff, Princeton Gamma - Tech. Inc. (Private communication).

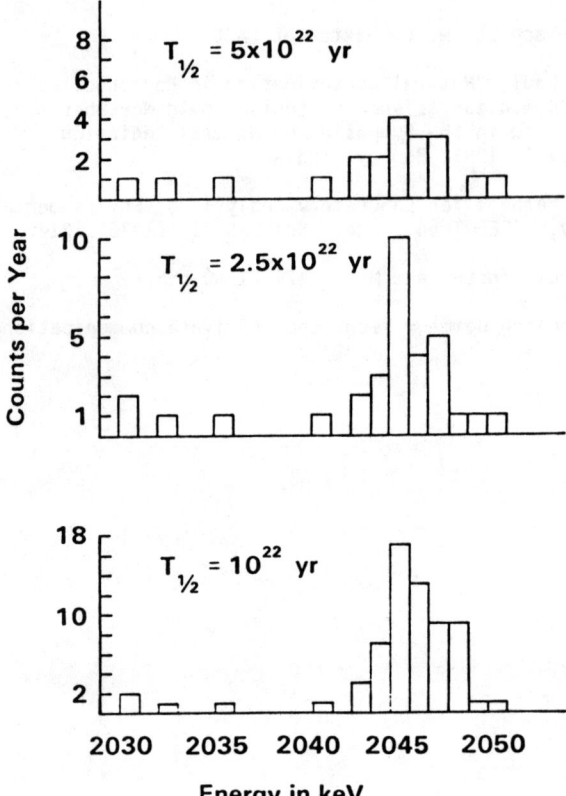

Fig. 1

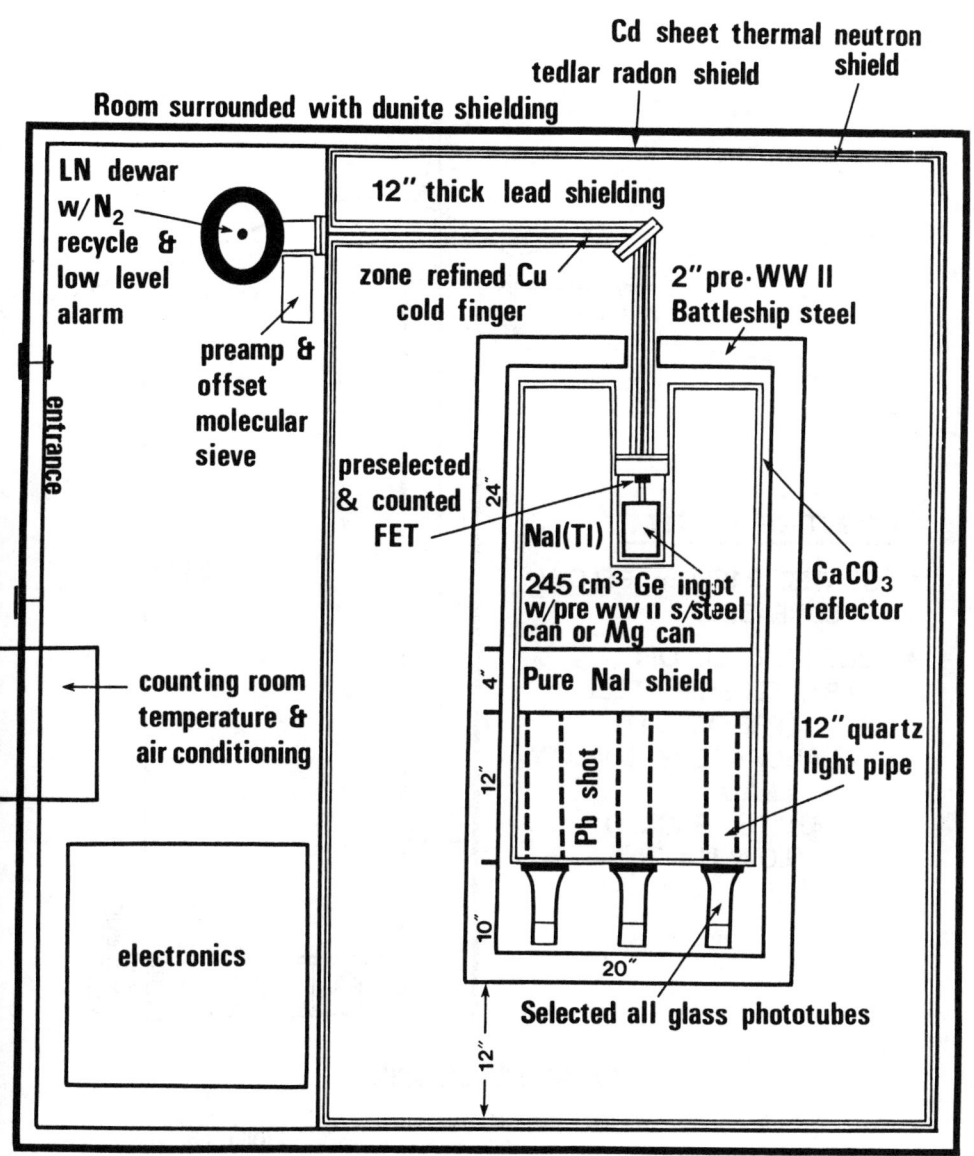

Fig. 2

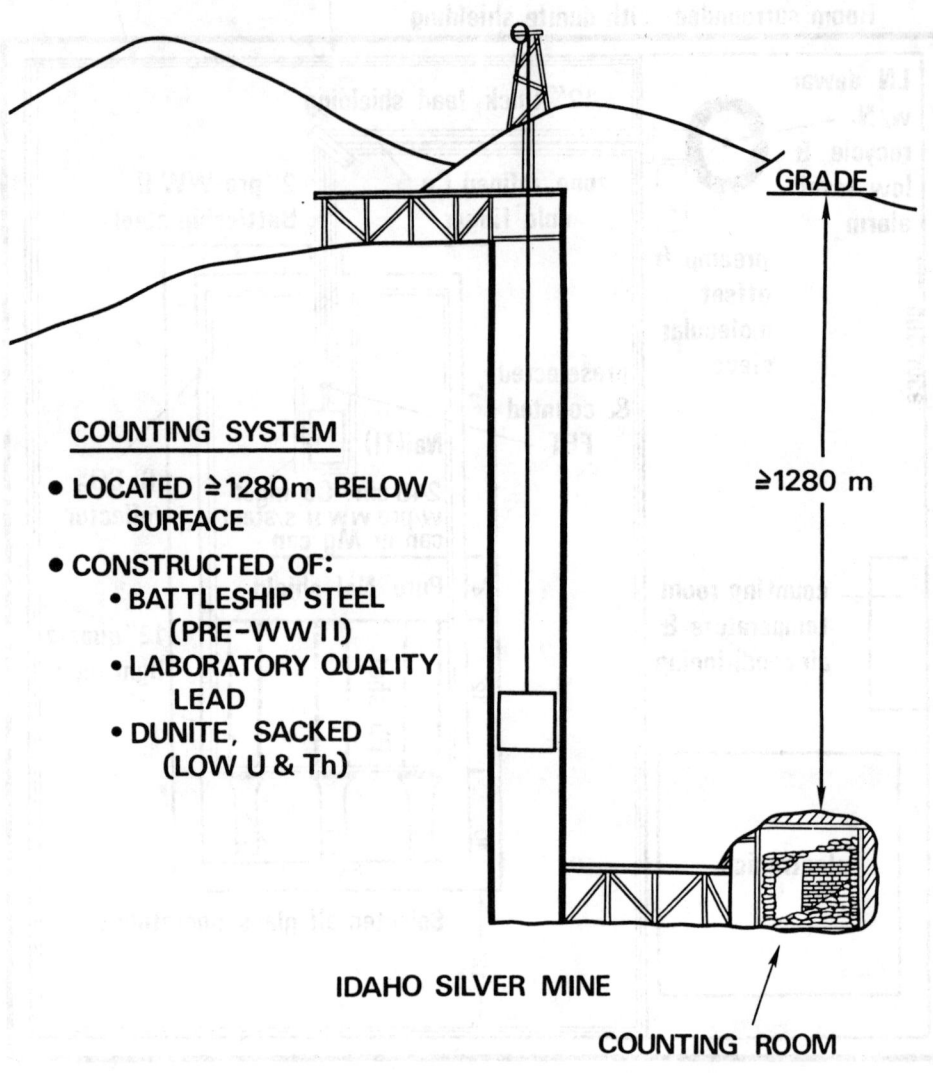

Fig. 3

DISCUSSION

WU: How will you protect your experiment from the background caused by Radon decaying into bismuth, which has killed previous experiments?

AVIGNONE: You saw our apparatus but it probably went by too quickly, but it was enclosed in tedlar. We have noticed this terrible Rn background and certainly it gives you that ^{214}Bi. We realize that it has got to be hermetically sealed and blown out with some nitrogen.

RUSHTON: I'm afraid Frank you have to go beyond hermetically sealing it even. What Mike Mo found - his data were taken before the radioactive velvet was put in--is that Rn will leak out of the walls of the construction materials and decay in the experimental volume. It then leaves a charged ion which is attracted to the source. In fact, Mo's data are completely consistent with all of the observed ^{214}Bi coming from the external environment, or from his chamber and plating onto his source and not being in the source.

AVIGNONE: We do intend to flush with gas if we must.

RUSHTON: Mo found that if you have the gas, in his case helium, flowing continuously, it purges the radon out, and it would kill the rate from that source. So it's easily manageable, you just have to know it is going to happen.

POSSIBLE DEVIATIONS FROM (V-A) CHARGED CURRENTS:
PRECISE MEASUREMENT OF MUON DECAY PARAMETERS

Mark Strovink*
University of California, Berkeley, CA 94720

INTRODUCTION

This short review examines the experimental limits on possible deviations from (V-A) charged weak currents, as would occur at some mass scale, for example, in manifestly left-right-symmetric electroweak theories. I shall consider both present and anticipated limits, emphasizing muon-decay experiments but including other experimental input where convenient.

At the outset, I shall take this opportunity to present a slightly pertinent result from the Berkeley-Fermilab-Princeton muon-scattering group[1], who obtain limits on the mass of a possible *neutral* muon which couples to right-handed currents. Turning to the parameters describing muon decay, I shall summarize too briefly the already precise experimental results of the 1960's. The major new experimental input to this field, nearly dormant through the last decade, is the measurement of longitudinal and transverse polarization of decay positrons performed by the ETH/Zurich-SIN-Mainz group at SIN[2]. After describing their results I shall mention several of the new experiments aiming to push further the measurement of muon decay parameters. Technologically, the most ambitious new effort is the Time Projection Chamber at LAMPF, under construction by the Los Alamos-Chicago-NRC/Canada-Carleton group. It is described in detail by J.D. Bowman in his contribution to these Proceedings. After brief remarks on the effects of possible charged Higgs boson exchange, I shall conclude by discussing present and possible future experimental constraints on the existance of a right-handed gauge boson W_R.

I. MASS LIMITS ON A HEAVY NEUTRAL MUON

The limit has recently been published[3] by the Berkeley-Fermilab-Princeton muon experiment, based on a 1978 exposure to the now-extinct Fermilab muon beam. The apparatus, shown in Fig. 1, used a distributed magnetized-steel calorimeter for high luminosity and efficient identification of muons in the final state. During the course of this experiment, the study of lepto-produced "extra" final-state muons progressed from observation of a few tens of multi-muon events to quantitative study of high-statistics samples -- in this case, some 10^5 events with two muons in the final state, due primarily to charmed-quark-pair production followed by semileptonic decay.

---

*This work was supported by the Director, Office of Energy Research, Office of High Energy and Nuclear Physics, High Energy Physics Division of the U.S. Department of Energy under Contract No. W-7405-ENG-48.

ISSN:0094-243X/81/720046-25$1.50 Copyright 1981 American Institute of Physi

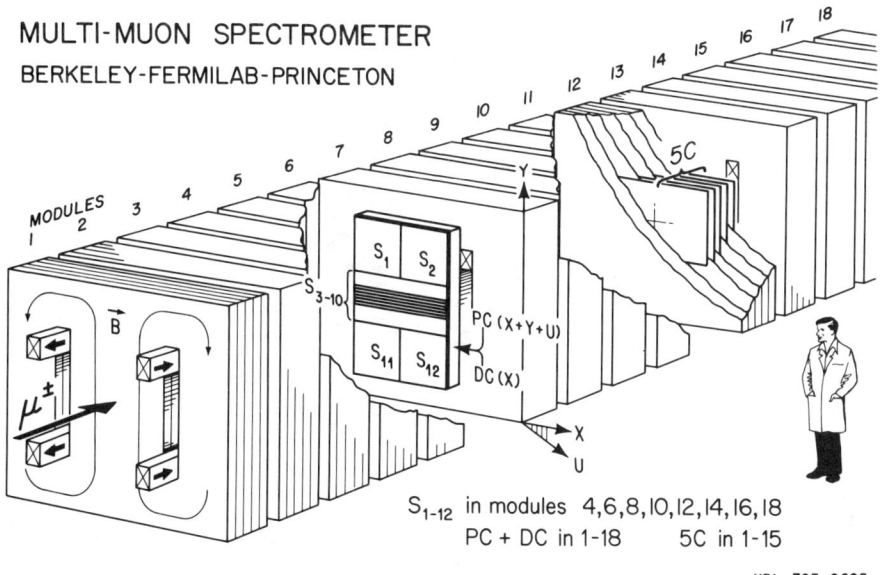

Fig. 1. Sketch of the Berkeley-Fermilab-Princeton multimuon spectrometer. The spectrometer magnet, serving also as a target and hadron absorber, reaches 19.7 kG within a 1.8×1×16-m^3 fiducial volume. Over the central 1.4×1×16 m^3, the magnetic field is uniform to 3% and mapped to 0.2%. Eighteen pairs of proportional (PC) and drift chambers (DC), fully sensitive over 1.8×1 m^2, determine the muon momenta typically to 8%. The PC's register coordinates at 30° (u) and 90° (y) to the bend direction (x) by means of 0.5-cm-wide cathode strips. Banks of trigger scintillators (S_1-S_{12}) occupy eight of the eighteen magnet modules. Interleaved with the 10-cm thick magnet plates in modules 1-15 are 75 calorimeter scintillators resolving hadron energy E_{had} with rms uncertainty $1.5 E_{had}^{1/2}$ GeV. Not shown upstream of module 1 are 1 PC and DC, 63 beam scintillators, 8 beam PC's, and 94 scintillators sensitive to accidental beam and halo muons.

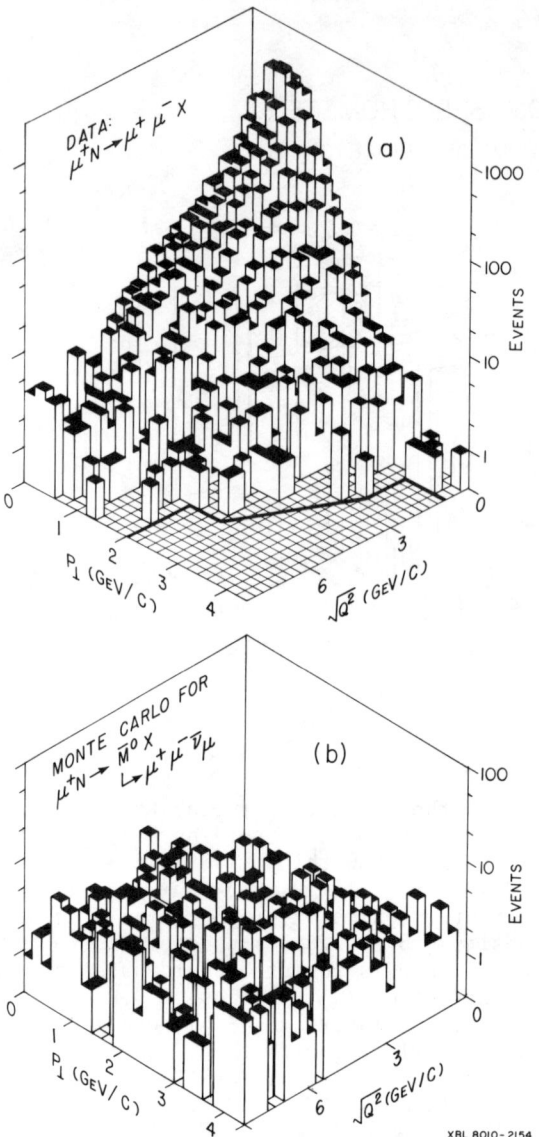

Fig. 2. Two-dimensional distributions of dimuon-final-state events vs. $\sqrt{Q^2}$ and $p_\perp$, the daughter muon momentum transverse to $\vec{Q}$. For this analysis, Q^2 is defined by taking the highest-energy beam-sign final state muon to be a scattered beam muon. The vertical scale is logarithmic; bin populations range from 0 to 450. Distribution (a) shows the data and an empirically chosen contour within which these events are contained. Distribution (b) is 77.4× the simulated population from production and decay of a 6 GeV/c² $\bar{M}^0$, with the assumptions described in the text. The 3.5 events in (b) lying outside the contour in (a) give the quoted σB limit at this mass.

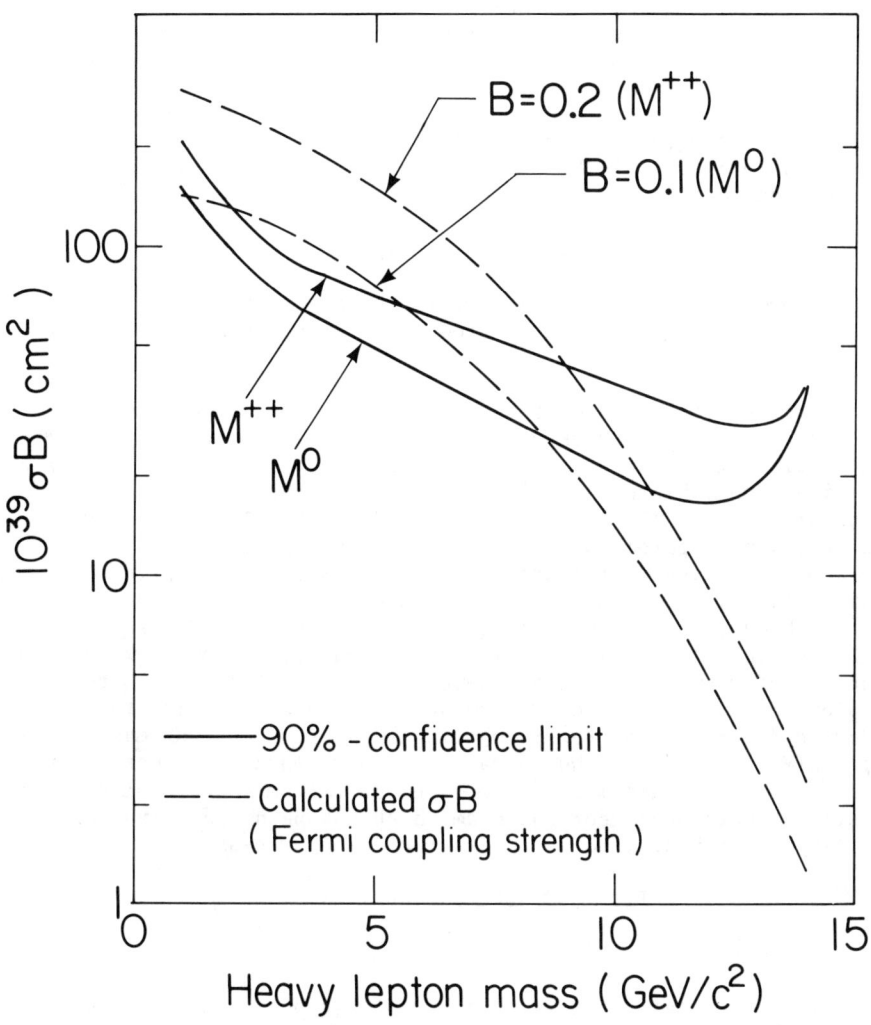

Fig. 3. Mass-dependent limits on the product of cross section and $\mu\mu\nu$ branching ratio (σB) for $\bar{M}^0$ and M^{++} production. Also indicated are the calculated σB for the production of $\bar{M}^0$'s and M^{++}'s, where the branching ratio is assumed to be 0.1 and 0.2 for $\bar{M}^0$ and M^{++}, respectively. To 90% confidence the data exclude the production of an $\bar{M}^0$ or M^{++} coupled with Fermi strength to a right-handed current in the mass range $1 < m_M < 9$ GeV/c^2.

A primary motivation for the search for a heavy neutral muon (M^0) had been the "hybrid" gauge model which placed the M^0 in a right-handed doublet with the μ^-. In the electron sector, this model has since been ruled out by the polarized-electron scattering experiments at SLAC[4]. Nevertheless, the M^0 has surfaced from time to time in other gauge models [5,6] which are not yet phenomenologically defunct. In most cases, the M^0 is made to couple with near Fermi strength to muons via a right-handed current. This is a good match to the large "unnatural" polarization of muon beams derived from forward π decay. The Berkeley-Fermilab-Princeton group searched for the reaction

$$\mu^+(L.H.)N \to \bar{M}^0 X; \quad \bar{M}^0 \to \mu^+\mu^-\bar{\nu}_\mu.$$

The two-muon-final-state signature is shared by "background" processes like charm or other hadron production with a subsequent decay into μ^-.

After various cuts were applied to enhance the sensitivity to $\bar{M}^0$ events relative to background, the data were analyzed as though the final-state μ^+ were a scattered beam muon, and accumulated on a two-dimensional histogram (Fig. 2) of $\sqrt{Q^2}$ vs. $p_\perp$, the μ^- momentum transverse to $\vec{Q}$. The background (Fig. 2(a)) has low Q^2 because of the photon propagator, and low $p_\perp$ because of the small charmed quark mass. Using a standard parton model with logarithmic scale-noninvariance, the simulated $\bar{M}^0$ events are found to have larger Q^2 and $p_\perp$, because of the W_R propagator and the higher $\bar{M}^0$ mass (6 GeV/c^2 in Fig. 2(b)). Simulated $\bar{M}^0$ populations in the background-free region result in the mass-dependent 90%-confidence limits on σB plotted in Fig. 3. For M^0 production, these lie below the levels expected for B=0.1 and Fermi coupling strength in the M^0 mass interval 1<$M(M^0)$<9 GeV/c^2. No comparable experimental information on the M^0 exists in this mass range. In fact, I am unaware of another experiment outside a neutrino beam which has been sensitive, on an event-by-event basis, to any weak production process.

II. MUON DECAY EXPERIMENTS

A. Baroque Era

Figure 4 recalls the (V-A) shape of the positron spectrum in μ^+ decay, and Table 1 reproduces its dependence upon the usual parameters ρ, η, δ, and ξ. The forward-backward asymmetry (where "forward" is defined to be opposite to the μ^+ spin direction) is complete at the energy endpoint x=1, where the forward decay rate vanishes in the (V-A) limit. Note that all experiments sensitive to ξ actually measure the product ξP_μ, where P_μ is the polarization along its direction of motion of a μ^+ which arises from π^+ decay at rest. According to (V-A), and/or if there exists only one muon-neutrino and it is massless, $P_\mu \equiv -1$. Measurement of this product can enhance the sensitivity of such experiments to departure from (V-A).

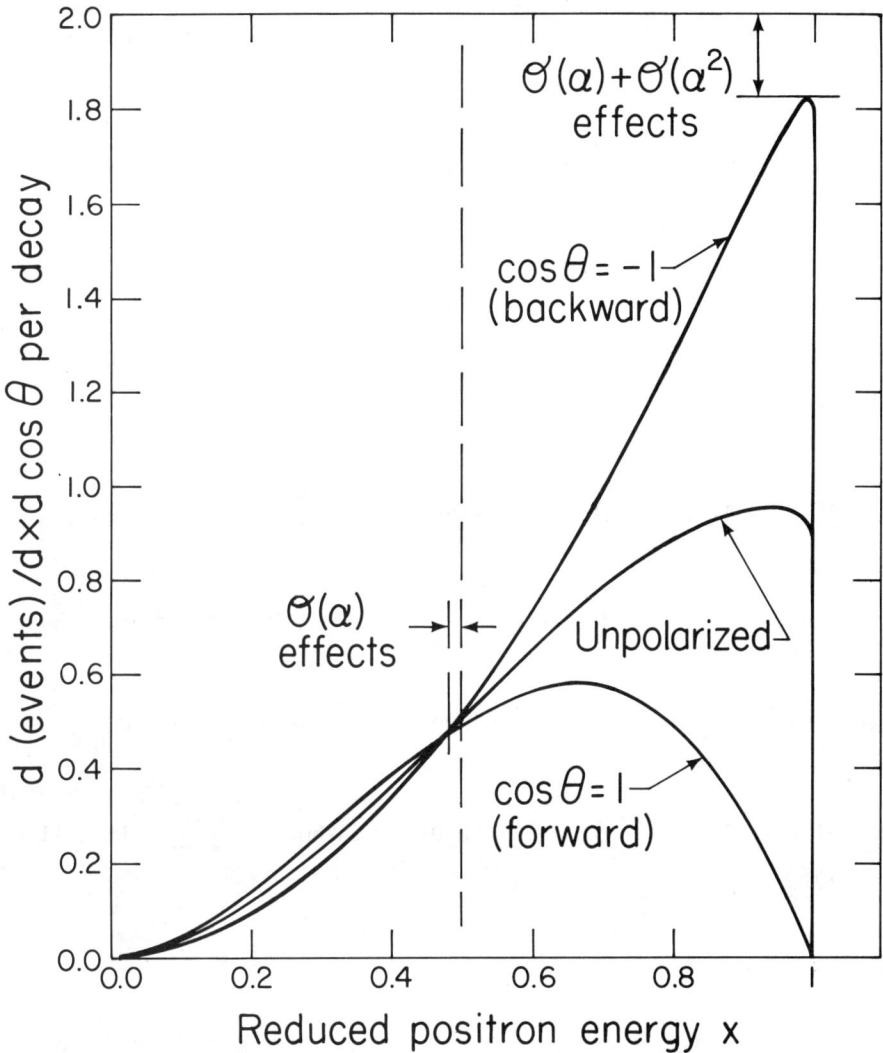

Fig. 4. Distribution in $x=E_{e^+}/E_{e^+}(\max)$ for μ^+ decay, according to (V-A). The parameter is $\cos\theta$, where θ is the angle between the forward direction (opposite to the μ^+ spin direction) and the positron momentum. In the forward direction near $x=1$, the lowest curve approaches $(1-4x)$ before radiative corrections are applied. The curves are radiatively corrected to order α (α^2 near the endpoint), and take into account the finite positron mass. The decay parameters ρ, δ, η, and ξ parameterize, respectively, the shape of the unpolarized-μ curve, the shape of the difference between polarized-μ curves, the low-energy shape of the unpolarized-μ curve, and the magnitude of the difference between polarized-μ curves.

Table 1. Definition and "classical" measurements of the muon-decay parameters. The defining equation sums over final-state helicities and is written to lowest order in m_e/m_μ and α.

$$\frac{d\Gamma^+}{x^2 dx d\cos\theta} \propto (3-2x)+(\tfrac{4}{3}\rho-1)(4x-3)+12\frac{m_e}{xm_\mu}(1-x)\eta$$

$$+\left[(2x-1)+(\tfrac{4}{3}\delta-1)(4x-3)\right]\xi P_\mu \cos\theta$$

Parameter	World average	Primary experiment(s)
"symmetric shape"	$\rho=0.7518\pm0.0026$	Bardon et al. (Ref. 7) Peoples et al. (Ref. 8)
"low energy"	$\eta=-0.12\pm0.21$	Derenzo (Ref. 9)
"asymmetric shape"	$\delta=0.755\pm0.009$	Fryberger (Ref. 10)
"polarization"	$\xi P_\mu=0.972\pm0.013$	Akhmanov et al. (Ref. 11)

Table 1 recounts the experimental successes [7,8,9,10,11] of the 1960's (see Ref. 12 for a complete review). Consistency with the (V-A) predictions was achieved, save for a 2.2-standard-deviation discrepancy in ξ. The symmetric and asymmetric shape parameters ρ and δ were extremely well-measured; one wonders if modern technology can produce a large further reduction in the systematic uncertainties. The error on the low-energy parameter η can be improved with higher statistics and positron energy resolution. To keep abreast of new η measurements, however, the radiative corrections should eventually be carried to second order. The classical measurement of ξ, performed in emulsion at 140 kGauss, did not have access to precise positron energy information. The result was based on the mean energy-independent front-back asymmetry. The fact that the rate near x=1 is nearly proportional to (1-ξ) in the forward direction should make possible a considerably more precise result in a future experiment.

B. New Measurements of Decay Positron Polarization at SIN

Preliminary results from the ETH/Zurich-SIN-Mainz group were reported at the 8th ICOHEPANS (Vancouver '79)[2]. The experiment analyzes the polarization of positrons from μ^+ decay by detecting their annihilation and Bhabha scattering in a magnetized Fe foil. Both the longitudinal and transverse e^+ polarization are analyzed; the latter is measured both within and normal to the plane formed by the e^+ momentum and μ^+ spin. This is made possible by precession of the μ^+ spin in a plane parallel to the foil (Fig. 5).

The preliminary results of this experiment are summarized in Table 2. The longitudinal e^+ polarization is

$$P_L^e = 0.94 \pm 0.08.$$

This is a substantial improvement upon the world average, but not a strong constraint on theory, compared for example to the 1.3% error on ξP_μ (Table 1). The transverse e^+ polarization is a function of the parameters α, α', β, and β', which, like the aforementioned muon decay parameters, depend on the scalar, pseuduscalar, vector, axial vector, and/or tensor coefficients in, for example, the charge-retention Hamiltonian (see Ref. 13 for full details). The primed parameters are T violating, corresponding to e^+ polarization out of the $\vec{p}_e$-$\vec{\sigma}_\mu$ plane. Under various conditions of physical interest, $\alpha=\alpha'=0$. If so, the experiment yields

$$\beta/A = -0.004 \pm 0.033$$
$$\beta'/A = -0.003 \pm 0.033,$$

where A is a sum of squares of coefficients approximately equal to 16.

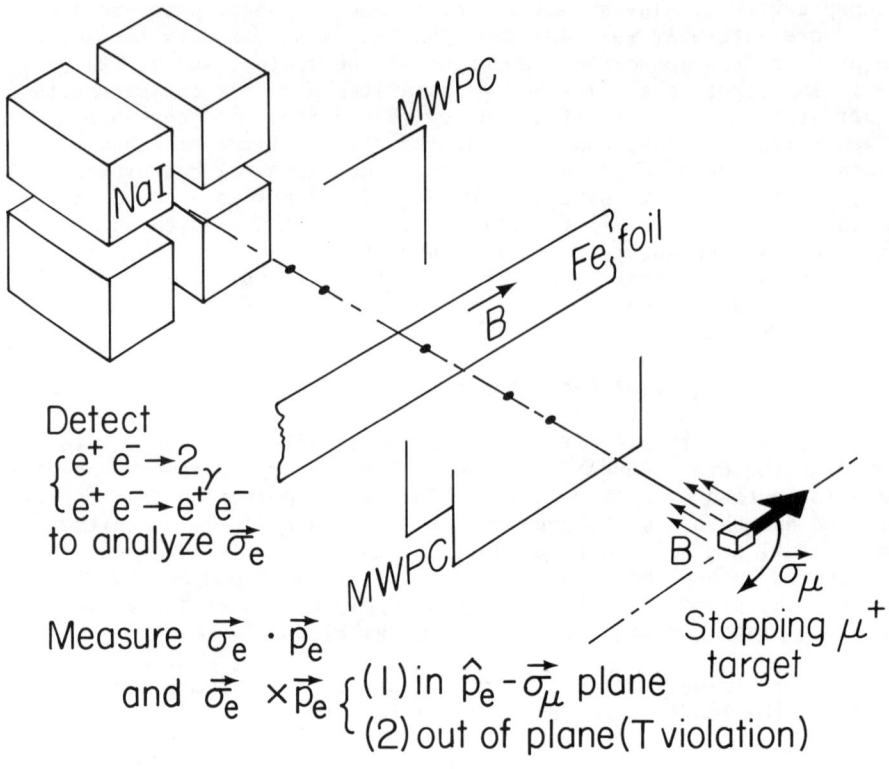

Fig. 5. Apparatus of Corriveau et al. (ETH/Zurich-SIN-Mainz collaboration, Ref. 2). The experiment measures the longitudinal and transverse polarization of μ^+-decay positrons. Both interesting types of particle pairs produced in the magnetic foil (2γ or e^+e^-) are detected; the magnetization is reversed at 15-minute intervals. Plastic scintillators (not shown) and 3 multiwire proportional chambers record the charged particle trajectories; energies of e^+, e^-, and γ are measured in four NaI crystals. The plane formed by the positron momentum and the muon spin rotates continuously as the latter precesses in the magnetic field near the stopping target. Transverse polarization both within and perpendicular to this plane is analyzed.

Table 2. Preliminary results of Corriveau et al. (ETH/Zurich-SIN-Mainz collaboration, Ref. 2), on the longitudinal polarization P_L^e and transverse polarization $\vec{P}_T^e$ of positrons from polarized μ^+ decay. Components of transverse polarization both within and perpendicular to the muon spin-positron momentum plane are measured. The notation is that of Scheck (Ref. 13).

Result for longitudinal polarization:

$$P_L^e = 0.94 \pm 0.08$$

$[P_L^e \text{ (previous world average)} = 1.00 \pm 0.14]$

$\vec{P}_T^e = \vec{P}_T^e(x, \alpha, \alpha', \beta, \beta')$, where

$x \equiv$ reduced positron energy

$\alpha \equiv |C_S|^2 + |C_S'|^2 - |C_P|^2 - |C_P'|^2$

$\beta \equiv |C_V|^2 + |C_V'|^2 - |C_A|^2 - |C_A'|^2$

$\alpha' \equiv 2\, \text{Im}(C_S C_P'^* + C_S' C_P^*)$
$\beta' \equiv 2\, \text{Im}(C_V C_A'^* + C_V' C_A^*)$ } T violating

$C_i \equiv$ coefficient of $(\bar{e}\Gamma_i\mu)(\bar{\nu}_\mu \Gamma^i \nu_e)$

$C_i' \equiv$ coefficient of $(\bar{e}\Gamma_i\mu)(\bar{\nu}_\mu \Gamma^i \gamma_5 \nu_e)$
   in charge-retention Hamiltonian.

Results for $\alpha, \alpha', \beta, \beta'$ free:

$\beta/A = -0.057 \pm 0.057$ $\qquad$ $\alpha/A = 0.16 \pm 0.12$

$\beta'/A = -0.049 \pm 0.057$ $\qquad$ $\alpha'/A = 0.14 \pm 0.14$,

where $A \approx 16$.

Results for $\alpha \equiv \alpha' \equiv 0$, β and β' free:

$\beta/A = -0.004 \pm 0.033$

$\beta'/A = -0.003 \pm 0.033$

In the same notation, the muon decay parameter η is

$$\eta = (\alpha - 2\beta)/A.$$

The parameter α is zero if $|C_S|=|C_P|$ and $|C_S'|=|C_P'|$ in the charge-retention Hamiltonian (see Table 2 and Ref. 13). Alternatively, one may construct[14] a "phenomenological gauge-theory Lagrangian"

$$\mathcal{L} \propto \bar{e}\gamma^\mu(1-\gamma_5)\nu_e \bar{\nu}_\mu \gamma_\mu (1-\gamma_5)\mu + \delta \bar{e}\gamma^\mu (1+\gamma_5)\nu_e \bar{\nu}_\mu \gamma_\mu (1+\gamma_5)\mu + h\bar{e}\nu_e \bar{\nu}_\mu \mu + h'\bar{e}\gamma_5 \nu_e \bar{\nu}_\mu \gamma_5 \mu$$

allowing for an arbitrary combination of left, right, scalar, and pseudoscalar couplings, e.g. exchange of W_L and W_R (without mixing), plus charged Higgs. In this construction, again, $\alpha=0$[14]. Therefore, in these interesting cases,

$$\eta = -2\beta/A,$$

and the SIN result may be interepreted as

$$\eta_{\alpha=0} = 0.008 \pm 0.066.$$

This is an important advance over the world average

$$\eta = -0.12 \pm 0.21.$$

Possibly, the error would shrink further if T invariance were assumed.

C. New Experiments Measuring Muon-Decay Parameters

Three new experiments of which I am aware are listed in Table 3. (Apologies are extended to those pursuing other initiatives, or whose institutional affiliations are incorrectly reproduced). I have already mentioned the large commitment being devoted to the Time Projection Chamber at Los Alamos[15]. It is expected to record $\gtrsim 10^8$ decay positrons from positive muons stopped in the methane TPC gas, with good momentum analysis over 4π solid angle except along (or opposite to) the (axial) direction of muon polarization. Muon depolarization (due both to epithermal and to thermal muonium formation in the methane) is expected to occur at the 1-2% level within the magnetic field. The *statistical* errors on all four muon decay parameters are calculated to be ~one order of magnitude smaller than existing combined uncertainties (Table 3). For discussion of the expected *systematic* errors, the reader is referred to J.D. Bowman's presentation.

Experiment 134/176 (Berkeley-British Columbia) at TRIUMF[16] utilizes a classical short-focussing solenoid as a single-channel positron-momentum analyzer. Despite the concomitant sacrifice in event rate it is quite useful for measurement of the low-energy parameter η, because the positrons encounter only vacuum between the μ^+-stopping scintillator and the annular momentum slit at the focal

Table 3. Experiments in progress which measure the muon-decay parameters. Anticipated <u>statistical</u> errors are shown in parentheses; anticipated <u>overall</u> errors are shown without parentheses.

	ρ	σ (overall) (σ(statistical))		
	ρ	η	ξP_μ	δ
World average (Ref. 12)	0.0026	0.21 / 0.066 [a]	0.012	0.009
LASL #455 (Ref. 15) LASL/Chicago/NRC/Carleton (Anderson/Bowman)	(0.00023)	(0.006[b])	(0.001)	(0.006)
TRIUMF #134/176 (Ref. 16) Berkeley/British Columbia I (Crowe)		(<0.1)	≤0.005 (≈0.001)	
TRIUMF #185 (Ref. 17) Berkeley/British Columbia II			≈0.001 (≈0.0003)	

[a] from measurement by Corriveau <u>et al</u>. (Ref. 2) of decay positron transverse polarization, assuming $\alpha=0$ (see Table 2).

[b] calculated statistical error does not include effects of radiative corrections.

plane. The spectrometer is to be adapted for use in measuring ξP_μ by precessing the μ^+ spin in a plane containing the solenoid axis. It will operate in a manner similar to that of a standard "muon spin rotation" apparatus, with the ability to select a positron momentum band corresponding to nearly complete time-modulation of the observed decay rate. This technique must sacrifice the advantage of a longitudinal magnetic field to "hold" the spin; muon depolarization is to be suppressed by chemical means.

Experiment 185 at TRIUMF[17], under construction by a second Berkeley-British Columbia group, is aimed at a definitive measurement of ξP_μ (statistical and systematic error $\lesssim 0.1\%$). Most of the running will be devoted to precise (0.5%) momentum measurement of decay positrons emitted opposite to the stopped muon spin direction, where (V-A) predicts that the rate must vanish. Only $\approx 2 \times 10^5$ events, obtainable in a few shifts, can provide the necessary statistical precision. The data sample thus should be highly complementary to that which will be collected by the Time Projection Chamber.

The group designing Experiment 185 feel it necessary to insure that uncertainties in correcting for muon depolarization, both in the beam transport and in the stopping medium, can be demonstrated *a priori* to be negligible at the 10^{-3} level. The experiment requires use of a "surface" μ^+ beam derived from decay of π^+ resting within a few mg/cm^2 of the surface of a thin carbon target illuminated by 520 MeV protons from TRIUMF. Each muon is to be tagged in position and angle by low-mass driftchambers, and in phase with respect to cyclotron RF (43 nsec period) to suppress contamination by "cloud" muons born promptly near the target. At TRIUMF, design and operation of such beams is highly developed[18]. Depolarization in the liquid He stopping target is to be suppressed by the uniquely high ionization potential of He: only $\leq 2\%$ of the thermalized muons will be bound in muonium. Depolarization of muons in this fraction will be suppressed by $\gtrsim \times 50$ in the longitudinal magnetic field. Table 4 lists the expected sources and levels of depolarization uncertainty in the beam and in the stopping target.

I shall take the opportunity to exhibit two sketches of the spectrometer being constructed. Figure 6 is a layout of the experiment, showing the upstream target solenoid and downstream 90° focussing positron spectrometer. Figure 7 exhibits the target solenoid (length 1 m) and nearby detectors in greater detail. The scale of the experiment is such that operation may be expected by early 1982.

III. REMARK ON CHARGED HIGGS LIMITS FROM μ DECAY

Up to this point I have not related the discussion of muon-decay parameters to any physical mechanism for departure from (V-A) predictions. The remainder of this review deals with two such mechanisms: charged Higgs exchange, and right-handed gauge boson (W_R) exchange (section IV).

Table 4. Anticipated depolarization of surface-beam muons in TRIUMF Experiment 185 (Ref. 17)

Depolarization of Beam

Source	Upper limit on error in correction
Coulomb scattering upstream of target	0.0005
Beam divergence at target	0.0006
Cloud muon contamination	0.0002
Jaw/slit scattering	?

Depolarization in Liquid He Target ($\vec{B}$≳7 kGauss)

Source	Upper limit on correction to polarization
Coulomb scattering	0.00001
Epithermal muonium formation	0.00001
Thermal muonium formation	0.00050
Impurities	0.00010
Wall stops	0.00010
Molecular ion rotation	0.00005
Total	0.00080

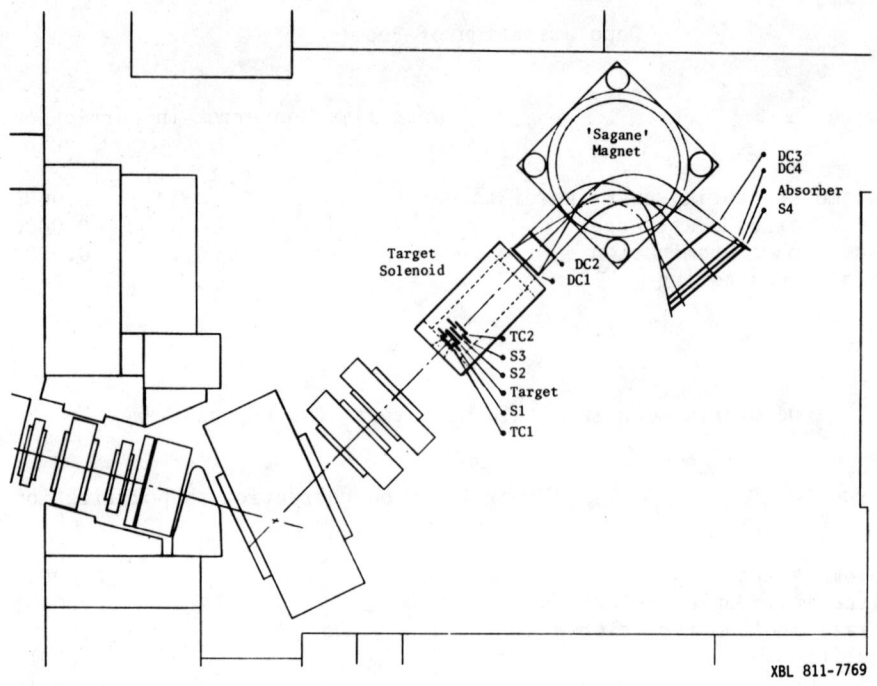

Fig. 6. Apparatus under construction for the Berkeley-British Columbia Experiment #185 at TRIUMF (Ref. 17). The experiment will be sensitive to the product of ξ, the polarization parameter describing μ^+ decay, and the polarization P_μ of the μ^+ from π^+ decay at rest. The anticipated sensitivity $\sigma(\xi P_\mu) \approx 10^{-3}$ would represent a factor 13 improvement over the current world average. It is to be achieved by measuring at the high-energy endpoint the rate of decay e^+ emission opposite to the μ^+ polarization direction. If this rate is found to be zero with the anticipated sensitivity, the mass of any right-handed gauge boson W_R must exceed 600 GeV/c^2. Two magnets are used in the apparatus: the target solenoid's axial field "holds" the stopped μ^+ spin and focusses the forward decay e^+, and the "Sagane" cylindrical dipole magnet is used in a 90° spectrometer with driftchambers DC2 and DC3 at the foci.

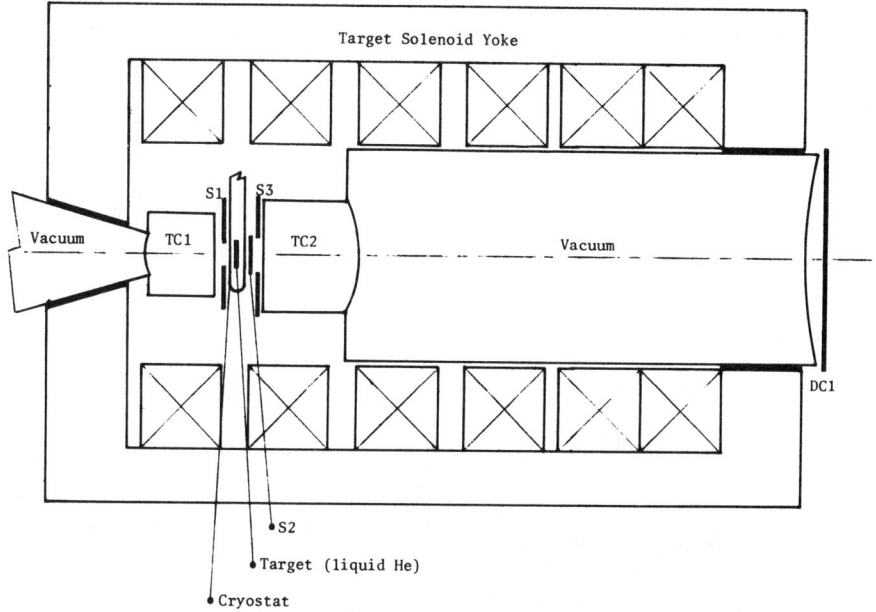

XBL 811-7820

Fig. 7. Enlarged view of apparatus in the vicinity of the target, now being constructed for the Berkeley-British Columbia Experiment #185 at TRIUMF (Ref. 17). (The full layout is shown in Fig. 6). The experiment will exploit the uniquely high ionization potential of (liquid) He in order to suppress depolarization due to muonium formation by stopping μ^+. Depolarization in the small residual muonium fraction will be made negligible by the Paschen-Back effect in the ≥ 7 kGauss longitudinal "holding" field. In the figure, TC1, TC2, and DC1 are low-mass driftchambers; S1-S3 are scintillators.

Haber, Kane et al.[19], and McWilliams and Li[20], have introduced a general charged-Higgs coupling which contributes to the Lagrangian a piece

$$\mathcal{L}^{C.H.} = 2^{3/4} [G_F^{1/2} M_{H^+}] \{ \bar{\psi}_f [\alpha_{ff'}^R \cdot \tfrac{1}{2}(1+\gamma_5) + \alpha_{ff'}^L \cdot \tfrac{1}{2}(1-\gamma_5)] \psi_{f'} \cdot \bar{H}(x) + h.c. \},$$

where f' and f are the initial and final fermions, and M_{H^+} is the charged Higgs mass. I have used the notation of Ref. 20; in the notation of Ref. 19,

$$\beta_{Kane} = 2^{1/2} M_\ell / (M_{H^+} \alpha_{\ell\nu}^L),$$

where ℓ is a charged lepton and ν is its neutrino. With a coupling of this form, $\alpha=0$ (see section II.B), and $\eta=-2\beta/A$; the SIN positron polarization experiment[2] constrains $\alpha_{ff'}$ just as does η:

(SIN $\vec{\sigma}_e \times \vec{p}_e$) $-0.20 < \alpha_{\mu\nu}^L \alpha_{e\nu}^L < 0.24.$ (III-1)

By comparison, the existing measurement of ξ constrains

(ξ) $|\alpha_{\mu\nu}^L| \{ (\alpha_{e\nu}^L)^2 + (\alpha_{e\nu}^R)^2 \}^{1/2} < 0.33.$ (III-2)

If a future experiment measures $\eta = 0 \pm 0.016$, or $\xi = 1 \pm 0.001$, the magnitude of the numbers in (III-1) and (III-2) will be reduced to 0.063.

Lest optimism be encouraged by this prospect, let me repeat a point emphasized by the authors of Refs. 19 and 20. The neutral (and by inference, the charged) Higgs mass is expected to be at least of order 3-10 GeV/c^2, and the coupling parameter $\alpha_{\ell\nu}$ is expected to be of order m_ℓ/m_{H^+}. If so, these low-energy experiments are hopeless! If $\alpha_{\ell\nu}$ turns out not to be proportional to m_ℓ/m_{H^+}, measurement of the branching ratio $(\pi^+ \to e^+ \nu_e)/(\pi^+ \to \mu^+ \nu_\mu)$ can be expected, for typical experimental accuracy, to produce a 1-2 order-of-magnitude greater sensitivity to (some) $\alpha_{ff'}$ than can be expected from the μ decay experiments. Unless there exists a mechanism to suppress quark-Higgs relative to lepton-Higgs coupling, there remains little motivation to search for effects of charged Higgs exchange in muon decay.

IV. CONSTRAINTS ON THE EXISTENCE OF W_R

The possible existence of one or more right-handed gauge bosons would be of great consequence to selection of a gauge group for grand unification. Moreover, considerable aesthetic appeal is held out by the possible restoration of "manifest left-right symmetry" to the electroweak interaction above some mass scale. A general discussion of the phenomenological constraints on right-handed currents is available from Bég, Budny, Mohapatra, and Sirlin[21], as appended by Holstein and Treiman[22].

This discussion will be phrased in terms of the physical variables $\delta \equiv (M(W_L)/M(W_R))^2$, where $M(W_L)$ [$M(W_R)$] is the mass of the left- [right-] handed gauge boson, and ζ, the angle by which W_L and W_R mix. This angle is the same as in Ref. 21. In the (V-A) limit, $\delta = \zeta \equiv 0$. Near this limit, the variables used in Refs. 21 and 22 are:

(Ref. 21) $\begin{cases} \eta_{AV} \approx -1+2(\delta-\rho+4\delta\zeta-\delta^2) \\ \eta_{AA} \approx 1+4(\zeta+\zeta^2-2\delta\zeta) \end{cases}$

(Ref. 22) $\begin{cases} x \approx \delta-\zeta \\ y \approx \delta+\zeta \end{cases}$

In these terms, the present experimental situation is summarized by the two-standard-deviation limits in Fig. 8. At present, the primary limit on the mixing angle ζ is set by the ρ parameter in μ decay; the primary limit on the mass-square ratio δ is set by the electron polarization in Gamow-Teller β decay [23]. Note that the scales of these contours are proportional to the square root of the corresponding experimental error. The elliptical contour in Fig. 8 arising from measurement of ξP_μ would have provided the most severe constraint on δ, if the central value had not fallen well below the (V-A) prediction. The remaining muon decay parameters retain their (V-A) values even if δ and ξ are nonzero[21].

Turning to future experimental constraints, Fig. 9 exhibits the effect of improvements envisaged by the Princeton group[24] upon their previous measurements[25] of the asymmetry parameter A(0) in ^{19}Ne β decay. Figure 10 shows limits which may be obtained by new measurements of ρ and ξ, as well as by comparison of electron helicity in Fermi and Gamow-Teller β decay. If the new experiments do remain relatively consistent, and if their relative sensitivities are distributed as Fig. 10 would suggest, most of the new information on possible right-handed currents will come from the measurement of ξP_μ. This helps to explain our enthusiasm at LBL and TRIUMF for undertaking so exacting a measurement.

I have appended in Fig. 11 and its caption the rate estimates for a conceivable future search for W_R effects in the reaction $e^- p \to \nu_e X$ in 30 GeV $e^- \times$800 GeV p collisions at the proposed HERA facility. It would be an enormously challenging experiment. The 500-GeV mass scale seems to go a long way toward equalizing in difficulty even so disparate a collection of experiments as Figs. 9-11 represent.

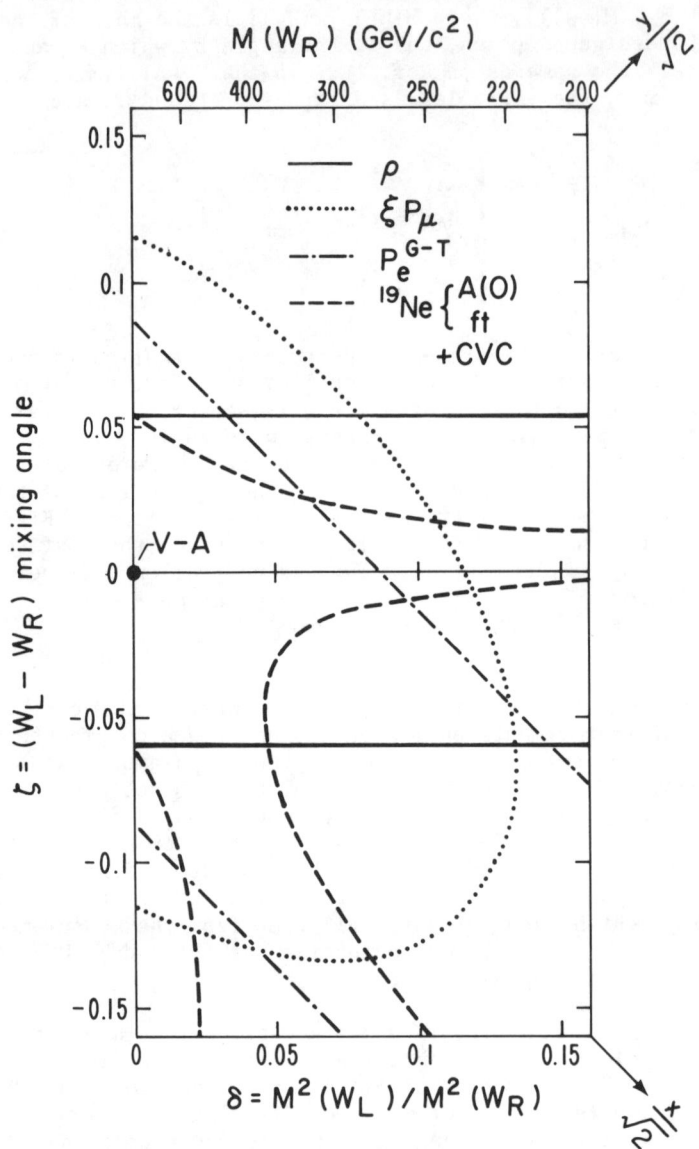

Fig. 8. Existing two-standard-deviation limits on the parameters δ (square of $M(W_L)/M(W_R)$ mass ratio) and ζ (W_L-W_R mixing angle) describing a possible right-handed gauge boson W_R. See Ref. 21 for the definition of ζ. If only a single left-handed gauge boson exists, $\delta \equiv \zeta \equiv 0$. The mixing angle is limited by the muon decay parameter ρ (Ref. 8); and also by the asymmetry parameter A(0) in ^{19}Ne β decay (Ref. 25), combined with decay rate measurement and calculations using the conserved-vector-current hypothesis (Ref. 22). The W_R mass is limited by the electron polarization measured in Gamow-Teller β decay (Ref. 23), and also by measurement of the product ξP_μ of the polarization parameter ξ describing muon decay and the polarization P_μ of a μ^+ from π^+ decay at rest.

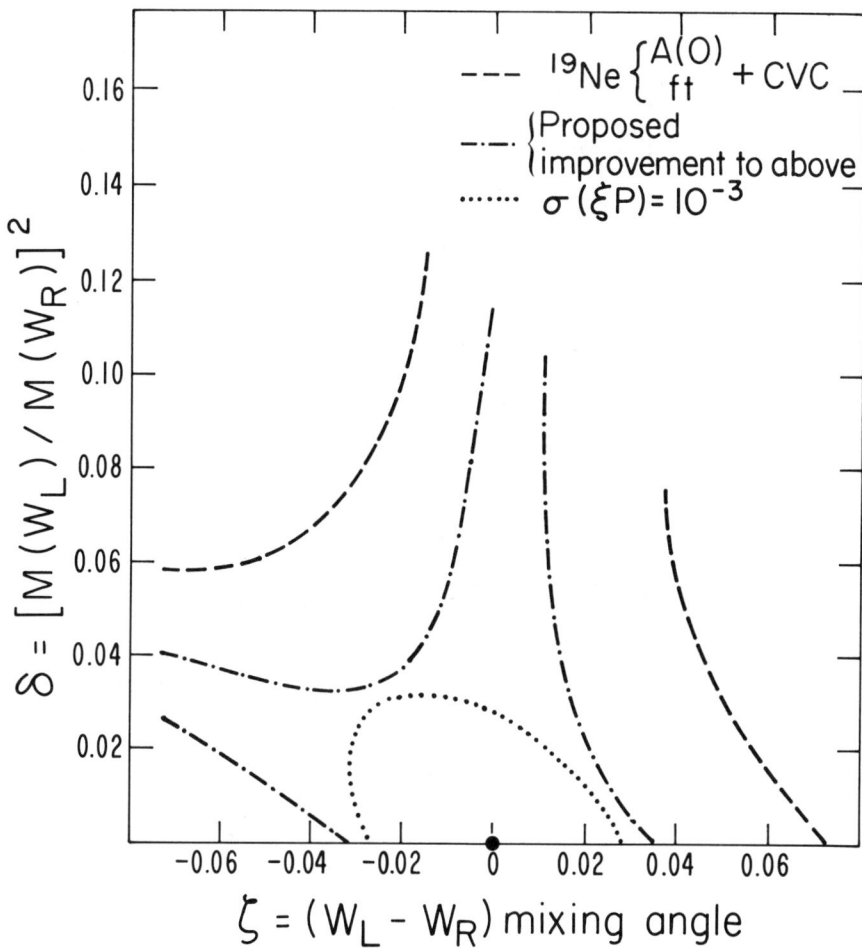

Fig. 9. Proposed improvement by the Princeton group (Ref. 24) upon the two-standard-deviation limits on ξ and δ based in part on their previous measurement of the asymmetry parameter A(0) in ^{19}Ne decay (Refs. 22, 25). The existing limit represented by the dashed line is the same as that in Fig. 8 (note reversal of axes). For comparison, the restriction on δ and ξ which would be obtained by a 0.1% measurement of ξP_μ (see Fig. 10) is given by the dotted ellipse.

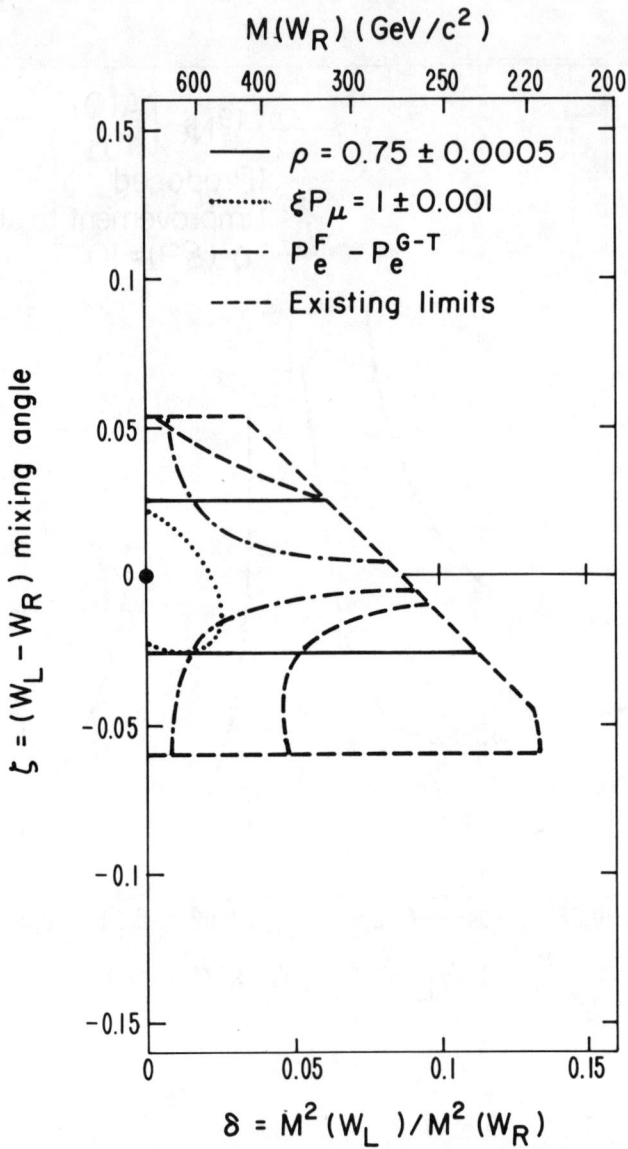

Fig. 10. Anticipated two-standard-deviation limits on the mass-squared ratio δ and mixing angle ζ parameterizing a possible right-handed W, to be obtained by experiments proposed or in progress. The dashed contours reproduce the most restrictive of the existing limits in Fig. 8. The error attached to the muon decay parameter ρ is twice the *statistical* error anticipated from 10^8 events in the Los Alamos-Chicago-NRC/Canada-Carleton experiment under construction at LAMPF. A comparison of electron polarization in Fermi and Gamow-Teller transitions proposed by the Michigan group (Ref. 26) will limit the product $\zeta\delta$ (dot-dashed contour). The dotted ellipse corresponds to the sensitivity anticipated for the Berkeley-British Columbia measurement of ξP_μ (Ref. 17; Figs. 6 and 7).

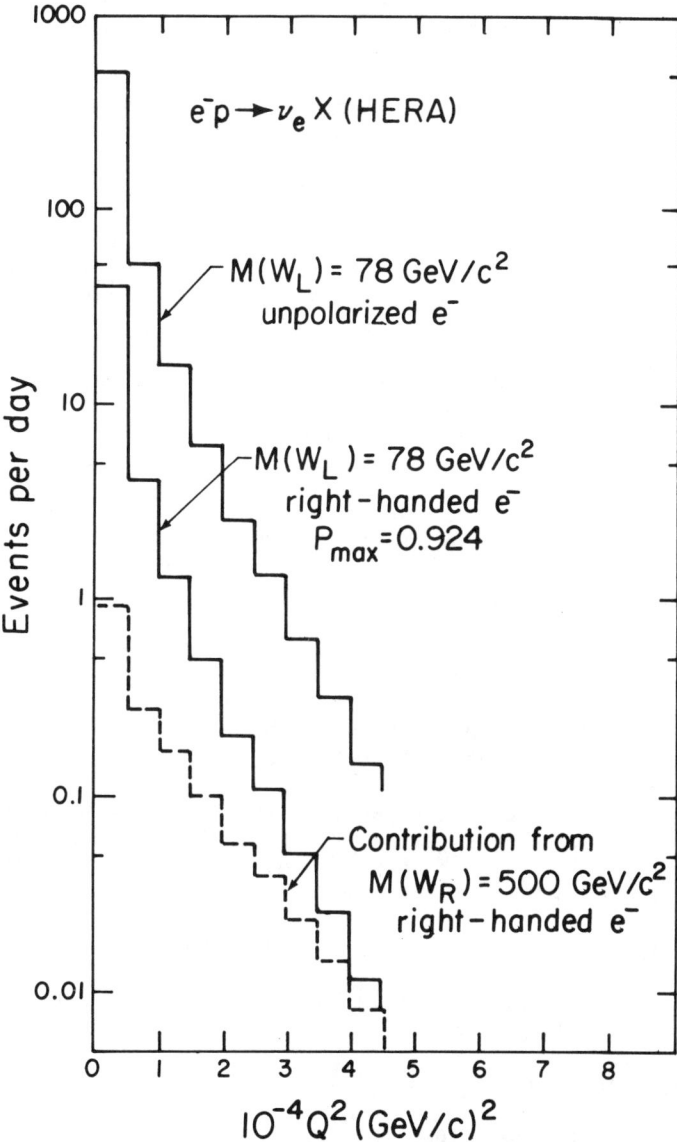

Fig. 11. Conceivable search for effects of a right-handed gauge boson W_R at a future e^-p colliding-beams facility. Rate estimates for HERA, at 30 GeV e^- ×800 GeV p the most ambitious of the proposed e^-p rings, are adapted from the ECFA workshop proceedings (Ref. 27). The top curve is the event rate per interval $\Delta(Q^2)=5000$ $(GeV/c)^2$ for the as-yet-unobserved reaction $e^-p \to \nu_e X$, assuming *unit* detection efficiency and *average* luminosity of 10^{32} cm^{-2} sec^{-1}. The contribution from W_R exchange ($M(W_R)$=500 GeV/c^2) is undetectable against this "background". The latter may be reduced by longitudinal electron polarization (0.924 is the quantum limit). In this limit, for example, for $Q^2>10^4$ $(GeV/c)^2$ W_R exchange makes a 15% difference in an event rate of 2 per day, using the ideal luminosity and detection efficiency mentioned above.

V. ACKNOWLEDGMENT

I wish to express my appreciation to the organizers of this splendid Workshop, and to Bob Cahn of LBL for calculations and advice on a number of the issues I have discussed. This work was supported by the Director, Office of Energy Research, Office of High Energy and Nuclear Physics, High Energy Physics Division of the U.S. Department of Energy under Contract No. W-7405-ENG-48.

REFERENCES

[1] A.R. Clark, K.J. Johnson, L.T. Kerth, S.C. Loken, T.W. Markiewicz, P.D. Meyers, W.H. Smith, M. Strovink, and W.A. Wenzel (Berkeley/LBL); R.P. Johnson, C. Moore, M. Mugge, and R.E. Shafer (Fermilab); G.D. Gollin, F.C. Shoemaker, and P. Surko (Princeton).

[2] F. Corriveau et al., submitted to 8th International Conference on High Energy Physics and Nuclear Structure, Vancouver, Canada, August 1979. See also J. Egger, in the Proceedings of the above Conference, edited by D.F. Measday and A.W. Thomas, Nucl. Phys. A335, 91-94 (1980).

[3] A.R. Clark et al., Phys. Rev. Lett. 46, 299 (1981).

[4] C.Y. Prescott et al., Phys. Lett. 77B, 347 (1978), and Phys. Lett. 84B, 524 (1979).

[5] F. Wilczek and A. Zee, Nucl. Phys. B106, 461 (1976); T. Cheng and L. Li, Phys. Rev. D16, 1425 (1977); D. McKay and H. Muczek, Phys. Rev. D19, 985 (1979); M. Abud and A. Bottino, Nuovo Cimento 51A, 473 (1979); Z. Hioke, Prog. Theo. Phys. 58, 1859 (1977).

[6] S. Weinberg and B.W. Lee, Phys. Rev. Lett. 38, 1237 (1977); Y. Achiman and B. Stech, Phys. Lett. 77B, 384 (1978). These models and those of Ref. 5 specify μ^+-$\bar{M}^0$ couplings of Fermi strength with $m_{M^0} \approx 4$-5 GeV/c^2 and $B(\bar{M}^0 \to \mu^+\mu^-\nu) \approx 0.1$-$0.2$.

[7] M. Bardon et al., Phys. Rev. Lett. 14, 449 (1965).

[8] J. Peoples, Ph.D. Thesis, Columbia University, Nevis Cyclotron Laboratory Report No. 147 (1966).

[9] S. Derenzo, Phys. Rev. 181, 1854 (1969).

[10] D. Fryberger, Phys. Rev. 166, 1379 (1968).

[11] V.V. Akhmanov et al., Sov. J. Nucl. Phys. 6, 230 (1968).

[12] A.M. Sachs and A. Sirlin, in Muon Physics, edited by V.W. Hughes and C.S. Wu (Academic Press, New York, 1975), Vol. II, pp. 49-81.

[13] F. Scheck, Physics Reports 44, 187 (1978). See also Ref. 12.

[14] C.J. Martoff, "Reconsideration of a General Amplitude for Muon Decay from a Gauge-Theory Style Lagrangian", University of California, Berkeley report, 1978 (unpublished); C.J. Martoff, private communication.

[15] H.L. Anderson, J.D. Bowman, C.M. Hoffman, H.S. Matis, R.J. McKee, D.E. Nagle, W.W. Kinnison, C.K. Hargrove, H. Mes, A.L. Carter, and D. Kessler, "High Precision Study of the μ^+ Decay Spectrum" (H.L. Anderson and J.D. Bowman, spokesmen), Proposal #455 to the Los Alamos Meson Physics Facility, November 1978 (unpublished); R.J. McKee, "Extraction of the Michel Parameters by Maximum Likelihood", LAMPF Experiment #455 internal report, 1980 (unpublished); J.D. Bowman, these Proceedings.

[16] J.H. Brewer, C.M. Clawson, K.M. Crowe, C.J. Martoff, J.M. Miller, and W.A. Zajc, "Measurement of the Eta Parameter in Muon Decay" (K.M. Crowe, spokesman), Proposal #134 to TRIUMF, 1979 (unpublished); J.H. Brewer, C.M. Clawson, K.M. Crowe, J.A. Jansen, C.J. Oram, and M. Salomon, "Measurement of the Parameter ξ in the Muon Decay" (K.M. Crowe, spokesman), Proposal #176 to TRIUMF, 1980 (unpublished).

[17] G. Gidal, C. Oram, H.M. Steiner, M. Strovink, and R.D. Tripp, "Precise Measurement of the Polarization Parameter ξ: A Search for the Effects of a Right-handed Gauge Boson in μ^+ Decay", Proposal #185 to TRIUMF, 1980 (unpublished).

[18] C.J. Oram, J.B. Warren, G.M. Marshall, and J. Doornbos, Nucl. Inst. and Meth. 179, 95 (1981).

[19] H.E. Haber, G.L. Kane, and T. Sterling, Nucl. Phys. B161, 493 (1979).

[20] B. McWilliams and L.-F. Li, Carnegie-Mellon University preprint COO-3066-146 (April, 1980).

[21] M.A.B. Bég et al., Phys. Rev. Lett. 38, 1252 (1977).

[22] B.R. Holstein and S.B. Treiman, Phys. Rev. D16, 2369 (1977).

[23] J. Van Klinken, Nucl. Phys. 75, 145 (1966).

[24] D. Schreiber and F.P. Calaprice, "Beta Asymmetry ($A\vec{J}\cdot\vec{p}_e$) of ^{19}Ne and Its Relationship to SCC - CVC and Right-handed Currents", Princeton University research proposal, 1980 (unpublished); F.P. Calaprice, private communication.

[25] F.P. Calaprice et al., Phys. Rev. Lett. 35, 1566 (1975).

[26] D. Newman and A. Rich, "Precision Positron Polarimetry - A New Technique in Weak Interaction Studies", University of Michigan

Technical Report DE-AC02-79ER10451, 1980 (unpublished); A. Rich, private communication.

[27] <u>Study</u> <u>on</u> <u>the</u> <u>Proton</u>-<u>Electron</u> <u>Storage</u> <u>Ring</u> <u>Project</u> <u>HERA</u> (Report of the Electron Proton Working Group of ECFA), edited by U. Amaldi, Report No. ECFA 80/42 - DESY HERA 80/01, 17 March 1980, Section II.2, Fig. 2.4 (unpublished).

COMMENTATOR

J. D. Bowman
Los Alamos Scientific Laboratory

I am going to briefly describe an experiment that is being prepared at Los Alamos to measure the spectral shape and the polarization dependence of electrons from normal µ decay. The obvious virtue of the study of this decay is that the particles are only weakly, or at worst electromagnetically, interacting and for a (V-A) theory the radiative corrections can be calculated very well. Our present knowledge of the Mischel parameters which determine the shape of this spectrum is summarized in Table 1: ρ is the high energy polarization-independent shape parameter, η is the low energy polarization-independent shape parameter, ξ multiplies the polarization dependent part, and δ is the polarization dependent shape parameter. These values have been analyzed by Dorenzo and they give values for the couplings shown in Table 2. As you see rather large amounts of non(V-A) couplings are allowed.

The proposed experiments aim at achieving the limits shown in Table 2 on the coupling constants, roughly a factor of 3 or 4 improvement on existing experiments. The existing experiments were done in the early sixties, they were limited by two considerations: all of the experiments observed about 10^6 µ-decays, and they used thick stopping targets and relatively massive detectors. The features of the present experiment are that we aim at statistics of 10^8 events. That would be obtained in taking one event per macropulse at LAMPF for 10^6 sec. We use a time projection chamber which has low mass, it measures down to 1 MeV, which is particularly important for the determination of η. It also has a large solid angle. We use a surface muon beam which provides nearly 100% polarization and we measure the 4 Mischel parameters simultaneously, which gives one certain advantages in their simultaneous determination. The muon beam comes from π^+'s decaying at rest to $\mu^+\nu_\mu$ in the surface of the production target. The μ^+ has a momentum of 30 MeV/c and is 100% polarized. For the particular beam that we have the e:µ ratio is about 1:1; ΔP/P is 2.5% which gives about a 10 mg/cm^2 range straggling; and we have (6 x 10^5) µ's/sec. We may trade off one or another of these things in the final beam.

Now to say a little about our time projection chamber. The surface muon beam enters as shown in Fig. 1. We detect the µ coming in with some beam counters. If we think we get a µ that we like, we turn off the beam by an electrostatic deflector; so we are guaranteed that we only study the decay of 1 µ at a time. There is an electric field that is anti-parallel to the direction of the muon. A magnet borrowed from SLAC provides a field parallel to the electric field. Ionization electrons from the µ which is slowing down drift to a series of multiplication detectors. The µ stops and then decays. The electron spirals around in the magnetic field and its ionization electrons drift to the detectors. We look at every segment of the track of the muon and the track of the electron. X and Y are given by the place where the electrons arrive at the array of detectors. Z is given by the length of time that they take to arrive. We display in Fig. 2 some tracks that are reconstructed from a prototype detector with the array of detectors that detect the arrival of electrons. We use a flash encoder and a shift register to acquire and analyze this data. The flash encoder sequentially analyzes the signals from each of the pad detectors. We display in Fig. 3 the data collection sequence. A µ stops and the electrons drift towards the pad. After about 600 ns there is a pulse produced on the pad by the arrival and multiplication of the electron signals. Every 40 ns the ADC or flash encoder digitizes the signal that is present in that interval of time. One builds up a digital representation of this charge arrival profile which is stored in an 8-bit x 256 shift register. At any time this shift register contains the information on the 20 microseconds of history of charge arrival profile over the whole area of the TPC. We presently have the magnet in place and the fiber glass vacuum tank is being made. We are working on various parts of the instrumentation, the largest open problem is to get all the electronics working. We expect to have test runs in the Summer of '81 and data runs in the Fall of '81.

MICHEL PARAMETER ERRORS

Parameter	$\sqrt{e_{kk}/N}$	Current Accuracy
ρ	0.00023	0.003
η	0.0061	0.21
ξ	0.00099	0.014
δ	0.00064	0.009

Table 1

LIMITS ON THE COUPLING CONSTANTS

COUPLING CONSTANT	OUR LIMITS	DERENZO'S LIMITS
Axial vector	$0.988 \leq x_A \leq 1.052$	$0.76 \leq x_A \leq 1.20$
Tensor	$x_T \leq 0.027$	$x_T \leq 0.28$
Scaler	$x_S \leq 0.048$	$x_S \leq 0.33$
Pseudoscaler	$x_P \leq 0.048$	$x_P \leq 0.33$
Vector axial vector phase	$0° \leq \Phi_{va} \leq 2.6°$	$0 \leq \Phi_{va} \leq 15°$

Table 2

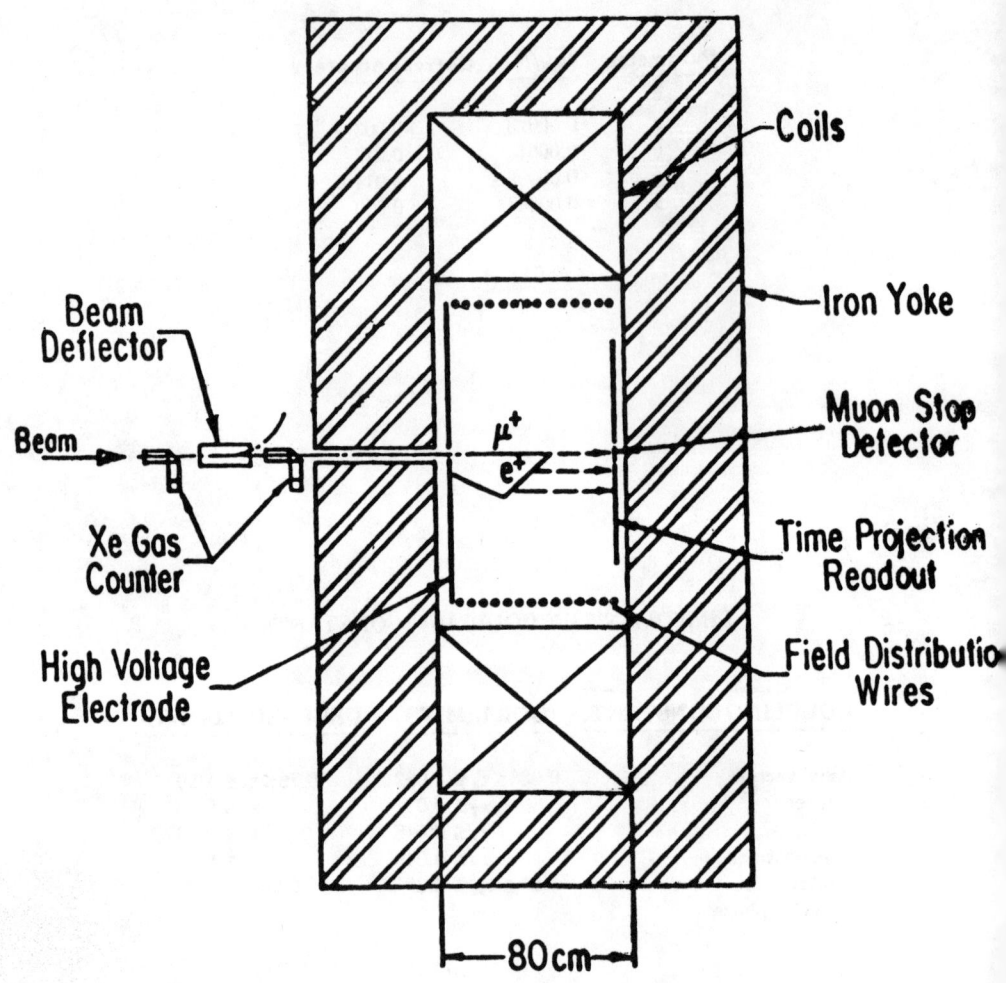

Figure 1

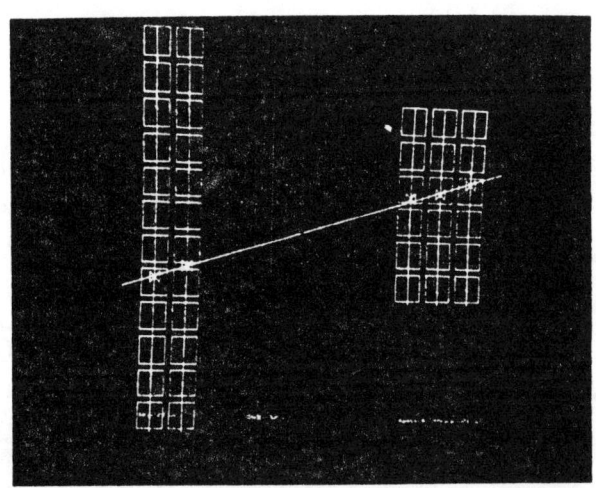

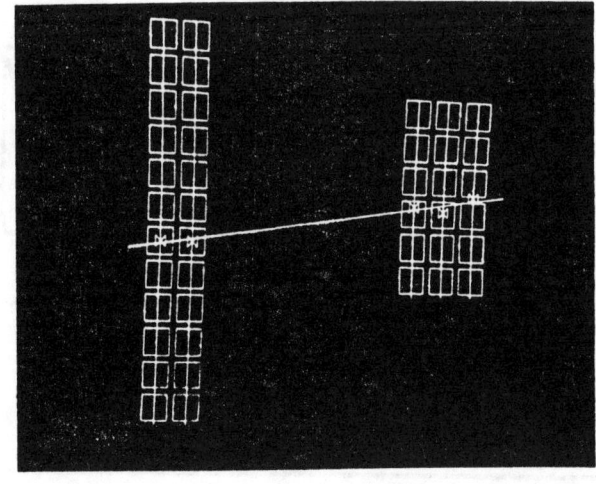

Figure 2

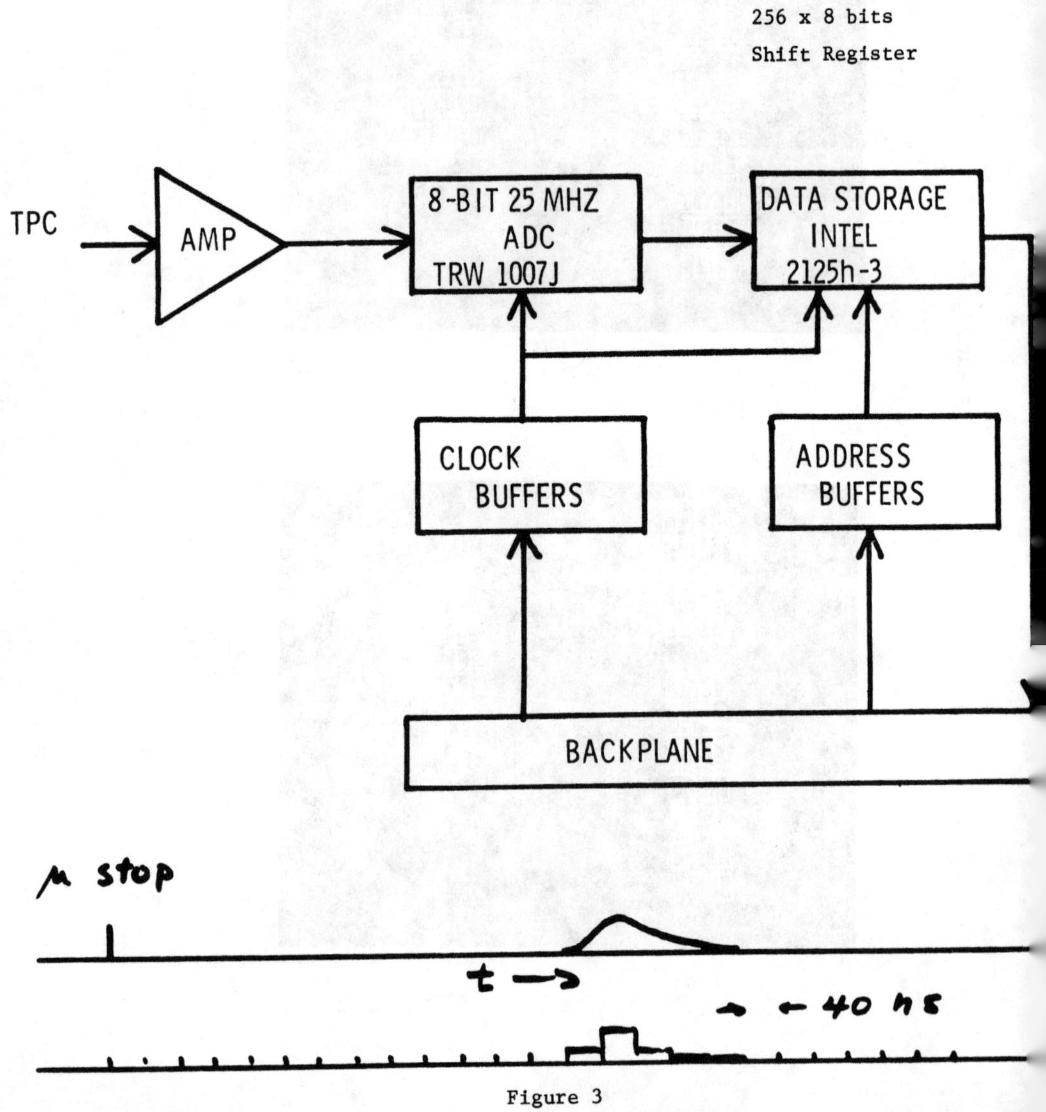

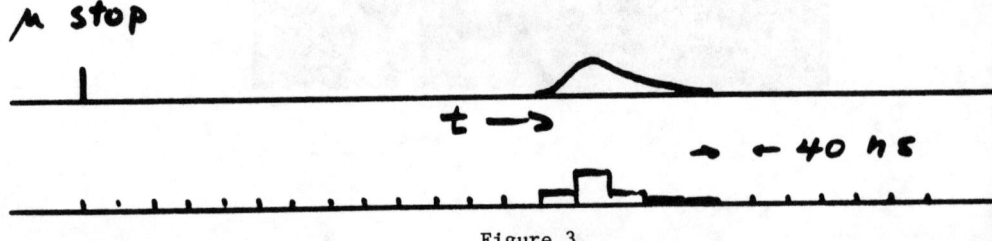

Figure 3

DISCUSSION

SIRLIN: It seems to me that at the level of precision of the experiments you are discussing, it may be necessary to include the radiative corrections of $O(\alpha^2)$, i.e. two-loop corrections. For example, at the one-loop level the radiative corrections increase the effective value of ρ by about 0.04, so it is conceivable that the two-loop corrections may induce changes of $O(0.04\alpha) = O(3 \times 10^{-4})$ and perhaps larger. Is there anybody calculating these two-loop corrections?

BOWMAN: We don't know if anybody is doing them. Our point of view is that they can be done and we hope somebody does them. Since we believe they can be done for V-A, we plan to do them.

SIRLIN: The second point I wanted to mention is that major accurate measurements of the Mischel parameters will provide some information on neutrino mixing, particularly the neutrino mixing of the neutrino of the muon and the neutrino of the tau, because the mass of the tau neutrino is sufficiently small to participate in the decay. It is very, very interesting.

WANG: I just want to comment on the first point by Dr. Strovink. MARK J at PETRA has set a very stringent limit on the existence of heavy leptons, I think with larger mass than you have. It's just another source of experimental input.

STROVINK: Not on neutral leptons.

WANG: Their limits are on charged leptons.

A PLANNED EXPERIMENT ON PARITY VIOLATION FOR ATOMIC HYDROGEN

V. Hughes
Yale University
New Haven, CT

Maurice Goldhaber has the view that theories should always be discussed in the present tense, accelerators in the future tense and experiments in the past tense. I think he is absolutely right, but let me discuss some experiments which are presently in progress. The phenomenological hamiltonians are shown here:

$$H_{PNC}^{quarks} = \sum_{quarks} \frac{G}{\sqrt{2}} [\varepsilon_{AV}(e,u)\bar{e}\gamma^\lambda \gamma_5 e \bar{u}\gamma_\lambda u + \varepsilon_{AV}(e,d)\bar{e}\gamma^\lambda \gamma_5 e \bar{d}\gamma_\lambda d$$
$$+ \varepsilon_{VA}(e,u)\bar{e}\gamma^\lambda e \bar{u}\gamma_\lambda \gamma_5 u + \varepsilon_{VA}(e,d)\bar{e}\gamma^\lambda e \bar{d}\gamma_\lambda \gamma_5 d] \tag{1a}$$

$$H_{PNC}^{atoms} = \sum_{nucleons} \frac{G}{\sqrt{2}} [c_{1p}\bar{e}\gamma^\lambda \gamma_5 e \bar{p}\gamma_\lambda p + c_{2p}\bar{e}\gamma^\lambda e \bar{p}\gamma_\lambda \gamma_5 p$$
$$+ c_{1n}\bar{e}\gamma^\lambda \gamma_5 e \bar{n}\gamma_\lambda n + c_{2n}\bar{e}\gamma^\lambda e \bar{n}\gamma_\lambda \gamma_5 n] \tag{1b}$$

$$c_{1p} = 2\varepsilon_{AV}(e,u) + \varepsilon_{AV}(e,d)$$
$$c_{1n} = \varepsilon_{AV}(e,u) + 2\varepsilon_{AV}(e,d)$$
$$c_{2p} = 2F\,\varepsilon_{VA}(e,u) + (F-D)\,\varepsilon_{VA}(e,d) \tag{1c}$$
$$c_{2n} = (F-D)\varepsilon_{VA}(e,u) + 2F\varepsilon_{VA}(e,d)$$

$$g_A = F+D$$
$$F = 0.42$$
$$D = 0.92$$

There are four coupling constants involved in the electron-nucleon or electron-quark sector, with the usual notation for the vector and axial vector coupling for the leptons and quarks. For atoms, you usually write it the second way with the c's for proton and neutron, which bear the relationship shown in (1c) to the four coupling constants. In principle, from hydrogen, one can determine all of these. In the heavy atom parity experiments that have been done thus far, one essentially determines a weak charge-type quantity $Q_W(Z,A) = 2[Zc_{1p} + (A-Z)c_{1n}]$. Of course, the Weinberg-Salam theory relates all of these to a single parameter $\sin^2\theta_W$:

$$c_{1p} = \frac{1}{2}(1-4\sin^2\theta_W); \quad c_{1n} = -\frac{1}{2}$$

$$c_{2p} = \frac{g_A}{2}(1-4\sin^2\theta_W); \quad c_{2n} = -\frac{g_A}{2}(1-4\sin^2\theta_W)$$

$$\varepsilon_{AV}(e,u) = \frac{1}{2}(1 - \frac{8}{3}\sin^2\theta_W)$$

$$\varepsilon_{AV}(e,d) = -\frac{1}{2}(1 - \frac{4}{3}\sin^2\theta_W)$$

$$\varepsilon_{VA}(e,u) = \frac{1}{2}(1 - 4\sin^2\theta_W)$$

$$\varepsilon_{VA}(e,d) = -\frac{1}{2}(1 - 4\sin^2\theta_W)$$

The aim of our experiment is to determine these constants without relation to the Weinberg-Salam theory. From the high energy e-D experiment, as you know, one measures the asymmetry in the scattering of polarized electrons from unpolarized deuterons:

$$A = \frac{d\sigma^+ - d\sigma^-}{d\sigma^+ + d\sigma^-}.$$

The principal quantity determined is the value

$$A^{eD}/Q^2\alpha\, a = 2\varepsilon_{AV}(e,u) - \varepsilon_{AV}(e,d).$$

The SLAC-Yale e-D experiment gave $a = (-9.7 \pm 2.6) \times 10^{-5}$.

Now the hydrogen atom experiments, none of which have been done yet (although several are underway) involve the familiar Lamb shift diagram shown in Fig. 1. The energy levels in the magnetic field, the Lamb shift at 6000 megahertz separation and the hyperfine intervals are shown in Fig. 2. The main reason for using the n=2 state, of course, is that the 2P state and 2S state are close by, so a parity-violating term in the Hamiltonian, which would mix S and P states, is associated with a small energy denominator; hence there is a relatively large admixture of the other parity state if the parity non-conserving term is present. The experiment essentially involves making a transition within the 2S state, say a magnetic dipole transition, and for the moment I will discuss the transition 1 → 4. You then note that if you have the parity-violating term in the Hamiltonian and apply not only the magnetic rf field but also an electric rf field, you can get a transition through the two-step process of the parity-violating term in the Hamiltonian and the rf electric field. It is the interference between these two amplitudes with regard to the intensity of the transition that people hope to look at.

To make this a little more transparent, let me consider one simple case. Let's consider the transition 1 → 4. The experiments are often considered at the magnetic field where levels cross, where of course you have a smaller energy denominator, zero in fact, but you have the decay rate of states that must be considered. So we consider transition 1 → 4 by a normal parity-conserving magnetic dipole transition. Now the state 4 will have have admixed a bit of state 8; that will be the most important admixture at the crossing field of about 1100 G. The admixture will be given as follows: let the parity-conserving (PC) M1 amplitude induced by $B_x \cos \omega t$ be b_M. Let the mixed state be $|4'\rangle = |4\rangle + \delta_{4,8}|8\rangle$. The admixture will be given by:

$$\delta_{4,8} = \frac{\langle 8|H_{PNC}|4\rangle}{(E_4-E_8) - \frac{i\hbar}{2}(\Gamma_4-\Gamma_8)},$$

which, if you evaluate it, turns out to be about $-2.6 \times 10^{-10}(c_{1p} + 1.1\ c_{2p})$. So, if you have not only an rf magnetic field present but also an rf electric field, say $E_y \cos(\omega t + \phi)$, then you will get an amplitude for this transition 1 → 4 also of this same form. Let the parity non-conserving E1 amplitude induced by E_y be b_E. The overall transition probability will be $\propto |b_M + b_E|^2$. Now what one hopes to be able to see is, of course, the interference term which is $\propto b_M b_E$. The interference depends on the phase angle ϕ between the electric and magnetic field. It is that quantity which one should be able to vary with the least probability of systematic errors. So, if you say you are going to change ϕ from $+\pi/2$ to $-\pi/2$, the fractional change in the intensity of this transition turns out to be

$$\delta P/P \propto \frac{|b_E|}{|b_M|} = 5 \times 10^{-7}(c_{1p} + 1.1\ c_{2p})\frac{E_y}{B_x}.$$

The final ratio here depends on the microwave arrangement you have.

The above is a typical order of magnitude, so one is forced to look for a very small change in intensity. And, of course, that is where the difficulty in the experiment arises. You basically are looking for a pseudoscalar-type term. There is a diagramatic way people talk about this. Fig. 1(b) would be the normal magnetic dipole transition between the initial 2S state and the final 2S state induced by an rf field B. Fig. 1(a) would be the parity-violating transition involving the two-step process, the parity-violating term and the applied electric field. One can also induce the primary transition by the so-called induced Stark effect by applying a static electric field, and then an rf electric field as well (cf. Fig. 1(c)). Measuring the interference between these two, is the aim of several experiments.

Now, what is the situation? At Yale, we've built the apparatus, we've seen the basic M_1 transitions. What we are trying to do is look for the transition near zero field for certain technical reasons, and that's where we stand. There are two other groups seriously working on this problem, at Michigan and at Seattle. The big aims are to determine all four constants ultimately and, ideally with sufficient accuracy, to measure virtual radiative corrections and test renormalizability. As far as our own work is concerned, it is very hard to project when things will happen. I would be very surprised if we had anything interesting for this kind of group in less than a year.

One other experiment I would just like to mention, that is quite active, involving an MIT-Yale-CCNY-Harvard collaboration, is to look for parity-nonconservation in essentially a SLAC-type experiment, using polarized electrons on carbon. Because carbon is a $J = 0$, $I = 0$ nucleus, the asymmetry is independent of form factor. The projected asymmetry, taking the Weinberg-Salam theory, is about 10^{-6}. The hope is to get to the order of 10^{-7} in the measurement, as has been done in polarized pp scattering.

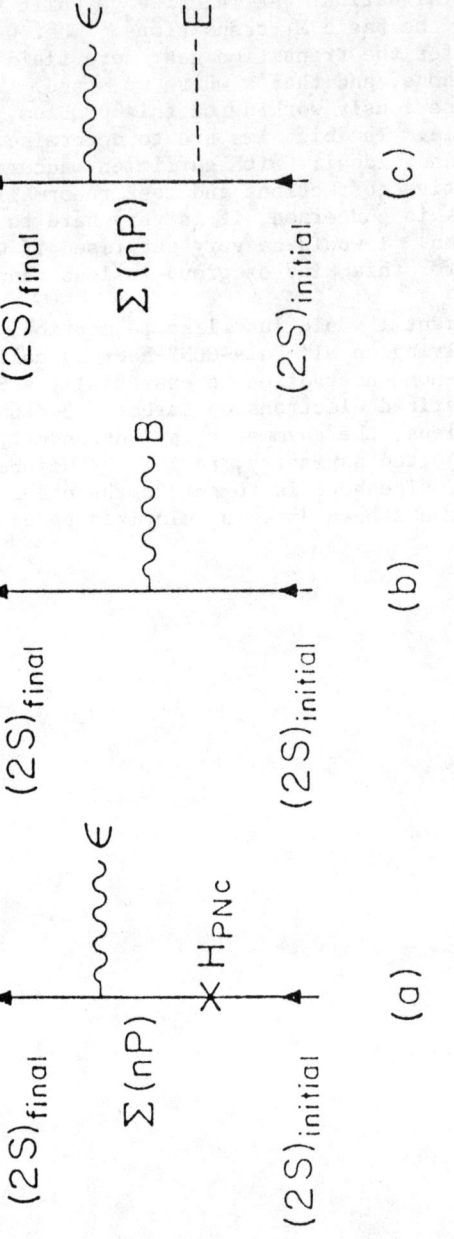

Fig. 1

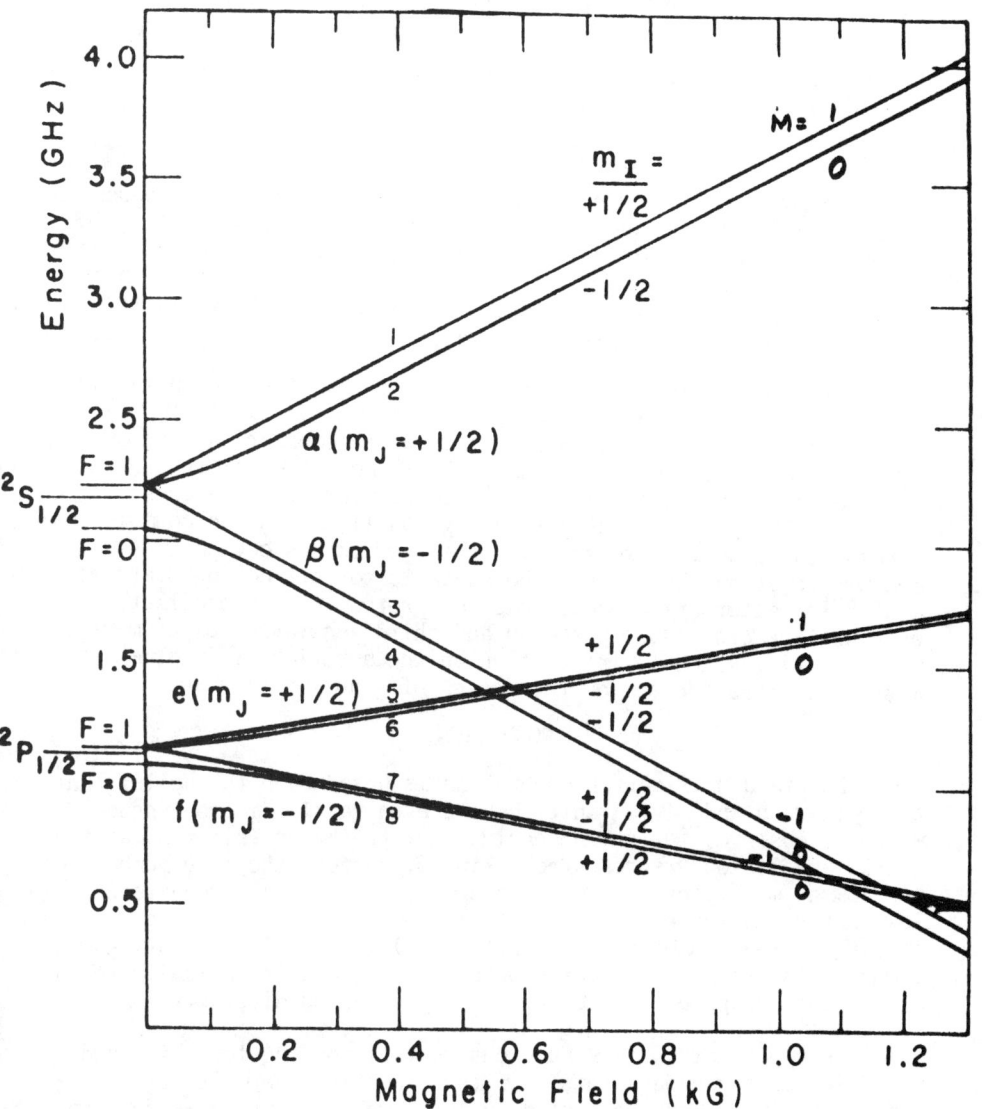

Fig. 2. The relevant energy levels of hydrogen in a magnetic field, drawn approximately to scale. The zero field hyperfine splittings are $a(2\,^2S_{1/2}) = 177.56$ MHz and $a(2\,^2P_{1/2}) = 59.2$ MHz. The Lamb shift is 1057.9 MHz.

Baryon and Lepton Number Non-Conservation

PROBING THE DESIGN OF GRAND UNIFICATION
THROUGH CONSERVATION LAWS

Jogesh C. Pati
Center for Theoretical Physics, Department of Physics and Astronomy
University of Maryland, College Park, Maryland 20742

ABSTRACT

The purpose of this talk is to note a few special consequences of gauging "maximal" quark-lepton symmetries such as SU(16), which is the maximal symmetry for a single family of fermions. Within these symmetries, violations of B, L and F are spontaneous rather than explicit. Furthermore these symmetries as a rule permit intermediate mass scales $\sim (10^3-10^6$ GeV) and $(10^8-10^{11}$ GeV) filling the so-called grand plateau between 10^2 and 10^{15} GeV. It has been shown in earlier papers that within these symmetries proton may decay via <u>four</u> alternative modes: i.e. proton $\to$ one or three leptons or antileptons plus mesons; some of which can coexist. It is now observed that even n-$\bar{n}$ oscillations can coexist with (B-L) conserving proton-decays of the type $p \to e^+\pi^0$ etc. without posing any conflict with the cosmological generation of baryon-excess; both these processes can possess measurable strengths so as to be amenable to forthcoming searches. Search for alternative decay modes of proton and n-$\bar{n}$ oscillations, even as rare processes in second and third generation experiments, would provide valuable information on the question of intermediate mass-scales and thereby on the design of grand unification.

I. INTRODUCTION

The grand unification-hypothesis is based upon the belief that there lies an underlying unity behind even vastly diverse phenomena which we observe.[1-3] At the basic level the hypothesis assumes that quarks and leptons represent one kind of matter; they are members of one common multiplet; the weak, electromagnetic and strong forces are low energy manifestations of a single force, which is generated by a gauging of the symmetry group of this multiplet. The primary gauge symmetry is assumed to break spontaneously and hierarchically to an appropriate subgroup to account for the observed diversity.

The hypothesis offers numerous tests involving low, high and ultrahigh energy experiments. Of these perhaps the most striking -- and the one which can be tested in the immediate future -- concerns the question of conservations of baryon number, B, lepton number, L and fermion number F. Fermion number is defined as just the sum of quark number B_q plus the lepton number L: $F \equiv B_q + L = 3B + L$. With quarks and leptons in one multiplet of a spontaneously broken local gauge symmetry G, one is inevitably led to question the conservations of these quantum numbers.[2] I shall begin by outlining briefly the circumstances which lead to this questioning.

The violations of B, L and F may arise in two alternative ways:

(i) Either spontaneously
> In this case the quantum numbers B, L and F are well defined and conserved in the limit of the unifying symmetry. Their conservations are lost through spontaneous symmetry breaking at low energies and low temperatures. They are fully recovered at sufficiently high temperatures and thus in the very early universe.

(ii) or explicitly
> Here B, L and in general F are violated explicitly in the basic gauge interactions. In other words they cannot even be defined in the limit of the unifying symmetry. They are fortuitously nearly conserved only at low energies and low temperatures. But their violations are effectively strong (of order α) at sufficiently high energies and high temperatures, where the full gauge symmetry is restored.

As a rule spontaneous violations of B, L and F arise if one gauges the "maximal" symmetry defined by the space of quarks and leptons, while explicit violations of these quantum numbers arise only provided one chooses to gauge some specific subgroups of the maximal symmetry. The two cases may be demonstrated by the following illustrative examples.

Consider a fermionic system consisting of three four component[5] quarks plus one four component lepton; the three quarks define three colors of a given flavor:

$$F = (q_r, q_y, q_b, \ell) \tag{1}$$

The <u>maximal</u> symmetry of this system of four four component fermions <u>is $\overline{SU(4)}_{color}$</u>. Let us first choose to gauge this maximal symmetry.[2] There would then arise a <u>massless</u> spin-1 gauge particle V_μ^{15} associated with the canonical 15th generator of SU(4), whose coupling would be given by

$$g V_\mu^{15} J_\mu^{15} = g V_\mu^{15} [(2\sqrt{6})^{-1} \{ \sum_{i=r,y,b} \bar{q}_i \gamma_\mu q - 3\bar{\ell}\gamma_\mu \ell\}$$
$$+ \text{(gauge spin-1-current)}] \tag{2}$$

From the fermionic content of the current J_μ^{15}, we see that the corresponding charge is proportional to $(B_q - 3L) = 3(B-L)$.

This illustrative example serves to demonstrate the general result that <u>once quarks and leptons are put in one multiplet and</u>

their maximal symmetry is gauged, then some linear combination of baryon and lepton-numbers (in this example B-L) must be among the generators of the local symmetry; coupling to a massless spin-1 gauge particle. It is thus conserved in the basic lagrangian.

Likewise if we put left-handed fermions $F_L = (q_r, q_y, q_b, \ell)_L$ and left-handed antifermions $F_L^C = (q_r^C, q_y^C, q_b^C, \ell^C)_L$ in one multiplet and gauge their maximal symmetry[6] -- for this illustrative example the symmetry would be $SU(8)$[7] -- fermion number $F \equiv B_q + L = 3B+L$ representing a second independent linear combination[8] of baryon and lepton numbers must also be among the generators of the local symmetry. There would then be a second massless spin-1 particle V_μ^F whose coupling to fermions is given by,

$$g J_\mu^F V_\mu^F = g V_\mu^F (1/4) [(\Sigma \bar{q} \gamma_\mu q + \bar{\ell} \gamma_\mu \ell)_L - (q \to q^C, \ell \to \ell^C)] \tag{3}$$

In 1955 Lee and Yang[9] showed that if the conservation of baryon number (which was generally assumed to hold at that time) was due to a massless gauge particle coupled to baryon number, then the effective coupling generated through the exchange of such a massless particle must be much weaker than gravitational in order that the limits from Eötvos type experiments may be satisfied. It is easy to see that as long as the Eötvos type experiments involve different kinds of matter with varying baryon to lepton-ratios, the same conclusion would follow for the effective coupling of a massless gauge particle coupled to any linear combination of baryon and lepton numbers.

So long as one did not wish to postulate the grand unification hypothesis unifying quarks and leptons, however, one did not have to introduce a massless spin-1 particle coupled to any linear combination of baryon and lepton-numbers in the first place; and in the second place if one did introduce such a particle, one was free to associate with it an arbitrarily small effective coupling. Now the crucial point is that within the grand unification-hypothesis, with a maximal gauging of the quark-lepton-symmetry, massless spin-1 gauge particles coupled to linear combinations of baryon and lepton-numbers would have to be introduced[2,6] as we saw before. Furthermore, within this hypothesis, the effective coupling of a gauge particle can not be arbitrarily weak, since all effective gauge interactions are of order $\alpha \approx 1/137$ at the grand unification mass-scale. There is then just one choice left in order for the hypothesis to be consistent with Eötvos-type experiments. Such spin-1 gauge particles coupled to linear combiantions of B and L must acquire mass spontaneously. It then follows that the associated charges (for the illustrative example B-L and F must be violated spontaneously. Later I exhibit several specific mechanisms illustrating such spontaneous violations.

Instead of gauging the maximal symmetry, one might have gauged a subgroup of the maximal symmetry still retaining the requirement that quarks and leptons are in the same irreducible multiplet of the subgroup. In this case one need not have any generator within the gauged subgroup representing a linear combination of baryon and lepton numbers; but then these quantum numbers would be violated explicitly by the basic gauge interactions. To see this, consider again the illustrative example of four fermions (q_r, q_y, q_b, ℓ) with the maximal symmetry being $SU(4)$. One may now choose to gauge a subgroup $SU(2)$ such that it treats (q_r, q_y) and (q_b, ℓ) as doublets. In this case, there would be no linear combination of B and L among the generators of the gauged symmetry; both B and L would indeed be violated here explicitly even prior to spontaneous symmetry breaking, because one and the same gauge particle would couple for example to both $\bar{q}_r \gamma_\mu q_y$ as well as to $\bar{q}_b \gamma_\mu \ell$ currents. An analogous situation is in fact what happens in some realistic models of grand unification such as $SU(5)^3$ and^{10} $SO(10)$, both of which are subgroups of the maximal one family symmetry,^{6,11,12} $SU(16)$ (see elaborations later).

Thus the violations of B and L may well be explicit. This would be the case if there was a good reason why Nature would choose a specific subgroup of the maximal symmetry. The point of the remark made, however, is that even if baryon and lepton numbers are conserved in the basic lagrangian (and this holds automatically if we gauge the "maximal" symmetry of the quark-lepton-multiplet), either B or L or both must be violated spontaneously as the gauge particles of the quark-lepton symmetry acquire masses; this is in order that the theory may be compatible with limits from Eötvos type experiments. This line of reasoning had led in 1973 to a questioning of baryon and lepton-number conservations with the prediction that the lightest baryon -- the proton -- must decay.^{2}

The main point of my talk would be to bring out certain special features of gauging "maximal" symmetries such as $SU(16)$ or its family-extensions. The interest in such symmetries stems from the fact that as a rule they permit several intermediate mass-scales^{13} lying within the so-called "grand plateau" between 10^2-10^{15} GeV^{14} and thus offer new experimental possibilities for future high energy accelerators (including those to be built in the 1-10 TeV range). As a consequence of these intermediate mass-scales, they permit four major decay-modes for protons, some of which can coexist.

(i) p $\longrightarrow$ 3 leptons + mesons $(\Delta F=0; \Delta(B+\frac{L}{3})=0)$
  (F=3)   (F=3)

(ii) p $\longrightarrow$ lepton + mesons $(\Delta F=-2; \Delta(B+L)=0)$
  (F=3)   (F=1)

(iii) p → antilepton + mesons ($\Delta F=-4; \Delta(B-L)=0$)

(iv) p → 3 antileptons + mesons ($\Delta F=-6; \Delta(B-\frac{L}{3})=0$)   (4)

While this had been observed in earlier papers,[6] recently Salam, Strathdee and I observed[12] that proton-decay especially of the $\Delta F = -4$-variety (p → $e^+\pi^0$) can coexist with $\Delta B=2, \Delta L=0$ n-$\bar{n}$-oscillations,[15] both having measurable strength. Such a coexistence does not pose any difficulty with the comological generation of baryon-excess.

II. MAXIMAL SYMMETRY AND ITS SPONTANEOUS DESCENT

Let me specify what I mean by "maximal" gauged symmetry. This corresponds to gauging all fermionic degrees of freedom[6] with fermions consisting of quarks and leptons. Thus with n two component left-handed fermions F_L plus n two component right-handed fermions F_R (which may be replaced by the left-handed charge conjugate fields $(F^c)_L$), the maximal symmetry is SU(2n). As an example, for a single family of eight left-handed fermions (six quarks and two leptons) plus their antiparticles, the maximal symmetry is SU(16). As stated before[7] such symmetries generate triangle anomalies, which are avoided by postulating that there exists a conjugate mirror set of fermions[16] $F_{L,R}^m$ which couples to the gauge mesons through the helicity flip coupling ($F_{L,R} \leftrightarrow F_{R,L}^m$). Thus by "maximal" symmetries, we shall mean symmetries which are maximal up to the discrete mirror symmetry.

The discrete mirror symmetry applies to the gauge interactions only. The gauge invariant Yukawa couplings of the basic $F_{L,R}$ and the mirror $F_{L,R}^m$ fermions to scalar multiplets can still be independent. This would in general assign differing diagonal mass-scales to the basic versus the mirror fermions. A direct mass-mixing between the basic and the mirror fermions can be avoided, if so desired, through the imposition of a discrete symmetry in the basic lagrangian: ($F_{L,R} \leftrightarrow F_{L,R}$, $F_{L,R}^m \leftrightarrow - F_{L,R}^m$ plus the corresponding symmetry transformations on the associated gauge and Higgs particles). Since the diagonal mass terms of the mirror fermions violate physical $SU(2)_L$, like those of the basic fermions, however, it follows that the mirror fermions can not be superheavy. Their masses -- assuming that the associated Yukawa coupling constant is ≲ e -- can not be much heavier than about 100 GeV. The LEP machine may thus be very suitable for testing the mirror hypothesis.[17] The characteristic signature of the mirror fermions is their helicity flip coupling compared to the basic fermions. If the mirror fermions do not mix at all, or mix very little with the basic fermions, there would be a stable or semistable heavy mirror-matter.[17]

For maximal gauging of the three families (e, µ and τ) one would need to gauge[11] $[SU(16)]^3$ or the still extended symmetry SU(48). (As expressed elsewhere[19] such gigantic symmetries should be viewed only as <u>effective gauge symmetries</u> with the associated gauge particles, quarks, leptons, mirror fermions and Higgs bosons being composites of a more economical system of preons.) Spontaneous symmetry breaking can permit the descent of SU(48) or $[SU(16)]^3$ to the familiar low-energy symmetry $SU(2)_L \times U(1) \times SU(3)_C$ (via for example the diagonal symmetry $SU(16)_{e+\mu+\tau}$). As mentioned before, such extended maximal symmetries permit signals for grand unification at low and intermediate mass scales ($\sim 10^4$ to 10^5 GeV and 10^8-10^{10} GeV)[20] and thereby offer richer experimental possibilities than for example SU(5) or SO(10). Furthermore, depending upon the pattern of SSB of SU(48) or $[SU(16)]^3$, the observed interfamily universality (e↔µ↔τ) can appear, for example in the electroweak-sector, only below an energy scale $\sim 10^5$ GeV. Above such energies one may see fundamental distinctions between e, µ and τ-families. In what follows I shall (for simplicity) use SU(16), in much of my discussion, <u>as a language</u> for maximal symmetries, though it will ultimately be viewed as part of an extended maximal symmetry such as SU(48) or $[SU(16)]^3$).

A possible chain of extending the observed low energy symmetry $SU(2)_L \times U(1) \times SU(3)^{col}$ towards maximally gauged symmetries is outlined in Fig. 1.

$$SU(2)_L \times U(1) \times SU(3)^{col}_{L+R}$$
$$\downarrow$$
$$SU(2)_L \times SU(2)_R \times U(1)^{(B-L)}_{L+R} \times SU(3)^{col}_{L+R}$$
$$\downarrow$$
$$SU(2)_L \times SU(2)_R \times SU(4)^{col}_{L+R}$$
$$\downarrow$$
$$SU(8) \times SU(8)$$
$$\downarrow$$
$$SU(16)_{e+\mu+\tau}$$
$$\downarrow$$
$$[SU(16)]^3, SU(48)$$

Fig. 1. Towards a maximal gauging

The chain, exhibited here, is based upon the assumption that Nature is intrinsically left-right symmetric;[8] the symmetry breaks spontaneously such that low energy weak interactions utilise only left-handed particles. This chain inevitably passes through the subunification-symmetry[2] $SU(2)_L \times SU(2)_R \times SU(4)^{col}$, which embodies left-right symmetry as well as quark-lepton unification. A few desirable intermediate symmetries, which are not exhibited here, are [17] $[SU(4)]^4$ and [21] $[SU(6)]^4$ each of which contain the subunification-symmetry $SU(2)_L \times SU(2)_R \times SU(4)^{col}_{L+R}$.

The maximal one family symmetry SU(16) puts eight left-handed fermions consisting of six quarks plus two leptons and their left-handed antiparticles in one multiplet. For the electron-family, the multiplet is:

$$F_e = \{(u,d)_{r,y,b}, \nu, e^- | (d^c, u^c)_{r,y,b}, (e^c, \nu^c)\}_L \qquad (5)$$

The 255 gauge particles of SU(16) contain (63+63+1)=127 gauge particles belonging to the subgroup $SU(8) \times SU(8) \times U(1)_F$ where $U(1)_F$ denotes fermion number plus (8x8) + (8x8)=128 gauge particles belonging to the coset, which take fermions to antifermions and vice versa. The former 127 carry fermion number zero and the latter 128 carry fermion number ±2. The gauge particles of SU(16) are symbolically listed in Fig. 2.

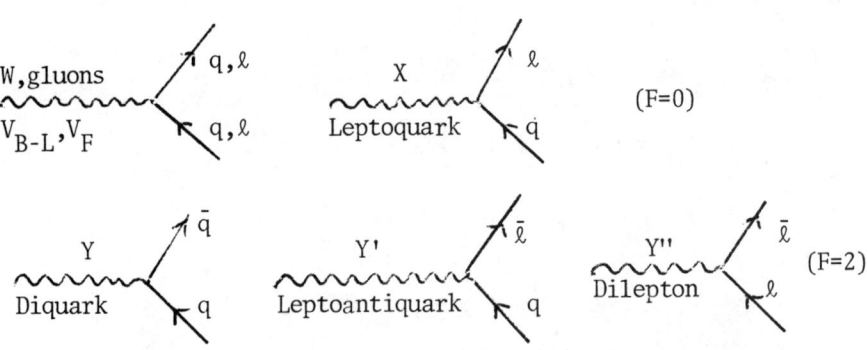

Fig. 2. Gauge particles of a maximal symmetry

Note that each of the gauge particles prior to SSB carry definite baryon, lepton and fermion numbers. Furthermore, among the gauge particles there exist two (i.e. V_{B-L} and V_F), which couple to B-L and fermions number F respectively. Following the argument presented in Sec. I, it follows that both B-L and F must be violated spontaneously as the associated massless spin 1 particles acquire masses. They are violated by two distinct mechanisms:

(i) <u>Gauge mixing</u>: Gauge particles carrying different B, L and G get mixed through SSB; this induces violations of B, L and F. In particular the mixing of Y with $\bar{Y}'$ will induce proton-decay into antilepton while that of Y with $\bar{X}$ will induce proton decay into lepton as illustrated in Figs. 3(a) and (b) respectively at the end.

Mixings	Selection Rules	Transitions
$Y \leftrightarrow \bar{Y}'$	$\Delta(B-L)=0; \Delta F=-4$	$p \to \bar{\ell}+\text{mesons}$
$X \leftrightarrow \bar{X}$	$\Delta(B+L)=0; \Delta F=-2$	$p \to \ell+\text{mesons}$

The mixing of Y and $\bar{Y}'$ occurs as SU(16) descends spontaneously to $SU(2)_L \times SU(2)_R \times SU(4)^C$, while that of Y with $\bar{X}$ occurs at a lower stage of SSB when $SU(2)_L \times U(1)$ breaks to $U(1)_{em}$. Both these mixings can arise through vacuum expectation values of a Higgs multiplet of the type $\Omega \begin{Bmatrix} \{AB\} \\ \{CD\} \end{Bmatrix} \sim \{16 \times 16\}_{symm} \times \{16^* \times 16^*\}_{symm}$ (see Ref. 12 for details).

The amplitudes for the corresponding processes will depend upon the mixing masses versus the diagonal masses, which in turn depend upon the route of spontaneous descent (see later):

$$\text{Amp}(p \to e^+\pi^0)_{\Delta F=-4} \approx g^2(\Delta^2_{YY'}/m^2_Y m^2_{Y'}) \qquad (6)$$

$$\text{Amp}(p \to e^-\pi^+\pi^+)_{\Delta F=-2} \approx g^2(\Delta^2_{XY}/m^2_X m^2_Y) \qquad (7)$$

For the simplest pattern of SSB, $m^2_Y = m^2_{Y'} = \Delta^2_{YY'}$; and thus the amplitude for $p \to e^+\pi^0$ reduces to (g^2/m^2_Y). The mixing (mass)^2 Δ^2_{XY} must be proportional to a scale representing $SU(2)_L$-breaking.[22] Thus $\Delta^2_{XY} \lesssim M m_{W_L}$, where M represents the heavier of the two masses: m_X and m_Y.

(ii) <u>Spontaneously Induced Yukawa Transitions</u>: Alternatively the violations of B, L and F may also arise through spontaneously induced Yukawa-transitions of the type:

$$q \to \ell + \phi$$
$$q \to \bar{\ell} + \phi'$$
$$q \to \bar{q} + \phi'' \tag{8}$$

Two typical mechanisms for the first and the third transitions are exhibited in Figs. 4a and 5a at the end. The effective Yukawa-transition $q \to \ell + \phi$ shown in Fig. 4a, taken in third order followed by an invariant $\lambda\phi^4$ scalar quartic coupling, subject to $\langle\phi\rangle \neq 0$, induces proton-decay into three leptons plust mesons (see Fig. 4b)

$$\text{proton} \to 3 \text{ leptons} + \text{mesons} \quad (\Delta(B+\tfrac{L}{3}) = 0; \Delta F=0) \tag{9}$$

Likewise the effective transition $q \to \bar{\ell} + \phi'$ taken in third order induces proton decay into three antileptons plus mesons:

$$p \to e^+ \bar{\nu} \bar{\nu} \text{ etc.} \quad (\Delta(B-\tfrac{L}{3})=0; \Delta F=-6) \; . \tag{10}$$

And the effective transition $q \to \bar{q} + \phi''$ taken in third order induces $n \leftrightarrow \bar{n}$-oscillations (see Fig. 5b):

$$n \leftrightarrow \bar{n} \quad (\Delta B=\pm 2, \Delta L=0, \Delta F=\pm 6) \; . \tag{11}$$

Details of the processes (8)-(11), including the types of Higgs multiplets which are relevant for these processes, may be found in Refs. 23 and 12.

Two common features of both mechanisms (i) and (ii), are worth noting:

(a) They utilise only spontaneous rather than explicit violations of B, L and F.

(b) None of these mechanisms is tied in any way to the nature of quark-charges. They hold for quark-charges being either integral or fractional.

The Subgroups SU(5) and SO(10)

Instead of gauging the maximal symmetry SU(16), one might have gauged[3,10] one of its subgroups SU(5) or SO(10). The subgroup SU(5) contains neither B-L nor F among its generators; while SO(10) contains B-L but not F. Both lead to explicit violations of B, L as well as F. This comes about as follows: While Y and Y' coupled to $\bar{q}^c \gamma_\mu q$ and $\bar{\ell}^c \gamma_\mu \ell$-currents are distinct gauge particles within the maximal symmetry SU(16) (as shown in Fig. 2), for the subgroups SO(10) and SU(5) only the symmetric combinations $Y_S \equiv (Y+\tilde{Y}')/\sqrt{2}$ are present;[4] the antisymmetric combinations

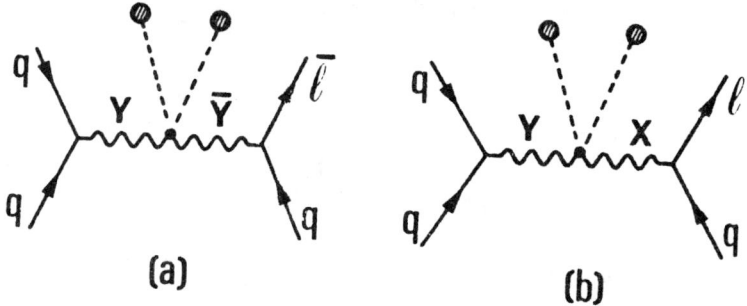

Fig. 3a,b: Spontaneously Induced Gauge Mixings leading to $3q \to \bar{\ell}$ and $3q \to \ell$-transitions. For definition of the canonical gauge fields, see Fig. 2.

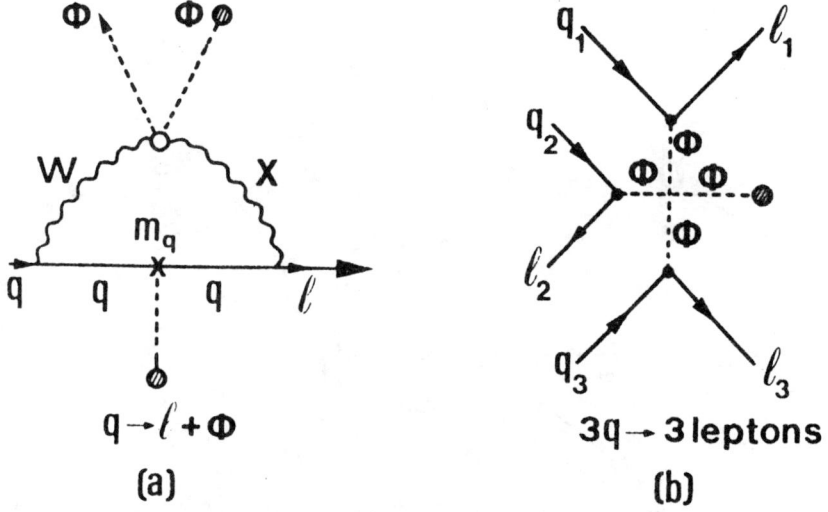

Fig. 4a,b: Spontaneously induced $q \to \ell + \phi$ and $3q \to 3\ell$ transitions.

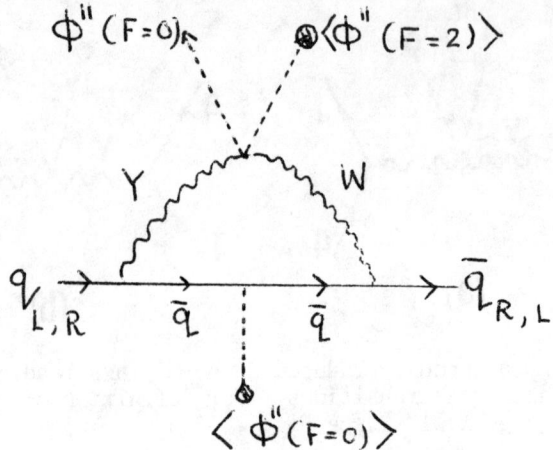

Fig. 5a. A typical diagram for the induced Yukawa-transition $q \to \bar{q} + \phi''$ (F=0), which violates fermion number by two units.

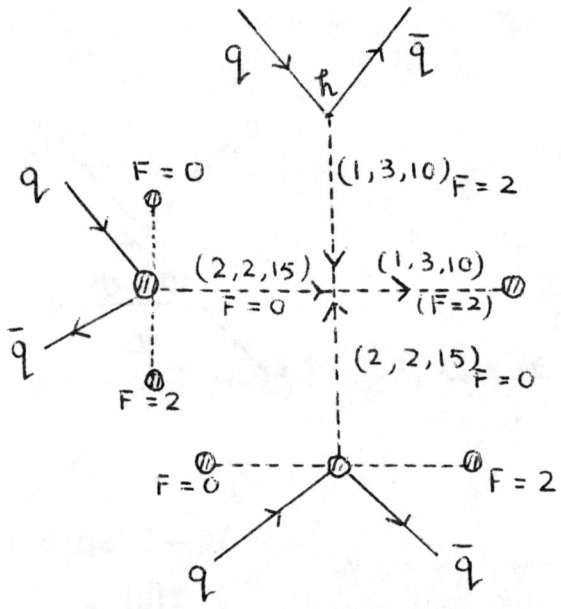

Fig. 5b. The effective transition $3q \to 3\bar{q}$ which induces n-$\bar{n}$ oscillations. The intermediate scalar fields denoted by dotted lines all belong to the single multiplet $\underline{136}$ of SU(16). They are specified in terms of their transformation properties with respect to the subsymmetry $SU(2)_L \times SU(2)_R \times SU(4)^C \times U(1)_F \subset SU(16)$, see text. Only one of the three Yukawa vertices, labelled by h, is SU(16) invariant, the other two are not.

$Y_a \equiv (Y-\tilde{Y}')/\sqrt{2}$ do not appear. The exchange of Y_S in the second order of gauge interactions would quite clearly induce transitions of the type $3q \to \bar{\ell}$ (i.e. $p \to e^+\pi^0$ etc.) leading to explicit violations of B, L and F.

These two cases, spontaneous versus explicit violations of B, L and F may agree in their predictions for proton-decay[25] and for that matter in all low energy-properties. After all SO(10) and SU(5) are subgroups of SU(16); consequently, if SU(16) descends for example via SO(10) rather than via alternative routes (see discussion later), its low energy predictions would coincide with those of SO(10). The two cases, spontaneous versus explicit violations of B, L and F, nevertheless differ from each other conceptually as well as in their physical consequences. Spontaneous violation would disappear at high temperatures exceeding the masses of the relevant gauge particles, while explicit violation would acquire its maximum strength at such temperatures, where the gauge particles would be massless. This distinction plays its most obvious role in the early stage of the universe.[25]

Route of Spontaneous Descent of SU(16)

It is now of interest to see the alternative routes of spontaneous descent of SU(16) down to the low energy symmetry $SU(2)_L \times U(1) \times SU(3)_C$. The three most obvious routes are:

(I) via $SU(8) \times SU(8) \times U(1)_F$, where the two SU(8)'s operate in the spaces of the eight fermions and the eight antifermions, respectively, and $U(1)_F$ represents fermion number.

(II) via SO(10) with respect to which the 16-plet remains irreducible.

(III) via $SU(2)_q \times SU(4)_\ell \times U(1)_{|B|-|L|}$, where $SU(12)_q$ operates on six quarks plus six antiquarks, $SU(4)_\ell$ on two leptons plus two antileptons and $U(1)_{|B|-|L|}$ represents $|B|-|L|$.

These three alternative routes together with the Higgs multiplets, whose VEV could induce spontaneous descent of SU(16) via these routes, are exhibited in Fig. 6. Vacuum expectation values of three scalar multiplets suffice to give the desired descents:[12] the adjoint 255, the fourth rank tensorial field

$\phi^{\{AB\}}_{\{CD\}} \sim \{16 \times 16\}_{symm} \times \{16^* \times 16^*\}_{symm}$ and $136 = (16 \times 16)_{symm}$. The multiplet 136 is used to give masses to fermions as well as to gauge bosons.

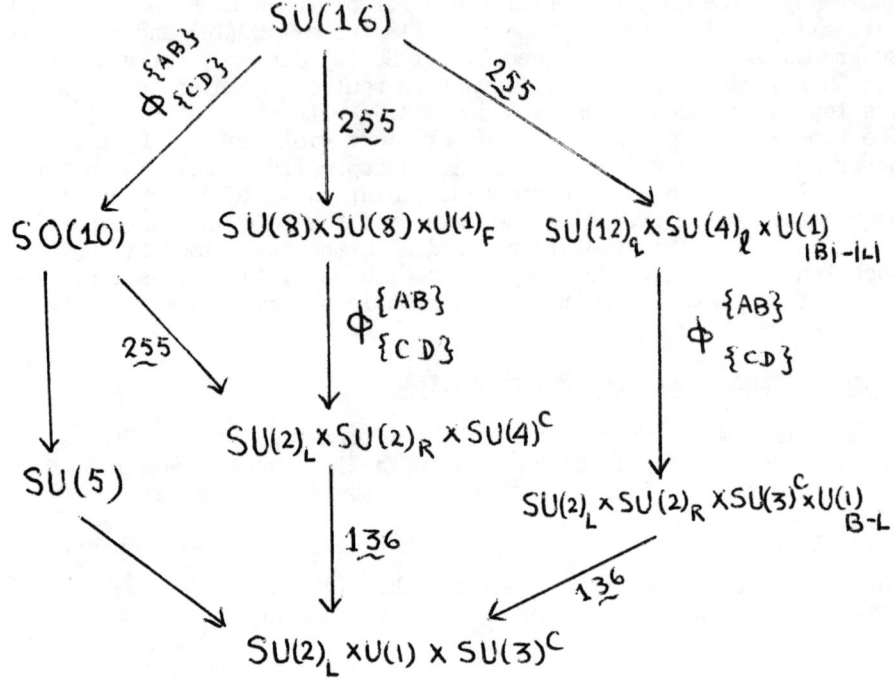

Fig. 6 Alternative routes of SSB of SU(16). Here $\phi\{AB\}_{\{CD\}}$, 255 and 136 denote scalar multiplets transforming like $(16 \times 16)_{symm} \times (16^* \times 16^*)_{symm}$, adjoint and $(16 \times 16)_{symm}$ - representations of SU(16) respectively.

VEV of the adjoint $\underline{255}$ by itself can take SU(16) into SU(8) x SU(8) x U(1)$_F$; while that of the tensorial field $\phi^{\{AB\}}_{\{CD\}}$ can break SU(16) in general straight to SU(2)$_L$ x SU(2)$_R$ x SU(4)^C and for a special case[12] to SO(10); simultaneously it breaks B, L and F. Thus if <$\underline{255}$> far exceeds <$\phi^{\{AB\}}_{\{CD\}}$>, the descent via SU(8) x SU(8) x U(1)$_F$ would be prominent. In this case it turns out[12] that the proton may decay via a variety of modes:
(i) $p \to e^+ + \pi^0$ etc. ($\Delta F=-4, \Delta(B-L)=0$); (ii) $p \to e^- + \pi^+\pi^+$ etc. ($\Delta F=-2, \Delta(B+L)=0$); and (iii) $p \to e^- + \nu\nu + \pi^+\pi^+$ etc. ($\Delta F=0, \Delta(B+L/3)=0$). Here the $\Delta F=-2$ and $\Delta F=0$-modes can even far supercede the $\Delta F=-4$ mode.

The relevant characteristic mass-scales for the three processes turn out to be $\approx (10^{12}-10^{14})$GeV, (10^8-10^{11})GeV and (10^4-10^5) GeV respectively, the corresponding proton-lifetime being $10^{31\pm2}$ years (see Ref. 12 for details).

If on the other hand <$\phi^{\{AB\}}_{\{CD\}}$> far exceeds <$\underline{255}$> and in particular the special SO(10)-chain materialises, then only the $\Delta F=-4$-mode (i.e. $p \to e^+ + \pi^0$ etc.) would be prominent. Thus, as mentioned before, predictions of SU(5) and SO(10) can emerge as special cases of the "maximal" symmetry SU(16).

The third route involving descent via SU(12)$_q$ x SU(4)$_\ell$ x U(1) can result from an alternative pattern of VEV of the adjoint $\underline{255}$ with the subsequent breakdowns taking place through <$\phi^{\{AB\}}_{\{CD\}}$> and <$\underline{136}$>. In contrast to the first two routes, the third route separates quarks from leptons at the first stage of SSB and thus assigns the heaviest masses to the lepto-antiquark Y' as well as the leptoquark X (see Fig. 2); the diquark Y in this case can, however, be much lighter than Y' and X, as it acquires a mass only at the second stage of SSB. We now observe that for a gauge mass-pattern given by

$$m_X \approx m_{Y'} \approx 10^{14} \text{ GeV}$$
$$m_Y \approx \Delta_{YY'} \sim 10^4-10^5 \text{ GeV} \qquad (12)$$

which is permissible[26] for the third route, $\Delta B=2, \Delta L=0$ n-$\bar{\text{n}}$ oscillations as well as proton-decay of the $\Delta F=-4$ variety (i.e. $p \to e^+ \pi^0$ etc.) can coexist, each with measurable strength. This is explained below.

The amplitude for $3q \to 3\bar{q}$ transition leading to n-$\bar{\text{n}}$ oscillations utilises third order of the effective Yukawa interaction $q \to \bar{q}+\phi''$

(see Figs. 5a and 5b). Noting that the amplitude for $q \to \bar{q} + \phi''$ is proportional to $(\alpha^2/m_Y^2)\ln(m_Y^2/m_W^2)$, a value of $m_Y \sim (1-3) \times 10^4$ GeV for the diquark mass yields an amplitude $\sim (10^{-29}-10^{-33})$ GeV^{-5} for the $3q \to 3\bar{q}$-transition for reasonable choice of the relevant scalar coupling constants (see Ref. 12). This in turn gives rise to an n-$\bar{n}$ mixing mass $\delta m(n\bar{n}) \sim 10^{-29}-10^{-33}$ GeV, the corresponding n-$\bar{n}$ oscillation period for free neutrons being $\approx (10^{+5}-10^{+9})$ sec with a deuteron lifetime $\approx 10^{30}-10^{38}$ years. (The $\Delta B=2, \Delta L=0$ deuteron-decay into pions was considered in Ref. 6). Needless to say, the above should be regarded only as a rough estimate of the order of magnitude.

The important point to note is that for a mass-scale $\approx (1-3) \times 10^4$ GeV characterising the mass-scale of the diquark Y, n-$\bar{n}$ oscillations begins to acquire a strength, which should be measurable in the near future.

Now with $\Delta_{YY'}^2 = m_Y^2$ (see eq. (12)), the amplitude for $\Delta F=-4$ proton decay modes (i.e. $p \to e^+\pi^0$ etc.) given by eq. (6) reduces simply to $g^2/m_{Y'}^2$. This has the canonical strength $\approx 10^{-29}$ GeV^{-2} for a lepto-antiquark mass $m_{Y'} \approx 10^{14}$ GeV, which yields a proton lifetime in the range of $10^{30\pm2}$ years. In other words, for this third route, not only n-$\bar{n}$ oscillations but also proton decay (of the $\Delta F=-4$ variety) can occur, each with observable strength. What is worth noting is that proton does not decay unduly rapidly despite the ultralight mass of the diquark Y($m_Y \approx 10^4$ GeV). This is something which was possible only because of the spontaneous nature of the B, L, F violations occuring within a maximal symmetry (for which Y-$\tilde{Y}'$ mixing vanishes as $m_Y \to 0$).

In contrast to the case presented here, it may be noted that Mohapatra and Marshak[15] working within the subunification symmetry $SU(2)_L \times SU(2)_R \times SU(4)^c$, proposed in Ref. 2, have constructed a model which permits n-$\bar{n}$-oscillation, but not proton-decay.[27] Most grand unification models (e.g. these based on SU(5) or SO(10) -- at least in their simplest version) permit proton-decay but not n-$\bar{n}$ oscillation.[28] The case arising here within the maximal symmetry SU(16) permits both.

Finally, it may be observed that in this third route of spontaneous descent, not only the lepto-antiquark Y' but also the leptoquark X becomes superheavy ($m_X \sim m_{Y'} \sim 10^{14}$ GeV). This has the consequence that the $\Delta F=0$, $\Delta F=-2$ and $\Delta F=-6$ proton-decay modes, i.e. $p \to 3$ leptons + mesons, $p \to \ell$ + mesons and $p \to 3\bar{\ell}$ + mesons, will all be strongly suppressed since these processes involve exchange of either the leptoquark X or the leptoantiquark Y' (see e.g. Figs. 4a, 4b and 3b).

The selection rules for B, L, F violations associated with the three alternative routes of spontaneous descent and the corresponding characteristic mass-scales are summarised in Table I.

Table I

Route of Spontaneous Descent	Selection Rules*	Characteristic Mass Scales
I SU(16)	$\Delta F=-4\,(p \to e^+\pi^0)$ --Yes	10^{12}-10^{14} GeV
$\to$ SU(8)xSU(8)xU(1)$_F$	$\Delta F=-2\,(p \to e^-\pi^+\pi^+)$ --Yes	10^{8}-10^{11} GeV
$\to$ SU(2)xSU(2)xSU(4)	$\Delta F=0\,(p \to e^-\nu\nu\pi^+\pi^+)$ Yes	10^{4}-10^{5} GeV
$\langle 255 \rangle \gg \langle \phi_{\{CD\}}^{\{AB\}} \rangle$	$\Delta F=-6\,(p \to 3\bar{\ell})$ No	
	$\Delta B=2\,(n \leftrightarrow \bar{n}\text{ oscill.})$ No	
II SU(16)$\to$SO(10)	$\Delta F=-4\,(p \to e^+\pi^0)$ Yes	10^{14} GeV
$\to$ SU(2)xSU(2)xSU(4)	$\Delta F=0,-2,-6$ No	
$\langle \phi_{\{CD\}}^{\{AB\}} \rangle \gg \langle 255 \rangle$	$\Delta B=2\,(n \leftrightarrow \bar{n}\text{ oscill.})$ No	
with special condns. on VEV-parameters (see Ref. 12)		
III SU(16)	$\Delta F=-4\,(p \to e^+\pi^0)$ Yes	10^{14} GeV
$\to$ SU(12)$_q$xSU(4)$_\ell$U(1)	$\Delta B=2\,(n \leftrightarrow \bar{n}\text{ oscill.})$ Yes	10^{4}-10^{5} GeV
$\to$ SU(2)xSU(2)xSU(3)xU(1)	$\Delta F=0,-2,-6$ No	

*For the first route of spontaneous descent, there exist subpossibilities under which $\Delta F=0$ and $\Delta F=-2$-proton decay modes can supercede the $\Delta F=-4$-decay mode. For this and for the case permitting $\Delta F=-6$ proton-decay (i.e. $p \to 3\ell$+mesons), see Ref. 12.

We see that some of the alternative routes of spontaneous descent require the existence of an intermediate mass-scale as low as about 10^4 GeV with or without a higher intermediate scale in the range of 10^8-10^{11} GeV. It has been shown elsewhere[20] that (consistent with the renormalisation group equations for the running coupling constants and the observed values of $\sin^2\theta_W$ and α_s, such intermediate mass-scales are indeed permissible, if we allow basic family distinctions for quarks and leptons and gauge their maximal symmetries such as $[SU(6)]^3$ or $SU(48)$ as mentioned in the introduction.

We now remark that the coexistence of the alternative proton-decay modes ($\Delta F=0,-2,-4$ and -6) with or without n-$\bar{\text{n}}$ oscillations, symbolising different characteristic mass-scales does not pose any conflict with the cosmological generation of baryon excess for a universe, assumed initially to be matter-antimatter symmetric. To put the matter in perspective, Weinberg[29] observed that there would be a conflict of the sort mentioned above if the violations of B, L and F are explicit. The argument is briefly this: since the $\Delta F = 0, -2$ and -6 proton decays as also $\Delta B = 2$ n-$\bar{\text{n}}$ oscillation are mediated by intermediate mass scales $M_I (\sim 10^4$-10^{11} GeV$) \ll M \sim 10^{14}$GeV, such processes would have relatively fast rates $\sim \alpha T$ or $\alpha^2 T$ in the early Universe at temperatures in the range $M_I \ll T \ll M$. Thus these processes (with rates exceeding the rate of expansion of the Universe) would be in thermodynamic equilibrium and would wipe out any baryon excess generated in earlier epochs at temperatures $\sim M$ due to the $\Delta F = -4$, $\Delta(B-L) = 0$-process, unless a specific linear combination $B + aL$ is conserved ($a \neq 0$). This would imply that only one of the three modes $\Delta F = 0, -2$ or -6 can coexist with the $\Delta F = -4$ mode -- not more than one -- and furthermore there should be no $\Delta B = 2$ n-$\bar{\text{n}}$ oscillation.

The conflict mentioned above does not arise, however, if the violations are spontaneous[27,11] rather than explicit. This is because for the spontaneous case the violations ($\Delta F = 0, -2, -6$ or $\Delta B = 2$) associated with a mass scale M_I disappear for temperatures $> M_I$. These violations appear only at temperatures $T \leq M_I$. However, the associated gauge particles acquire their masses at the same time that the violations appear. The rates of B, L and F violating processes ($\Delta F = 0, -2, -6$, etc.) damped by the associated gauge masses are now lower than the expansion rate of the Universe at this epoch. This keeps all these processes out of equilibrium whenever they are operative. Consequently, baryon excess generated in an earlier epoch $T > M_I$ by for example the $\Delta F = -4$ process is not wiped out by processes occurring at later times. We thus see that there is no conflict between the coexistence of $\Delta F = 0, -2, -4$ and -6 proton decay modes (with even $\Delta B = 2$ n-$\bar{\text{n}}$ oscillation) and the generation of the baryon excess, if the violations of B, L, F are spontaneous. This is one of the crucial differences between explicit versus spontaneous violations of B, L, F.

Finally, it may be observed that in the event SU(16) descends spontaneously through a <u>single mass-scale</u> to the low energy symmetry $SU(2)_L \times U(1) \times SU(3)^C$,

(i) there would arise a long extending "desert" between 10^2-10^{15} GeV.[14]

(ii) one would obtain[14,30] $\sin^2\theta_W \approx 0.21$.

(iii) Proton would decay predominantly via the $\Delta F=-4$, $\Delta(B-L)=0$-mode (i.e. via $e^+\pi^0$, $\bar{\nu}\pi^+$-modes etc.). All other decay modes of proton as well as n-n̄ oscillations would be strongly supperssed, and

(iv) the "standard" calculations[30] carried out in the context of the minimal SU(5)-model for proton-decay rate together with radiative corrections for $\sin^2\theta_W$ would apply, barring perhaps small corrections owing to the presence of mirror fermions in SU(16). Thus proton would be expected to have a lifetime $\tau_p \approx 10^{30.5 \pm 2}$ years.

In short, the physics of SU(16) descending spontaneously via a single stage hierarchy would coincide essentially with that of SU(5) at low energies. The <u>extra</u> righthanded neutrino would have a superheavy majorana mass,[31] as also the gauge particles coupling to V+A flavor-currents. They would be unobservable. Proton-decay would perhaps be the only observable striking consequence of grand unification. This situation, while logically permissible within SU(16), is no doubt rather dull compared to what it would be if SU(16) descends hierarchically in multiple stages permitting intermediate mass-scales. It would be exciting to see traces of exotic leptoquark-interactions and of left-right symmetry through manifest V+A-interactions at an energy-scale of order 10 TeV, which might be achievable in the near future. But then, who knows what Nature has indeed chosen for the hierarchy of spontaneous descent?

What one has argued[11,12,20] is that exciting new physics at low (1 GeV) and moderate energies ($\lesssim$ 100 TeV) can legitimately be expected, consistent with presently known phenomena, if we believe that the underlying unity of quarks and leptons is based upon a maximal symmetry. Ongoing searches for proton-decay and n-n̄ oscillations can help us judge better in the near future as to whether some of these expectations are likely to be borne out. In particular, the observation of the $\Delta F = 0$, or -6 mode (i.e. p → 3ℓ or 3ℓ̄ + mesons) and/or n-n̄ oscillation at any level within conceivable future will strongly suggest the existence of new physics at 10 to 100 TeV region and thereby motivate building high energy machines in this range. For this reason, second and third generation experiments for proton decay and n-n̄ oscillation must be planned to look for all modes listed here as possible rare processes, in case they are not found in the first generation experiments. As we now sit at low energies, this might be the best way to probe into the design of grand unification.

It is a pleasure to thank Abdus Salam and John Strathdee for stimulating discussions. The research was supported in part by a grant from the U. S. National Science Foundation.

REFERENCES AND FOOTNOTES

1. J. C. Pati and Abdus Salam, "Lepton hadron unification" (unpublished), reported by J. D. Bjorken in the Proceedings of the 15th High Energy Physics Conference held at Batavia, Vol. 2, p. 304, September (1972); J. C. Pati and Abdus Salam, Phys. Rev. D8, 1240 (1973).

2. J. C. Pati and Abdus Salam, Phys. Rev. Letters 31, 661 (1973); Phys. Rev. D10, 275 (1974); Phys. Letters 58B, 333 (1975).

3. H. Georgi and S. L. Glashow, Phys. Rev. Letters 32, 438 (1974).

4. Quark number B_q is +1 for quarks and -1 for antiquarks. Since baryon number is +1/3 for quarks, $B_q = 3B$.

5. For simplicity we do not introduce here chiral gauge transformations operating on massless left and right handed fermions in this illustrative example, since such a gauge structure would not still alter the main point of our discussion. See later for introduction of chiral gauges.

6. J. C. Pati, Abdus Salam and J. Strathdee, Il Nuovo Cimento 26A, 77 (1975); J. C. Pati, Proceedings of the Second Orbis Scientiae, Coral Gables, Florida, p. 253, January (1975); J. C. Pati, S. Sakakibara and Abdus Salam, ICTP, Trieste, preprint IC/75/93 (1975), unpublished.

7. These symmetries would ordinarily involve triangle anomalies, which are avoided however through the introduction of conjugate mirror fermions $F_{L,R}^m$. The gauge symmetry would then be maximal up to the mirror discrete symmetry ($F_{L,R} \leftrightarrow F_{R,L}^m$). The introduction of the mirror fermions would still retain well defined baryon, lepton and fermion numbers, which would be the sums of the corresponding numbers for the basic fermions $F_{L,R}$ and the mirror fermions $F_{L,R}^m$ (i.e. $B = B_F + B_{F^m}$ etc.).

8. Note that the first combination (Bq - 3L) corresponding to the current exhibited in (2) is traceless in the space of (quarks and leptons), while the second combinations (Bq + L) is traceless in the full space of (fermions and antifermions).

9. T. D. Lee and C. N. Yang, Phys. Rev. $\underline{98}$, 1501 (1955).

10. H. Fritzch and P. Minkowski, Ann. Phys. (NY) $\underline{93}$, 193 (1975); H. Georgi, Proc. AIP Conf., Williamsburg (1974).

11. J. C. Pati and Abdus Salam, "Quark lepton unification and proton decay," ICTP, Trieste, IC/80/72, Invited talk presented by J. C. Pati at the Grand Unification Workshops held at Erice (March 1980) and New Hampshire (April 1980), to appear in the Proceedings.

12. J. C. Pati, Abdus Salam and J. Strathdee, Trieste Preprint "Probings through proton-decay and n-$\bar{n}$ oscillations," IC/80/183; Nuclear Physics (to appear).

13. V. Elias, J. C. Pati and Abdus Salam, Phys. Rev. Lett. $\underline{40}$, 920 (1978).

14. H. Georgi, H. Quinn and S. Weinberg, Phys. Rev. Lett. $\underline{33}$, 451 (1974).

15. V. A. Kuzmin, Pisma Zh. Eksp. Teor. Fiz. $\underline{13}$, 335 (1970); S. L. Glashow, Cargese Lectures (1979); R. N. Mohapatra and R. E. Marshak, Phys. Rev. Lett. 44, 1316 (1980). L. N. Chang and N. P. Chang, CCNY-HEP-80/5 ($\overline{1}$980); J. C. Pati, Abdus Salam and J. Strathdee, work reported at the 20th International Conference on High-Energy Physics, Madison, Wisconsin (July 1980).

16. It is possible to avoid anomalies by introducing multiplets other than conjugate multiplets. An example is $\bar{5}$ + 10 for SU(5) (Ref. 3). However in such a theory no generator of the local symmetry can be associated with a linear combination of B, L or F, so that their breaking will necessarily be of the explicit variety.

17. J. C. Pati and Abdus Salam, Phys. Lett. $\underline{58B}$, 333 (1975). J. C. Pati, "An introduction to unification" in Topics in Quantum Field Theory and Gauge Theories, Proc. Salamanca 1977, page 262; ed. by J. A. de Azcarraga, published by Springer-Verlag.

18. J. C. Pati and Abdus Salam, Phys. Rev. $\underline{D10}$, 275 (1974); R. N. Mohapatra and J. C. Pati, Phys. Rev. $\underline{D11}$, 566, 2558 (1975); G. Senjanovic and R. N. Mohapatra, Phys. Rev. $\underline{D12}$, 1502 (1975).

19. J. C. Pati, "Magnetism as the origin of preon binding," Phys. Lett. January 1, 1981; J. C. Pati, Trieste preprint IC/79/80; to appear in the Proc. of the 1980-ν Conference, Erice (June 1980). J. C. Pati, Abdus Salam and J. Strathdee, "A preon model with hidden electric and magnetic type charges," IC/80/180, to appear in Nuclear Physics.

20. B. Deo, J. C. Pati, S. Rajpoot and Abdus Salam, "On proton decay" (unpublished), 1979. This work is reported briefly in Ref. 11.

21. J. C. Pati, Proc. Scottish Univ. Summer School Lecture notes, held at St. Andrews (1976). V. Elias and S. Rajpoot, Phys. Lett. (1979).

22. S. Weinberg, Phys. Rev. Lett. $\underline{43}$, 1566 (1979); F. Wilczek and A. Zee, Phys. Rev. Lett. $\underline{43}$, 1571 (1979).

23. B. Deo, J. C. Pati, S. Rajpoot and Abdus Salam, "Multiple stage descent of maximal symmetries," unpublished 1979; see Ref. 11 for a summary of the results.

24. To be specific SU(16) contains gauge particles transforming as $(2,2,6)_{+2}$ as well as $(2,2,6)_{-2}$; the transformation is with respect to the subsymmetry $SU(2)_L \times SU(2)_R \times SU(4)^C \times U(1)_F$. By contrast the subgroup SO(10) contains only the 24 real symmetric combinations "Y_s" $\equiv (1/\sqrt{2}) [(2,2,6)_{+2} + (2,2,6)_{-2}]$, but not the corresponding antisymmetric combinations "Y_a" $\equiv (1/\sqrt{2}) [(2,2,6)_{+2} - (2,2,6)_{-2}]$. Since any member belonging to the set "Y_s" couples to a mixture of F = +2 and F = -2 currents, its exchange leads to an explicit violation of F, B and L.

25. J. C. Pati, "Grand unification and proton-stability," Proc. of Workshop held at Madison, Wisconsin (Dec. 1978).

26. Descent of maximal symmetries such as $[SU(16)]^3$ via routes separating quarks and leptons at the first stage (i.e. via route III and variants thereof) permit mass-patterns of the type exhibited in eq. (12). See remarks in Ref. 12. Details of these investigations are being carried out with M. Özer.

27. It is instructive to compare the mechanism of n-n̄ oscillation presented here (Fig. 5a and 5b) with that of Mohapatra and Marshak (Ref. 15). The two mechanisms have an apparent resemblance in that both utilize third order of Yukawa like transitions (see Fig. 5b); nevertheless they differ intrisically from each other as explained below. The distinction stems primarily from the fact that the subunification symmetry[2] SU(2) x SU(2) x SU(4), which Mohapatra and Marshak use, does not contain fermion number F as a generator, while SU(16) does. Thus for the subunification symmetry fermion number need not be conserved in the basic lagrangian even as a global symmetry; but for SU(16) it must.

Now M and M introduce the Higgs multiplets (3,1,10) and (1,3,10) together with invariant Yukawa and quartic interactions, which these multiplets can posses. In particular they utilize a quartic coupling for the $3q \to 3\bar{q}$-transitions, in which all four scalars entering the vertex are either $(3,1,\overline{10})$ or $(\overline{1,3,10})$.

Such a quartic coupling, though allowed by the subunification symmetry, is not allowed by SU(16). This follows by noting that both (3,1,10) and (1,3,10) belong to the single multiplet $\underline{136}$ of SU(16), [which incidentally must be introduced[12] within $\overline{SU}(16)$ to give masses to the fermions]. These two submultiplets are thus assigned definite fermion number (i.e. F = +2) in SU (16). A quartic scalar interaction-vertex, for which each of the four scalars entering the vertex brings in two units of fermion number, as above, can not clearly conserve fermion number, and is therefore not allowed within SU(16).

By contrast, for the mechansim exhibited in Fig. 5b, it turns out (see Ref. 12) that two of the scalars entering the quartic interaction-vertex must be (2,2,15) with F = 0, while the other two must be either (1,3,10) and (1,3,10)* or (3,1,10) and (3,1,10)* with F = +2 and -2 respectively, all these being submultiplets of $\underline{136}$ of SU(16). Such vertices needless to say conserve fermion number, as they must being invariant under SU(16). This illustrates the intrinsic difference between the two mechanisms referred to above.

28. See however remarks by N. P. Chang at this conference.

29. S. Weinberg, Phys. Rev. D$\underline{22}$, 1694, 1980.

30. A. J. Buras, J. Ellis, M. Gaillard and D. V. Nanopoulos, Nucl. Phys. B$\underline{135}$, 66 (1978); T. J. Goldman and D. A. Ross, Cal. Tech. Preprint 68-759 (1980); J. Ellis, M. K. Gaillard, D. V. Nanopoulos and S. Rudaz, CERN-Annecy preprint, LAPP-Th-14, CERN TH 2833; W. Marciano and A. Sirlin, Phys. Rev. Lett. 1981. Other references may be found here.

31. M. Gell-Mann, P. Ramond and R. Slansky, Rev. Mod. Phys. $\underline{50}$, 721 (1978).

DISCUSSION

ANDERSON: Could you just give us your most optimistic number for the mixing time in $n\bar{n}$ in your most favorable theories.

PATI: As I said, from the renormalization group equation, one finds the solution of about 10^5 GeV for the diquark mass, which corresponds to roughly an oscillation time in the range (these predictions cannot be taken literally but within a factor of 100) of the order of about 10^5 to 10^8 or 10^9 seconds. It comes right in this range. That is the important point - between 10^5 to 10^8 seconds so I think the thing to extract out of this is that the prediction is not that tight. But the complexion can exist, consistent with all the theoretical requirements.

SU(5) PREDICTIONS: τ_p, $\sin^2\theta_W^{exp.}$, m_W and m_Z

W.J. Marciano

Department of Physics and Astronomy
Northwestern University
Evanston, Illinois 60201

and

A. Sirlin

Department of Physics
New York University
New York, New York 10003

TABLE OF CONTENTS

I. The Georgi-Glashow SU(5) Model

II. Proton Decay

III. Renormalization Group Analysis

IV. Radiative Corrections to Neutrino Scattering and e-D Asymmetries

V. Precise SU(5) Predictions: $\sin^2\theta_W^{exp.}$, m_W and m_Z

VI. Summary

I. The Georgi-Glashow SU(5) Model

The SU(5) model introduced by Georgi and Glashow[1] is the most economical of all grand unified theories (GUTS). It is minimal in that only the standard fermions are required. They belong to three sequential generations, each of which is composed of a $\underline{5} + \underline{10}$ representation of SU(5). The lightest generation assignments are

$$\begin{pmatrix} d_1 \\ d_2 \\ d_3 \\ e^+ \\ \bar{\nu}_e \end{pmatrix}_R \qquad \frac{1}{\sqrt{2}} \begin{pmatrix} 0 & \bar{u}_3 & -\bar{u}_2 & -u_1 & -d_1 \\ -\bar{u}_3 & 0 & \bar{u}_1 & -u_2 & -d_2 \\ \bar{u}_2 & -\bar{u}_1 & 0 & -u_3 & -d_3 \\ u_1 & u_2 & u_3 & 0 & -e^+ \\ d_1 & d_2 & d_3 & e^+ & 0 \end{pmatrix}_L \qquad (1)$$

Similarly, there is a second generation containing μ, ν_μ, s, c and a third composed of τ, ν_τ, b, t with Cabibbo mixing among generations.

There are 24 gauge bosons in the SU(5) model. Twelve of these are the usual gauge fields $W^\pm$, Z^0, γ and 8 gluons of the standard $SU(3)_c \times SU(2)_L \times U(1)$ model, while the other twelve are exotic in that they carry color and have fractional electric charges $\pm 4/3$ and $\pm 1/3$. The latter bosons, $X^{\pm 4/3}$ and $Y^{\pm 1/3}$, belong to $SU(3)_c$ color triplets and together form an $SU(2)_L$ weak isodoublet; hence they are degenerate up to very small $SU(2)_L$ breaking, $m_X \approx m_Y \approx m_S$. These gauge fields and the pattern of symmetry breaking in the SU(5) model are illustrated in Fig. 1.

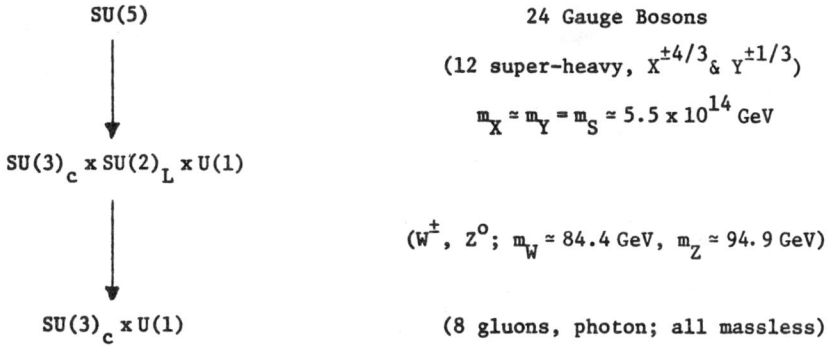

Fig. 1. Pattern of symmetry breaking and vector boson mass scales in the SU(5) model. ($\Lambda_{\overline{MS}}$ = 0.4 GeV used as input to obtain vector boson masses.)

The minimum Higgs scheme required to break the gauge symmetry and provide fermion masses is a real 24-plet and a complex 5-plet (there may be several 5-plets). In addition, to get the fermion masses right, a 45-plet of scalars may be required.[1] Although the masses of the physical scalar particles are somewhat arbitrary, we will assume when necessary that all physical scalars originating from the 24-plet and all fractionally charged scalars coming from 5-plets (and 45's) have super-heavy mass $\simeq m_S$, while the N_H isodoublets under weak $SU(2)_L$ have mass $\simeq m_W$.

A nice feature of embedding the standard model in a simple grand unified gauge group G, is that the bare couplings associated with the low energy subgroups $SU(3)_c \times SU(2)_L \times U(1)$ are constrained to be equal

$$g_{G_0} = g_{3_0} = g_{2_0} = g_{1_0} \tag{2}$$

In such a theory the observed unequal strengths of strong, weak and electromagnetic interactions is a consequence of performing present-day experiments at relatively low energies, so that higher order radiative corrections effectively enhance or diminish the various interaction strengths.[2] They become equal only at superhigh energies above all mass scales in the theory, i.e. at very very short distances.

With regard to the weak mixing angle, an appealing consequence of grand unification is that $\sin^2\theta_W^o$ is elevated from an infinite adjustable counterterm parameter (its role in the $SU(2)_L \times U(1)$ model) to a definite rational number. There is a nice formula relating $\sin^2\theta_W^o$ to the values of T_{3i} (third component of weak isospin) and Q_i (electric charge) of the members of any representation R of G (whether it be fermions, gauge bosons or scalars)

$$\sin^2\theta_W^o = \frac{e_0^2}{g_{2_0}^2} = \frac{\sum_{i \varepsilon R} T_{3i}^2}{\sum_{i \varepsilon R} Q_i^2} \tag{3}$$

So, in a theory such as SU(5) which requires only sequential fermions, one finds from the 5 + 10 representation in Eq. (1)

$$\sin^2\theta_W^o = \frac{4(1/2)^2 + 4(-1/2)^2}{2(-1)^2 + 6(2/3)^2 + 6(-1/3)^2} = \frac{3}{8} \tag{4a}$$

$$\frac{e_0^2}{g_{3_0}^2} = \frac{e_0^2}{g_{2_0}^2} = \frac{3}{8} \tag{4b}$$

The relationships between the bare quantities in Eq. (4) are natural (in the technical sense); hence the same relationships hold for the renormalized parameters, up to (large) finite calculable higher order corrections which will be described in Section III.

II. Proton Decay

The exotic super-heavy bosons $X^{\pm 4/3}$ and $Y^{\pm 1/3}$ mediate baryon number violating processes such as proton decay.[1,2,3] Some of the SU(5) proton decay modes are schematically illustrated in Fig. 2.

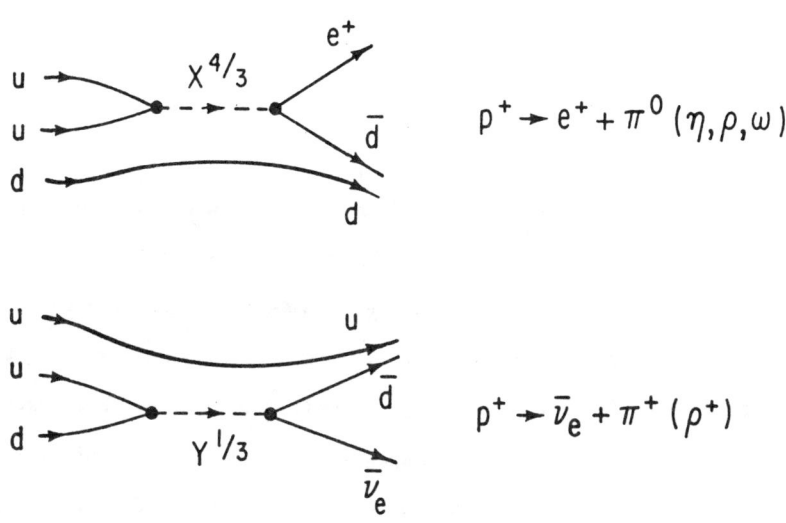

Fig. 2. Some proton decay channels in the SU(5) model.

There exist several calculations of the predicted proton lifetime, τ_p, as a function of m_S. The details of those calculations and their estimated uncertainties have been thoroughly reviewed.[4,5] At present the quoted values of τ_p are in the range[3,4,5,6]

$$\tau_p \simeq (2 - 38) \times 10^{-29} (m_S \text{ in GeV})^4 \text{ yr.} \qquad (5)$$

for the SU(5) model. We note, however, that given the uncertainties in the individual calculations, it is perhaps better to employ a more conservative range of predictions such as

$$\tau_p \simeq 10^{-28\pm1} (m_S \text{ in GeV})^4 \text{ yr.} \qquad (6)$$

in comparing theory and experiment.

III. Renormalization Group Analysis

The higher order radiative corrections to the natural relationships in Eqs. (2) and (4) are sizeable because they contain large logarithms of the form $\alpha^n \ln^n(m_S/\mu)$, $n=1,2\ldots$; μ = low energy mass scale. Fortunately, such terms can be summed by using renormalization group techniques; an approach initiated by Georgi, Quinn and Weinberg.[2] Their procedure has been extended to include next to leading logs[7] (by employing two terms in the effective beta functions) and simplified by using modified minimal subtraction, $\overline{MS}$, in the framework of dimensional regularization to define the renormalized parameters.[4,8,9]

In the $\overline{MS}$ prescription (as applied to GUTS), the renormalized running couplings $\alpha_i(\mu) = g_i^2(\mu)/4\pi$, $i=1,2,3$ are defined by subtracting out all $(\frac{1}{n-4} + \frac{1}{2}\gamma - \ln\sqrt{4\pi})$ loop contributions such that they satisfy the boundary conditi

$$\alpha_1(m_S) = \alpha_2(m_S) = \alpha_3(m_S) \qquad (7)$$

Further subtracting a $-1/22$ contribution associated with super-heavy vector boson loops, so that a complete decoupling of all super-heavy particles occurs (i.e. $\frac{1}{n-4}+\frac{1}{2}\gamma - \ln\sqrt{4\pi} - \frac{1}{22}$ is subtracted from super-heavy vector boson loops), one obtains a slightly different set of couplings $\hat{\alpha}_i(\mu)$ which satisfy the SU(5) boundary condition[8]

$$\frac{1}{\hat{\alpha}_1(m_S)} = \frac{1}{\hat{\alpha}_2(m_S)} - \frac{1}{6\pi} = \frac{1}{\hat{\alpha}_3(m_S)} - \frac{1}{4\pi} = \frac{1}{\alpha_1(m_S)} - \frac{5}{12\pi} \qquad (8)$$

Next, using the relationship

$$\frac{1}{\hat{\alpha}(\mu)} = \frac{5}{3\hat{\alpha}_1(\mu)} + \frac{1}{\hat{\alpha}_2(\mu)} \qquad (9)$$

for the effective electromagnetic coupling, one finds from Eq. (8)

$$\frac{\hat{\alpha}(m_S)}{\hat{\alpha}_3(m_S)} = \frac{3}{8} + \frac{3\hat{\alpha}(m_S)}{16\pi} \qquad (10a)$$

$$\sin^2\hat{\theta}_W(m_S) = \frac{\hat{\alpha}(m_S)}{\hat{\alpha}_2(m_S)} = \frac{3}{8} + \frac{5\hat{\alpha}(m_S)}{48\pi} \qquad (10b)$$

To obtain the running couplings $\hat{\alpha}_i(\mu)$ for $\mu < m_S$, the beta-function equations

$$\mu\frac{\partial}{\partial\mu}\hat{\alpha}_i(\mu) = \beta(\hat{\alpha}_i) = b_i\hat{\alpha}_i^2 + b_{ij}\hat{\alpha}_i^2\hat{\alpha}_j; \quad i,j = 1,2,3 \qquad (11)$$

are integrated using the boundary condition in Eq. (8). The effective b_i and b_{ij} in Eq. (11) receive contributions only from particles with masses $< m_S$; so in the SU(5) model with n_g generations of fermions and N_H light Higgs doublets ($m_H \simeq m_W$)[2,4,10]

$$b_1 = -\frac{1}{2\pi}(-\frac{4}{3}n_g - \frac{1}{10}N_H) \tag{12a}$$

$$b_2 = -\frac{1}{2\pi}(\frac{22}{3} - \frac{4}{3}n_g - \frac{1}{6}N_H) \tag{12b}$$

$$b_3 = -\frac{1}{2\pi}(11 - \frac{4}{3}n_g) \tag{12c}$$

The b_{ij} are given in refs. 7 & 9. Integrating Eq. (11) down to $\mu = m_W$ and using the boundary condition in Eq. (8), one finds

$$\frac{\hat{\alpha}(m_W)}{\hat{\alpha}_3(m_W)} = \frac{3}{8}[1 - (\frac{66+N_H}{6})\frac{\hat{\alpha}(m_W)}{\pi}\ell n\frac{m_S}{m_W} + \frac{\hat{\alpha}(m_W)}{2\pi} + \ldots] \tag{13}$$

$$\sin^2\hat{\theta}_W(m_W) = \frac{3}{8}[1 - (\frac{110-N_H}{18})\frac{\hat{\alpha}(m_W)}{\pi}\ell n\frac{m_S}{m_W} + \frac{5\hat{\alpha}(m_W)}{18\pi} + \ldots] \tag{14}$$

where ... represents the effect of the b_{ij} in Eq. (11). Those contributions are of order $(\hat{\alpha}(m_W)/\pi)^2 \ell n(m_S/m_W)$. Although not explicitly displayed, they are included in all subsequent numerical estimates. Their inclusion in Eq. (13) is particularly important. (Note, the ... in Eqs. (13) and (14) depend on n_g; we take $n_g = 3$.) The effective coupling $\hat{\alpha}(m_W)$ is related to the usual fine structure constant $\alpha = 1/137.036$ by[11]

$$\frac{1}{\hat{\alpha}(m_W)} = \frac{1}{\alpha} - \frac{2}{3\pi}\sum_f Q_f^2 \ell n(m_W/m_f) + \frac{1}{6\pi} + \ldots \tag{15}$$

where the sum is over all fermions with $m_f < m_W$, (Q_f is their electric charge). Using the three known generations of fermions with $m_t \simeq 18$ GeV and $m_W \simeq 84.4$ GeV, we estimate $\hat{\alpha}(m_W) = 1/127.66$. The quantity $\hat{\alpha}_3(m_W)$ is the conventional[12] $\overline{MS}$ running coupling of quantum chromodynamics (QCD) evaluated at $\mu = m_W$. Parameterizing $\hat{\alpha}_3(m_W)$ in terms of $\Lambda_{\overline{MS}}$, the QCD mass scale in the running coupling

obtained using two terms in the effective 4 flavor QCD beta-function
(4 flavors because experimental determinations of $\Lambda_{\overline{MS}}$ are made at $Q^2 \simeq 10 GeV^2$,
i.e. below the b flavor threshold), we find[11] from Eq. (13), including the
b_{ij} terms (for $N_H \ll 68$)

$$\frac{m_S}{m_W} = 6.5 \times 10^{12} \times 10^{-0.19(N_H-1)} \times (\Lambda_{\overline{MS}}/0.4 GeV)^{1.07} \quad (16)$$

Taking $m_W \simeq 84.4$ GeV and employing the proton lifetime estimate in Eq. (6) gives

$$\tau_p = 9 \times 10^{30 \pm 1} \times 10^{-0.76(N_H-1)} \times (\Lambda_{\overline{MS}}/0.4 GeV)^{4.28} \, yr \quad (17)$$

For $\Lambda_{\overline{MS}} = 0.4$ GeV and $N_H = 1$, we see that the SU(5) model predicts $\tau_p \simeq 10^{31 \pm 1}$ yr; not far from the present experimental bound $\tau_p > 10^{29}$ yr and within reach of ongoing experiments[13] which plan to search up to about $\tau_p \simeq 10^{33} \sim 10^{34}$ yr. Note that if $N_H > 1$, m_S and τ_p decrease; so for $\Lambda_{\overline{MS}} = 0.4$ GeV, the SU(5) model can accommodate at most 3 or 4 light Higgs doublets before running into trouble with the proton lifetime bound.

Using Eq. (16) in Eq. (14), we obtain the SU(5) prediction for $\sin^2\hat{\theta}_W(m_W)$

$$\sin^2\hat{\theta}_W(m_W) = 0.2087 + 0.004(N_H-1) + 0.006\ln(0.4 GeV/\Lambda_{\overline{MS}}) \quad (18)$$

How does this value compare with experiment? To make a precise comparison, one must consider the effect of ordinary $O(\alpha)$ $SU(2)_L \times U(1)$ radiative corrections on the particular experiment considered. That is the subject of our recent work[11,14,15] which we report on in the next two sections.

IV. Radiative Corrections to Neutrino Scattering and e-D Asymmetries

At present, the most precise determination of $\sin^2\theta_W^{exp.}$ comes from deep-inelastic ν_μ-hadron scattering. In those experiments the ratio of neutral current (NC) and charged current (CC) cross-sections is measured and compared with the theoretical prediction of the Weinberg-Salam $SU(2)_L \times U(1)$ model which depends rather sensitively on $\sin^2\theta_W^{exp.}$. To extract a precise value of $\sin^2\theta_W^{exp.}$, one must include the $O(\alpha)$ radiative corrections to both cross-sections. Contact between theory and experiment is most easily made by performing a phenomenological renormalization; i.e. we replace $(g_{2_0}^2/8m_Z^2\cos^2\theta_W^o)^2$ $(\rho_{NC} G_\mu/\sqrt{2})^2$ and $\sin^2\theta_W^o$ by $\sin^2\theta_W(\nu_\mu;h)(q^2)$ in the neutral current cross-section formula and $(g_{2_0}^2/8m_W^2)^2$ by $(\rho_{CC} G_\mu/\sqrt{2})^2$ in the charged current cross-section formula ($G_\mu = 1.16632 \pm 0.00002 \times 10^{-5} \text{GeV}^{-2}$ is the muon decay constant[16]). We then compute $\rho_{NC}^2 = 1 + O(\alpha)$ and $\rho_{CC}^2 = 1 + O(\alpha)$ directly. A detailed analysis has been carried out.[14,15] Here we present only the large and potentially large contributions to $\rho^2 = \rho_{NC}^2/\rho_{CC}^2$ for the case in which the x and y variables of deep-inelastic scattering are fully integrated

$$\rho^2 = \rho_{NC}^2/\rho_{CC}^2 = 1 - \frac{\alpha}{\pi}[\ln(\frac{m_Z^2}{2<-q^2>}) + 2] + \frac{3\alpha}{8\pi\sin^2\theta_W}\frac{m_t^2}{m_W^2} - \frac{3\alpha}{8\pi\cos^2\theta_W}\ln(\frac{m_\phi^2}{m_Z^2}) \qquad (19)$$

+ smaller contributions

where m_t = top quark mass, m_ϕ = Higgs scalar mass and $<-q^2>$ represents the average momentum transfer in deep-inelastic ν_μ-hadron scattering. The first log in Eq. (19) comes from the photonic corrections to the CC cross-section; it is complete analogous to the large logarithm encountered in β-decay and its coefficient is essentially model independent (i.e. unaffected by strong interactions).[17] The constant term accompanying the log represents a leading photonic contribution evaluated in the parton model. The term proportional to m_t^2/m_W^2 is potentially important for $m_t > m_W$, a feature pointed out by Veltman.[18] The last log will be significant only if $m_\phi \gg m_Z$. (We have displayed only the dominant part of a single Higgs multiplet contribution for large m_ϕ.) Included in the smaller contributions are mass singularity terms proportional to $\ln(<-q^2>/m_i^2)$ where m_i is the mass of the initial quark; they are discussed in ref. 14. Assuming $m_t < m_W$, $m_\phi \approx m_Z$ and using $<-q^2> = 20 \text{GeV}^2$, we find

$$\rho^2 \approx 0.98 \qquad (20)$$

for the ratio of <u>fully integrated inclusive</u> cross-sections.

A detailed phenomenological analysis of the world-wide data on ν_μ-hadron scattering has been carried out by Kim, Langacker, Levine and Williams.[19] Performing a two parameter fit to the data, they found

$$\rho = 0.999 \pm 0.025, \qquad \sin^2\theta_W^{(\nu_\mu;h)}(-20\text{GeV}^2) = 0.232 \pm 0.027 \qquad (21)$$

A one parameter fit was made only for $\rho=1$; however, using that result[19] and including the main dependence on ρ^2, we find[15]

$$\sin^2\theta_W^{(\nu_\mu;h)}(-20\text{GeV}^2) = 0.234 \pm 0.012 - 0.50(1-\rho^2) \qquad (22)$$

(Note that a 1% shift in ρ leads to about a 4% shift in $\sin^2\theta_W^{(\nu_\mu;h)}(-20\text{GeV}^2)$.) For $\rho^2 = 0.98$ as estimated in Eq. (20), the result in Eq. (22) implies

$$\sin^2\theta_W^{(\nu_\mu;h)}(-20\text{GeV}^2) = 0.224 \pm 0.012 \qquad (23)$$

This is the experimental value that we will subsequently compare with the SU(5) prediction.

In the case of ν_μ-e and $\bar{\nu}_\mu$-e scattering, the $O(\alpha)$ corrections were previously computed.[14,20] One finds $\rho_{NC}^{(\nu_\mu;e)}$ and $\rho_{NC}^{(\bar{\nu}_\mu;e)}$ are close to one. Unfortunately, existing experiments have a rather large uncertainty in their neutrino flux (the absolute neutral current cross-section rather than a ratio is measured). Presently, there is about a 20% uncertainty in the value of

$\sin^2\theta_W^{(\nu_\mu;e)}(0)$ extracted from experiment.[21] In the future, proposed high statistics experiments should be able to measure $\sin^2\theta_W^{(\nu_\mu;e)}(0)$ with an accuracy of about 0.01. They will then provide a test of GUTS.

The SLAC deep-inelastic e-D asymmetry measurements presently yield a value[22]

$$\sin^2\theta_W^{(e;D)}(-1.6\text{GeV}^2) = 0.223 \pm 0.020 \qquad (24)$$

where ±0.005 of the quoted uncertainty is theoretical.[19] The theoretical uncertainty includes an estimate of the box diagram radiative corrections,[23] which if considered alone would increase the central value in Eq. (24) by about 0.004. Other still larger corrections come from γ-Z mixing through fermion loops. Our procedure is to absorb the γ-Z mixing effect into the theoretical definition of $\sin^2\theta_W^{(e;D)}(-1.6\text{GeV}^2)$. In the next section we show that the γ-Z mixing brings the SU(5) prediction for $\sin^2\theta_W^{(e;D)}(-1.6\text{GeV}^2)$ very close to the central value in Eq. (24).

V. Precise SU(5) Predictions: $\sin^2\theta_W^{\text{exp.}}$, m_W and m_Z

The SU(5) effective low energy parameter $\sin^2\hat\theta_W(m_W)$ given in Eq. (18) is not precisely the same as the quantity $\sin^2\theta_W^{\text{exp.}}(q^2)$ defined in the previous section and presently measured. (They differ by O(α) terms.) Indeed, $\sin^2\theta_W^{\text{exp.}}(q^2)$ should vary somewhat with q^2 and depends on the particular experiment considered. To compare experimental results with the predictions of any GUT, $G \supset SU(2)_L \times U(1)$, we have computed the O(α) corrections to neutrino-induced neutral current scattering within the framework of an arbitrary G.[11] Our procedure is as follows: Given a $\sin^2\hat\theta_W(m_W)$ determined by the $\overline{\text{MS}}$ renormalization group analysis outlined in Section III, we calculate directly the remaining O(α) radiative corrections to neutrino neutral current scattering

In doing so, we employ an effective $SU(2)_L \times U(1)$ theory with heavy particles ($m > m_Z$) decoupled (left-out) and replace $\sin^2\theta_W^0$ by $\sin^2\hat{\theta}_W(m_W)$. (Similarly, α_{i_0} is replaced by $\hat{\alpha}_i(m_W)$.) Modified minimal subtraction is used to eliminate the divergences encountered in the one loop calculations, while μ, the dimensional regularization unit of mass, is set equal to m_W. In that way we find for ν_ℓ ($\ell = e, \mu, \tau$) neutral current scattering at $-q^2 \ll m_W^2$, the GUT prediction[11]

$$\sin^2\theta_W^{exp.}(q^2) = \sin^2\hat{\theta}_W(m_W) - \frac{\hat{\alpha}(m_W)}{2\pi}B - \frac{\alpha(-q^2)}{2\pi}[\frac{c^2}{3} + \frac{1}{2} - 2J_\ell(q^2) + \sum_f (C_{3f}Q_f - 4s^2 Q_f^2)J_f(q^2)] \quad (25)$$

where $c^2 \equiv \cos^2\hat{\theta}_W(m_W)$, $s^2 \equiv \sin^2\hat{\theta}_W(m_W)$, $J_f(q^2) \equiv \int_0^1 dx\, x(1-x) \ln[(m_f^2 - q^2 x(1-x))/m_W^2]$, ($m_f$ is the mass of fermion f), C_{3f} is twice the third component of weak isospin (eg. $C_{3e} = -1$, $Q_e = -1$) and

$$\alpha^{-1}(-q^2) = \alpha^{-1} + \Pi_{\gamma\gamma}(q^2) - \Pi_{\gamma\gamma}(0) \quad (26a)$$

$$\alpha^{-1}(-q^2) \simeq \alpha^{-1} - \frac{2}{\pi} \sum_f Q_f^2 \int_0^1 dx\, x(1-x) \ln[(m_f^2 - q^2 x(1-x))/m_f^2] \quad (26b)$$

In Eq. (25) the contributions denoted by B arise from box diagrams and are given by $B = (5/2 - 61s^2/20 - 9s^4/10 + 14s^6/9)/(2c^4)$ for ν-hadron scattering and $B = (19/8 - 17s^2/4 + 3s^4)/c^2$ for ν-lepton scattering. In Eq. (25) we included higher order vacuum polarization, $\Pi_{\gamma\gamma}(q^2)$, corrections to the photon propagator which replaced $\hat{\alpha}(m_W)$ by $\alpha(-q^2)$ in some of the terms. The hadronic contribution to $\alpha(-q^2)$ in Eq. (26a) can be obtained from measurements of $\sigma(e^+e^- \to hadrons)$ via a dispersion relation,[16] thereby avoiding the uncertainty associated with quark masses in Eq. (26b).

We note that there are two sources of large logarithms in Eq. (25). The first one in $J_\ell(q^2)$ comes from the neutrino charge radius, while the logs in

$J_f(q^2)$ are due to γ-Z mixing through fermion loops.[24] These two distinct contributions tend to cancel; a feature which keeps the bracketed expression in Eq. (25) small and fairly insensitive to variations in q^2.

Our formula in Eq. (25) is applicable to any GUT. For the SU(5) model we merely insert the result for $\sin^2\hat{\theta}_W(m_W)$ in Eq. (14) into Eq. (25). Then, by an algebraic rearrangement valid to $O(\alpha)$ one can rewrite the resulting expression as[11]

$$\sin^2\theta_W^{exp.}(q^2) = \frac{3}{8}[1 - \frac{(110-N_H)}{18}\frac{\alpha(-q^2)}{\pi}\ln(\frac{m_S}{m_W})+\ldots] - \frac{\hat{\alpha}(m_W)}{2\pi}B$$

$$- \frac{\alpha(-q^2)}{2\pi}[\frac{1}{2} - 2J_\ell(q^2) + \sum_f (C_{3f}Q_f - \frac{3}{2}Q_f^2)J_f(q^2)] \quad (27)$$

In this form the bulk of the γ-Z mixing goes into replacing $\hat{\alpha}(m_W)$ by $\alpha(-q^2)$ in the coefficient of the large $\ln(m_S/m_W)$ terms. The prediction for $\sin^2\theta_W^{exp.}$ in Eq. (27) exhibits the following features: Since the hadronic contribution to $\alpha(-q^2)$ can be obtained from experimental e^+e^- annihilation data and $C_{3f}Q_f - \frac{3}{2}Q_f^2 = 0$ for the u, c and t quarks, our result is independent of m_u, m_c and m_t. It is also independent of the ordinary neutral Higgs mass m_ϕ. The ... in Eq. (27) can be estimated by including the b_{ij} terms in the beta-functi They increase $\sin^2\theta_W^{exp.}(q^2)$ by about 0.22%.[11] For $\sin^2\hat{\theta}_W(m_W) \simeq 0.21$ and $m_u \simeq m_d \simeq$ $m_s \simeq 0.1 GeV$, we find that $\sin^2\theta_W^{exp.}(q^2)$ monotonically decreases by $\simeq 0.9\%$ over th interval $0 \leq -q^2 \leq 100 GeV^2$ in the case of ν_μ scattering.

Using the estimate in Eq. (16) for m_S/m_W, we find from Eq. (25) or (27)

$$\sin^2\theta_W^{(\nu_\mu;h)}(-20GeV^2) = 0.2098 + 0.004(N_H-1) + 0.006\ln(0.4GeV/\Lambda_{\overline{MS}}) \quad (28)$$

for deep-inelastic ν_μ-hadron scattering at $q^2 = -20GeV^2$. This SU(5) prediction is in good agreement with Eqs. (21) and (23) for $\Lambda_{\overline{MS}} \simeq 0.4 GeV$ and $N_H = 1 - 4$.

Instead of predicting $\sin^2\theta_W^{(\nu_\mu;h)}(-20\text{GeV}^2)$ as a function of $\Lambda_{\overline{MS}}$, one can consider the alternative strategy of ignoring the QCD sector. Then using the experimental result $\sin^2\theta_W^{(\nu_\mu;h)}(-20\text{GeV}^2) = 0.224\pm0.012$ as input and allowing for a variation of 1 standard deviation, we find from Eq. (27) for $N_H=1$

$$m_S < 3.6 \times 10^{14} \text{ GeV} \rightarrow \tau_p < 1.8 \times 10^{31} \text{ yr} \qquad (29)$$

Note, however, that these restrictive bounds would increase considerably if a larger variation of $\sin^2\theta_W^{(\nu_\mu;h)}(-20\text{GeV}^2)$ is allowed and/or if $N_H>1$.

In the case of ν_μ-e scattering at $q^2 \approx 0$, we find from Eq. (27)

$$\sin^2\theta_W^{(\nu_\mu;e)}(0) = 0.2104 + 0.004(N_H-1) + 0.006\ell n(0.4\text{GeV}/\Lambda_{\overline{MS}}) \qquad (30)$$

which for $\Lambda_{\overline{MS}} = 0.4\text{GeV}$ and $N_H=1-4$ is consistent with existing data[21] (given the large experimental uncertainties).

For the SLAC deep-inelastic e-D asymmetry measurements, we have not yet completed studying the $O(\alpha)$ radiative corrections. However, we can make the following observations: The extraction of $\sin^2\theta_W^{(e;D)}(-1.6\text{GeV}^2)$ is not very sensitive to deviations in $\rho^{(e;D)}$ from 1. A 1% change in $\rho^{(e;D)}$ produces only about a 0.8% shift in $\sin^2\theta_W^{(e;D)}(-1.6\text{GeV}^2)$. Box diagrams if considered alone[23] increase $\sin^2\theta_W^{(e;D)}(-1.6\text{GeV}^2)$ by about 0.004; however, as mentioned in the previous section, the theoretical uncertainty in Eq. (24) includes this effect. Since there are other theoretical uncertainties of comparable magnitude but opposite in sign,[19] we leave for now the box diagram effects in the quoted uncertainty rather than correcting for them. Finally, for this process, charge radii contributions are much smaller and therefore less important than in neutrino scattering. On the other hand, γ-Z mixing through fermion loops is the same as in neutrino scattering. Therefore, for the e-D asymmetry experiment, we expect the prediction in Eq. (25) will be modified by the inclusion of new individually small $O(\alpha)$ corrections and the suppression of the large charge radius contribution (the $J_\ell(q^2)$ term). For the SU(5) model one should find[25] (see Eq. (27))

$$\sin^2\theta_W^{(e;D)}(-1.6\text{GeV}^2) = \frac{3}{8}[1 - \frac{110-N_H}{18}\frac{\alpha(1.6\text{GeV}^2)}{\pi}\ln(\frac{m_S}{m_W}) + \ldots] \quad (31)$$

$$+ \text{ individually small } O(\alpha) \text{ corrections}$$

Assuming that the individually small $O(\alpha)$ corrections don't add up to a large effect, we expect the bracketed expression in Eq. (31) to represent a good approximation for the SU(5) model prediction. Using $\alpha(1.6\text{GeV}^2) = 1/136.2$, we estimate from Eq. (31) and Eq. (16)

$$\sin^2\theta_W^{(e;D)}(-1.6\text{GeV}^2) \simeq 0.22 + 0.004(N_H-1) + 0.006\ln(0.4\text{GeV}/\Lambda_{\overline{MS}}) \quad (32)$$

For $\Lambda_{\overline{MS}} \simeq 0.4\text{GeV}$ and $N_H=1-4$, this SU(5) prediction is in good agreement with the experimental result $\sin^2\theta_W^{(e;D)}(-1.6\text{GeV}^2) = 0.223 \pm 0.020$ quoted in Eq. (24).

We have also combined the result in Eq. (27) with our earlier analysis[14] of the renormalized parameter $\theta_W \equiv \cos^{-1}(m_W/m_Z)$ to obtain the SU(5) prediction

$$\sin^2\theta_W \equiv 1 - m_W^2/m_Z^2 = 0.2100 + 0.004(N_H-1) + 0.006\ln(0.4\text{GeV}/\Lambda_{\overline{MS}}) \quad (33)$$

This can be used[11] in conjunction with our previously obtained radiatively corrected mass formulas[10,16,26]

$$m_W = \frac{38.66 \text{ GeV}}{\sin\theta_W}, \qquad m_Z = \frac{77.32 \text{ GeV}}{\sin 2\theta_W} \quad (34)$$

to obtain precise SU(5) predictions for m_W and m_Z as functions of $\Lambda_{\overline{MS}}$ and N_H. Alternatively, when m_W and m_Z are accurately determined, one can turn our formulas around to obtain SU(5) predictions for $\Lambda_{\overline{MS}}$, m_S, τ_p and $\sin^2\theta_W^{\text{exp}}$.

For example, (if $N_H=1$)

$$\Lambda_{\overline{MS}} = 0.4\text{GeV} \exp[35(1 - (\frac{84.36\text{GeV}}{m_W})^2)] \quad (35a)$$

$$= 0.4\text{GeV} \exp[47.7(1 - (\frac{94.91\text{GeV}}{m_Z})^2)] \quad (35b)$$

The results of this section are illustrated in table I for $N_H = 1$ and $\Lambda_{\overline{MS}}$ in the range 0.1GeV to 0.8GeV using the conservative estimate for τ_p given in Eq. (6).

$\Lambda_{\overline{MS}}$ (GeV)	$\sin^2\theta_W$	m_W (GeV)	m_Z (GeV)	m_S (GeV)	τ_p (yr)
0.1	0.218	82.74	93.58	1.3×10^{14}	$3 \times 10^{28 \pm 1}$
0.2	0.214	83.53	94.23	2.6×10^{14}	$5 \times 10^{29 \pm 1}$
0.3	0.212	84.01	94.62	4.0×10^{14}	$3 \times 10^{30 \pm 1}$
0.4	0.210	84.36	94.91	5.5×10^{14}	$9 \times 10^{30 \pm 1}$
0.5	0.209	84.63	95.13	7.0×10^{14}	$2 \times 10^{31 \pm 1}$
0.6	0.208	84.86	95.32	8.5×10^{14}	$5 \times 10^{31 \pm 1}$
0.7	0.207	85.05	95.48	1.0×10^{15}	$1 \times 10^{32 \pm 1}$
0.8	0.206	85.22	95.63	1.2×10^{15}	$2 \times 10^{32 \pm 1}$

Table I. SU(5) predictions for $N_H=1$. For a given $\Lambda_{\overline{MS}}$, each increase of N_H by one Higgs doublet increases $\sin^2\theta_W$ by 0.004, decreases m_W by 0.80GeV, decreases m_Z by 0.66GeV, decreases m_S by a factor of 0.65 and decreases τ_p by a factor of 0.17.

VI. Summary

For $\Lambda_{\overline{MS}} = 0.4\text{GeV}$ as suggested by electroproduction experiments,[12] we find that the SU(5) model predicts

$$\sin^2\theta_W^{(\nu_\mu;h)}(-20\text{GeV}^2) = 0.2098 + 0.0040(N_H-1) \tag{36a}$$

$$\sin^2\theta_W^{(\nu_\mu;e)}(0) = 0.2104 + 0.0040(N_H-1) \tag{36b}$$

$$\sin^2\theta_W^{(e;D)}(-1.6\text{GeV}^2) \simeq 0.22 + 0.0040(N_H-1) \tag{36c}$$

$$m_W = 84.36 - 0.80(N_H-1) \text{ GeV} \tag{36d}$$

$$m_Z = 94.91 - 0.66(N_H-1) \text{ GeV} \tag{36e}$$

$$m_S = 5.5 \times 10^{14 - 0.19(N_H-1)} \text{ GeV} \tag{36f}$$

$$\tau_p = 9 \times 10^{30 \pm 1 - 0.76(N_H-1)} \text{ yr} \tag{36g}$$

The prediction for the proton lifetime in Eq. (36g) is within easy reach of ongoing experiments. When compared with the present experimental bound $\tau_p > 10^{29}$ yr it implies that the number of allowed light Higgs doublets $N_H \leq 4$. For $N_H = 1-4$, the SU(5) prediction for $\sin^2\theta_W^{(\nu_\mu;h)}(-20\text{GeV}^2)$ is in good agreement with the experimental value 0.224 ± 0.012 (see Eq. (23)), which we obtained after taking radiative corrections into account. The prediction for $\sin^2\theta_W^{(e;D)}(-1.6C$ in Eq. (36c) (which includes only the γ-Z mixing effect) is also in good agreement with the experimental value 0.223 ± 0.020.

We strongly advocate precise measurements of all the quantities in Eq. (36 In particular, a determination of m_W and m_Z by the next generation of high energy accelerators will critically test the SU(5) model and provide important new input for deciphering the structure of the fundamental interactions.

References

1. H. Georgi and S. Glashow, Phys. Rev. Lett. $\underline{32}$, 438 (1974).
2. H. Georgi, H. Quinn and S. Weinberg, Phys. Rev. Lett. $\underline{33}$, 451 (1974).
3. A. Buras, J. Ellis, M. Gaillard and D. Nanopoulos, Nucl. Phys. B$\underline{135}$, 66 (1978).
4. W. Marciano, Proceedings of "Orbis Scientiae 1980".
5. P. Langacker, "Grand Unified Theories and Proton Decay" SLAC-PUB-2544. An up to date discussion of proton decay is given in this review.
6. Calculations of τ_p as a function of m_S are given in ref. 3 and in M. Machacek, Nucl. Phys. B$\underline{159}$, 37 (1979); C. Jarlskog and F. Yndurain, Nucl. Phys. B$\underline{149}$, 29 (1979); J. Donoghue, Phys. Lett. $\underline{92B}$, 99 (1980); A. Din, G. Girardi and P. Sorba, Phys. Lett. $\underline{91B}$, 77 (1980); E. Golowich, Phys. Rev. D$\underline{22}$, 1148 (1980).
7. T. Goldman and D. Ross, Phys. Lett. $\underline{84B}$, 208 (1979); preprint CALT-68-759 (1980).
8. S. Weinberg, Phys. Lett. $\underline{91B}$, 51 (1980). The source of the -1/22 term and the need to subtract it in order to obtain complete decoupling is discussed in this paper.
9. P. Binetray and T. Schucker, preprints CERN-TH-2802 (1979) and TH-2857 (1980); L. Hall, Harvard preprint HUTR-80/A024 (1980); N. Chang, A. Das and J. Perez-Mercader, CCNY-HEP-75/25; J. Ellis, M. Gaillard, D. Nanopoulos and S. Rudaz, CERN preprint TH-2833 (1980).
10. W. Marciano, Phys. Rev. D$\underline{20}$, 274 (1979); AIP Conference Proceedings No 59, Particles and Fields - 1979 edited by B. Margolis and D. Stairs p373.
11. W. Marciano and A. Sirlin, "Precise SU(5) Predictions for $\sin^2\theta_W^{exp.}$, m_W and m_Z" Northwestern Univ. preprint (1980).

12. A. Buras, Rev. Mod. Phys. $\underline{52}$, 199 (1980).

13. See L. Sulak, Proceedings of the International Conference on Neutrino Phys. Bergen 1979.

14. W. Marciano and A. Sirlin, Phys. Rev. D$\underline{22}$, 2695 (1980).

15. W. Marciano and A. Sirlin, manuscript in preparation.

16. A. Sirlin, Phys. Rev. D$\underline{22}$, 971 (1980).

17. A. Sirlin, Rev. Mod. Phys. $\underline{50}$, 573 (1978).

18. M. Veltman, Nucl. Phys. B$\underline{123}$, 89 (1977).

19. J. Kim, P. Langacker, M. Levine and H. Williams, Univ of Penn. preprint UPR-158T to be published in Rev. of Mod. Phys.

20. P. Salomonson and Y. Ueda, Phys. Rev. D$\underline{11}$, 2606 (1975); N. Byers, R. Ruckl and A. Yano, Physica $\underline{96A}$, 163 (1979).

21 L. Mo, "Review of Purely Leptonic Interactions of Weak Neutral Currents" VPI preprint HEP-80/8.

22. C. Prescott et al., Phys. Lett. $\underline{77B}$, 347 (1978); $\underline{84B}$, 524 (1979).

23. W. Marciano and A. Sanda, Phys. Lett. $\underline{77B}$, 383 (1978); Phys. Rev. D$\underline{17}$, 305 (1978); see also E. Derman and W. Marciano, Ann. of Physics, $\underline{121}$, 147 (19

24. Details of the calculations in Section V are given in Refs. 11, 14, 15 and

25. A detailed discussion of the radiative corrections to e-D asymmetries will be presented in a subsequent publication.

26. Alternative calculations of m_W and m_Z are given in: M. Veltman, Phys. Let $\underline{91B}$, 95 (1980); Erratum, M. Green and M. Veltman, Nucl. Phys. B$\underline{169}$, 137 (1980); F. Antonelli et al., Phys. Lett. $\underline{91B}$, 90 (1980) and Erratum to be published.

DISCUSSION

STROVINK: At the Madison Conference, the European Muon Collaboration, which has the most precise data on scale-non-invariance in high energy charged lepton scattering, presented a variety of results in which the central value of Λ was always less than 0.1 and the upper limit on Λ was always less than 0.2, which was the lower limit you chose. If I carry out your arithemtic I find that if one believes their central value of, let's call it 0.1, one would obtain an upper limit of 10^{29} years on the proton lifetime. If the European Muon Collabortion results were substantiated, would SU(5) be ruled out?

MARCIANO: The simplest version of SU(5) that I've used for these calculations predicts a proton lifetime of $\tau_p \simeq 3 \times 10^{28 \pm 1}$ yr for $\Lambda_{\overline{MS}} = 0.1$ GeV. Given the present experimental bound $\tau_p > 10^{29}$ yr, I would say that the simplest version of the SU(5) model might be on the verge of being ruled out if $\Lambda_{\overline{MS}}$ is really 0.1 GeV and proton decay isn't seen pretty soon in the ongoing experiments. You could, however, modify the prediction somewhat.

STROVINK: That is the question.

MARCIANO: By adding a fourth fermion generation you could raise τ_p by a factor of $\simeq 6$. Or you might complicate the Higgs structure. If, however, you want to retain the nice prediction for $\sin^2\theta_W$, you are more constrained than you might think. A lower value for $\Lambda_{\overline{MS}}$ does generally imply a smaller prediction for τ_p and therefoer makes the proton decay experiments even more exciting.

Nucleon Decay Experiments[1]

M. Cherry, K. Lande, C. K. Lee and R. I. Steinberg

Physics Department

University of Pennsylvania

Philadelphia, PA 19104

A considerable number of proton decay experiments is now being planned, under construction or running.[2] In this talk we will briefly review these experiments with emphasis on the types of backgrounds that will be encountered and the rejection efficiencies which will be necessary to carry out meaningful proton decay searches.

The common characteristic of all proton decay experiments is the requirement of very large active detector masses, $10^2 - 10^4$ tons. In order to reduce the rate of cosmic ray induced signals in such large masses it is necessary to locate these detectors deep within the earth. The two sources of cosmic ray induced background are muons, whose contribution is depth dependent and neutrinos, which give a background that is independent of depth. The various proton decay experiments fall into three different depth groups, Fig. 1: those at about 1.5×10^3 hg/cm^2, those at 4.5×10^3 hg/cm^2 and one at 7×10^3 hg/cm^2. The depth determines not only the background, but also the size of the cavity that can be excavated.

In order to understand the background rejection factors necessary to be able to observe proton decay with $\tau_N \geq 10^{30}$y, it is useful to consider the various background processes as if they were nucleon decays and to calculate the apparent nucleon lifetime produced by these processes. The largest backgro te is due to throughgoing muons, Fig. 2. We thus calculate the apparent nucleon cay rate which will be produced by misidentifying each throughgoing muon as nucleon decay. This rate depends on the depth and the surface area of the tector. The apparent nucleon decay rate ($1/\tau_A$) is the background rate divided by the mber of nucleons in the detector.

$$\frac{1}{\tau_A} = \frac{\text{(muon flux)} \times \text{(detector area)}}{N_o \times \text{(detector volume)} \times \text{(density)}}$$

$$\alpha \; 1/\text{(detector volume)}^{1/3}$$

nce (detector volume)$^{1/3}$ is the same (within a factor of 2) for most the experiments under consideration, we can characterize the various periments at a given depth by a single apparent lifetime. For throughgoing ons at 1500 hg/cm^2 we have $\tau_A \sim 10^{25}$ y; at 4500 hg/cm^2, $\tau_A \sim 5 \times 10^{27}$ y; at 7000 hg/cm^2, $\tau_A \sim 2 \times 10^{28}$ y. Thus, to observe ucleon lifetime of 10^{32} y, for example, it will be necessary to ieve background rejection factors of 10^7, 2×10^4, and 5×10^3 for three depths considered above.

Similar considerations can be raised for a nucleon decay signal t requires a large initial pulse followed by a $\mu \to e$ decay. For pping muons, we have

$$\frac{1}{\tau_A} = \frac{\text{(muon flux)} \times \text{(muon stops/gm cm}^{-2})}{N_o}$$

The cosmic ray muons have average energies of 200-300 GeV and thus pping probabilities of about 10^{-5} per gm cm^{-2}. The apparent nucleon lifetime g. 3) for a muon induced signal followed by a $\mu \to e$ decay is therefore about 10^2

larger than that of a throughgoing muon alone.

The muon backgrounds have a well defined vertical orientation which helps in rejecting them. There is also a background due to cosmic ray neutrinos which is isotropic. The neutrino interaction rates, of course, are depth independent and give an apparent nucleon lifetime of $\sim 10^{31}$ y.

The various proton decay experiments can also be described by experimental approach. The largest volume experiments involve totally active detectors that utilize water Cerenkov counters. These range from the completely open detector at Fairport Harbor to the highly modularized Homestake detector. A second set of experiments employs a fine grain calorimeter approach - a large mass of inert absorbing material interspersed with proportional or flash tubes.[3] The latter approach has the advantage of achieving some topology reconstruction but is not able to measure energy release.

Homestake Results[4]

We began operation of the Homestake Detector at the beginning of 1979. Electronics and apparatus appropriate to the various experimental aims of the detector were turned on during the year. We here describe and summarize data relevant to proton decay acquired in 1979.

We have examined the data of 127 days observation. During this period we observed 89 $\mu \rightarrow e$ decays. The events are characterized by 3 parameters: number of modules involved, i.e. spatial extent of energy deposition; magnitude of energy deposition; and presence or absence of veto counts. We display these events in Fig. 4, where the energy deposition is shown along the x-axis and the module number along the y-axis. From a Monte Carlo simulation of proton decay for the various decay modes, we know that nucleon decay events will deposit 150 to 400 MeV of energy and that the decay secondaries will, with high probability, be confined to one or two modules. Module numbers ≥ 3 are inconsistent with nucleon decay. About half of the events with $n \geq 2$ are also accompanied by one or more veto scintillator pulses. This fraction of scintillator involvement is consistent with that expected for cosmic ray background events.

Our interest is focussed on the one- and two-module events (Fig. 5). Of the 16 two-module events, 4 also involve scintillation counters and thus are labeled as cosmic ray events. Eleven of the remaining 12 events involve two vertically related modules with the muon decay in the lower module. This is exactly the topology expected for stopping cosmic ray particles. Proton decays, however, should have an isotropic distribution of module pairs. In order to avoid confusion with stopping cosmic ray particles, we will restrict our two-module proton decay candidates to those that have horizontal or upward pion-muon trajectories. The one remaining two-module event involves a prompt energy deposit of 980 MeV, a value that is above that expected for nucleon decay.

Of the nine one-module events, three involve scintillation counters and thus can be labeled as cosmic ray induced. For the remaining six events it is necessary to compare the energy deposit distribution with that predicted by our Monte Carlo simulations. The single module events should show less than 400 MeV energy deposit. Very few of these events should be below 200 MeV. Of the remaining six single-module events only two are close to the acceptable range for proton decay. They both have a prompt energy deposit of 150 MeV. Since both of these events are at the low end of the energy spectrum, an end with very low occurrence probability, it is reasonable to claim that we have no remaining candidates for proton decay.

We will determine the effective running time by using the 89 cosmic ray induced muon decays. Since each such decay corresponds to 2×10^{29} nucleon-y, our running time is $89 \times 2 \times 10^{29} = 1.8 \times 10^{31}$ nucleon-y. From the various branching ratio calculations and the behavior of pions in the intranuclear region and in water, we determine that 27% of all nucleon decays will result in muon decays. Since almost all nucleon decay channels contribute with similar strength to this signature, the 27% value is insensitive to most modifications of the branching ratio predictions. Finally, we introduce an acceptance factor of 0.85 to take account of the exclusion of downward muon trajectories. Combining these numbers, we find the effective running time to be $1.8 \times 10^{31} \times 0.85 = 4 \times 10^{30}$ nucleon-y. No nucleon decays in 4×10^{30} y corresponds to an upper limit, at the 90% confidence level, of 2.3 events, or a lower limit for the total nucleon lifetime of 1.8×10^{30} y at the same 90% confidence level.

Other Recent Experimental Results

At the 1980 International Neutrino Conference Miyake reported on two proton decay candidates that were observed in detectors operated in the Kolar Gold Field laboratory during the past 15 years. These events give a lifetime of $(1.7 \pm 1.2) \times 10^{30}$ years. Confirmation of this observation would provide us with the substance of many more pleasant meetings in Blacksburg.

Footnotes and References

1. Research supported in part by the U.S. Dept. of Energy under contract DE-AC02-81ER40012 and by the U.S. National Science Foundation under contract AST-79-08670.
2. Recent thorough reviews of proton decay experiments have been given by J.C. van der Velde, SLAC Summer Institute on Particle Physics, August 1980 and by M. Goldhaber, Neutrino '80, Erice, June 1980.
3. Communicated by M. Conversi, to be published in Proc. of 1980 International DUMAND Symposium.
4. M. Deakyne et al., Proc. XVth Rencontre de Moriond, March 1980; Proc. First Workshop on Grand Unification, Durham, N.H., April 1980.

NUCLEON DECAY EXPERIMENTS

	DEPTH/LOCATION	FIDUCIAL MASS (TONS)	DETECTION TECHNIQUE	STATUS
A)	1500 hg/cm^2			
	1) Fairport Harbor (I.M.B.)	$6-9 \times 10^3$	Water Cerenkov	Construction
	2) Park City (H.P.W.)	10^3	Water Cerenkov	Construction
	3) Soudan (Minn.-Argonne)	$1.5 \sim 10^2$	Calorimeter	Construction
B)	4500 hg/cm^2			
	1) Homestake (Penn-BNL)	3×10^2	Water Cerenkov	Running
	2) Mt. Blanc	1.5×10^2	Calorimeter	Construction
	3) Frejus	10^3	?	Planning
C)	7000 hg/cm^2			
	1) Kolar	1.5×10^2	Calorimeter	Construction

Fig. 1

BACKGROUND RATES AND APPARENT NUCLEON LIFETIMES
FOR A DETECTOR CONTAINING 10^{33} NUCLEONS*

		DEPTH	1500 hg/cm^2	4500 hg/cm^2	7000 hg/cm^2
1)	Muons				
	a) Throughgoing	rate	$\sim 10^8$/y	$\sim 2 \times 10^5$/y	$\sim 5 \times 10^4$/y
		τ_A	$\sim 10^{25}$ y	$\sim 5 \times 10^{27}$ y	$\sim 2 \times 10^{28}$ y
	b) Stopping ($\mu \to e$ decay)	rate	$\sim 10^6$/y	$\sim 10^3$/y	$\sim 2 \times 10^2$/y
		τ_A	$\sim 10^{27}$ y	$\sim 10^{30}$ y	$\sim 5 \times 10^{30}$ y
2)	Neutrinos				
	a) Interactions	rate	$\sim 10^2$/y		
		τ_A	$\sim 10^{31}$ y		
	b) $\mu \to e$	rate	~ 30/y		
		τ_A	$\sim 3 \times 10^{31}$ y		

Fig. 2

*As explained in the text, for the detectors listed in Fig. 1, at a given depth, the apparent lifetimes are nearly independent of detector volume.

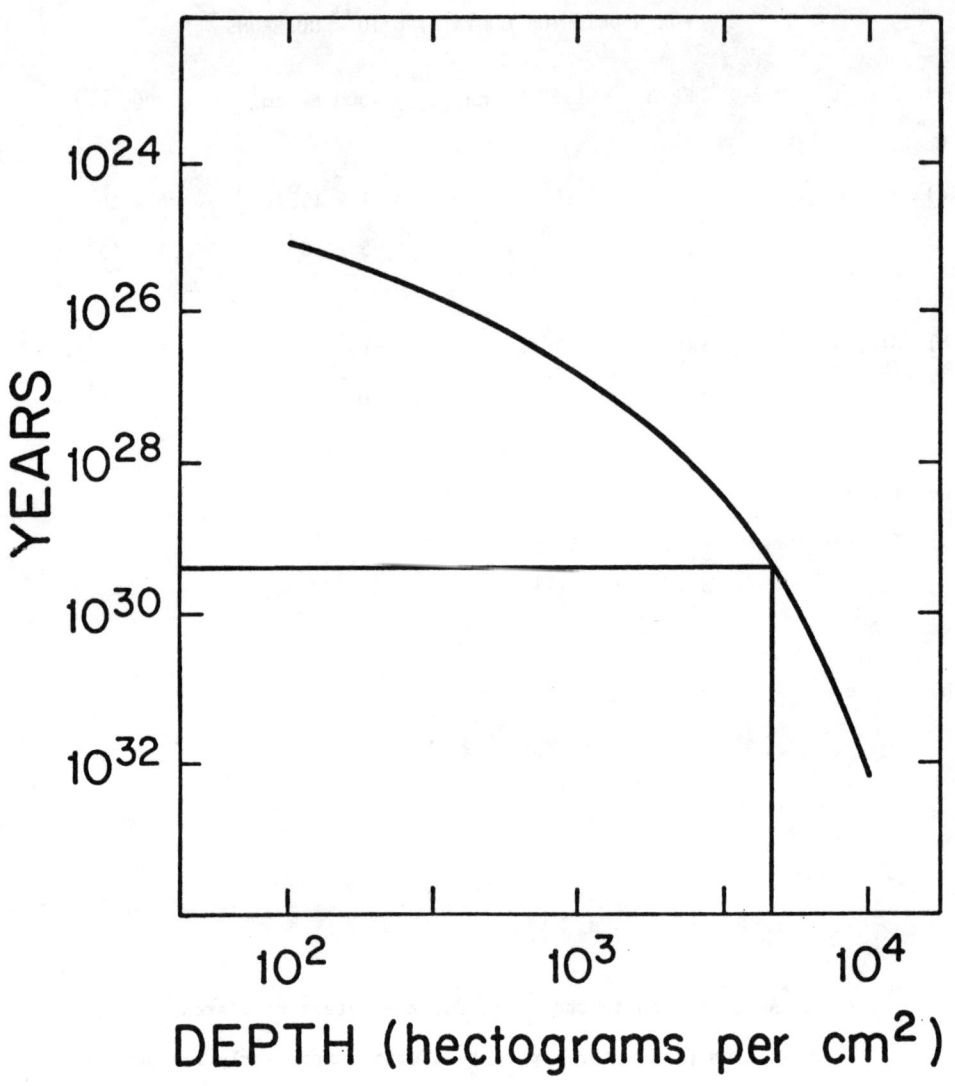

Fig. 3 Apparent Nucleon Lifetime as a Function of Depth

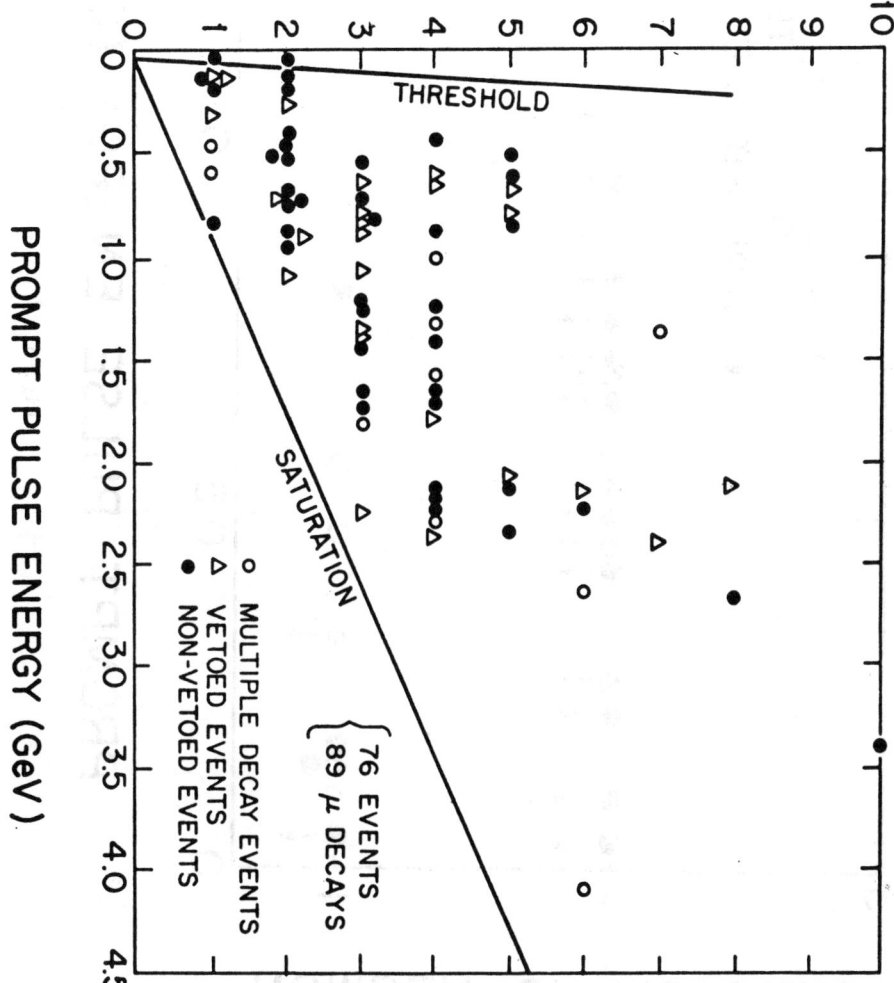

Fig. 4

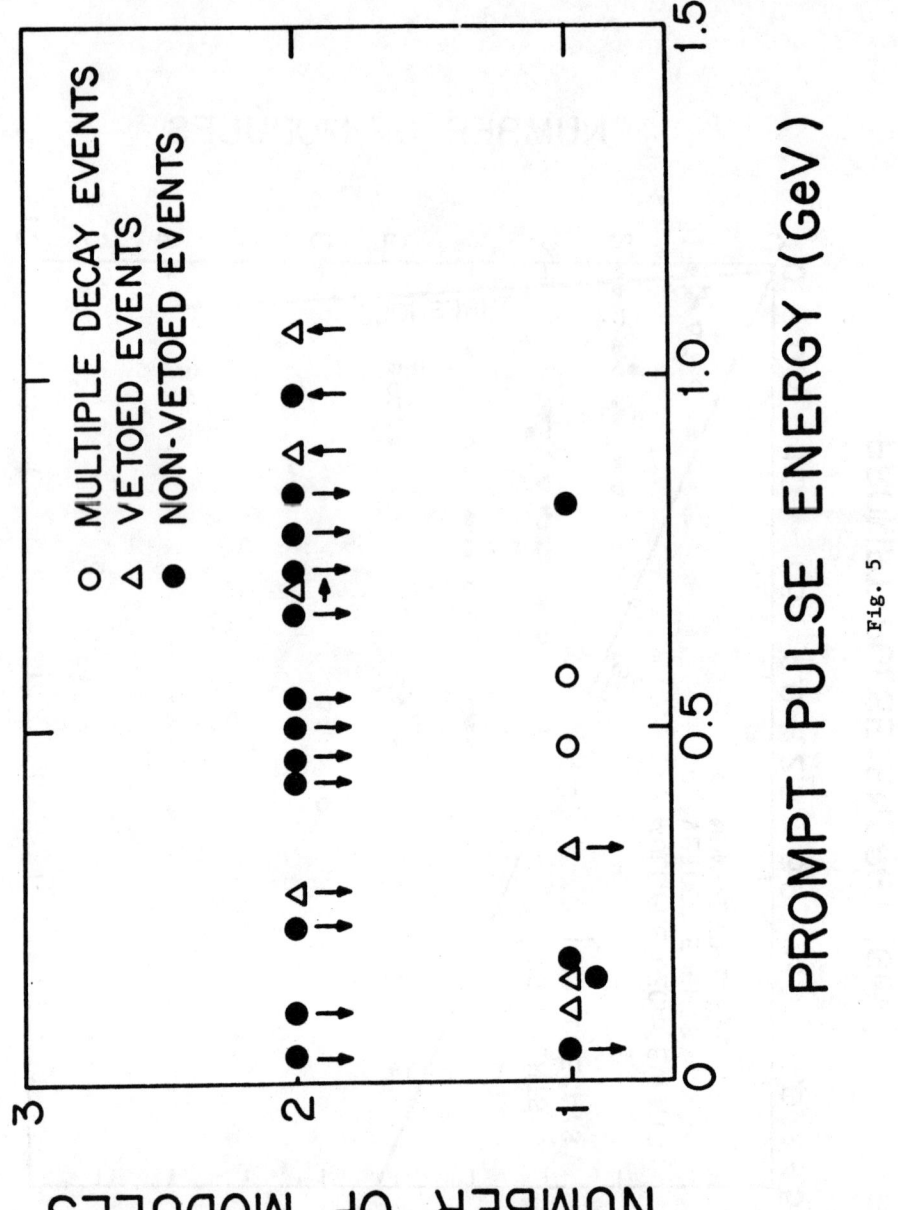

Fig. 5

DISCUSSION

ABASHIAN: The single best known imput in the whole experiment seems to be the mass of the proton, which is very well known. But I don't surmise from your presentation any indication as to how well the total energy of the outgoing products is determined in any of the experiments. If you can actually show that the energy deposited is equal to the mass of the proton, that would be a very sensitive test. Do you want to say something about that?

LANDE: There are several limitations on the reconstruction of the nucleon mass from the pulse heights observed in the detector. There is the energy lost by the pion secondaries in escaping from the nucleus. According to calculations by D. Sparrow, one half of the pions are absorbed in a nucleus, such as oxygen. Of the remaining half, one third will suffer a nuclear scattering on the way out of the nucleus. There are also energy losses suffered by pions because of scattering in the internuclear domain. All of these losses are missed by Cerenkov detectors and cannot be corrected for.

We have carried out Monte Carlo calculations of the visible energy for various nucleon decay modes and we see some possibility of using the fraction of visible energy to distinguish between different decay modes. This approach may permit a measurement of branching ratios, if nucleon decay is ever observed.

MARSHAK: Could you also see $\Delta B = 2$ events with these experiments?

LANDE: The $\Delta B = 2$ decays ($N \to \bar{N}$ transitions) followed by a $\bar{N}$-N annihilation would result in almost 2 GeV of visible energy. The annihilation process would produce 4-6 charged pions which would result in multiple $\mu \to e$ decays. We are actively examining our data for such events.

PATI: Could you comment on the two Kolar gold mine events and what kind of decays are present. If you say they are not proton decays, then what can they be?

LANDE: I'll certainly make the first comment.

PATI: The second one is more important.

LANDE: It is hard for me to decide whether or not the two Kolar events are indeed nucleon decay. I have only seen transparencies of them. One event has a strong resemblance to $p \to \pi^0 e^+$. I certainly hope that they are indeed examples of nucleon decay and that Miyake finds more of them.

PATI: Do you feel that this interpretation is reasonable?

LANDE: It is a reasonable interpretation.

PATI: Does anything conflict with this interpretation?

LANDE: For that event, I know of nothing. Now, the other event is much more complicated.

PATI: Could it be e^+ or τ?

LANDE: They have reported a number of unusual events in their cosmic ray neutrino experiment. I thought that some of these might be τ decays and from the angular distribution of those events it even seemed possible that the ν_τ might have resulted from neutrino oscillation.

PATI: How about $e^-\pi^+\pi^+$?

LANDE: The pattern is less suggestive of $e^-\pi^+\pi^+$ than of $e^+\pi^o$, but without additional information I cannot distinguish among these possibilities.

ROSEN: I wonder if I could make a cautionary remark along the following lines. We heard earlier this morning about how difficult it is to see the process of double β decay which has a lifetime 10 orders of magnitude greater than the expected lifetime of the proton. It is certainly true that in the proton decay much more energy is released and so you get a bigger bang for your bucks as it were, but on the other hand you do have those 10 orders of magnitude to fight. Now, one conceivable way in which you might be able to gain something is not to think in terms of experiments that last for one, two or even possibly 10 years, but rather experiments that last the age of the universe. That is to say, the same kinds of geochemical experiment as people have done in the case of double β decay. If, inside a nucleus, a nucleon disappears, you may end up with a different nucleus which, if stable, can sit around for a very long time. You could then gain a great deal in the scale of your experiments. And you would also have an advantage in that you would not be relying as heavily as you appear to do on a presumed decay mode. You would have a decay mode-independent experiment.

LANDE: I think there are two points to be made. One is the question of whether one can do experiments where signals are very small. The second question is how could one try to model something like the various decay modes and their various branching ratios. The first question, I think is that the signal isn't as weak as for double β decay because what I tried to show in my earlier transparency was that with the scales of the background it is possible to look for something in the range of 10^{30}–10^{32} years, and the backgrounds for certain of these decay processes are on the scale of 10^{30} years to begin with, so you are only trying to affect the background by factors of 10 or 100 in order to be able to do that. Now that isn't as painful as the double β decay exercise. It isn't just the energy it's **also** the pattern of the thing that you are looking for. Now the question of branching ratios in decay modes - that is a very different story. We were very worried about that question and the idea of choosing a single channel which may have a big number, but could also have a small number and possibly by next year be totally forbidden depending upon how the fates determine things, its a frightening prospect. So we felt that what was important to say was that there are a collection of particles that are lighter in mass than the proton. Any of them is a possible candidate. We said what is a characteristic of those. The thing that was most characteristic is that the μ-e decay channel was common to most of these secondary particles. And they often ended up as ρ's or ω's which decayed to π's and then gave μ-e's. And so by having many channels, our thought was that if the branching ratio calculations were to make one channel become more popular and one less popular the sum wouldn't change and therefore one could become more model independent. Now there is the possibility that in fact it would end up with many neutrinos and one π^+ which doesn't go anywhere. Then you could really be trapped. The geochemical approach looks feasible for lifetimes of 10^{27} years, but I am not sure that it is possible to achieve 10^{30} years with this technique.

ON NEUTRON OSCILLATIONS

N.P. Chang
Department of Physics, The City College, CUNY, New York, NY 10031

1. Introduction

Grand unification has given rise to renewed speculations on the nature of conservation of fundamental quantum numbers. When quarks and leptons are put in the same multiplet it is natural to expect the ultimate conversion of quarks into leptons, so that the hitherto intrinsic stability of the proton against such decays as [1]

$$p \to e^+ \pi^0 \tag{1}$$

is no longer assured. The final experimental verdict on this crucial question is yet to come, with an early verdict likely to be quite revolutionary.

In this talk I will dwell on a related question, viz. that of $\Delta B = 2$ transitions, a prominent example of which being the oscillation of n into $\bar{n}$. If protons are not to be forever, can $n - \bar{n}$ oscillations be far behind?[2]

Actually, already in 1955 Lee and Yang studied the question of B-conservation in physics in an elegant, little paper.[3] Their conclusion, based on the result of Eotvos experiment, was that if B-number is a local gauge symmetry, then the B coupling constant, η, must satisfy the constraint

$$\frac{\eta^2}{\hbar c} \lesssim 5.9 \times 10^{-44} \tag{2}$$

In modern language, one common way out is to spontaneously break the B-gauge symmetry with the consequence that the force due to B-gauge boson exchange is short ranged, and no constraint on η results. The other way is simply to let B not be a conserved quantum number <u>at all</u>, although it will turn out for low energies to have like a global quantum number that is approximately conserved. This latter is quite easily realised in grand unification as we shall show.

2. B-L

In SU(5) grand unification, the proton decay (such as eq.(1)) conserves B-L. This B-L conservation turns out, totally within the context of SU(5), to be quite accident. The puzzle arises because B-L is not a generator of SU(5) as is clear from a listing for the fundamental representation

	B	L	B-L
d_{iR}	$\frac{1}{3}$	0	$\frac{1}{3}$
e_R^+	0	-1	1
ν_{eR}^c	0	-1	1

$$Tr(B-L) = 3$$

so that B-L conservation would appear to be a mystery.

As Wilzek and Zee have noted,[4] however, there is a global number, X, which can be assigned to $\underline{5}_R$ so that it, together with the internal hypercharge, will mimic B-L. To wit, for the $\underline{5}_R$, we have

	Y_W	X	B-L
d_{iR}	-2/3	3	$\frac{1}{3}$
e_R^+	1	3	1
ν_R^c	1	3	1

where $\frac{2}{5} Y_W + \frac{1}{5} X = B-L$ (3)

Given this, it is easy to show that for the $\underline{10}_L$ representation the same relation (eq. (3)) requires $X(\underline{10}_L) = +1$.

So far we have not brought in the Higgs. Because fermions can couple to the Higgs and self-couplings of Higgs occur, X assignments have to be given also to the Higgs. If a consistent X assignment can be given to all the Higgs, then X would be a good, global quantum number. This would obviously be true when SU(5) is unbroken. Even when SU(5) is broken, however, X will remain to a good approximation, a conserved global quantum number.

To see this, recall that in a broken SU(5) theory, processes mediated by the gluons, $W_\mu^\pm$, Z_μ and A_μ preserve Y_W and, by eq.(3), conserve B-L. Processes mediated by the leptoquarks and Higgs have the properties (i = 1,2,3; a = 4,5):

Mediated by		Mass	ΔY_W	ΔX	Δ(B-L)
Leptoquarks	X_r^{ia}	$\sim 10^{14}$ GeV	$-\frac{5}{3}$	0	$-\frac{2}{3}$
$\underline{5}$ Higgs	H^i	$\sim 10^{12}$ GeV	$-\frac{2}{3}$	-2	$-\frac{2}{3}$
	H^a	$\sim 10^3$ GeV	1	-2	0

Since the $\underline{24}$ Higgs does not couple to the fermions they will not directly cause Δ(B-L) $\neq$ 0 transitions.

From the above table it is clear that ΔB = 2 processes are necessarily suppressed by the large mediating masses,$^{.5}$ viz. matrix elements are $\sim 1/M^6$, with $M \geq 10^{12}$ GeV.

The minimal SU(5) model has only $\underline{5}$ and $\underline{24}$ Higgs present in the Lagrangian. The Lagrangian turns out to possess an extra global invariance with the assignmen

		X - quantum number
gauge bosons		0
$\underline{5}_R$	fermion	3
$\underline{10}_L$	"	1
$\underline{5}$	Higgs	- 2
$\underline{24}$	"	0

This global invariance is not a consequence of nor is it related to the fundament SU(5) invariance.

3. Non-minimal SU(5)

Minimal SU(5) therefore allows the proton to decay but does not permit the neutron to oscillate. But, as I have mentioned earlier, grand unification could allow $n - \bar{n}$ oscillation. How does this come about? Let us introduce a $\underline{15}$ Higgs and see what happens.

The two allowed, renormalisable interactions

$$h \, \tilde{\psi}_R^\alpha \, C^\dagger \, \psi_R^\beta \, H^\dagger_{\alpha\beta} \qquad (4)$$

$$m_i H^\alpha \, H^\beta \, H^\dagger_{\alpha\beta}$$

have mutually exclusive X - assignments with $X(H^{\alpha\beta}) = + 6$ and $- 4$ respectively. Thus, if we fix $X(H^{\alpha\beta})$ to be $+ 6$ by the Yukawa interaction, the trilinear Higgs interaction will no longer be invariant under global X transformation but transforms on the whole as $X = - 10$ tensor.

Lay Nam and I introduced the $\underline{15}$ Higgs simply to prove a point with respect to B-L. Phemenologically we know that the minimal Higgs system is not enough to generate the correct fermion mass spectrum nor is the minimal Higgs adequate for baryon asymmetry. If we argue that the Higgs system should be enlarged, then ac cording to prevailing views of dynamical symmetry breakdown, we should probably all the way to the maximal set of Higgs that can couple to fermions of a given generation and include all of $\underline{5}$, $\underline{15}$, $\underline{45}$ and $\underline{50}$. Of this set only $\underline{15}$ can and wi break B-L conservation.

A measurement of neutron oscillation is thus a direct probe of the Higgs system, uniquely confirming the presence of the 15. Whether or not we need, for phenomenological purposes, all of the rest of the Higgs (45 and 50) we cannot of course as yet tell. What we have found is the need for 45 in order to maximize the $n - \bar{n}$ oscillation and still not conflict with proton decay.

The scenario for a non-minimal SU(5) grand unification is thus quite rich and interesting. In contrast with the touted "desert" of minimal SU(5), we have the full panoply of intermediate mass scales in SU(5) that populate the desert ("Let the desert bloom"), and among the buried ruins are such old absolute quantum numbers as B and L.

A severe constraint on the possible range and hierarchy of mass scales comes from a study of the $\sin^2\theta_W$ parameter. We have found that the following mass spectrum is an allowed optimum scenario:[6] (here $(8,2)_{45}$ refers to SU(3) 8-plet, SU(2) 2-plet member of 45, etc).

χ^{ia}	leptoquarks	$\sim 10^{15}$ GeV
$(3,1)_5$		$\sim 10^{13}$ GeV
$(3,1)_{45}$		
$(3^*,1)_{45}$		$\sim 10^{10}$ GeV
$(3^*,2)_{45}$		
$(8,2)_{45}$		
$(6,1)_{15}$		
$(6^*,1)_{45}$		$\sim 10^{4-5}$ GeV
$(3,2)_{15}$		
$(3,3)_{45}$		
$(1,2)_5$		
W,Z		$\sim 10^3$ GeV
$(1,3)_{15}$		
$(1,2)_{45}$		

With this scenario, the result of the renormalisation group analysis[8] is $\sin^2\theta_W$ .232. This is an improvement over minimal SU(5) prediction of .211, a value that may be somewhat low compared with the experimental result .23 ± .02.

A further word about how this wide range of mass scale was settled upon is helpful at this point. In the renormalisation group analysis of M_X and $\sin^2\theta_W$, we have used as input, besides the usual α_s and α_{em}, the additional constraints that $(3,1)_{45}$ be $\sim 10^{10}$ GeV and $(6,1)_{15}$, $(6^*,1)_{45}$ be $\sim 10^{4-5}$ GeV. The former constraint is in order not to conflict with proton decay lifetime limit and the latter is simply so as to make n-n̄ oscillation as large as possible.

With this choice, proton decay proceeds predominantly via leptoquark exchange and therefore will not exhibit B-L violation in its decays. (i.e. $p \to e^- + \pi^+\pi^+$ suppressed relative to $p \to e^+\pi^0$ etc). The n-n̄ matrix element proceeds via the diagram

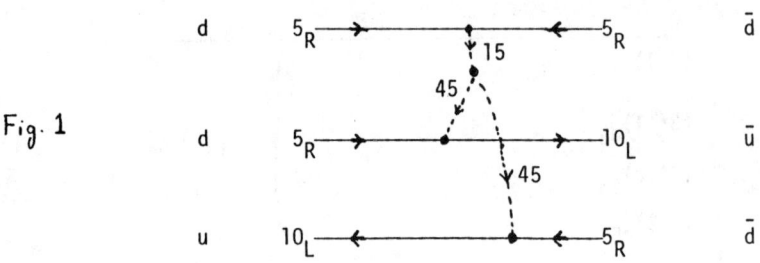

Fig. 1

with $(6,1)_{15}$, $(6^*,1)_{45}$ as the mediating fields. The matrix element is (m = dimensional coupling of 15-45-45)

$$\frac{h' h^2 m}{M_6^2 M_{6*}^4} \qquad (5)$$

where M_6 is mass of $(6,1)_{15}$ and M_{6*} is mass of $(6^*,1)_{45}$ and h' and h are the respective Yukawa couplings of 15 and 45 to the light fermions. Since the light fermions will acquire mass through the final SU(2) breaking of the 5 and 45, h must be $\sim 10^{-4}$. The 15 does not acquire vacuum expectation value at the tree level so h' is not constrained. From baryon asymmetry considerations, however, we prefer h' to be $\sim 10^{-1}$. Finally we assume for simplicity that $M_6 = M_{6*} = m$ so that the n-n̄ matrix element is

$$\frac{h' h^2}{\cdot m^5} \sim \frac{10^{-9}}{m^5} \equiv \varepsilon \tag{6}$$

What then should m be? We must at this point turn to nuclear stability constraints to get some limit on ε.

4. Nuclear Stability Constraint

For this we consider the effective field theory for a B- nonconserving neutron

$$\mathcal{L} = - \bar{\psi}\gamma\partial\psi - m\bar{\psi}\psi - \varepsilon \bar{\psi} C^{-1} \psi + \varepsilon \psi C \tilde{\bar{\psi}} + \mathcal{L}_{strong} + \mathcal{L}_{weak} \tag{7}$$

Kuzmin et al (1980)[9] made an estimate of ε based on the following argument. An (A,Z) nucleus may by the ε mixing term turn into a metastable nucleus with Z protons, A-Z-1 neutrons and one $\bar{n}$. This metastable nucleus will have a different binding energy than the original nucleus, with the difference, ΔM, being of order tens of MeV or so. The $\bar{n}$ in the nucleus has a Γ_{ann} for annihilating with a p or n into pions so that

$$(A,Z) \xrightarrow{\varepsilon} (A,Z)^*$$

$$\xrightarrow{\Gamma_{ann}} \begin{cases} (A-2,Z) + \pi^+\pi^- \\ (A-2,Z-1) + \pi^+\pi^0 \end{cases}, \text{ etc.}$$

The total rate is given by[10]

$$\Gamma = (A-Z) \varepsilon^2 \frac{1}{2\pi} \frac{\Gamma_{ann}}{(\Delta M)^2 + \Gamma_{ann}^2/4} \tag{8}$$

Since, by a Fermi gas model, one can directly estimate Γ_{ann} to be ~ 135 MeV $\gg \Delta M$, we find

$$\Gamma = (A-Z) \varepsilon^2 \frac{2}{\pi} \frac{1}{\Gamma_{ann}} \tag{9}$$

With a nuclear stability limit of 10^{30} years/A, the limit on ε becomes

$$\varepsilon \lesssim 10^{-31} \text{ GeV} \tag{10}$$

In turn, the mass scale m has to be $\gtrsim 10^{4.4}$ GeV.

The shocking thing about neutron oscillation is not that they are allowed to oscillate but that a <u>free</u> neutron oscillation into anti-neutron can occur in as short a time scale as

$$t_{n\bar{n}} \equiv \frac{1}{\varepsilon} \sim 10^7 \text{ sec} \tag{11}$$

and still not conflict with nuclear stability limits. It is for this reason that neutron oscillation has caught the fancy of nuclear physicsts who are now busy proposing and setting up experiments.

The physical reason for this paradox is that in free space, n and $\bar{n}$ are completely degenerate. By analogy with K-$\bar{K}$ system, the Hamiltonian for the system is, in the rest frame,

$$H = \begin{pmatrix} m & \varepsilon \\ \varepsilon & m \end{pmatrix} \qquad (12)$$

From this it is easy to obtain

$$P_{\bar{n}n}(t) = [\sin \varepsilon t]^2 \qquad (13)$$

In the nuclear medium, however, the nuclear forces between n-n and n-$\bar{n}$ are not identical and the degenracy between the n and $\bar{n}$ is lifted. If 2Δ is the energy level splitting between n and $\bar{n}$ in the medium, then old fashioned perturbation theory gives

$$P_{\bar{n}n}(t) = \frac{\varepsilon^2}{\Delta^2} [\sin \sqrt{\Delta^2 + \varepsilon^2}\, t]^2 \qquad (14)$$

Actually, there is an interesting pairing effect[11] which modifies somewhat this argument. Time does not permit me to dwell on this effect except to note that the true ground state for a B-non-conserving world is a BCS ground state with paired neutrons of opposite momentum and separately paired antineutrons. The practical result of this pairing is unfortunately too small to affect the oscillation results based on usual K-$\bar{K}$ analog. For details please refer to our paper.

In the experimental context, a neutron propagating in truly free space is hard to get. The reason is the presence of Earth's magnetic field which causes the n↑ and $\bar{n}$↑ to precess in opposite directions, and thus cause a Larmor splitting.[12] The result for n-$\bar{n}$ oscillation in the presence of a perpendicular magnetic field is ($\Delta \equiv \sqrt{B^2 + \varepsilon^2 m^2/\omega^2}$)

$$(\sin \Delta t)^2 \left[\frac{\varepsilon m}{\omega B} + \frac{p^2}{m\omega} \frac{B}{(\omega^2 - B^2)}\right]^2 + (\cos \Delta t)^2 \left[\frac{\varepsilon p^2}{m} \frac{1}{\omega^2 - B^2}\right]^2$$

$$(15)$$

Earth's magnetic field gives rise to a $B \sim 3 \times 10^3$ sec^{-1}. In order to enhance the probability for $\bar{n}$ in an initial slow n-beam, the earth's field must be degaussed. If B is degaussed to ~ 1 sec^{-1}, then an optimistic $\dot{\epsilon} \sim 10^{31}$ GeV gives at t = 1 sec

$$P_{\bar{n}n} \sim 10^{-14} \tag{16}$$

5. Baryon Asymmetry

An important question concerning baryon asymmetry[13] may be raised with regard to the class of theories that allow for n oscillation. In the standard big bang cosmology, according to the Weinberg scenario, baryon excess is produced by the decay of the big X ($(3,1)_5$ member of Higgs family) at $T_D \sim 10^{11}$ GeV, small compared with its $\sim 10^{13}$ GeV mass. In standard theory this baryon excess is not thermalized by subsequent collisions because baryon number nonconserving matrix elements involving virtual X exchange are severely suppressed relative to baryon-conserving collisions.

The existence of an intermediate mass scale [the small X] with its baryon-nonconserving decays will act to wash out the baryon asymmetry produced by the first generation. The question thus is how the second generation of baryon asymmetry may be produced.

Let us call the $(6,1)_{15}$ member of Higgs family the new X with the mass M ($\sim 10^{4.4}$ GeV) and $\Gamma_x (= h'^2 M/4\pi)$ The temperature T_D where $\Gamma_x \sim H$ (Hubble expansion rate) is at $\sim 10^{10}$ GeV, which is very high compared with M. Therefore unlike the Weinberg scenario this new X will not immediately decay away. Inverse decays will be energetically allowed to replenish the X. Baryon excess, produced at the first generation, will thus be wiped out by this new thermal equilibrium

$$X \rightleftarrows d_R + d_R \tag{17}$$

The condition of equilibrium requires that

$$\mu(X) = 2\mu(d_R) \tag{18}$$

This reaction coexists with the reaction $d + \bar{d} \rightleftarrows 2\gamma$. In fact since the rate for the annihilation into two photons is $\sim 10^{-5}$ T for T down to about 10^2 M the annihilation channel forces

$$\mu(X) = \mu(d_R) = 0 \tag{19}$$

As T further cools down to M = $10^{4.4}$ GeV, however, Γ_x remains at $\sim 10^{1.4}$ GeV while Γ_{ann} is now $10^{-.6}$ GeV. In this domain, $d\bar{d}$ annihilation and dd scattering may be neglected relative to $X \rightleftarrows d\,d$ and thermal equilibrium condition now reads

$$\mu(X) = 2\mu(d_R) \neq 0 \tag{20}$$

The physical meaning of this condition is that in this range of T, there is a new number conservation. As T cools down still further the inverse decays become unfavorable and the X decays away, creating baryon excess.

According to this new scenario, therefore, for $T \gtrsim M$, the number density of X and d_R are given by thermal distributions with zero chemical potential, viz.

$$n_X = .122\ T^3$$
$$n_{d_R} = .092\ T^3 \qquad T \gtrsim M \qquad (21)$$

For $T < M$, the total number of X's will decrease and the total number of d_r increases such that $(N_X + \frac{1}{2} N_{d_R})$ is constant. The chemical potential that maintains this evolves as follows:

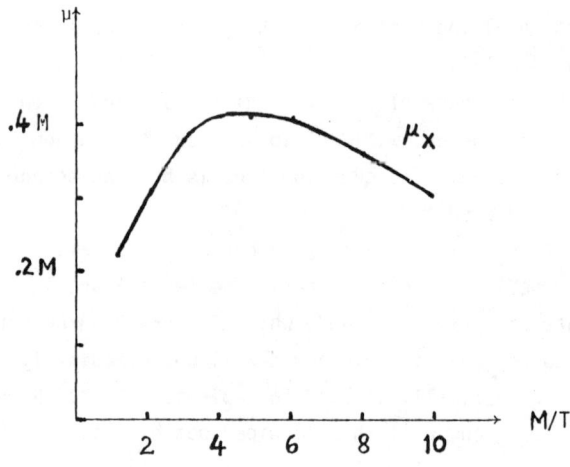

For $T < M$, this chemical potential translates into a sharp decrease in N_X (the total number of X), reflecting upon the fact that as T drops below M, the inverse decays begin to fail to replenish X. By $T \sim M/10$ or so, the X will have essentially decayed away, producing baryon excess.[14]

To see this scenario work out mathematically,[15] we can write down the Boltzmann equation for N_X, N_{d_R}, $\bar{N}_X$, $\bar{N}_{d_R}$, in terms of z ($\equiv M/T$) and the reduced chemical ν ($\equiv \mu/T$)

$$N_X \equiv \int \frac{d^3p}{e^{\sqrt{p^2+z^2}-\nu_X}-1}\ \frac{1}{(2\pi)^3} \equiv \int \frac{d^3p\ n_X(p,z)}{(2\pi)^3} \qquad (22)$$

$$N \equiv N_{d_R} = \int \frac{d^3p}{e^{p-\nu}+1}\ \frac{1}{(2\pi)^3} \equiv \int \frac{d^3p\ n(p,z)}{(2\pi)^3} \qquad (23)$$

$$\frac{d}{dz} N_x = - z^2 K \left[1 - (1-\tfrac{\eta}{2}) e^{2\nu-\nu_x}\right] \int \frac{d^3p}{\sqrt{p^2+z^2}} \, n_x(p,z) \tag{24}$$

$$\frac{d}{dz} N = 2K z^2 \left[1 - (1-\tfrac{\eta}{2}) e^{2\nu-\nu_x}\right] \int \frac{d^3p}{\sqrt{p^2+z^2}} \, n_x(p,z) \tag{25}$$

$$\frac{d}{dz} \bar{N}_x = - K z^2 \left[(1 - \tfrac{\eta}{2}) - e^{2\nu-\nu_x}\right] \int \frac{d^3p}{\sqrt{p^2+z^2}} \, \bar{n}_x(p,z) \tag{26}$$

$$\frac{d}{dz} \bar{N} = 2K z^2 \left[(1 - \tfrac{\eta}{2}) - e^{2\nu-\nu_x}\right] \int \frac{d^3p}{\sqrt{p^2+z^2}} \, \bar{n}_x(p,z) \tag{27}$$

Here K is a measure of Γ_x relative to Hubble expansion rate, viz ($m_p \equiv$ Planck mass $\sim 10^{19}$ GeV)

$$K = \frac{m_p \Gamma_x}{a M^2} \tag{28}$$

where $dT/dt = - aT^3/m_p$. In our scenario, $K \gg 1$.

Taking advantage of this fact, we can develop a $1/K$ expansion for the solutions, with the leading order satisfying

$$\begin{aligned} \nu_x &= 2\nu - \tfrac{\eta}{2} \\ \bar{\nu}_x &= 2\bar{\nu} + \tfrac{\eta}{2} \end{aligned} \tag{29}$$

$$N_x + \tfrac{1}{2} N = \text{constant} = \bar{N}_x + \tfrac{1}{2} \bar{N} \tag{30}$$

The solutions for ν and ν_x are as exhibited in Fig. 1. Note that as T drops below M/10, ν approaches an asymptotic value.

The baryon excess may be obtained from

$$N - \bar{N} = \eta \int d^3p \, n(p,z) \left[1 - n(p,z)\right]_{z \to \infty} \tag{31}$$

Throughout, η refers to the CP violating parameter that distinguishes between $\to QQ$ and the $\bar{Q}\bar{Q} \to X$ rates and may be arranged to be $\sim 10^{-9}$.

6. Partial Unification Theory

So far I have talked about neutron oscillation only in the context of grand unification. The proposal by Mohapatra and Marshak, however, is entirely in the context of the partial unification. Partial unification (or petit unification) professes to be less ambitious and permits strong interactions to be a separate direct product group from the electro-weak unification group. They propose the partial unification group of nature.

$$SU(2)_L \times SU(2)_R \times SU(4)$$

where $SU(4')$ unifies $SU(3)_C \times U(1)_{B-L}$. Their essential observation is that the analog of the Gell-Mann-Nishijima formula now reads

$$Q = I_{3R} + I_{3L} + \frac{B-L}{2} \qquad (33)$$

and B-L is now a generator of local gauge symmetry. To avoid the Lee-Yang conclusion they assume B-L is spontaneously broken in the physical ground state.

The particle content in this theory may be written [16]

		$SU(2)_L$	,	$SU(2)_R$	,	$SU(4')$
Fermions:	$\psi_L \sim$ (	2	,	1	,	4)
	$\psi_R \sim$ (	1	,	2	,	4)
Higgs:	$\phi \sim$ (	2	,	2	,	1)
	$\Delta \sim$ (	3	,	1	,	10)
	$\Delta \sim$ (	1	,	3	,	10)
	$\Sigma \sim$ (	1	,	1	,	15)

The hierarchy of breaking is arranged so as to get $m_X \gg m_{W_R} \gg m_{W_L} \gg m_q$, viz. the vacuum expectation values are

$$\left. \begin{array}{ll} \langle \Delta_{R,44}^{1+i2} \rangle = v & m_{W_R} = gv \\[6pt] \langle \Delta_{L,ij}^a \rangle = 0 & \\[6pt] \langle \phi \rangle = \begin{pmatrix} k & 0 \\ 0 & k' \end{pmatrix} & m_{W_L} = gk \sim 100 \text{ GeV} \\[6pt] \langle \Sigma \rangle = v \begin{bmatrix} 1 & & \\ & 1 & \\ & & 1 \\ & & & -3 \end{bmatrix} & m_X \geq 10^4 \text{ GeV} \end{array} \right\} \qquad (34)$$

The graph responsible for neutron oscillation in PUT is the analog of Fig. 1,

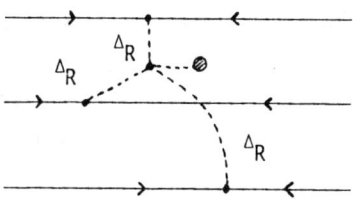

with matrix element

$$\simeq \frac{\lambda h^3 <\Delta_{R,44}>}{(m_{\Delta_R})^6} \approx 10^{-29} \text{ GeV} \qquad (35)$$

A unique feature of the PUT model is the absence of proton decay. They would root for neutron to oscillate but would forbid such decays as

$$p \to e^+ \pi^0 \quad \text{(B-L conserving nucleon decay)}$$
$$n \to e^- \pi^0 \quad \text{(B-L violating nucleon decay)}$$

This does not, however, mean that the present salt mine experiments would, according to PUT, be negative experiments. The <u>very</u> existence of neutron oscillation would imply the reaction

$$(A,Z) \to (A-2,Z) + \pi^+\pi^- \qquad (36)$$

If ε is as big as their estimate, the rate for this reaction would be large enough to give an unmistakable (and unexpected) signal in the salt mine experiments[17] under way. The PUT scenario would allow only very energetic pions in the decay debris, and <u>no</u> leptons of either charge.

Finally, let me conclude this section by adding a note on the existence of a heavy Majorana neutrino, N, in PUT which is not called for in SU(5). The electron neutrino acquires a small mass as a result of N, whereas in SU(5) the electron can acquire a small mass through the radiatively acquired vacuum expectation value of <u>15</u>.

7. Conclusion

In this talk I have tried to describe the two main solutions to the dilemma posed in 1955 by Lee and Yang. One is to continue to gauge baryon number locally but break it spontaneously to avoid the constraint, eq.(2). The other is to not let B be a quantum number <u>at all</u>. Before the advent of grand unification it was difficult to envision a context for this second possibility. With grand unification it has become very easy to give up B as a classifying quantum number and end up with the startling possibility that not only is the free proton unstable but that a free neutron could also oscillate into a $\bar{n}$ in as short a time as 10^7 sec.

Present and contemplated experiments on nucleon decay <u>as well as</u> on neutron oscillation can serve to very clearly distinguish between the two solutions and show which is nature's choice to handle baryon number. To facilitate our understanding of the crucial implications of these two types of experiments, I have summarized their possible outcomes and interpretations in the following table:

	If Salt-mine exp't observe:	and if n-oscillation exp't observe:	Implications
(I)	$p \to e^+\pi^0$ $n \to e^+\pi^-$	null result i.e. $t_{n\bar{n}} > 10^{10}$ sec, say	i) PUT is ruled out ii) GUT (minimal SU_5) favored
(II)	$p \to e^+\pi^0$ $n \to e^+\pi^0$ $(A,Z) \to (A-2,Z) + \pi^+\pi^-$	$n \leftrightarrow \bar{n}$	i) PUT ruled out ii) GUT (min SU_5) out iii) extra-GUT favored
(III)	$p \to e^+\pi^0$ $\to e^-\pi^+\pi^+$ $n \to e^+\pi^-$ $\to e^-\pi^+$ $(A,Z) \to (A-2,Z) + \pi^+\pi^-$	$n \leftrightarrow \bar{n}$	Same as case II with altered pattern of mass breaking among 15 and 45 Higgs
(IV)	$(A,Z) \to (A-2,Z) + \pi^+\pi^-$	$n \leftrightarrow \bar{n}$	i) PUT favored
(V)	null result	null result	i) NUTS favored no unification beyond $SU(3) \times SU(2) \times U(1)$

Clearly we are poised on the threshold of potentially revolutionary discoveries of physics. Grand unification has appropriately led the way to new vistas in physics. As always, only further experiments can tell us where this new vista will lead us. At the very least, grand unification has already achieved a degree of unification of the many disciplines of physics that in itself is exciting and unprecedented.

References

. S. Glashow and H. Georgi, Phys. Rev. Lett. 32, 483 (1974).
. L.N. Chang and N.P. Chang, Phys. Lett. 92B, 103 (1980), (E) ibid 94B, 551 (1980). Phys. Rev. Lett. 45, 1540 (1980)
  Preliminary results reported by L.N. Chang in Guangzhou Conf. Jan. 1, 1980.
. S. Glashow, HUTP 79/A029, HUTP 79/A059; Proc. of Cargese Summer School (1979).
. T.K. Kuo and S. Love, Phys. Rev. Lett. 45, 93 (1980).
. R.N. Mohapatra and R. Marshak, Phys. Rev. Lett. 44, 1316 (1980); Phys. Lett. 94B, 183 (1980).
. J.C. Pati and A. Salam, see remarks by Pati in Proc. of First Workshop on a Grand Unification (1980).
. J. Schneider, Ann. Phys. Fr. 4, 221 (1979).
  For other suggestions on B-L violations, see
. F. Wilczek and A. Zee, Phys. Lett. 88B, 311 (1979); Phys. Lett. 93B, 389 (1980).
. T.D. Lee and C.N. Yang, Phys. Rev. 98, 1501 (1955).
. F. Wilczek and A. Zee, Phys. Lett. 88B, e11 (1979).
. S. Weinberg, HUTP 80/A023.
. F. Wilczek and A. Zee, Phys. Lett. 88B, 311 (1979).
. L.N. Chang and N.P. Chang, "On Neutron Oscillations" - CCNY-HEP 81/1.
. The h appropriate for the heavy quarks would be correspondingly higher ($\sim 10^{-2}$).
  The study is based on the work reported in
. N.P. Chang, A. Das, J. Perez-Mercader, Phys. Rev. D22, 1414 (1980).

. V.A. Kuzmin, Institute for Nuclear Research, Moscow, Report IYaI-P-0161 (1980).
. Eq.(8) here has an extra 2π to correct for the missing normalisation in Kuzmin et al formula.

11. L.N. Chang and N.P. Chang, Phys. Rev. Lett. $\underline{45}$, 1540 (1980).
12. S. Glashow first pointed out the importance of this effect.
13. M. Yoshimura, Phys. Rev. Lett. $\underline{41}$, 281 (1978);
    S. Weinberg, Phys. Rev. Lett. $\underline{42}$, 850 (1979);
    D. Toussaint, S.B. Treiman, F. Wilczek and A. Zee, Phys. Rev. $\underline{D19}$, 1036 (1
14. In our model, the $(6^*,1)_{45} \equiv X'$ can also decay into $\bar{u}\,\bar{d}$ with however a rat much smaller than Γ_x ($h \sim 10^{-3}\, h'$). If M' (mass of X') had been much smal than M (mass of $(6,1)_{15}$) then X' would act as an eraser of baryon excess produced by $(6,1)_{15}$. Fortunately, we have $M' \sim M$ so that at the same time that inverse decays of X fail to replenish the X, the same is happening to X'. Baryon excess produced by X and X' decays thus tend to add. This conclusion is unaffected by the baryon-conserving collision rate, Γ_c, which is $\sim 10^{-1}$ GeV in this temperature range.
15. L.N. Chang and N.P. Chang, Cosmological Production of Baryon Excess with Intermediate Mass Scale, CCNY-HEP-81/2.
16. R.N. Mohapatra and R. Marshak, Phys. Rev. Lett. $\underline{44}$, 1316 (1980);
    Phys. Lett. $\underline{94B}$, 183 (1980).
17. For an excellent review see M. Goldhaber, Proceedings of Rochester Conferer at Madison, Wisconsin (1980).

DISCUSSION

PATI: I would like to ask you something again. That is this question of consistency with the generation of baryon excess. Since you do have low mass scales for the n-n̄ oscillation, and since everything is explicit violation, how do you get around the problem?

CHANG: Well, I know you both work in this field and have been worried about this very important problem and I just happen to have prepared some slides which might address that matter. When we have cosmological expansion at very high temperatures, it is argued (and that is the starting assumption of everybody in this case) that the chemical potential of all these particles are equal, and they equal zero. Therefore, if you work it out, you will find that the number density, the total number divided by the volume is such and such. As the volume expands, the total number does not change, therefore the number density decreases like T^3. Now in the early moments of the expansion everything is in equilibrium. Around 10^{12} GeV, the heavy X's decouple from thermal equilibrium and since $T_D < M_X$ promptly decay away and your excess is produced. Now the crucial question is what happens after that. In a standard scenario, at lower temperatures the baryon number non-conserving collision rate is much, much smaller than the expansion rate so baryon excess is preserved.

What happens now when you begin to have intermediate scales? Well, they wash out. At lower temperatures you have the new 15 plet which can now decay into two quarks. At the place where the decay rate becomes relevant as far as the expansion rate goes, this decoupling temperature is much higher than its mass. Now according to Weinberg that, therefore, is a no-no. Why? Because things are still in thermal equilibrium at this T_D. According to a little theorem of Yang and Yang, Treiman, Wilczek, Zee and Weinberg, if it goes in thermal equilibrium initially, it will be frozen in thermal equilibrium even at this temperature T_D, and therefore there will be no baryon excess produced. However, the important point is these particles are still there, they have not yet decayed. They will decay when the temperature comes down to the order of M_X itself. During this interval, this is really a dissocation.

PATI: What particle is?

CHANG: The new 15 plet, which is the one that will produce the second generation baryon excess. It has baryon number-producing decays. So long as inverse decays can replenish the new X's, no net baryon excess will be produced. The chemical potential must be maintained in the process. Things are still in the thermal equilibrium.

What is this chemical potential? Is it still the original chemical potential (which is zero), or is is something different? If we assume it is the same chemical potential of zero, you find that the number density of the X, as the temperature cools down to order of its mass, is .023 T^3. That is, the total number of X is then proportional to .023 vs. the original .122 at higher temperatures.

What happend to all those missing X's? Part of them may have gone into an increase in the number of d's, but you can check the actual balance; it does not work out unless the chemical potential rearranges itself, and that is what happens. If you check what the chemical potential does, you find that the chemical potential rearranges itself in this interval (it is no longer zero) and that the net number of d's balance each other in accordance with the dissociation, as discussed in my text.

When the inverse decay no longer is energetically favored, you get your baryon excess generation. You can estimate the number through the number of particles that are present at that point. The baryon excess is therefore now generated at a lower temperature than what you thought at first. After this new X has decayed, the baryon-violating collisions are again much smaller, and the baryon excess will not be washed out by subsequent evoluation.

ROSEN: Your theory violates actually $\Delta(B-L)$ in a certain way -- say $\Delta(B-L) = 2$. The neutron oscillation is a $\Delta B = 2$ term. Presumably the same interaction contains terms where $\Delta B = 1$ and $\Delta L = -1$ as well. Therefore, you could get proton decay out of the very same term as gives you the neutron oscillation. However, the mass scale appears to be much smaller than is consistent with the limits of the proton lifetime. You have a 10^5 GeV boson instead of a 10^{15} GeV one.

CHANG: In arranging my mass scenario, I have been careful to insist on the (3,1) member of 45 plet to be heavy enough not to spoil proton decay. The point is, it is a different member of the 45 plet that appears in neutron oscillation.

A PROPOSED EXPERIMENT FOR A SENSITIVE SEARCH FOR $\Delta B = 2$ TRANSITIONS VIA n-n̄ MIXING

G. R. Young, H. O. Cohn, T. A. Gabriel, R. A. Lillie, P. D. Miller,
and
R. R. Spencer

Oak Ridge National Laboratory, Oak Ridge, Tennessee 37830*

and

M. S. Goodman and R. Wilson

Harvard University, Cambridge, Massachusetts 02128

and

W. M. Bugg, G. T. Condo, T. Handler, and E. L. Hart

University of Tennessee, Knoxville, Tennessee 37916

Baryon number has long been assumed to be an absolutely conserved quantity; no strong experimental evidence suggests otherwise. Experimental observation of its violation would have important implications for our understanding of nature and for efforts to describe all forces in one unified theory. Extensive theoretical effort is being expended at present to extend the unification of electromagnetic and weak forces demonstrated by Weinberg and Salam to include strong interactions as well.[1] An almost universal prediction of these grand unified theories (GUT) is the prediction of the existence of processes which do not conserve baryon number, $\Delta B \neq 0$, notably the prediction of the decay of the proton. Once one considers possible extensions of minimal GUT to include a non-minimal Higgs sector,[2,3] or if one considers partial unification schemes which do not contradict present data on electroweak interactions,[4] it is found that $\Delta B = 2$ processes may exist, notably the process $n \leftrightarrow \bar{n}$. Experimental verification of the existence of $n \to \bar{n}$ transitions would constitute a most fundamental discovery about the

---

*Operated by Union Carbide Corporation under contract W-7405-eng-26 with the U. S. Department of Energy.

behavior of matter and would serve to discriminate among general classes of unification schemes. We describe the general considerations for such an experiment and report on an experiment being proposed to search for these transitions using the Oak Ridge Research Reactor.

If one considers the following mass matrix for n and $\bar{n}$ states

$H = \begin{pmatrix} X & \alpha \\ \alpha & Y \end{pmatrix}$, $\psi = \begin{pmatrix} n \\ \bar{n} \end{pmatrix}$, where $X = M_n + V_n$, $Y = M_{\bar{n}} + V_{\bar{n}}$, M_n the neutron mass, V_n the external potentials experienced by a neutron and $\alpha = \hbar/T_{n\bar{n}}$, $T_{n\bar{n}}$ being the $n - \bar{n}$ mixing time, one sees that diagonalizing yields eigenvalues $\lambda_{1,2} = M_n \pm \sqrt{V_n^2 + \alpha^2}$ and the resultant transition probability becomes $P(t)_{n\bar{n}} = 2(1 - \cos \omega t)\sin^2\theta \cos^2\theta$, where we have taken $M_n = M_{\bar{n}}$, $V_n = -V_{\bar{n}}$, and one has $\omega = 2\sqrt{V_n^2 + \alpha^2}/\hbar$, $\tan \theta = \alpha/V_n + \sqrt{V_n^2 + \alpha^2}$. Since one can estimate^2 from observed matter stability on the order of $\tau \geq 10^{30}$ years that $\alpha \leq 10^{-22}$ eV $<< V_n$, the eigenstates of the diagonalized Hamiltonian are n and $\bar{n}$ to good approximation and the expression for $P(t)_{n\bar{n}}$ can be expanded for small ω and θ to give

$$P(t)_{n\bar{n}} = \frac{2\alpha^2}{(2V_n)^2} (1 - \cos \omega t) \cong \frac{\alpha^2 t^2}{\hbar^2}.$$

This yields a maximum transition probability $P_{max} = \alpha^2/V_n^2$. A convenient figure of merit to employ for comparison of various experimental arrangements is Nt^2, where N is the number of neutrons observed and t is the observation time. The following table gives examples of the maximum transition probability and oscillation frequencies for various values of V_n, all for $\alpha = 10^{-22}$ eV. Since neutrons have a magnetic moment $\mu_n = -1.91 \mu_N$, and since $\vec{\mu}_n = -\vec{\mu}_{\bar{n}}$, any external magnetic fields remove the $n - \bar{n}$ degeneracy.

TABLE 1

V_n	P_{max}	ω
1 keV	10^{-50}	10^{19}/sec
$\mu B = 6 \times 10^{-12}$ eV (1 gauss ≅ earth field)	10^{-22}	10^{4}/sec
$\mu B = 6 \times 10^{-15}$ eV (1 mgauss)	10^{-16}	10^{1}/sec

An interesting experiment would thus consist of introducing as many free neutrons as possible into a region of space in which all perturbing (i.e. magnetic) fields have been reduced to as small a value as possible. After allowing these neutrons to remain in the field-free region as long as possible (however, see below), one would search for the presence of anti-neutrons. This is most easily done by allowing any anti-neutrons present to annihilate on matter and searching for the characteristic $2M_n c^2$ (≅ 2 GeV) of energy released. Experimental results for low energy $\bar{p}$ annihilation on carbon indicate that the vast majority of annihilation products are pions with a mean multiplicity of about 4.5 per annihilation.[5] Since the neutrons and thus anti-neutrons of interest will have thermal (or smaller) energies, the center-of-mass motion is zero to very good approximation. Thus, a detection apparatus should be able to identify a multipion event with net momentum zero and characteristic energy 2 GeV and distinguish this from any background. Various aspects of a method to accomplish this will be discussed below.

Writing $P(t)_{n\bar{n}} = \alpha^2/(\mu B)^2 \sin^2(\mu B t)$, i.e. assuming removal of the $n,\bar{n}$ degeneracy by magnetic field perturbations only, one sees a gain in $P(t)_{n\bar{n}}$ is realized by decreasing the perturbing magnetic field B until $\mu B t \ll 1$. At that point, expanding $\sin^2(\mu B t)$ yields $P(t)_{n\bar{n}} \cong \alpha^2/(\mu B)^2 (\mu B t)^2 = (\alpha t)^2$.

It should be observed that the neutron and anti-neutron wave functions are out of phase with the factor $e^{-i(V_n - V_{\bar{n}})t/\hbar}$, leading to the same experimental condition of $V_{nt} \ll 1$. For a neutron moving in a 1 gauss field (roughly that of the earth), $\mu Bt < 1$ for $t \leq 10^{-4}$ sec. Such a short time severely depresses the attainable values of Nt^2, indicating the desirability of degaussing the earth's field. The maximum field to attain $\mu Bt < 1$ for various neutrons traveling a 20m flight path is given in the following table

TABLE 2

Type	B field
Thermal n, $t \approx 10^{-2}$ sec	$B < 11$ mG
Cold n, $t \cong 10^{-1}$ sec	$B < 1$ mG
Neutron mean life 914 sec	$B < .1$ μG

As all present sources of free neutrons involve observing neutrons with non-zero lab velocities, it should be noted that increasing t^2 by means of increasing the length of the "quasi field-free region" for a fixed area target results in <u>no</u> gain in Nt^2 due to loss of solid angle subtended by the detector

$$Nt^2 = \Phi \frac{\pi R^2}{4\pi L^2} \frac{L^2}{V^2} = \frac{\Phi}{4} \frac{R^2}{V^2},$$

Φ is the number of neutrons, V is their velocity, and R is the detector radius. Thus increasing the source intensity, detector area and running time while decreasing the mean neutron velocity (i.e. cooling the neutrons) all help to increase the sensitivity of the experiment. The distance from the neutron source enters only in the forms of requiring better magnetic field degaussing for longer flight paths and in decreasing source-related background for that same increase in path.

There are several possible sources of large numbers of neutrons which can be placed in relatively field-free regions. These include large H_2O Čerenkov detectors, cosmic ray neutrons observed in space, nuclear bomb blasts, accelerator spallation sources, and continuous and pulsed mode nuclear reactors. Considering the requirements of magnetic field shielding, source strength, ability to repeat measurements, maximum instantaneous γ-ray fluxes from neutron capture on the $\bar{n}$ annihilation target and ability to thermalize neutrons leads to the choice of a continuous mode reactor as the preferred source of the neutrons. Desirable features of reactors, especially high flux research reactors (as opposed to high power but low flux power reactors which have a diffuse core structure) include:

1) The neutron flux is continuous (the duty factor at the detector is 100%).
2) The source of neutrons is renewable (with the addition of uranium and money).
3) The energies of thermalized neutrons are below, $<T_n> \cong .025$ eV.
4) Fluxes of $\geq 10^{14}$ n/cm^2/sec are obtainable at accessible research reactors.
5) Ports in the containment vessel of research reactors allow bringing neutrons into a quasi field-free region in a controlled manner (although one is in essence working in the core for the needed large ports).
6) The detector of antineutrons, which is proposed to search for the 2 GeV released in $\bar{n}$ annihilation, can have active veto counters to allow tagging and rejection of the cosmic ray background. This is necessary to reject cosmic ray events with high efficiency. Unfortunately, the present type of n - $\bar{n}$ experiments, as opposed to p decay experiments, cannot be done underground as no reactors exist underground.

7) The reactor can be turned off to identify reactor related signal and to study non-reactor background events alone.

An isometric view of the layout of the n - n̄ mixing experiment proposed for the Oak Ridge Research Reactor (ORR) is shown in Fig. 1. The ORR is a 30 MW tank type reactor which uses enriched uranium as fuel. It has a set of six small (∼7 cm radius) beam ports used for conventional neutron scattering experiments, a set of pipes leading into the core for irradiation experiments, an underwater facility near the core for high intensity experiments, and two large (≥ 1400 cm^2) ports for engineering test experiments of reactor components and designs (see Fig. 2). One of these latter holes will be used to bring out a large number of neutrons for the proposed experiment. Provision has been made to place a bismuth γ-ray shield and D_2O secondary moderator directly into this hole in order to shield against direct γ rays from the core and moderate most of the fast flux from the core. (The thermal and fast neutron fluxes at the edge of the containment walls for the ORR are nearly equal.) This section is followed by an H_2O-floodable section for neutron absorption and a removable lead γ-ray shutter to allow working on the apparatus without needing to shut down the reactor or having to wait for core activity to decrease after a normal refueling shutdown. This allows regular checks of background rates and detector performance in the absence of reactor related events. A 20 meter long evacuated flight tube is included after these sections. This pipe is surrounded by several layers of mu-metal and serves as the field-free region of the experiment. Surrounding the beam pipe and the reactor end of the beam pipe is additional concrete shielding for biological purposes. At the end of the field-free pipe is a thin n̄ annihilation target, together with two collars, one of borated plastic to shield the detector from direct reactor neutrons, and one of lead to shield the detector from reactor γ rays. Downstream of the target is a CD_2 + $^6Li_2CO_3$ beam dump.

The target is surrounded by a cylindrical detector consisting of a layer of Pb-glass Čerenkov counters, followed by several layers of Fe or Pb-scintillator sandwich. (See Fig. 3.) As the primary $\bar{n}$ annihilation products are pions with a total energy of 2 GeV, mean multiplicity of 4.5 and net momentum of very nearly zero in the lab frame, the detector must be able to detect the energy deposited by photons from π° decay and ionization from $\pi^\pm$ deceleration. In addition, the detector must be relative insensitive, at least in its inner section, to low energy (<8 MeV) photons produced from thermal neutron (n,γ) processes in the target. This motivates the choice of Pb-glass for the inner active layer of the detector. The detector will have layers of $^6Li_2CO_3$, and Pb inside the Pb-glass layer to absorb all scattered thermal neutrons and degrade some of the soft γ-ray flux produced. The detector is surrounded by 5 feet of concrete to stop charged pions which leave the detector and prevent them from reaching the veto detector and thus rejecting a valid event. This concrete also adds to the ~1 meter steel outer shield, which is used to stop slow (<1 GeV) cosmic ray muons and convert cosmic ray hadrons and prevent them from reaching the veto scintillator.

Table 3 gives relative figures of merit Nt^2 in (neutrons/sec)sec^2 for various reactors (MIT, ILL, ORR, BSR, HFIR) and accelerator neutron sources (LAMPF and ORELA). The MIT, BSR (Oak Ridge), HFIR (Oak Ridge) and ORR "Goat Hole" (the proposed facility) are all conventional research reactors. The ILL reactor in Grenoble, France, has cold neutrons and has curved neutron guide tubes which nearly eliminate the fast neutron and γ flux delivered to the detector region; however, the neutron flux of a curved guide tube is quite low relative to the original reactor flux. The LAMPF and ORELA (a very high current electron accelerator) numbers assume full use of those beams; effects of losses due to thermalization of the neutrons produced have not been included. The poolside facility at the ORR has the largest Nt^2 of all reactors we have surveyed; however, it adds the serious complication of having to do the experiment under water. The "Goat Hole" facility at the ORR appears to be the best choice after this.

TABLE 3. Possible Neutron Sources

Reactor/Accelerator	ϕ n/cm^2-sec $\times 10^{13}$	A cm^2	Ω Fraction 4π $\times 10^{-4}$	t^2 sec	Nt^2
MIT	10	100	10	1×10^{-5}	5×10^7
ILL	$I = 2 \times 10^9$/sec		10^4	6×10^{-4}	10^6
LAMPF ($^{100\ \mu A}_{10\ n/p}$)	$I = 6 \times 10^{15}$/sec		2	6×10^{-5}	7×10^7
ORR (Goat Hole)	15	1400	2	6×10^{-5}	1.3×10^9
ORR (Poolside in Tank)	15	4200	8	1.5×10^{-5}	4×10^9
BSR	3	2090	8	1.5×10^{-5}	4×10^8
HFIR			6	6×10^{-5}	1.2×10^8
ORELA	$I = 10^{14}$/sec		2	1.5×10^{-5}	3×10^5

Figure 4 gives representative plots of the results of a preliminary study of the neutron fluxes expected at various stages through the reactor region and a possible design of the Bi/D$_2$O attenuator/moderator section at the core. These were calculated using a large Boltzmann transport code written at Oak Ridge which is used in reactor shielding design.[6] Figure 5 gives the resulting neutron flux as a function of energy at the end of the Bi/D$_2$O section. Significant suppression of the fast neutron flux is achieved.

Even after attenuating the reactor associated γ-ray flux, a potentially severe problem remains in the annihilation target region resulting from soft (<8 MeV) γ-rays produced from thermal (n,γ) processes. As seen in Table 4, even for the best available ratios for solid targets of photon-less capture cross sections to photon producing capture, which occur for nuclei such as ^{10}B and ^{6}Li, absorbing all the thermal neutrons gives rise to a large number of γ rays released at the target. This problem exists even for clean beams of thermal neutrons alone. To circumvent this, we plan to use a transmission target selected for a very small (n,γ) cross section, such as Bi or CD$_2$.

TABLE 4. Data for ^{10}B and ^{6}Li Targets (Thermal Neutrons)

Target	Reaction	γ's/sec for 2 x 10^{13} n/sec incident
^{10}B	$\sigma(n,\alpha)$ = 3.5 kbarns $\sigma(n,\gamma)$ = .5 barn $Q_{n,\gamma}$ = 11.456 MeV	2.8 x 10^9
^{6}Li	$\sigma(n,\alpha)$ = 900 barns $\sigma(n,\gamma)$ = 38 mbarn $Q_{n,\gamma}$ = 7.252 MeV	8.4 x 10^8

Unfortunately, no data exist for thermal $\bar{n}$ annihilation cross sections so one is forced to rely on theoretical estimates of cross sections. A naive partial wave analysis assuming total absorption yields very large cross sections, ∼30 Mbarn, which are unrealistically large. A more realistic estimate[7] of the scattering length yields cross sections that scale as $1/\sqrt{T_{\bar{n}}}$

(the behavior observed for thermal neutron (n,γ) cross sections) and reach values on the order of several thousand barns at thermal energies, indicating quite thin targets, on the order of 1 mg/cm² or less, can be utilized.

One other important check on an observed signal is summarized in Fig. 6. Since

$$P(t)_{n\bar{n}} \cong \frac{\alpha^2}{(\mu B)^2} \sin^2(\mu B t) ,$$

by varying the magnetic field <u>inside</u> the mu-metal magnetic shielding around the flight tube, one can ascertain that an effect seen only with the reactor on and producing neutrons, decreases in magnitude as $1/B^2$ as a secondary magnetic field is increased, as is expected for the transition probability for neutrons to oscillate into anti-neutrons. This provides an important check of possible spurious signals due to the large number of neutrons incident on the detector region from the reactor.

In summary, a sensitive search for $\Delta B = 2$ $n - \bar{n}$ transitions is proposed. A high flux reactor and magnetically shielded flight path are used to provide a large value of Nt^2. Future improvements to the experiment would indicate addition of a liquid deuterium moderator at the reactor wall to cool the neutrons from ~300°K to ~20°K and enlargement of the target region. The proposed experiment searches for an effect not heretofore considered experimentall and provides an improvement of several orders of magnitude over theoretical estimates of the mixing time $T_{n\bar{n}}$ based solely on considerations of the stabilit of nuclei.

REFERENCES

1. S. Weinberg, Rev. Mod. Physics $\underline{52}$, 515 (1980) and references therein; A. Salam, ibid, p. 525; S. L. Glashow, ibid, p. 539.

2. S. L. Glashow, Harvard Preprints HUTP-59, HUTP-79, 1980.

3. N. D. Chang, this conference; J. C. Pati, this conference.

4. R. E. Marshak, R. N. Mahapatra, Riazuddin, VPI Preprint VPI-HEP-80/7.

5. W. M. Bugg et al., Phys. Rev. Lett. $\underline{31}$, 1475 (1973).

6. W. W. Engle, Code ANISN, Oak Ridge National Laboratory, (unpublished).

7. A. K. Kerman, private communication.

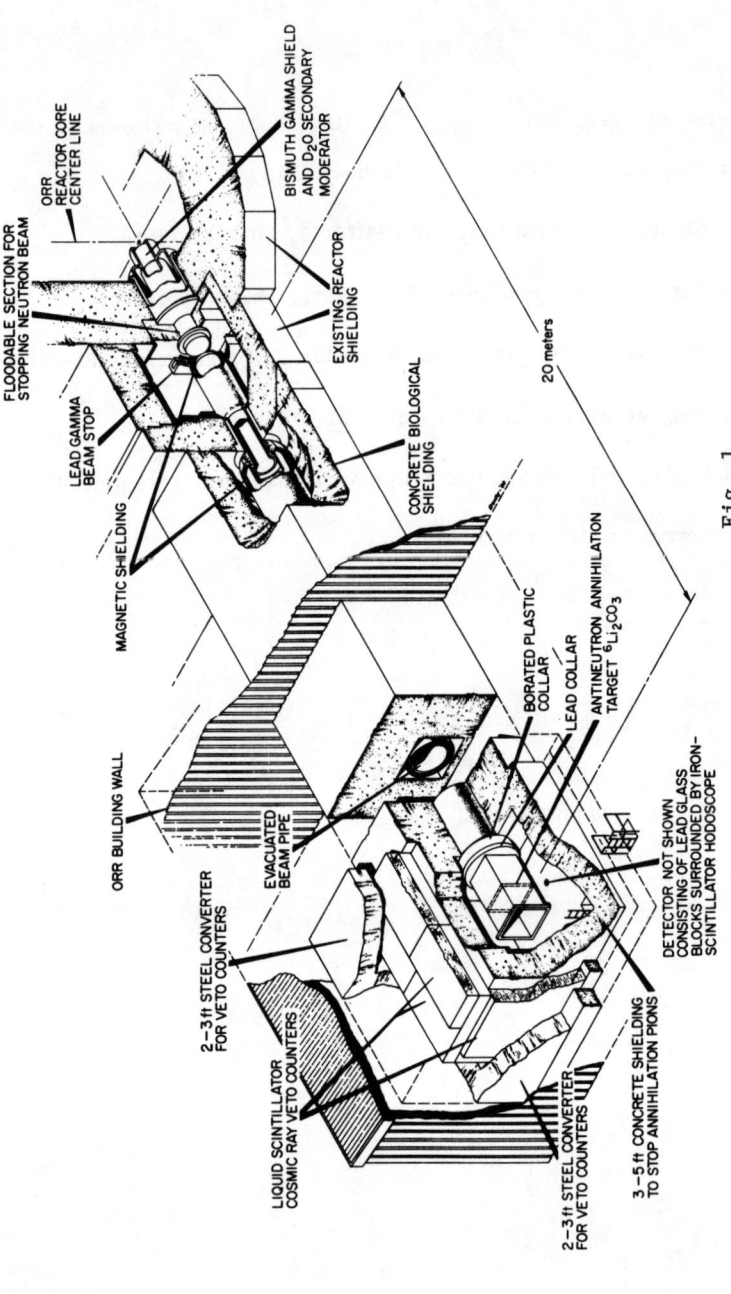

Fig. 1 Isometric view of the proposed n - n̄ mixing experiment at the ORR.

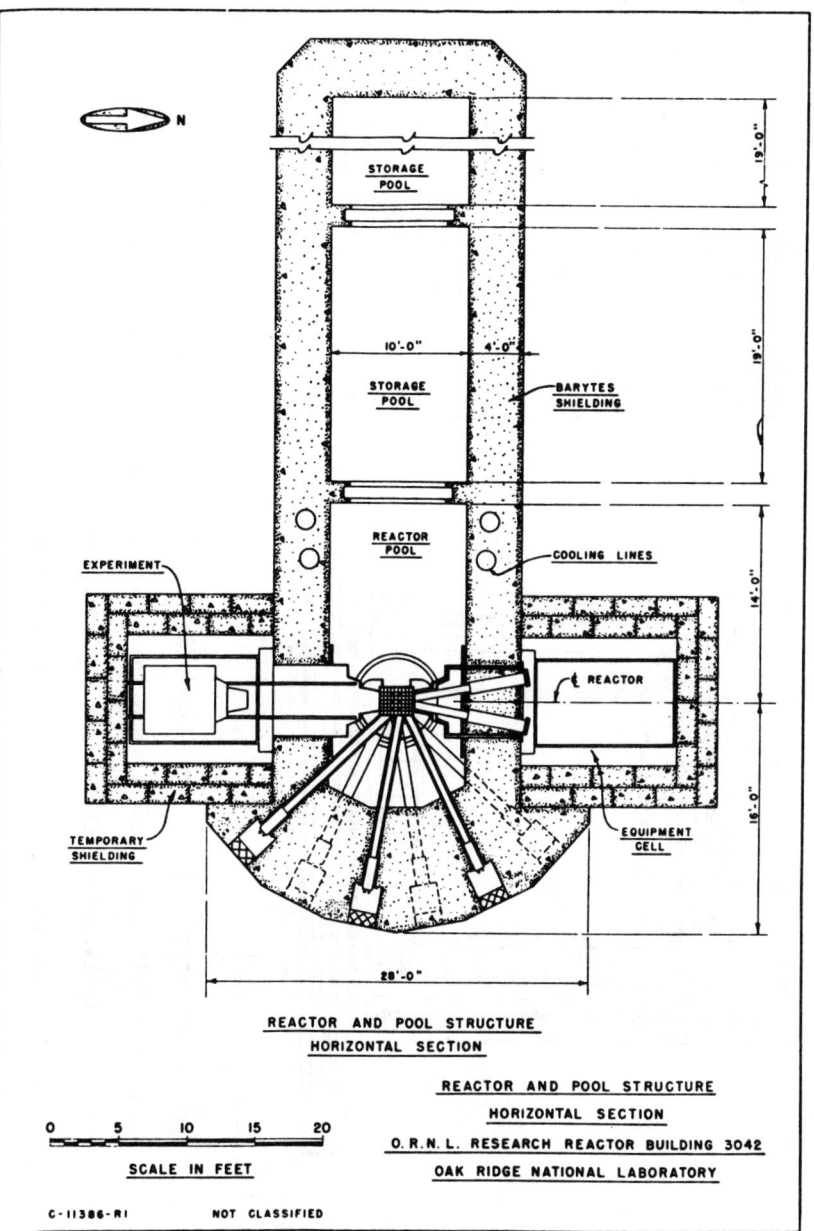

Fig. 2

Plan view of the ORR reactor and pool, showing experimental ports. The core is the small lattice at the center. The large apparatus on the left side marked "experiment" views the port to be used for the n - n̄ mixing experiment.

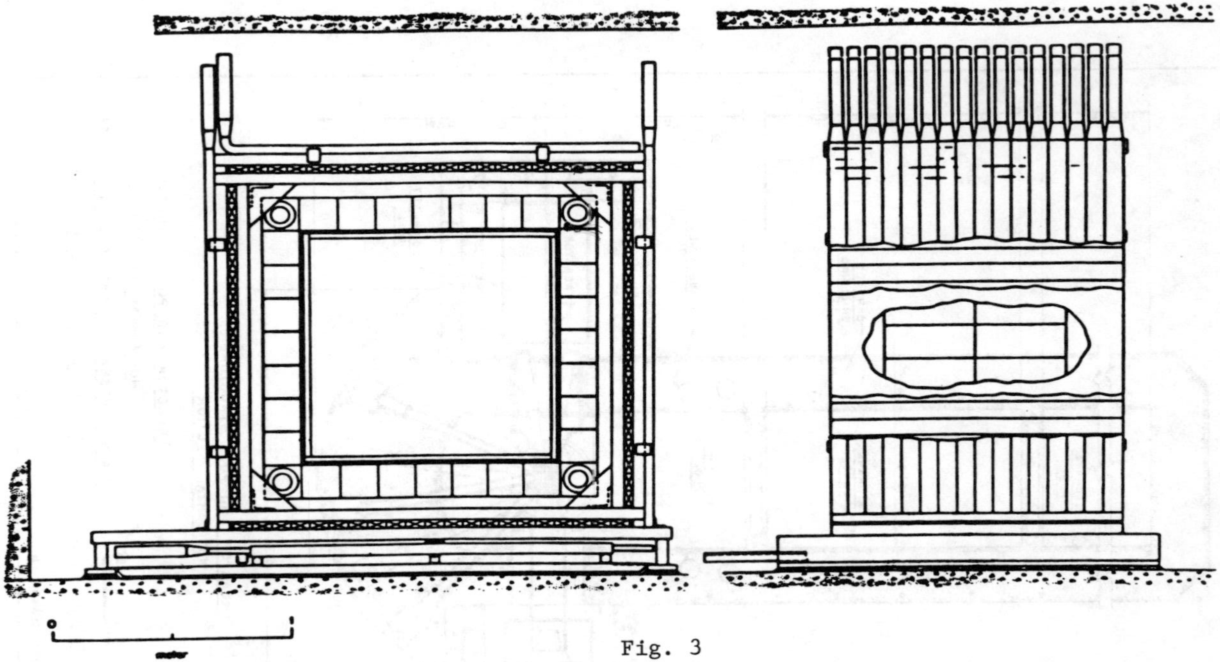

Fig. 3

Sketches of detector layout: beam view (left) and side view (right). A 1 m² n̄ annihilation target sits at the detector center, followed (radially) by layers of ^{6}Li$_2$CO$_3$ (to stop scattered thermal neutron flux) and Pb (for some γ-ray attenuation). Following this are the Pb-glass blocks and the Pb-scintillator sandwich.

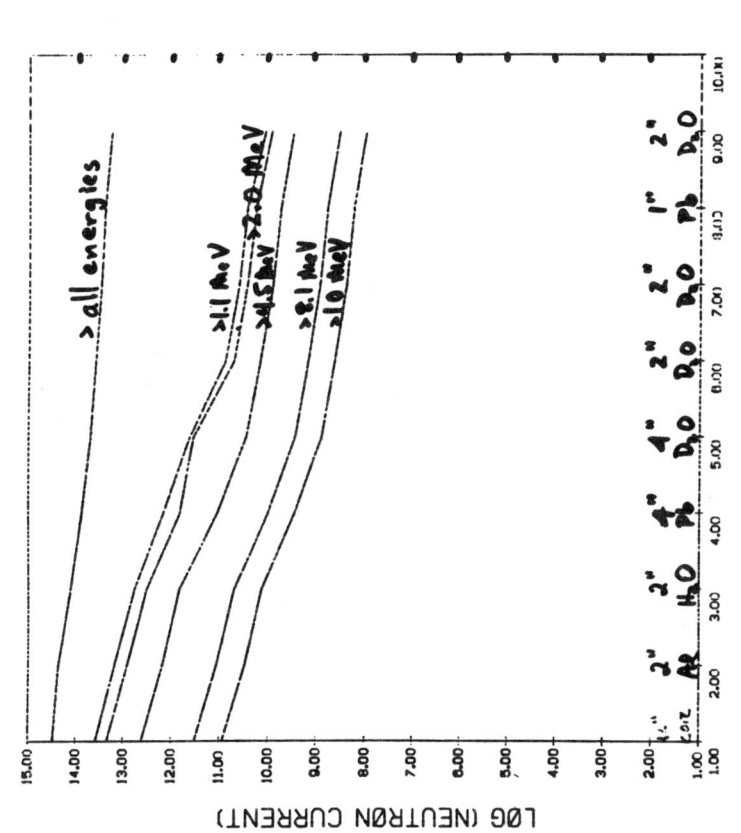

Fig. 4. Integral neutron currents, above a given energy cut, as a function of distance from the core in terms of zones of various materials. Plotted is the current after passing through the given zone and all previous zones.

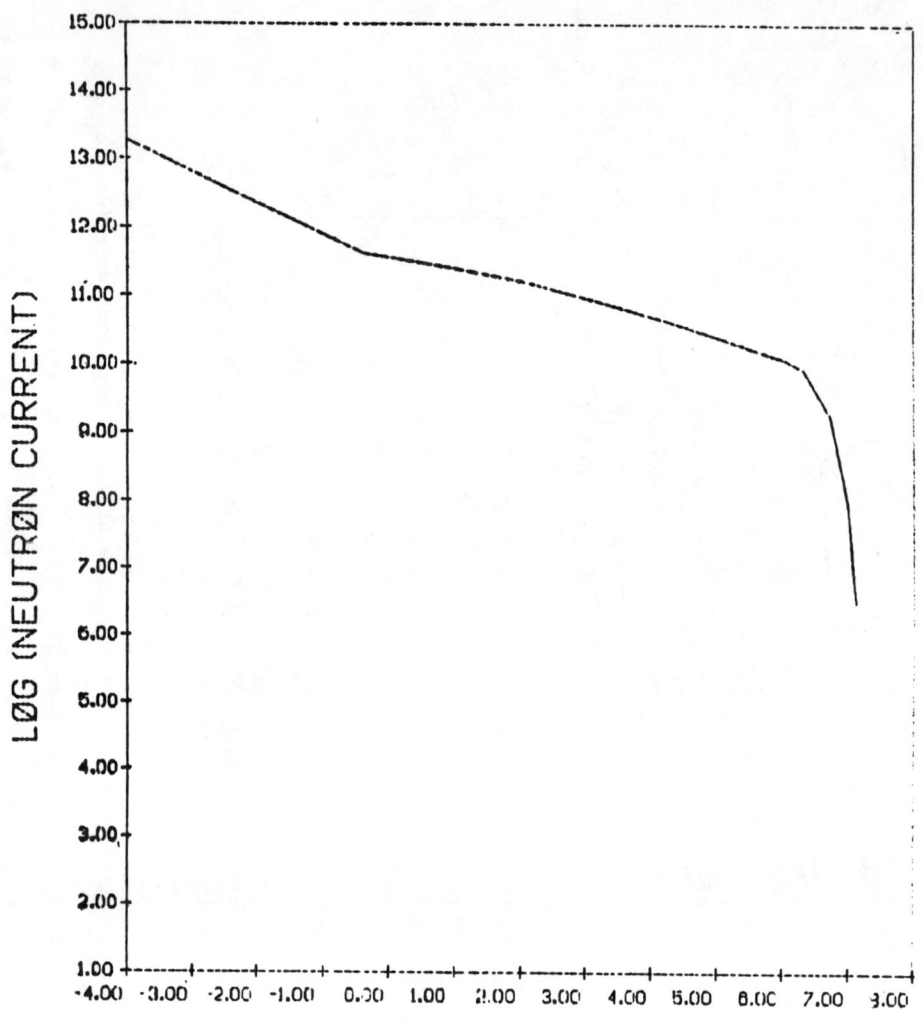

NEUTRØN FLUX - REACTØR ZØNE 9

Fig. 5

Log_{10}-Log_{10} plot of neutron current vs neutron energy for the last zone of Fig. 4.

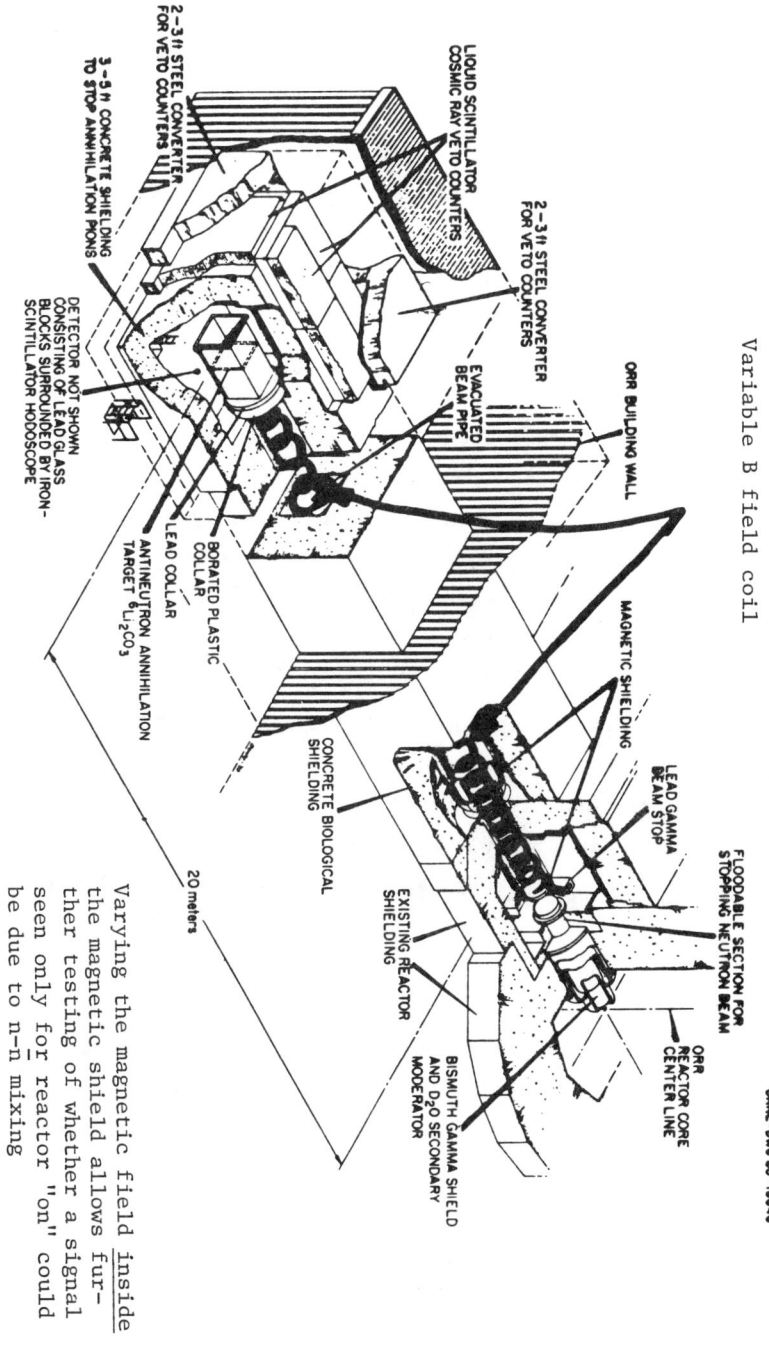

$$P(t)_{n\bar{n}} \simeq \frac{\sigma^2}{(\mu B)^2} \sin^2(\mu B t)$$

Fig. 6. Isometric view of n - n̄ experiment showing placement of magnetic field coil used to suppress observed n - n̄ transitions. This allows varying the magnetic field inside the magnetic shield allows further testing of whether a signal seen only for reactor "on" could be due to n-n̄ mixing

COMMENTATORS

Neutron Oscillation Experiment at Los Alamos

H. L. Anderson

Los Alamos Scientific Laboratory

The Omega West Reactor at the Los Alamos Scientific Laboratory operates at 8 MW and produces 2.5×10^{15} neutrons/s. Although this is less than the ILL reactor at Grenoble (57 MW), what counts is the fraction that can be brought to the detector. Figure 1 shows a cross-section of the reactor. It has a small core, $30 \times 70 \times 60$ cm^3 of enriched uranium metal and a large thermal column, 150×150 cm^2. By removing a section of the thermal column as shown a flux of 4×10^{14} neutrons per second over an area of 150×150 cm^2. If more of the thermal column is removed, namely that behind the boral curtain, this flux can be increased by a factor of 10.

The slow neutrons emerging directly out of the reactor are accompanied by a large flux of fast neutrons and γ-rays, making a difficult environment for any type of electronic detector. Besides the γ-rays coming from the reactor core, there are the γ-rays from neutron capture. The (n,γ) cross-section is appreciable in most nuclei, He^4 and He^3 are exceptions, so that care must be exercised to keep such processes to a minimum. Even in graphite and beryllium, materials most commonly used in slowing neutrons, the (n,γ) reactions are large enough to make it difficult to obtain neutron fluxes with less than 3% γ contamination. The event rate for the process $n \to \bar{n}$ is given by

$$\nu_{n \to \bar{n}} = \left(\frac{\ell/v}{\tau}\right)^2 \Phi \, T \, \epsilon$$

where ℓ is length of path travelled by slow neutrons of velocity v in a field free ($H < 10^{-3}$ gauss) vacuum before striking a target within the detector, τ is the mixing time, Φ is the flux of neutrons striking the target, T is the observation time, and ϵ is the efficiences for detecting the annihilation of $\bar{n}$ when it strikes the target and releases 2 GeV of energy in the form of pions. Since $\Phi \sim \frac{1}{\ell^2}$, the event rate is independent of ℓ for given detector area. Thus, the γ-ray flux can be reduced by setting the detector at as great a distance as possible from the source.

We have chosen to use a graphite reflector to redirect the slow neutrons to have them move down a 2.6 m diameter pipe at right angles to the axis of the thermal column, as shown in Figure 1. About 1/2 of the neutrons emerging from the reactor are redirected in this way. For reasons of economy we are proposing a phase I experiment with a detector much smaller than required if maximum sensitivity were to be obtained. The arrangement is shown in Figure 2. Our detector has an aperture of 1 meter. The target is covered with Li^6D pressings. We stop neutrons in Li^6D because Li^6 has a large absorption cross-section for slow neutrons via the process $n + Li^6 \rightarrow He^4 + H^3$. The (n,γ) process has a branching ratio of only 1/25,000. We also line the walls of the vaccuum pipe with Li^6D to keep low the (n,γ) processes which would otherwise occur if the neutrons were able to strike the walls of the pipe.

Neutron and γ-ray fluxes were calculated by R. J. McKee using a Monte Carlo transport program available at Los Alamos. Some of the neutron fluxes are indicated on Figure 2. In particular, 6.5 x 10" n/s strike the 1 m diameter target at the center of the detector. This gives an event rate of .7/day for a mixing time of 10^6 seconds. Here we have taken ℓ = 8.7 m, v = 2200 m/s, and ε = 0.8.

The detector is a cylindrical iron-scintillation hodoscope sandwich that surrounds the target. The detector cylinder is 2.6 m long, and 60 cm thick. It is a calorimeter made up of some 250 scintillation hodoscope elements distributed through the iron, more closely on the inside, further apart on the outside. Each hodoscope element, 2.6 m long, 1 cm thick, and from 15-30 cm wide is viewed by PM tubes at each end. We can reconstruct the pion shower from $\bar{n}$ annihilation by using time of flight out of both ends of the struck scintillation element for the Z coordinate, and its location in azimuth and radius for the θ and r coordinate. Although the space resolution will be only 10 - 15 cm, it should be sufficient to establish that particles detected, whether $\pi^{\pm}$ or the γ's from $\pi^°$, originate from a common vertex at the target.

For triggering we use a linear fan in to sum the pulses from all the PM tubes, looking for pulses that correspond to an energy of 100 -

200 MeV. This is a threshold well above the value that could result from pile up of the captured $\dot{\gamma}$'s entering the calorimeter. A managable γ flux is obtained by providing an inner Pb layer two radiation lengths thick. We calculate that 6×10^8 γ/s with average energy of about 1.2 MeV will enter our calorimeter. The average energy from this in a 10 ns gate is 7.2 MeV. Pile up effects can produce much large pulses, but only rarely (< 1/day) will pile up pulses exceed 25 MeV.

The trigger rate from cosmic ray muons and e.m. showers could be about 3200/sec. Thus, we need the cosmic ray veto outside the calorimeter as shown in Figure 2. The probability that the C.R. veto will veto an annihilation event is only a few percent. We reduce the number of neutron triggers by installing several meters of concrete shielding around the detector. This is important because a cosmic ray neutron could interact in the calorimeter and be mistaken for an annihilation event. An important aspect of this experiment, not possible in proton decay experiments, is that backgrounds can be separately and extensively studied with the annihilation events turned off.

The experiment can be extended to yield a higher sensitivity in mixing time. Thus in Phase II we would:

1) Remove more of the thermal column to increase Φ by 10 times.
2) Increase the diameter of the detector from 1 m to 2.6 m for an increase in Φ by a further factor of 6.7.
3) Cool the reflector to give 30°K neutrons. This decreases v and increases the event rate by a factor of 10.

In all, with an event rate of 470 events/day for $\tau = 10^6$ sec we can achieve a sensitivity of $\tau = 10^8$ sec by observing 4.7 ev/100 days.

Accompanying the increase in neutron flux and decrease in neutron velocity is a corresponding increase in γ-ray flux. To compensate for this we increase the drift pipe to 80 m. Figures 3 and 4 show how this could be done at the Omega West Reactor site at the bottom of a canyon. We propose to extend our 2.6 m diameter pipe outside the building, tunnelling into the canyon wall as shown. There is no special difficulty in doing this, only time and money.

I thank W. S. C. Williams, B. Barbiellini, S. Fukui, and Arne Lund for useful discussions. R. J. McKee and J. Solem are collaborating with me in this work.

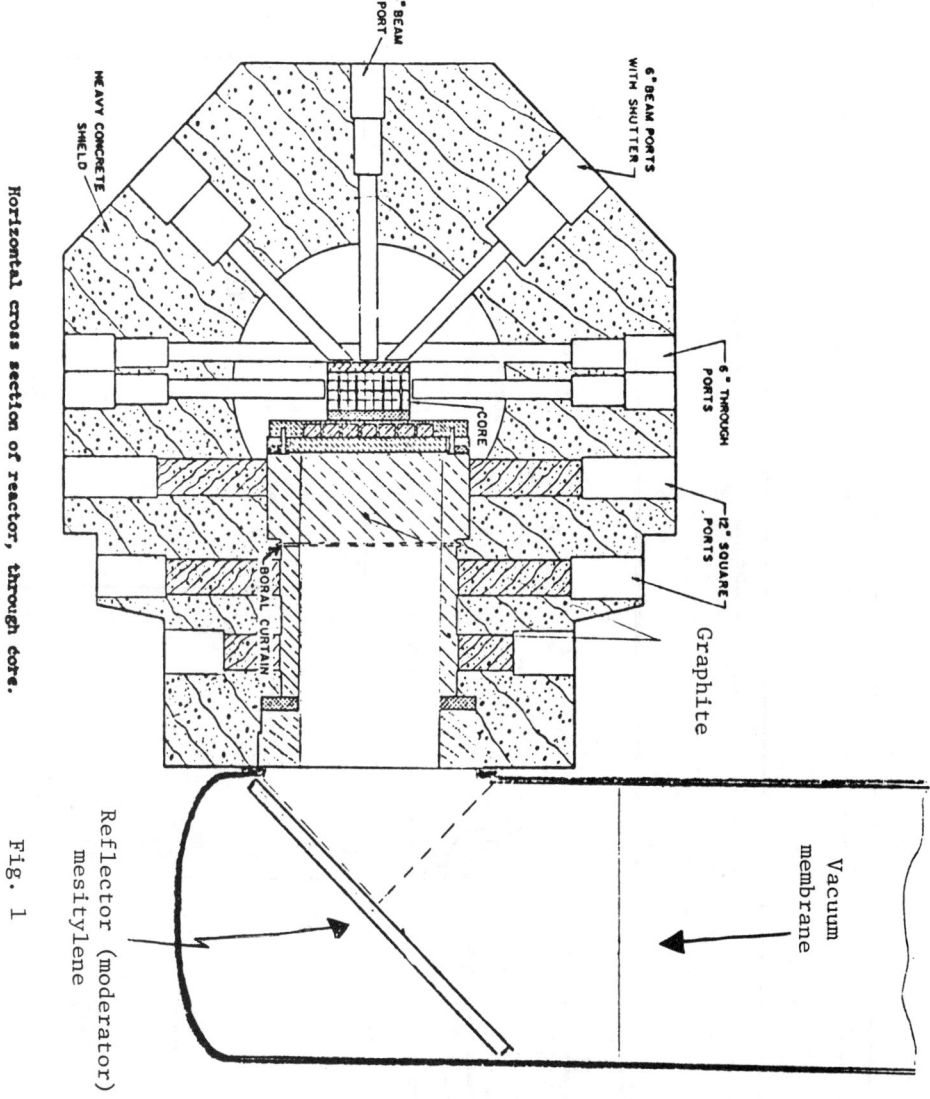

Horizontal cross section of reactor, through core.

Fig. 1

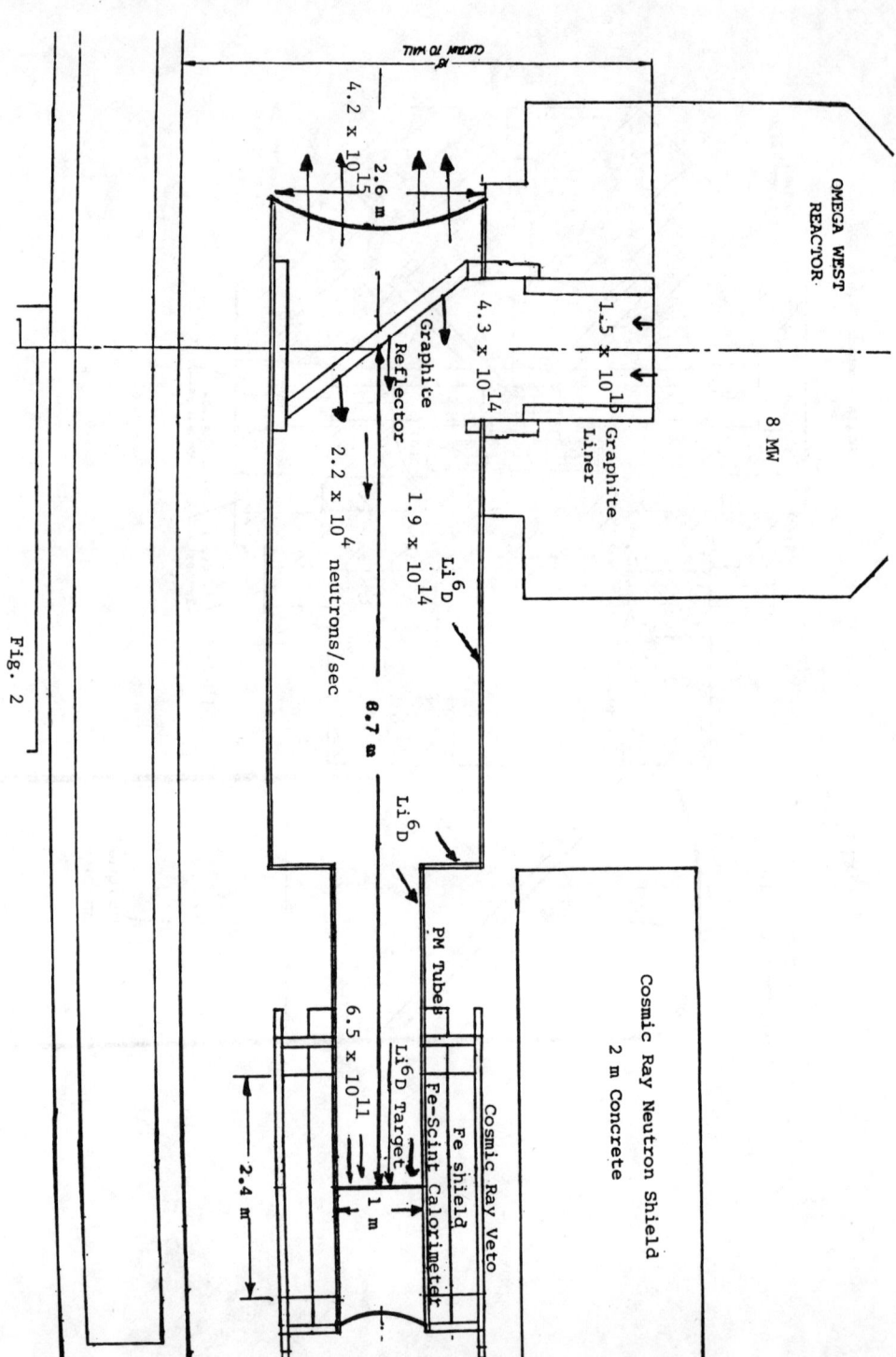

Fig. 2

Fig. 3

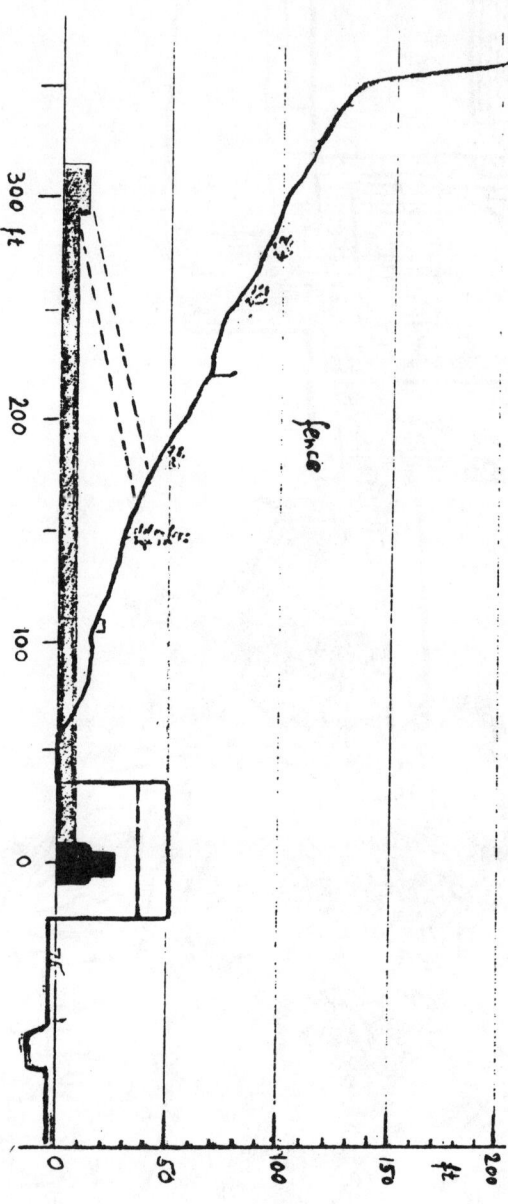

Figure 4

S. Ratti
Univ. di Pavia

We are preparing an experiment at the University of Pavia Reactor, a one quarter megawatt reactor, which is very similar to the experiment which is being planned by Professor Anderson. It has a thermal column, which is completely empty (Fig. 1) and our neutron source is half a meter from the core; so on the one meter square target we have roughly 5×10^{11} neutrons. We also have the option of operating the reactor in the pulsed mode and this should give us a chance to keep the cosmic ray background under control because we can pulse it for 30 milliseconds three or four times an hour. Our basic idea is not to have a calorimeter type detector but rather to use detectors like flash tubes which allow us to eventually reconstruct the interaction point. The spectrum of products coming from a Monte Carlo generation of neutron-antineutron annihilations seems to be equal to proton-antiproton; antineutron-proton annihilations are assumed to be equal to antiproton-deuteron neutron annihilations. These Monte Carlos (Fig 2,3) give us indications of relatively large multiplicity of particles; namely, if we believe the measured cross sections and assume we have a few eV anti-neutrons, we have an average multiplicity of five pions in the annihilation. By triggering on multiplicity and detecting the direction of the charged particles, and possibly the gamma rays if the gamma rays come from the π^o, we can select annihilation events from the background gamma rays which will be fairly numerous, I suppose, even though they are of very low energy. The minimum energy of gammas coming from π^o seems to be of the order of 20 MeV so they should be separable from the low energy gammas coming from all sorts of backgrounds (Fig. 4). This, of course, has to be studied experimentally. I think all these experiments will be well understood only when we understand the background, no matter who is doing what and where. The problem is to understand how the background behaves in all possible ways, it's probably the crucial point. Our sensitivity should be very similar to Prof. Anderson's because while our reactor has less power, the source is only 45 centimeters from the core, so we gain a great deal in this way.

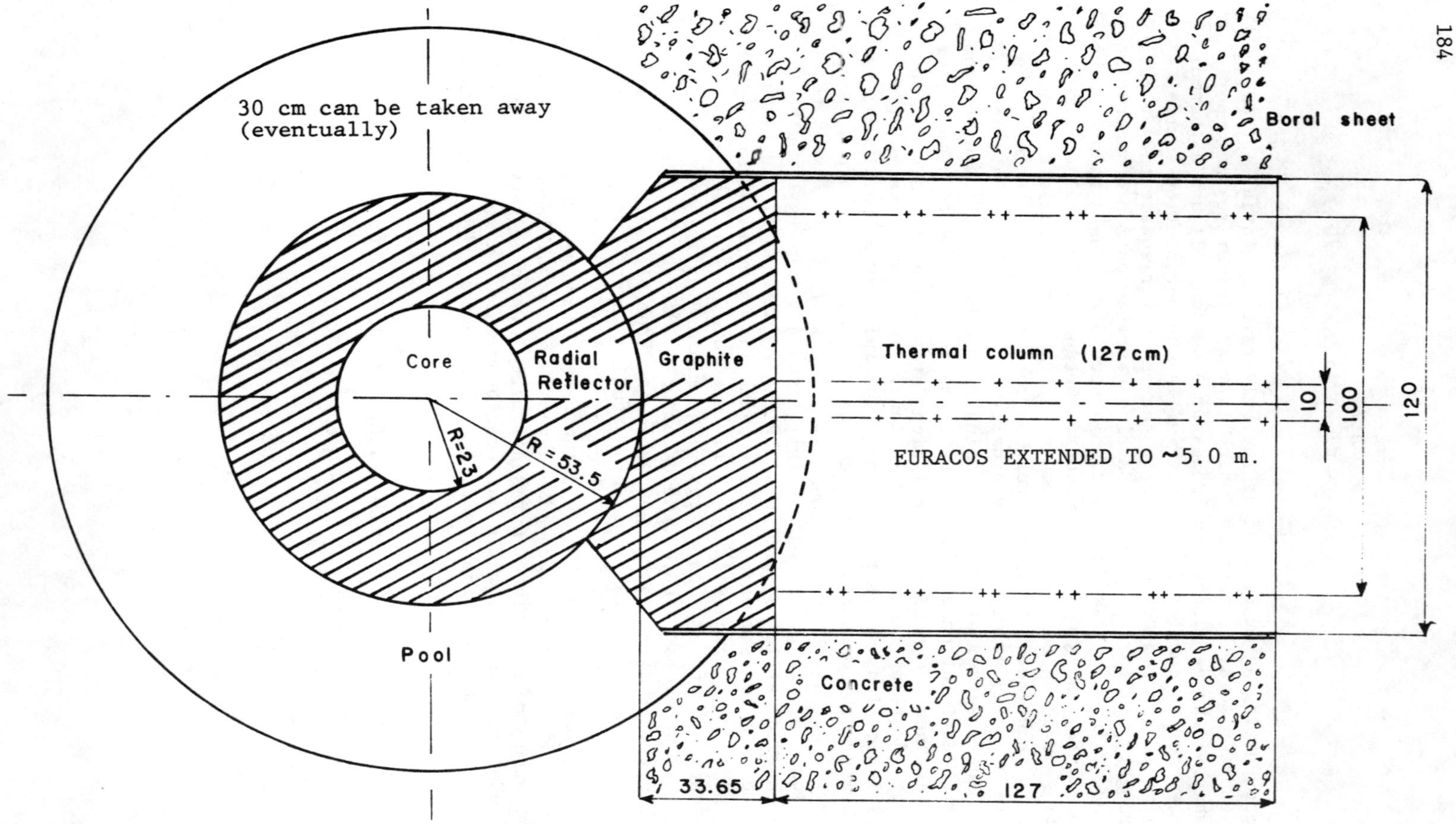

Fig. 1 PAVIA: Horizontal Section of the Triga Mark II Reactor

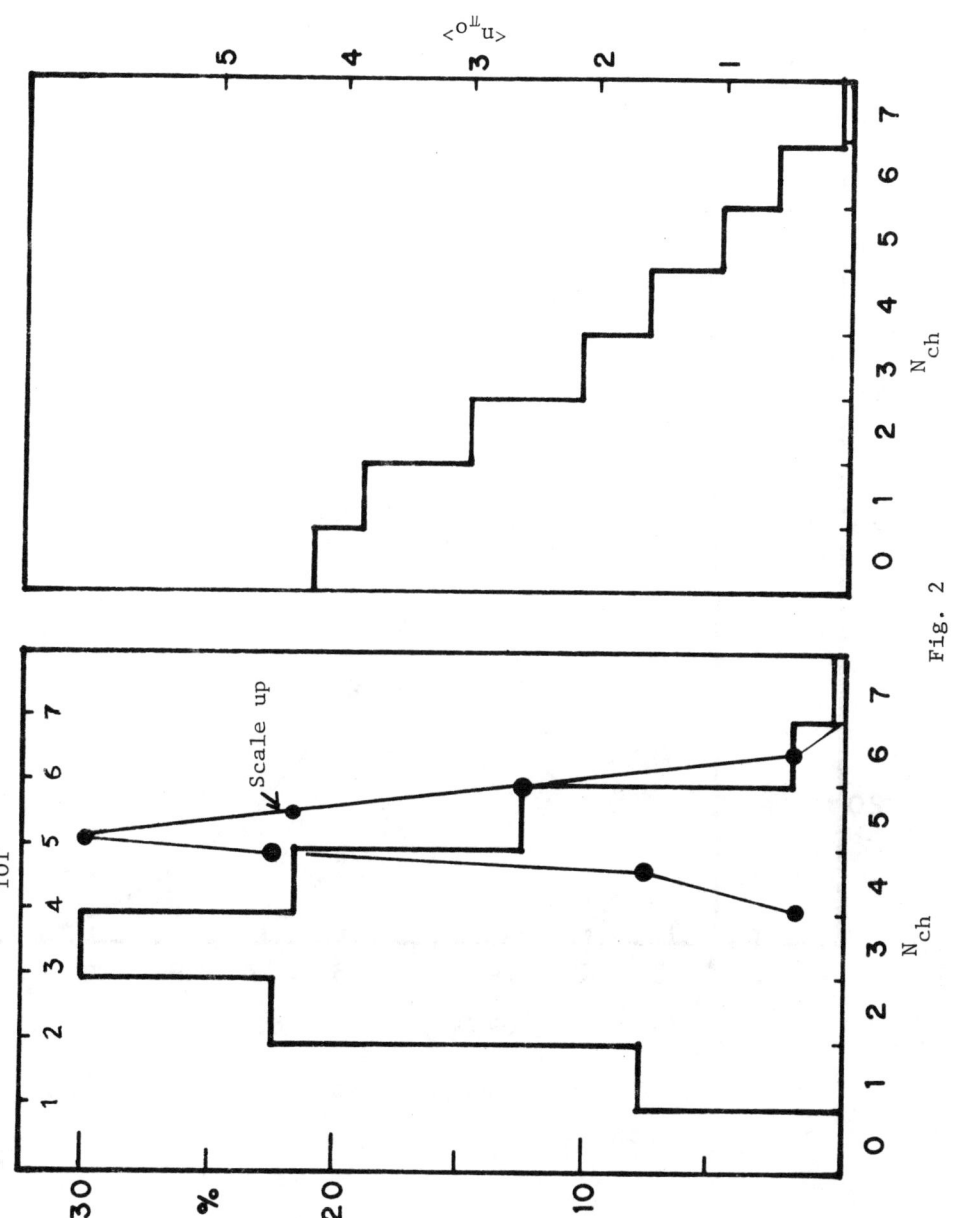

Fig. 2

Monte Carlo - Energy Distribution

120
100
80
60
40
20
# of Tracks
Charged Particles
.1
.2
.3
.4
.5
.6
.7
.8
.9
1.0
140 MeV
E$_{\pi^{\pm}}$ (GeV)

Fig. 3a

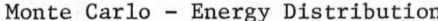

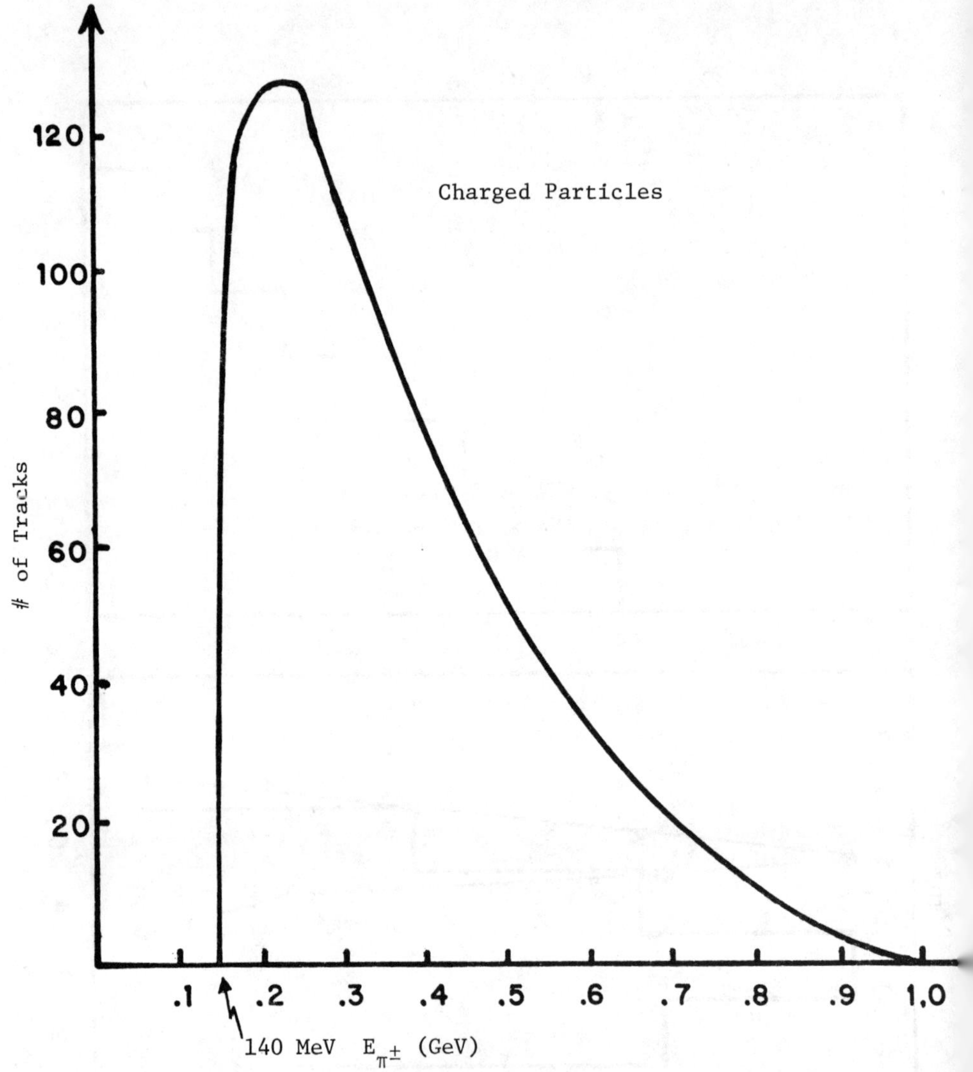

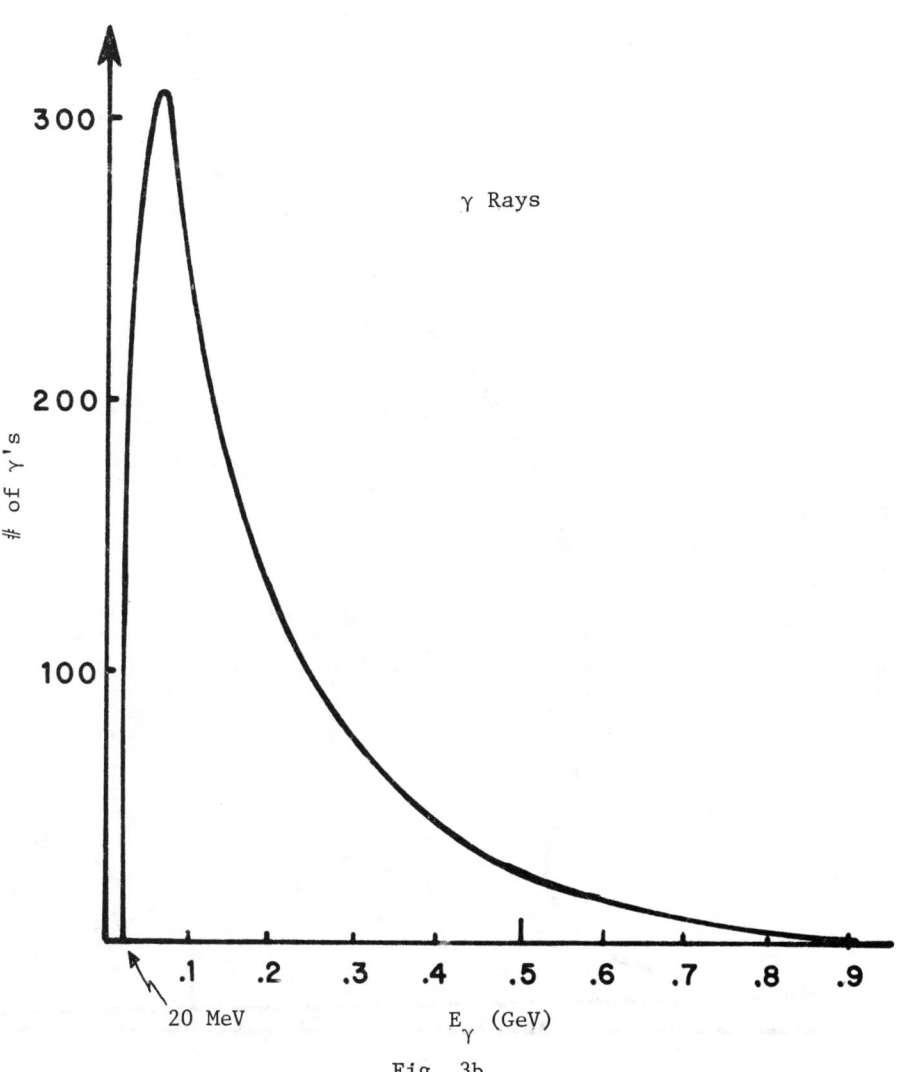

Fig. 3b

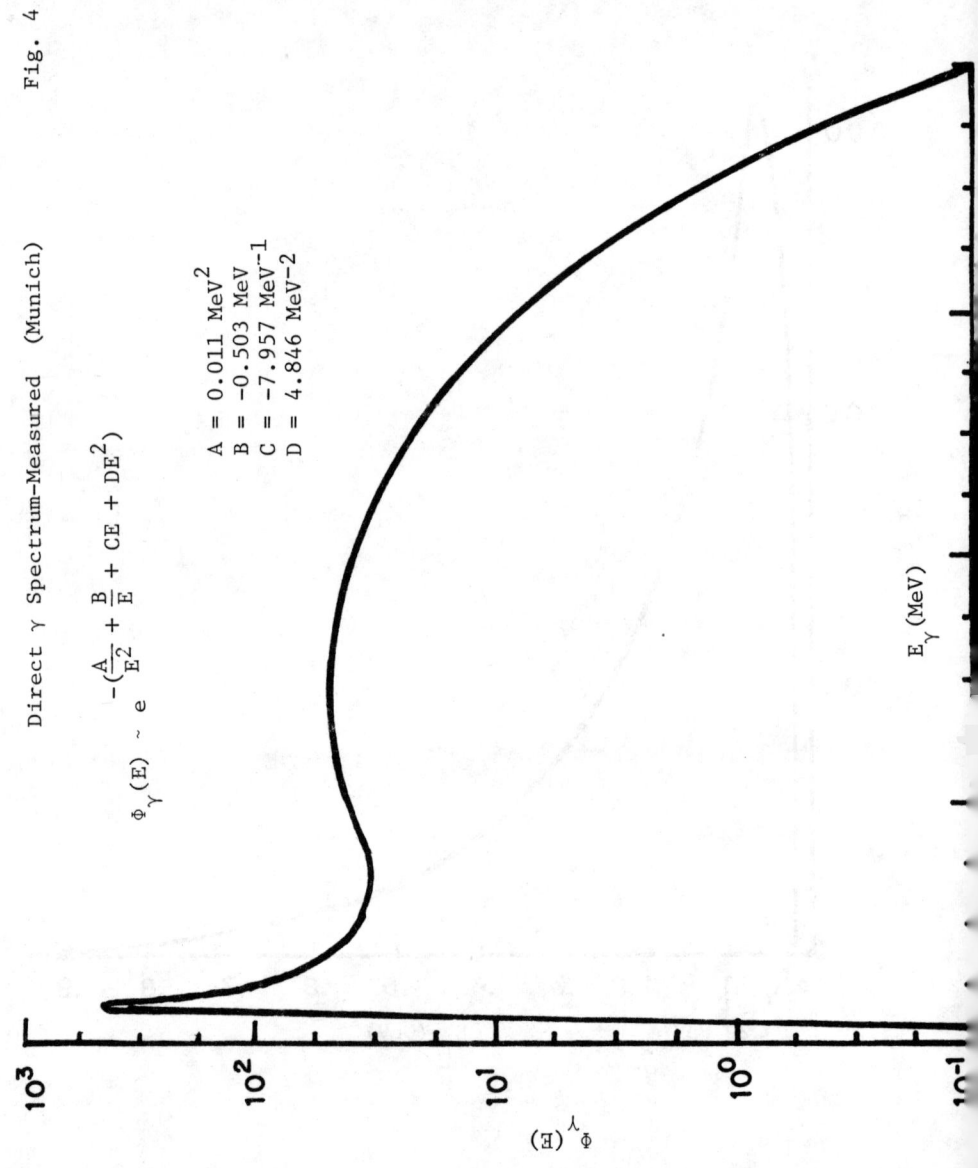

Fig. 4

Direct γ Spectrum-Measured (Munich)

$$\Phi_\gamma(E) \sim e^{-(\frac{A}{E^2} + \frac{B}{E} + CE + DE^2)}$$

A = 0.011 MeV^2
B = -0.503 MeV
C = -7.957 MeV^{-1}
D = 4.846 MeV^{-2}

T. Kamae
Lawrence Berkeley Laboratory

Since most of the stuff has already been discussed, I just want to describe what I have proposed to KEK in a very preliminary form. That is, accelerator beam dump experiments are as good as or even better, than high flux thermal neutron experiments at reactors. The essential point is that the cooling down to $10°$ Kelvin is readily available. As a matter of fact, Professor Ishikawa and his colleagues have built such a neutron source at KEK for solid state studies. In principle, even at the KEK booster accelerator (500 MeV, 10^{13} protons per second) we can obtain somewhat less than 5×10^{10} neutrons. In practice, 10^{10} may be a reasonable limit. However, these neutrons are at $10°K$, much colder than what you can obtain in most of the reactors. The existing cold neutron sources at KEK are shown in Fig. 1. A 500 MeV proton beam is dumped into a tungsten target, which can be replaced by uranium for a higher neutron yield. Solid methane cools down to $\sim 10°$ Kelvin. At this temperature, the velocity of the neutron is so slow that you gain a factor of, say, 10^3 to 10^4 relative to thermal neutrons. The present setup covers only a very minor part of the solid angle out of the moderator, but perhaps one can cover up to as much as 2π in solid angle. And the question that remains is how efficiently one can funnel neutrons into a small pipe. Perhaps one can use cold pipes which reflect back neutrons by total reflection. Such a neutron beam will be more manageable than the kind of beam pipe discussed by Herb Anderson and others. You may even be able to cool the neutrons further. For example, in a proposal by Yoshiki of KEK, you can cool down even to 5×10^{-5} electron volts, a scheme he has been trying out on a small scale. According to his scheme, one can contain the neutrons for a period of 200 seconds, in a SiD_2 box. At that wave length you may hope that the reflection off the wall is more or less coherent. If so, the change of phase is not random and the factor you gain over reactor experiments may be much bigger. So with the proton intensity offered by the Los Alamos Proton Linac one should be able to beat most of the reactor experiments. That is all I have to say.

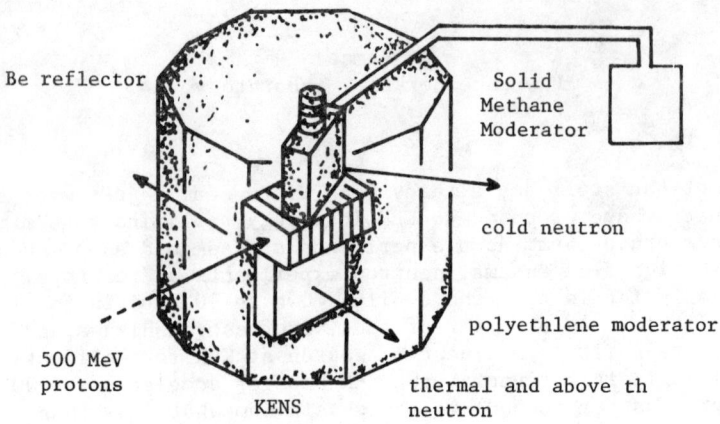

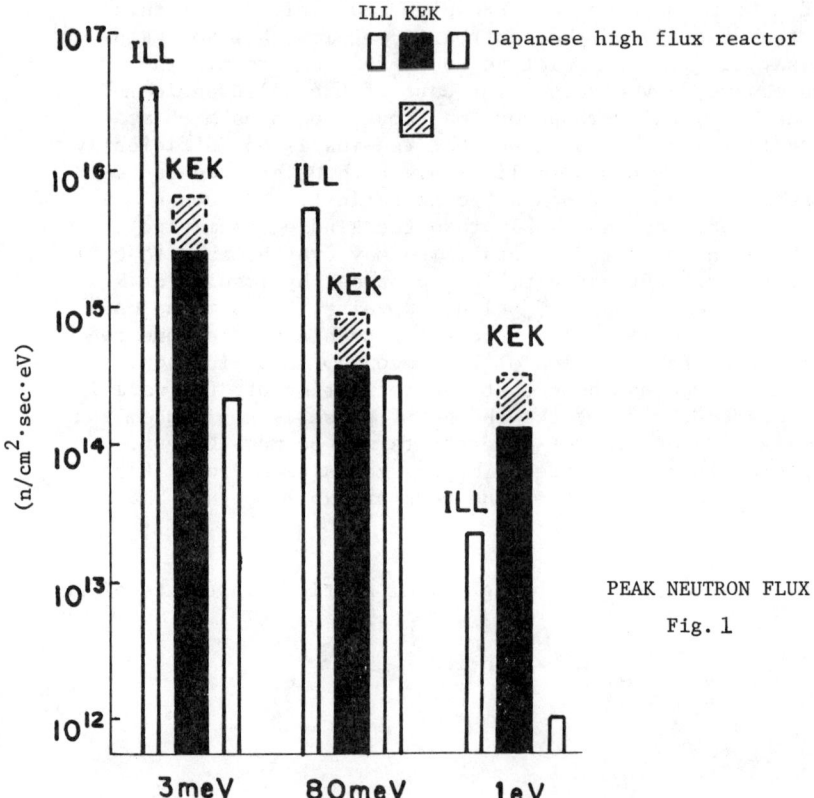

PEAK NEUTRON FLUX
Fig. 1

DISCUSSION

ANDERSON: Could Prof. Kamae say what he estimates now for the oscillation time that he can achieve in his experiment?

KAMAE: The mean collision time depends on the radius of the beam transport tube. A few milliseconds will be a reasonable estimate if the collision is incoherent. If coherence is retained, it can be an order of magnitude larger.

MARSHAK: Suppose one sees the neutron oscillations. The question then will be, is it an SU(5) theory with a complicated Higgs structure or is it PUT? PUT is the partial unification theory that can be expressed by the Pati-Salam group $SU(2)_L \times SU(2)_R \times SU(4)$ with B-L the fourth color. I should like to point out that there is a lot of physics in PUT. Namely, B-L is the weak hypercharge and the breakdown of B-L local symmetry is connected to spontaneous parity breakdown, and this means that you have both Majorana neutrinos ($\Delta L=2$) and neutron oscillations ($\Delta B=2$). Evidence for the heavy Majorana neutrino and neutron oscillations would clearly favor some version of PUT [and its possible generalization to SO(10) GUT].

CP VIOLATION AND THE DEVELOPMENT OF COSMOLOGICAL BARYON ASYMMETRY

Goran Senjanović
Brookhaven National Laboratory, Upton, N.Y. 11973

ABSTRACT

A discussion of the origin of the observed matter-antimatter asymmetry of the universe is presented in the context of the standard cosmological model. Except in the case of the minimal SU(5) theory, it is possible that grandunified theories predict the right order of magnitude for the ratio of baryon to photon number. The question of CP violation is addressed in detail and it is shown that, tied up with symmetry nonrestorationat high temperature, the soft CP violation does remain at $T \simeq 10^{15}$ GeV as to lead to the creation of baryon asymmetry in the very early universe.

The standard cosmological model seems to provide a rather successful picture for the evolution of the universe[1], at least up to the times of order of a second or so. We have been able to observe a very important relic of the early universe: a 2.7°K microwave, isotropic, blackbody radiation. However, the observed amount of matter in the universe and the lack of antimatter[3] (up to the cluster of galaxies) has been an outstanding puzzle for a long time. Recently, a resolution has been suggested[4] which attributes the baryon number of the universe $n_B/S \simeq 10^{-9} - 10^{-11}$ (S is the entropy) to baryon number and CP nonconserving decays of superheavy bosons (X) of grandunified theories during the very early stages of the universe. According to this picture at temperatures below the Planck mass and above the X boson masses the universe went through the epoch at thermal equilibrium during which time any previous matter-antimatter asymmetry would have been wiped out. When the temperature dropped below the superheavy boson masses and the inverse decay was blocked by $e^{-M/kT}$, the decays of these bosons into the channels with different baryon numbers presumably created a slight excess of particles over antiparticles. And finally, at much later times, when the baryons and antibaryons annihilated only matter was left, as the observations indicate. Quantitative analysis shows that for a large class of grandunified theories, the predicted baryon number does not disagree with the measurements.

The suggested scheme requires the extrapolation of the standard model, both of particle interactions and of cosmology, up to energies (temperatures) of order 10^{15} GeV or times up to the 10^{-35} sec. This, as outrageous as it seems, may not be totally unreasonable. Grandunified theories have offered us a long awaited explanation of why proton is so stable ($\tau_p \gtrsim 10^{30}$ years) and they also predict that we should, in near future, witness its decay. Their merits have been discussed at length[5] and they have become a respectable candidate for the old dream of unification of particle forces. On the other hand, we have no reason to suspect that the standard cosmological model will cease to be a valid description of the phenomena in the very early stages of the universe, as long as the temperature is below the Planck mass so that we can neglect the quantum gravitational effects. Of course, the whole picture may be totally wrong, but its simplicity and naturalness are highly suggestive. In this talk I will try to make the case for its validity.

For a baryon asymmetry to arise dynamically, independently of initial conditions, in addition to baryon number violation and the departure from the thermal equilibrium, CP cannot be a good symmetry. Otherwise, the X boson decay create an equal number of particles and antiparticles. More precisely, CP violation has to remain operative at $T \simeq 10^{15}$ GeV. It is important to see what constraints does that requirement imply on our present understanding of the origin of CP nonconservation. Now, in the context of gauge theories we have two basically different mechanisms[6] of CP violation, hard and soft, depending on the canonical dimension of the CP nonconserving piece of the Lagrangian $d(\bar{L})$. For $d(\bar{L}) = 4$ CP violation is hard, whereas for $d(\bar{L}) \leq 3$ CP violation is called soft. The most popular example of the first kind is the so called Kobayashi-Maskawa extension[7] of the standard model with complex Yukawa couplings. By soft CP violation we will, in what follows, assume CP to be spontaneously broken.

Hard CP violation, since it is characterized by complex couplings in the basic weak Lagrangian, will remain at all temperature. The case with soft CP violation is more subtle and it is tied up with the nature of symmetry breaking at high temperature. On the basis of the anology with ferromagnets and as confirmed[8] in the simplest, single Higgs model one would expect symmetry to be restored above the temperature of order of the scale of weak

interactions ($\simeq$ 300 GeV). But then soft CP violation would not present at
T $\simeq 10^{15}$ GeV, as is required from the considerations of baryon asymmetry.
However, the symmetry is not always restored at high temperature. Increased
complexity of the Higgs sector allows (at least for some) vacuum expectation
values to remain broken at high T.[9,10] Mohapatra and the author[9] have constructed a series of models in which soft CP violation is present at high T.
As we shall see, the resulting baryon asymmetry is in accord with observations.[11]
We should add that the main motivation for soft CP violation is not just aesthetic
or philosophical, but rather the fact that it seems mandatory in order to understand the smallness of strong CP violation.

In this talk I will discuss above issues in some detail. The rest of
the material is organized in the following manner: In section II we give a
brief discussion of the standard cosmological model with the aim to obtain the
expression for the expansion rate of the early universe. That will serve for
the comparison with decay rates given in section III, which will determine when
the universe went through the equilibrium epoch. There we also present a plausible
scenario for the development of matter-antimatter asymmetry. In section IV we
discuss the theories of CP nonconservation and the nature of high temperature
behavior of gauge theories. We also present some rough, qualitative estimates
of induced baryon number. Finally, in section V we offer some comments and
summarize the basic contents of this paper.

II. Standard cosmological model and the very early universe

The assumptions of isotropy and homogeneity, encouraged by observations,
lead to a unique form[1] for the metric of the space time (Robertson-Walker metric)

$$ds^2 = dt^2 - a^2(t) \left[\frac{dr^2}{1 - kr^2} + r^2(d\theta^2 + \sin^2\theta \, d\phi^2) \right] \tag{2.1}$$

where r, θ, ϕ are dimensionless parameters and $a(t)$ is the scale factor. By simple rescaling $r \to (|k|)^{-\frac{1}{2}} r$, $a \to (|k|)^{\frac{1}{2}} a$ it is obvious that k takes only three different values: $k = 0, +1, -1$. The distance between material points, one at the origin and other at $(r,0,0)$ is given by

$$R(t) = \int_1^2 \sqrt{-ds^2} = a(t) \int_0^2 \frac{dr}{(1 - kr^2)^{\frac{1}{2}}} \tag{2.2}$$

We then obtain the following results:

a) $k = 0 \Rightarrow R(t) = a(t)r$. The three-dimensional space is flat and infinite Such universe is called Eucledian.

b) $k = -1 \Rightarrow R(t) = a(t) \sin^{-1} r$. This is called an open universe, since for $r \to \infty$ $R(t)$ becomes infinite (infinite three dimensional space).

c) $k = 1 \Rightarrow R(t) = a(t) \sin^{-1} r$. Such models are called closed, since $0 \leq r \leq 1$ and $0 \leq R \leq \pi/2a$. It is still debated as to which is our universe.

The standard, or Friedmann model of the universe is obtained when Robertson-Walker metric (2.1) is combined with Einstein equations

$$R_{\mu\nu} - \frac{1}{2} g_{\mu\nu} R = 8\pi G T_{\mu\nu} \tag{2.3}$$

where the energy momentum tensor $T_{\mu\nu}$ has the form of the ideal fluid (which is forced by the dynamics)

$$T_{\mu\nu} = pg_{\mu\nu} - U_\mu U_\nu (p + \rho) \tag{2.4}$$

In the above p and ρ are the pressure and density. From (2.1), (2.3) and (2.4) one derives

$$\frac{1}{2} \left(\frac{da}{dt}\right)^2 - G\left(\frac{4\pi}{3} a^3 \rho\right) \frac{1}{a} = -\frac{k}{2} \tag{2.5}$$

which has a form of energy conservation, and

$$\frac{d\rho}{\rho} + 3(1 + p/\rho) \frac{da}{a} = 0 \tag{2.6}$$

which shows that the expansion of the universe is adiabatic.

The very early universe was radiation dominated, so that the, in addition needed, equation of state was $p = 1/3\rho$. Now, what we are after is the expansion rate of the universe, called Hubble constant

$$H \equiv \frac{1}{a} \frac{da}{dt} \qquad (2.7)$$

which is a function of time, or temperature. For a radiation dominated universe $\rho \simeq NT^4$, and since the term in the right hand size of (2.6) is negligible[13], one obtains

$$H \simeq \sqrt{N} \frac{T^2}{M_p} \quad \text{and} \quad t \simeq \frac{M_p}{\sqrt{N} T^2} \qquad (2.8)$$

What we will be discussing in this paper is the very early universe when the temperatures were between 10^{14} GeV and the Planck mass (10^{19} GeV). In terms of the age and the size of the universe

$$\begin{aligned} T_p &= 10^{19} \text{ GeV} \Rightarrow t_p = 10^{-43} \text{ sec} \quad r_p = 10^{-33} \text{ cm} \\ T &= 10^{15} \text{ GeV} \Rightarrow t = 10^{-35} \text{ sec} \quad r = 10^{-25} \text{ cm} \end{aligned} \qquad (2.9)$$

etc. For the sake of completeness, I have included a table of some, somewhat randomly chosen, important moments in the history of the universe.

$t = 10^{10}$ years	$T = 3°K$	PRESENT
$t = 10^6$ years	$T = 1$ eV	ATOMS
$t = 4$ minutes	$T \simeq 10$ keV	NUCLEO SYNTHESIS
$t = 10$ sec	$T = .5$ MeV	$e^+ e^-$ ANNIHILATION
$t = 1$ sec	$T = 1$ MeV	NEUTRINOS DECOUPLE
$t = 10^{-10}$ sec	$T = 100$ GeV	SU(2) × U(1) BREAKING
$t = 10^{-35}$ sec	$T = 10^{15}$ GeV	SU(5) BREAKING n_B CREATION MONOPOLE -11-
$t = 10^{-43}$ sec	$T_p = 10^{19}$ GeV	QUANTUM GRAVITY

T A B L E   1.

SOME IMPORTANT DATES IN THE HISTORY OF THE UNIVERSE

III. DEVELOPMENT OF BARYON ASYMMETRY

At enormously high temperature, much above all the known and conjectured particle masses it is natural to expect an equilibrium situation and therefore an equal number of particles and antiparticles. How did the universe then evolve into an asymmetric state, with mainly matter? Clearly, the answer could be that there are domains of matter and antimatter, large enough so that our observations are misleading. This picture still needs the explanation of the domain growth and of their separation. In any case, the magic number $n_B/S \simeq 10^{-9} - 10^{-11}$ would have to be postulated ad hoc, as an initial condition.

It was relaized,[14] a long time ago, that in order to have baryon number dynamically generated, we need baryon number nonconservation at some level, or otherwise the baryon number of the universe would be a fixed quantity, throughout a history of the universe. Now, grandunified theories, as a rule, predict baryon number violation and therefore provide a natural theoretical scheme that could explain, on the basis of fundamental, microscopic laws such a global property of the universe as its material content. In the last two years a scenario for the origin of matter-antimatter asymmetry has been developed. It, as we have mentioned, incorporates the basic and general aspects of grandunification and the standard cosmological model and provides a simple, and logically consistent picture. It is our task to show that. As we shall see, it requires, at this point, a lot of faith, but future tests (of the proton decay) will hopefully justify it.

We start by briefly recalling some of the basic features of grandunified theories.[15] Let us imagine the simplest possibility according to which a unifying group G is broken down to $SU(3)_C \times SU(2)_L \times U(1)$ at a single energy scale M_X. Following Georgi, Quinn and Weinberg,[16] we can trace the momentum dependence of coupling constants and derive the well known relations for the low energy parameters of weak and strong interactions

$$\sin^2\theta_W(M_W) = \frac{3}{8}\left[1 - \frac{55}{9}\frac{\alpha(M_W)}{\pi}\ln\frac{M_X}{M_W}\right]$$

$$1 - \frac{8}{3}\frac{\alpha(M_W)}{\alpha_S(M_W)} = 11\frac{\alpha(M_W)}{\pi}\ln\frac{M_X}{M_W} \tag{3.1}$$

Consistency of (3.1) predicts[17] then $M_X = 10^{14} - 10^{15}$ GeV. M_X corresponds roughly to the masses of superheavy or X bosons, which do not carry a fixed baryon number and whose enchange leads to baryon number violation. For example, one species of such bosons carry fractional charge $\pm 4/3$ and have interactions

$$L_{int} = g_X/\sqrt{2}\ X_{\mu\lambda}\left[\varepsilon_{ijk}\overline{u^C_{Lj}}\gamma^\mu u_{Lk} + \overline{d_i}\gamma^\mu e^+\right] \tag{3.2}$$

where C denotes charge conjugation and i, j, k stand for color. A tree level process in which X is exchanged leads to proton decay $p \to \pi^0 + e^+$ (see Fig. 1). From the prediction for $M_X \simeq 10^{14} - 10^{15}$ GeV, we can estimate $\tau_p \simeq 10^{31\pm 2}$ years, which is within reach of experiments now in progress.

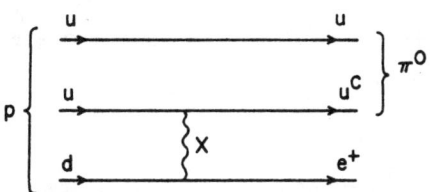

Fig. 1. Proton decay $p \to \pi^0 e^+$ as induced by exchange of X boson.

At very high temperature $T \gtrsim M_X$ the decays of X bosons should have played an important role, presumably being responsible for the observed baryon asymmetry. Namely, these bosons can decay into the channels with different baryon numbers

$$X \to \begin{cases} qq & B_1 = 2/3 \quad\text{branching ratio } r \\ \bar{q}\bar{l} & B_2 = -1/3 \quad\text{branching ratio } 1 - r \end{cases}$$

and

$$\bar{X} \to \begin{cases} \bar{q}\bar{q} & -B_1 = -2/3 \quad\text{branching ratio } \bar{r} \\ ql & -B_L = 1/3 \quad\text{branching ratio } 1 - \bar{r} \end{cases} \tag{3.3}$$

If the branching ratios r and $\bar{r}$ are different, then when the temperature dropped below M_X, we would expect that a small asymmetry should have developed. As the temperature dropped even further, much below M_X, the baryon violating processes gradually stopped playing an important role (for being very slow) and so the induced small baryon excess should roughly correspond to the amount of matter in the present universe.[18]

Let us discuss the above picture in some detail. First, the key (necessary) ingredients for the explanation of n_B/S are

(i)   <u>microscopic baryon number violation</u>

(ii) <u>departure from thermal equilibrium</u>

(iii) <u>CP violation</u>.

The condition (i) is automatically satisfied in most grandunified theories. Its necessity is obvious, unless we accept a baryon number of the universe as a mysterious initial condition.

The condition (ii) is also easy to understand.[19] If the baryon violating interactions are always in equilibrium, then the numbers of particles and antiparticles would be given by $e^{-m/kT}$ and $e^{-\bar{m}/kT}$ ($\bar{m}$ is the antiparticle mass), which are equal by CPT invariance: $m = \bar{m}$.

Finally, the condition (iii) comes about for the following reason. CP invariance implies the following equality between the amplitudes

$$M(i \rightarrow j) = M(\bar{i} \rightarrow \bar{j}) = M(j \rightarrow i) \tag{3.4}$$

where bars indicate, as before, CP conjugate states. Clearly, no asymmetry can be established.

We should make an important remark regarding the condition (iii). Namely, if we write for the amplitudes in the perturbation theory

$$\begin{aligned} M(i \rightarrow j) &= g_0 + g_1 F \\ M(\bar{i} \rightarrow \bar{j}) &= g_0^* + g_1^* F \end{aligned} \tag{3.5}$$

where g_0 denotes the tree level contribution and F denotes the Feynman amplitude, then[4]

$$|M(i \to j)|^2 - |M(\bar{i} \to \bar{j})|^2 = \text{Im} g_0 g_1^* \times \text{Im} F . \qquad (3.6)$$

Therefore, an absorptive part of the amplitude has to be nonvanishing (i.e., physical intermediate states) and also we need the interference between the lowest and higher orders. The latter is a very useful result, since it tells us that n_B/S is expected small, in some sense. How small precisely, will depend on the amount of CP violation at high temperature. Furthermore, one can also derive the following result:[20]

$$\sum_{j=B} |M(i \to j)|^2 = \sum_{j=-B} |M(\bar{i} \to \bar{j})|^2 \qquad (3.7)$$

which holds true to all orders in baryon conserving interactions and to the first order in baryon violating interactions. Therefore, $n_B/S \neq 0$ requires higher orders in baryon nonconserving forces.

Let us now follow the history of the universe from the earliest moments and present the scenario which meets all the necessary conditions (i), (ii), and (iii). We shall need to compare the expansion rate of the universe

$$H \simeq \sqrt{N}\, \frac{T^2}{M_p} \qquad (3.8)$$

and the decay rate of X bosons[21]

$$\Gamma_X \simeq \alpha_X N \frac{m_X^2}{\sqrt{T^2 + m_X^2}} \qquad (3.9)$$

where $\alpha_X \simeq 10^{-2}$ for gauge mesons and $\alpha_X \simeq 10^{-5} - 10^{-6}$ for Higgs boson interactions. In what follows we shall ignore the baryon violating collisions of light particles, since it can be shown that such processes cannot lead to baryon asymmetry.[4]

(1) $T \leq M_p$

Obviously, $\Gamma_X \ll H$ which means that the expansion rate of the universe is so fast that the decays do not occur. The baryon number is given now by the initial condition at $T = M_p$. Its value, as we shall see is irrelevant for future asymmetry.

(2) $T \simeq 10 \, m_X$

The main change that has occurred is that the expansion rate has showed down and $\Gamma_X \simeq H$. Therefore, X decays and inverse decays establish an equilibrium an so $n_B(T = 10 \, m_X) = 0$. That's a very important result and it means that irrespectively of the initial condition the universe is bound to go through an epoch of equilibrium during which any preexisting baryon number has to vanish. At these temperatures, we start naturally with a symmetric universe.

(3) $T < m_X$

Now, $\Gamma_X > H$. Decays are very important. However, inverse decays become more and more rare, due to Boltzmann suppression $e^{-m_X/kT}$. The needed departure from the equilibrium gets created. If X and $\bar{X}$ bosons do not decay equally fast, an excess of matter over antimatter will be created. What is required is CP violation.

(4) $T \ll m_X$

As the temperature drops down, X bosons will all decay. Created baryon excess should survive today. Of course, baryons and antibaryons will annihilate when temperatures is of order of their masses much later in the evolution of the universe, leaving only matter behind.

Now, in order to estimate the induced baryon to entropy ratio we need to know the density of X bosons n_X and the total entropy at $T \simeq m_X$. At such high temperature

$$n_X \simeq N_X T^3$$

$$S \simeq N T^3 \qquad (3.10)$$

where N_X is the number of X spin states and N is the number of all spin states, the assumption being that the baryon conserving collisions were in equilibrium so that (3.10) applies. But then[21]

$$\frac{n_B}{S} = \frac{N_X}{N} \Delta B \qquad (3.11)$$

with ΔB the net baryon number produced for the decay of X bosons. From (3.3) we evaluate ΔB

$$\Delta B = [rB_1 + (1-r)B_2 - \bar{r}B_1 - (1-\bar{r})B_2] = (r - \bar{r})(B_1 - B_2). \qquad (3.12)$$

Since $B_1 - B_2 \simeq 1$, we get

$$\frac{n_B}{S} \simeq \frac{N_X}{N} (r - \bar{r}) \simeq 10^{-2}(r - \bar{r}) \qquad (3.13)$$

In order to predict the correct amount of matter in the universe, we need $r - \bar{r} \simeq 10^{-7} - 10^{-9}$. In the next section we discuss the theories of CP violation and their predictions for $r - \bar{r}$. We shall, of course, need to discuss the high temperature behavior of gauge theories, in particular the theories of CP non-conservation.

IV. HIGH TEMPERATURE BEHAVIOR OF GAUGE THEORIES AND CP VIOLATION IN THE EARLY UNIVERSE

For the purpose of discussing phenomena which supposedly took place at $t \simeq 10^{-35}$ sec ($T \simeq 10^{15}$ GeV) we need to know the nature of baryon number and CP violating interactions at high temperature. That question is closely tied up to the origin of such interactions, namely whether they are intrinsic (that is, present in the basic symmetric Lagrangian) or the product of symmetry breaking.

If the interactions are intrinsic, then they <u>will</u> remain operative at any temperature, since the form of the Lagrangian is temperature independent. If they, however, result from symmetry breaking of the originally symmetric theory then the question of the form of such interactions depends on the nature of symme breaking at high temperature.

Now, baryon number violation is intrinsic in the minimal schemes, such as SU(5) or O(10), whereas in the Pati-Salam theory baryon number is spontaneously broken. For the sake of simplicity we assume intrinsic baryon number nonconservation, since there is no other reason, besides aesthetical and philosophical one, to assume otherwise.

In the case of CP nonconservation, we have similarly theories of intrinsic (hard) and spontaneously broken[22] (soft) CP violation. We describe first the minimal $SU(2)_L \times U(1)$ hard CP theory with 6 quarks, known as KM model.[7] One assumes a single Higgs doublet ϕ, which implies that $<\phi>$ can be made real by the use of gauge symmetry. In this case one requires complex Yukawa couplings in order to generate CP violation. The Yukawa interactions have the form

$$L_Y = \bar{\psi}_{iL} h_{ij} \phi n_{jR} + \bar{\psi}_{iL} \tilde{h}_{ij} \tilde{\phi} p_{jR} + h.c. \qquad (4.1)$$

where $\tilde{\phi} \equiv i\tau_2 \phi^*$, n_i and p_i stand for three up and down quarks and ψ_{iL} stands for left-handed doublets $\psi_{iL} = \binom{p_i}{n_i}_L$ (h, $\tilde{h} \epsilon C$). It turns out then, that when the quark mass matrices

$$M^n_{ij} = h_{ij} <\phi>$$
$$M^p_{ij} = \tilde{h}_{ij} <\phi> \qquad (4.2)$$

are diagonalized, so are the interactions of neutral, physical Higgs scalar with quarks. <u>The source of CP violation in this model is complex Cabbibo rotation</u>, which results from complex unitary matrices that diagonalize quark mass matrices The CP violation resides completely in the gauge meson interactions with quarks.

When this model is extended to SU(5) gauge theory, the doublet of SU(2) × U(1), gets replaced by a <u>5</u> dimensional Higgs multiplet

$$\phi_5 = \begin{pmatrix} H_1 \\ H_2 \\ H_3 \\ \phi^+ \\ \phi^0 \end{pmatrix} \qquad (4.3)$$

where H_i (i = 1,2,3) is a color triplet of fractionally charged, superheavy Higgs bosons. Due to the complex Yukawa couplings, their interactions with quarks will be CP nonconserving, and so their decays will not respect CP.

In the case of soft CP violation, one assumes all the couplings in the basic Lagrangian real, so that CP is a good symmetry prior to symmetry breaking. The motivation for these theories, besides the philosophical or aesthetical preference, is that they offer a natural resolution of the strong CP problem, as I will discuss below.

Let me first describe the simplest scheme, based on the two Higgs doublet $SU(2)_L \times U(1)$ model.[22] It turns that, consistent with a minimization of the potential, one can achieve, in a range of the free parameters of the potential that

$$<\phi_1> = \begin{pmatrix} 0 \\ v_1 \end{pmatrix}, \qquad <\phi_2> = \begin{pmatrix} 0 \\ e^{\pm i\alpha} v_2 \end{pmatrix} \qquad (4.4)$$

where v_1, v_2 and α are real numbers. Then the quark mass matrices, say for the down quarks

$$M^n_{ij} = h^1_{ij} v_1 + h^2_{ij} e^{i\alpha} v_2 \qquad (4.5)$$

become complex and, similar to the kM case, the Cabbibo rotation will be complex. The minimal soft CP model completely mimics kM scheme in the gauge meson sector. The extra physical Higgs scalars, present in the model, have CP nonconserving interactions with quarks which, if nothing else, tend to alter the KM prediction for the electric dipole moment of the neutron. One predicts $d_n^e \simeq (10^{-25} - 10^{-28})$ecm which serves as a distinguishing feature from the superweak[23] and kM model.

The question is, do we need soft CP violation? Since one measures only the resulting, physical effects why talk of spontaneous breaking of CP? The answer is: <u>natural explanation of the smallness of strong CP violation seems to favor soft CP violation</u>. Let us first review briefly what the strong CP problem is.

From the form of the QCD Lagrangain it was shown originally[24] that strong interactions conserve P and CP to order $G_F \alpha$ (and not only α). The result follows from the neglect of the allowed term (by the symmetry and renormalizability) $\varepsilon_{\mu\nu\alpha\beta} F^i_{\mu\nu} F^{i\alpha\beta}$, since such a term is a total divergence and was not expected to play a role in physical phenomena. However, from the work of 't Hooft and others we have learned that such a term cannot be ignored: through the perturbative, instanton effects it leads to an effective interaction

$$L_{eff} \sim c[e^{i\theta} \det|\bar{q}^0_L q^0_R| + e^{-i\theta} \det|\bar{q}^0_R q^0_L|] \qquad (4.6)$$

where $q^0_{L,R}$ denotes all the (weak eigenstates) quark flavors and c is a dimensional parameter. The interactions in (4.7) violate both P and CP. From the upper limit on the electric dipole moment of the neutron[26] $d_n^e \leq 10^{-24}$ ecm, one obtains a limit $\theta \leq 10^{-9}$. The burning question then becomes as why is θ so small? A trivial answer could be: set $\theta = 0$ and it will always remain such, since it cannot be induced perturbatively. Unfortunately, it doesn't work. As is well known, the quark mass matrices are in general arbitrary and complex, so that in the process of diagonalization

$$q^0_{L,R} \equiv U_{L,R} \, q_{L,R}$$
$$U^+_L M U_R = D \equiv \begin{pmatrix} m_1 & & 0 \\ & m_2 & \\ 0 & & \ddots \end{pmatrix} \qquad (4.7)$$

an additional complex phase in (4.6) will be induced, and the effective $\bar{\theta}$ parameter becomes

$$\bar{\theta} = \theta + i \ln \frac{\det M}{\det M^+} \qquad (4.8)$$

Well, maybe we should instead set $\theta_{tree} = 0$ and hope that perturbation theory keeps it finite and small. In the absence of any symmetry that of course won't work. Weak and electromagnetic interactions will induce infinities, since there is no reason for them not to (we know that all counterterms allowed by a symmetry must be present to ensure the renormalizability of gauge theories). For example, in kM scheme infinities were explicitly isolated[28] (albeit in high orders in perturbation theory).

An interesting suggestion has been made by Peccei and Quinn[29], who postulate an existence of extra $U_A(1)$ axial symmetry, which effectively removes $\bar{\theta}$ from the theory. Such a symmetry, as was realized by Weinberg and Wilczek[30], gets broken and so results in a pseudogoldstone boson, axion, which gets a tiny mass due to instanton effects. Axion seems to be ruled out experimentally[31], at least in the context of the standard model.

Another simple possibility is that some quark, presumably up quark, is massless so that the chiral symmetry eliminates $\bar{\theta}$. It seems to be disfavored by current algebra. In any case, question then just becomes: why is $m_u = 0$?

Finally, it has been suggested[32] that if CP is broken spontaneously, then the symmetry of the original Lagrangian may be used to set $\theta_{tree} = 0$ and then, hopefully, the same symmetry would keep θ finite to all orders in perturbation theory. Of course, it is $\bar{\theta}$ that has to be calculable and small.

One particular program[33] utilizes left-right symmetric gauge theories, derived by Pati, Salam, Mohapatra and the author[34], in order to explain parity violation in weak interactions. According to these theories parity violation is a low energy phenomenon (result of spontaneous symmetry breaking) which ought to disappear at high energies. In the case of the so called manifest left-right symmetric models[35], characterized by hermetian mass matrices $M = M^+$ (and $U_L = U_R$) $\bar{\theta}_{tree} = 0$. Infinities do not appear and $\bar{\theta}$ is a calculable quantity. Estimates[33] show $\bar{\theta}$ (1 loop) = 0 and also $\bar{\theta}$ (2 loop) $\leq 10^{-10}$. The strong CP violation becomes naturally small.

We hope to have convinced the reader that soft CP violation is a highly desirable tool in understanding the smallness of strong CP nonconservation. To make the whole program fully viable we have to make sure that soft CP violation remains at $T \simeq 10^{15}$ GeV[36], when the baryon excess was created. That is tied up with the whole question of symmetry behavior of gauge theories at high temperature, which we now address.

By the analogy with ferromagnetic systems, one would intuitively expect a phase transition at $T_c \simeq <\phi>$ as to lead to symmetry restriction for $T \gtrsim T_c$. This is exactly what the actual computations in the simple Higgs model demonstrated.[8] Let us take an example of the scalar model. The temperature dependent Higgs potential is[8]

$$V(T) - \left(-\frac{\mu^2}{2} + \frac{c}{2} T^2\right) \phi^+\phi + \frac{\lambda}{4} (\phi^+\phi)^2 \tag{4.9}$$

where

$$C = \frac{1}{8} \lambda \tag{4.10}$$

From the positivity of the potential at $T = 0$, $\lambda > 0$ and so $C > 0$. Therefore, the phase transition occurs at

$$T_c \simeq \sqrt{\mu^2/c} \tag{4.1}$$

and so

$$T > T_c : <\phi> = 0$$

$$T \leq T_c : <\phi> = \sqrt{\frac{-CT^2 + \mu^2}{\lambda}} = \sqrt{\frac{C}{\lambda}} \sqrt{T_c^2 - T^2} \tag{4.1}$$

Figure 2 shows the form at the potential for the two phase.

Fig. 2. The unbroken phase for $T > T_c$ in (a) and the broken phase for $T \leq T_c$ in (b). Actually, the picture (although essentially correct) is somewhat more subtle[37], when the one-loop terms for V(0) are included, which play a dominant role neat $T \simeq T_c$, as Coleman and Weinberg[38] have taught us. The phase transition becomes first order and schematically the situation looks as in Fig. 3.

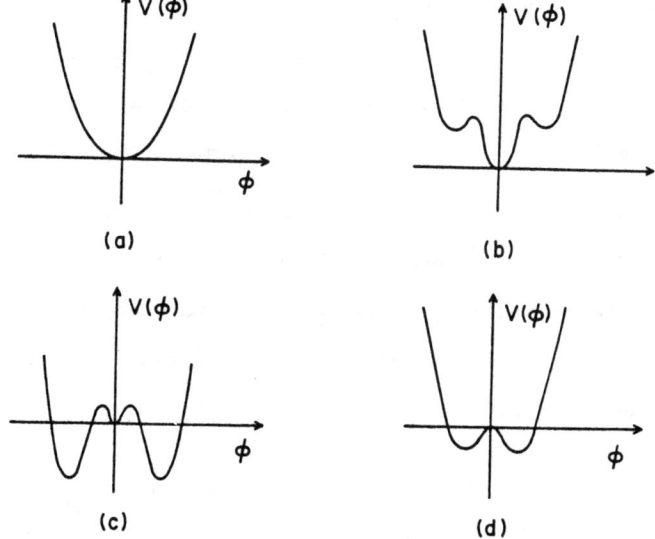

Fig. 3. Before going from unbroken phase a) to broken one d); the potential goes through phases b) and c) with broken and unbroken false vacua, respectively (in addition to true vacua).

We turn now to the implications of above analysis for the creation of baryon asymmetry in the early universe.

(i) **Hard CP violation: KM scheme**

As we have seen above $T_c \simeq 300$ GeV, $<\phi>$ vanishes. Therefore, for $T \geq T_c$, $M_q = 0$ and the Cabbibo rotation becomes identity: $U_c = 1$. At high temperature, gauge meson interactions conserve CP. Their interactions cannot, by themselves, induce a baryon asymmetry. Something else is needed and that, of course, are Higgs bosons. As we have seen in (4.3), the $\underline{5}$ dimensional Higgs of SU(5) consists of, besides the usual light Higgs particles, a color triplet of superheavy Higgs scalars whose interactions with quarks and leptons violate baryon number. Due to complex Yukawa couplings, these interactions violate CP as well, at all temperatures. Their decays, it turns out, play a dominant role in the generation of matter-antimatter asymmetry.

Still, the minimal, single Higgs scheme does not pass the test of predicting the correct n_B/S Namely, nonvanishing $r - \bar{r}$ appears only at the three loop level[39] and for ordinary quarks one gets a hopelessly small baryon number $n_B/S \leq 10^{-18}$. A way out is to postulate the existence of a rather heavy quark Q ($m_Q \simeq 100$ GeV), whose Yukawa couplings would not be small and so even the three-loop contribution would be nonnegligible.[40] Alternatively, one could imagine two $\underline{5}$'s of SU(5), in which case the one-loop diagram shown in Fig. 4 gives a nonvanishing contribution.[41]

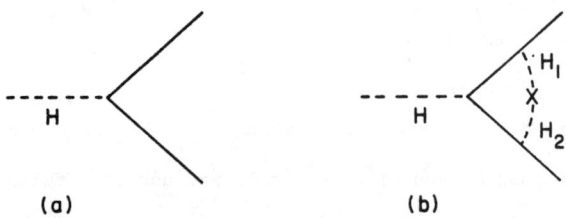

Fig. 4. Nonvanishing $r - \bar{r}$ which results from interference of the tree level graph a) and one-loop b).

Estimates give

$$r - \bar{r} \simeq \frac{1}{16\pi^2} h^2(m_X) \tag{4.14}$$

or

$$\frac{n_B}{S} \simeq (10^{-3} - 10^{-5}) h^2(m_X)$$

where $h(m_X)$ are the Yukawa couplings at $T = m_X$, which can be determined from $h(0)$ by the use of renormalization group equations. Yukawa couplings in gauge theories are asymptotically free and so they get smaller at high T. Roughly, for $h(0) \simeq 10^{-2}$ (b,t quarks), we should use $h(m_X) \simeq 10^{-3} - 10^{-2}$. Therefore,

$$\left(\frac{n_B}{S}\right)_{th} \simeq 10^{-11} - 10^{-7} \tag{4.15}$$

a value which does not contradict observation.

(ii) <u>Soft CP Violation</u>

If, as in our example discussed before, the symmetry gets always restored for $T > T_c \simeq 300$ GeV, then at high T, CP would become a good symmetry, since the underlying Lagrangian is CP conserving. That in turn implies $n_B/S = 0$. Do the global observations of the universe rule out spontaneous symmetry breaking as a mechanism for generating CP nonconservation? The answer, as we shall readily demonstrate, is no!

The essential point is that symmetry is not necessarily restored.[42] About a year ago, Mohapatra and myself[9] have redone the analysis of high temperature behavior of gauge theories, motivated by the desire to have a realistic model of CP nonconservation at $T \simeq 10^{15}$ GeV. It turns out that in the models with more complex Higgs sector, not all the vacuum expectation values vanish for $T > T_c$. We present a simple example of a model with two Higgs scalars, with a potential invariant under $\phi_i \to -\phi_i$,

$$V(T) = \tfrac{1}{2}(-\mu_1^2 + c_1 T^2)\phi_1^2 + \tfrac{1}{2}(-\mu_2^2 + c_2 T^2)\phi_2^2 +$$

$$+ \frac{\lambda_1}{4} \phi_1^4 + \frac{\lambda_2}{4} \phi_2^4 + \frac{\lambda_3}{2} \phi_1^2 \phi_2^2 \tag{4.16}$$

where

$$C_1 = \frac{1}{24} (3\lambda_1 + \lambda_3)$$

$$C_2 = \frac{1}{24} (3\lambda_2 + \lambda_3) \tag{4.17}$$

The positivity conditions at zero temperature, for the case of symmetry breaking $\langle\phi_1\rangle \neq 0 \neq \langle\phi_2\rangle$ are

$$\lambda_1, \lambda_2 > 0 \qquad \lambda_1\lambda_2 - \lambda_3^2 > 0 \tag{4.18}$$

It is clear from (4.17) then, that in the range of parameters

$$3\lambda_2 + \lambda_3 < 0, \quad \lambda_3 < 0$$
$$\lambda_1\lambda_2 - \lambda_3^2 > 0 \tag{4.19}$$

$C_1 > 0$ and $C_2 < 0$. Therefore, the temperature dependent mass term for the ϕ_2 field is negative and for the $T > T_c \simeq \sqrt{\mu_1^2/c}$

$$\langle\phi_1\rangle = 0, \quad \langle\phi_2\rangle^2 = \frac{1}{\lambda_2} (CT^2 - \mu_2^2) \tag{4.20}$$

As we promised, the symmetry remains partially broken. Along the same line one can easily construct an $SU(2)_L \times U(1)$ model with the two Higgs doublets ϕ_1 and ϕ_2, so that at high T $\langle\phi_1\rangle = 0$, $\langle\phi_2\rangle \neq 0$ and so the $SU(2) \times U(1)$ symmetry does not get restored at all. Since the gauge meson contribution to C_i terms, define in (4.16), is always positive, in the case of $SU(2) \times U(1)$ model we need some Higgs self-couplings to be not only negative, but also greater than g^2 in order to ensure that one of the C_i's is negative (see Ref. 9). We display below the t different phases, both which amount to the broken $SU(2) \times U(1)$ symmetry.

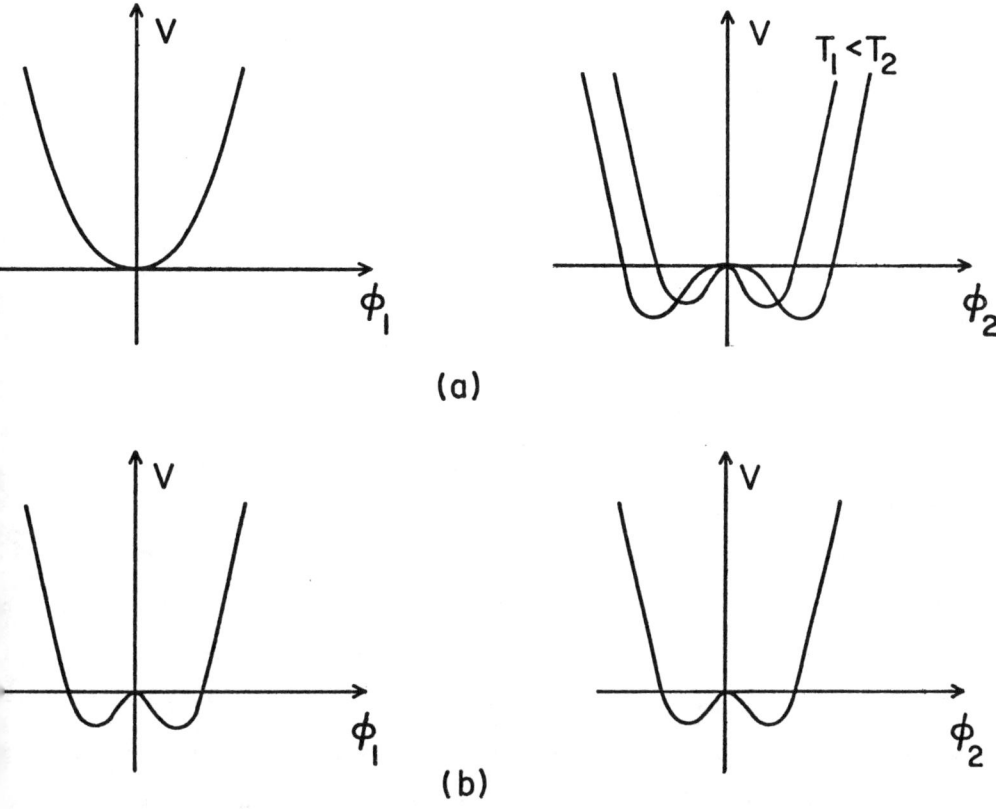

Fig. 5. High T (a) and low T (b) patterns of symmetry breaking in the model discussed above.

Of course, what we are after is CP breaking at high T. Then two Higgs doublets are not enough, since only one of vacuum expectation values remain broken and so by an SU(2) × U(1) rotation it can be made real. Hence the corollary: for soft CP broken at high T we need at least three Higgs doublets.

A realistic such grandunified theory in which CP is broken at $T \sim m_X$ is then easy to construct.[9] It is based on the SU(5) theory with the three sets of $\underline{5}$ dimensional Higgs multiplets. At $T \gg T_c \simeq 300$ GeV one can achieve the following pattern of symmetry breaking

$$<\phi_1> = \begin{pmatrix} 0 \\ 0 \\ 0 \\ 0 \\ v_1 \end{pmatrix}, \quad <\phi_2> = \begin{pmatrix} 0 \\ 0 \\ 0 \\ 0 \\ v_2 e^{i\alpha} \end{pmatrix}, \quad <\phi_3> = 0 \qquad (4.21)$$

where $v_i \alpha T$. As displayed in (4.5) quark mass matrices will be nonvanishing and complex, which induces complex Cabbibo rotation for six flavors. Similarly, the Higgs boson mass matrices will be complex, and as a result CP will be broken both in gauge meson and Higgs boson interactions with quarks and leptons. In principl both the superheavy gauge meson and Higgs boson decays will be inducing the baryo asymmetry. But first, it is important to make sure that quarks and leptons, whos mass depends on temperature are light enough so that the decays are possible. From

$$m_f(T) \simeq h(T) \, v(T) \qquad (4.22)$$

and using

$$v(T) \simeq \sqrt{c/\lambda} \, T \simeq T \qquad (4.23)$$

we obtain

$$m_f(T) \simeq h(T) \, T \qquad (4.24)$$

Now, the baryon asymmetry develops when T drops much below m_X so that t inverse decay is blocked, say at $T \simeq \frac{m_X}{10}$. Then

$$m_f \ (\frac{1}{10} m_X) \simeq 1/10 \ h(T) m_X \stackrel{<}{\sim} (10^{-3} - 10^{-4}) m_X \qquad (4.25)$$

Clearly, the phase space is enormous and so the previous analysis[4,39,40,41] which assumes massless fermions applies. It was shown[43] that the leading graphs which induce the baryon asymmetry are the following

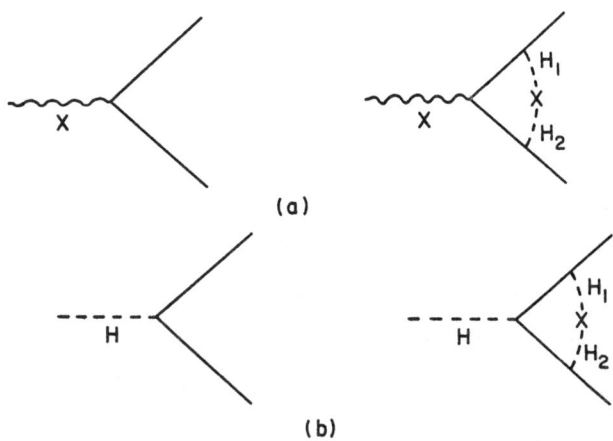

Fig. 6. Gauge meson (a) and Higgs boson (b) baryon number violating decays.

Except for a suppressed mixing between H_1 and H_2, which is of order $T^2/m_X^2 \simeq (10^{-1} - 10^{-2})$, the rest of the computation follows the conventional two Higgs model with the results given in (4.14) and (4.15). Therefore, we predict

$$\left(\frac{n_B}{s}\right)_{\text{soft CP}} \simeq 10^{-8} - 10^{-13} \qquad (4.26)$$

which is in agreement with observation.

In summary, due to increased symmetry breaking at high temperature soft CP violation in a large class of models doe remain at $T \simeq m_X$ as to ensure the non-vanishing dynamically originated baryon number. In terms of actual quantative prediction, we can say the same as for the other models of CP violation discussed before: they do not obviously fail. We clearly need a specific theory at our hands to be able to discuss these questions more precisely. As far as the discrimination between the various mechanisms of CP nonconservation, i.e., hard and soft, we have to patiently await future and more precise, low energy measurements.

One can wonder, in the context of soft CP models, what happens when the temperature increases beyond 10^{15} GeV. Well, it turns out that we cannot give a definite answer. Namely, ϕ^4 couplings are not asymptotically free and so the increase with temperature. Since at T = 0, at least some $\lambda_i \gtrsim g^2$, we expect $\lambda(T=m_X)/4\pi \simeq 1$, so that perturbation theory breaks down. It is hard to estimat precisely at what T it happens, but roughly $T \simeq 10^{17} - 10^{18}$ Gev. In other word we cannot say whether symmetry gets restored or remains broken. Of course, at $T = M_p$ quantum gravitational effects would start to play an important role, co fusing the situation even more.

An amusing comment can be made assuming that the symmetry does get restore at $T \gg m_X$. By the anology with ferromagnets one would expect domains with di ferent signs of CP phase to be formed as the universe cools down through the p transition. From (4.4) we see that both α and $-\alpha$ minimize the potential and s $(r-\bar{r})\alpha \pm \delta$, that would mean that universe consists of domains of matter and an matter.[44] This picture, in order to work, has to solve the questions of domai sizes and their separation, which may not be impossible in the case of the fir order phase transition. More work is called for. We should emphasize, howeve that symmetry may not be restored in which case the universe should be solely filled with matter.

Before closing, we would like briefly to mention the question of what get created in the early universe: matter or antimatter? At the first glance, th looks as a matter of semantics: whatever asymmetry develops, we can call it "matter". However, the situation is not completely arbitrary, and the "matter" is usually defined by the sign of CP violation in K meson decays, so what is needed is the relation between the signs of CP phases in X and K meson decays. Only if the signs are the same can we claim the fully successful understanding of the dynamical origin of the baryon number of the universe. At present the desired connection between the CP phases is still lacking.

Let me point out briefly what seems to be the problem in the case of the simplest grandunified theory, i.e., minimal SU(5) theory with a single $\underline{5}$ multiplet. At low energies, the only source of CP violation is the complex Cabbibo rotation, its complexity being the product of orginally complex Yukawa couplings. At high temperatures, Cabbibo rotation disappears since the quarks become massless and CP violation resides in Yukawa couplings only. In order to relate the CP phases in the light and the heavy section of the theory, one would need a one-to-one correspondence between the KM phase and the phases of Yukawa couplings, which, unfortunately, is obscured by the phase redefinitions used to simplify the form of KM matrix.

In my opinion, the hope of relating the phases in K and X meson decays lies in soft CP models. There, the symmetry will not be restored at high temperature, which by itself does not provide a solution. However, there is a class of models[45] where one forbids the flavor changing neutral Higgs current, which requires the existence of at least three $SU(2)_L \times U(1)$ doublets (or SU(5) $\underline{5}$'s) In such models the Cabbibo rotation becomes necessarily real[46], so that the only source of CP nonconservation, both at low and high T, are the complex Yukawa couplings. One has then to see what the connection between heavy and light Higgs couplings are. It should be done.

V. SUMMARY

Standard cosmological model and the grandunified theories offer plausible and logically consistent scenario for the dynamical development of matter-antimatter asymmetry of the universe. At the temperatures of order on 10^{16} GeV or s ($\simeq 10\ m_X$), the universe has undergone through an epoch of equilibrium which eras any previous baryon number. The present baryon asymmetry becomes therefore a ca culable quantity determined by the dynamics of the fundamental interactions pres in grandunified theories. These baryon violating forces which are responsible f the equilibrium situation were forced to go out of equilibrium as the universe grew older. Namely, as the temperature dropped below m_X, the inverse decays of superheavy bosons became more and more rare and so eventually all these boson decayed away. Due to CP violation, naturally present in these theories, the rates for X and $\bar{X}$ bosons were not the same and so the baryon asymmetry was created. Originally, there was only a tiny excess of particles over antiparticl which, when the baryons and antibaryons annihilated much later in the history o the universe, remained what we today observe as matter.

Unless there are heavy quarks (m $\simeq$ 100 GeV), the analysis of the induced baryon number tells us that the minimal SU(5) theory cannot account for the observed asymmetry. <u>We need at least two 5 Higgs multiplets</u>. It is an interes and amusing result, since our low energy phenomenology typically depends on the type of the scalar multiplets and not on their number.

Now, the predicted baryon number depends on whether the CP violation is of hard or soft origin, a question which has profound effect on our understanding of strong CP nonconservation. Hard CP violation, automatically present at high leads to $n_B/S \simeq 10^{-11} - 10^{-7}$. On the other hand, soft CP nonconservation requi theoretically perfectly acceptable symmetry restoration at high temperature in to remain at $T \simeq 10^{15}$ GeV and predicts $n_B/S \simeq 10^{-13} - 10^{-9}$. Both results do not disagree with observation. I should add that the computations are plagued by the uncertainties in the values of CP phases and of Yukawa couplings (or quark masses). At present, we just cannot make precise predictions, before all the quark masses are known.

The above picture appears appealing and plausible. However, clearly much more work remains to be done. In particular, in my opinion, there are two fundamental questions which need to be answered:

(i) What is the character of produced "matter", i.e., its sign?

(ii) Is the universe filled solely with matter or maybe there are large domains of matter and antimatter and we just happen to live in one of them? And if there are domains, how do we account for their size[47]?

References and Footnotes

1. See, for example S. Weinberg, Gravitation and Cosmology, Wiley and Sons, 1972, Chapter XV. For a more popular treat see S. Weinberg, The First Three Minutes, Basic Books, 1977.

2. A.A. Penzias and R.W. Wilson, Ap. J. 142, 419 (1965).

3. See, G. Steigman, Cargese Lectures (1971) and references therein.

4. M. Yoshimura, Phys. Rev. Lett. 41, 381 (1978); A. Yu. Ignatiev, N.V. Krasnikov, V.A. Kuzmin and A.N. Tavkhelidze, Phys. Lett. 76B, 436 (1978); S. Dimopoulos and L. Susskind, Phys. Rev. D18, 4300 (1978); D. Touissant, S. Treiman, F. Wilczek and A. Zee, Phys. Rev. D19, 1036 (1979); S. Weinberg, Phys. Rev. Lett. 42, 850 (1979); J. Ellis, M. Gaillard and D.V. Nanopoulos, Phys. Lett. 80B, 360 (1978).

5. For a review and an extensive set of references, see P. Langacker, SLAC - Pub-2544 (1980).

6. For a recent review and references, see G. Senjanović, invited talk at the XX International Conference on High Energy Physics, Wisconsin 1980 (to be published in the proceedings).

7. M. Kobayashi and T. Maskawa, Prog. Theoret. Phys. 49, 652 (1973).

8. D.A. Kirzhnitz and A.D. Linde, Phys. Lett. 42B, 471 (1972); S. Weinberg, Phys. Rev. D9, 3537 (1974); L. Dolan and R. Jackiw, Phys. Rev. D9, 2904 (1974); C. Bernard, Phys. Rev. D9, 3312 (1974). For a review, see A.D. Linde, Reports on Progress in Physics 42, 389 (1979).

9. R.N. Mohapatra and G. Senjanović, Phys. Rev. Lett. 42, 1657 (1979); Phys. Rev. D20, 3390 (1979); Phys. Lett. 89B, 57 (1979). For a summary, see Proceedings of the European Physical Society Meeting, 1979.

10. Alternative mechanism for symmetry breaking at high temperature was suggested by A.D. Linde, Phys. Rev. D14, 3345 (1976), but it requires an enormous lepton asymmetry of the universe, which we find rather unlikely.

11. R.N. Mohapatra and G. Senjanović, Phys. Rev. D21, 3470 (1980).

12. M.A.B. Bég and H.S. Tsao, Phys. Rev. Lett. 41, 278 (1978); R.N. Mohapatra and G. Senjanović, Phys. Lett. 79B, 283 (1978); H. Georgi, Hadr. J. 1, 155 (1978).

13. See, for example, Refs. 1 and 3.

14. S. Weinberg, in Lectures in Particles and Fields, ed. S. Deser and K. Ford, (Prentice-Hall, Englewood Cliffs, NJ 1964); A.D. Sakharov, Pis'ma Zh. Eksp. Teor. Fiz. 5, 32 (1967).

15. J.C. Pati and A. Salam, Phys. Rev. D10, 275 (1974); H. Georgi and S.L. Glashow, Phys. Rev. Lett. 32, 438 (1974). Se also Ref. 5 for further references.

16. H. Georgi, H. Quinn and S. Weinberg, Phys. Rev. Lett. 33, 451 (1974).

17. For a precise analysis, see W. Marciano, talk at this conference and references therein.

18. This requires rather subtle analysis of the thermalization of the asymmetry. See, S. Dimopoulos and L. Susskind, Phys. Lett. 81B, 416 (1979) and S. Treiman and F. Wilczek, Phys. Lett. 95B, 222 (1980).

19. For a detailed analysis, see Ref. 4.

20. D.V. Nanopoulos and S. Weinberg, Phys. Rev. D20, 2484 (1979).

21. S. Weinberg, Ref. 4.

22. T.D. Lee, Phys. Rev. D8, 1226 (1973); Phys. Reports C9, 148 (1974).

23. L. Wolfenstein, Phys. Rev. Lett. 13, 562 (1964).

24. S. Weinberg, Phys. Rev. Lett. 31, 494 (1973).

25. G. 't Hooft, Phys. Rev. D14, 3432 (1976); R. Jackiw and C. Rebbi, Phys. Rev. Lett. 37, 172 (1976); C. Callan, R. Dashen and D. Gross, Phys. Lett. 63B, 334 (1976).

26. N.F. Ramsey, W. Dress, P. Miller, P. Perrin and S. Pendlebury, Phys. Rev. D15, 9 (1977); V. Lobashev, et al., Leningrad Institute of Nuclear Physic preprint (1978).

27. V. Baluni, Phys. Rev. D19, 2227 (1979); R. Crewther, P. Di Vecchia, G. Veneziano and E. Witten, Phys. Lett. 88B, 123 (1979).

28. J. Ellis and M.K. Gaillard, Phys. Lett. (1978).

29. R. Peccei and H. Quinn, Phys. Rev. D16, 1791 (1977).

30. S. Weinberg, Phys. Rev. Lett. 40, 223 (1978); F. Wilczek, ibid. 40, 279 (1978).

31. See a talk by H. Tye at this Conference and references therein.

32. See Ref. 12. Also, see G. Segrè and A. Weldon, Phys. Rev. Lett. 42, 1181 (1979); S. Barr and P. Langacher, Phys. Rev. Lett. 42, 1659 (1979). For alternative attempts in the context of technicolor gauge theories, see E. Eichten, K. Lane and J. Preskill, Phys. Rev. Lett. 45, 225 (1980).

33. Bég and Tsao; Mohapatra and Senjanović, Ref. 12.

34. J.C. Pati and A. Salam, Phys. Rev. D10, 275 (1974); R.N. Mohapatra and J.C. Pati, Phys. Rev. D11, 566; 2588 (1975); G. Senjanović and R.N. Mohap Phys. Rev. D12, 1502 (1975).

35. M.A.B. Bég, R.V. Budny, R.N. Mohapatra and A. Sirlin, Phys. Rev. Lett. 38 1252 (1977).

36. For the possibility of baryon creation at much lower temperature, A. Masi and R.N. Mohapatra, Max-Planck preprint (1980).

37. A.D. Linde, Phys. Lett. 70B, 306 (1977). See also A. Guth and E. Weinber Phys. Rev. Lett. 45, 1131 (1980) and M. Sher, Phys. Rev. D22, 2989 (1980)

38. S. Coleman and E. Weinberg, Phys. Rev. $\underline{D7}$, 1888 (1973).

39. S. Barr, G. Segrè and A. Weldon, Phys. Rev. $\underline{D20}$, 2494 (1979); D.V. Nanopoulos and S. Weinberg, Ref. 20; A. Yildiz and P. Cox, Phys. Rev. $\underline{D21}$, 906 (1980), T. Yanagida and M. Yoshimura, Nucl. Phys. $\underline{B168}$, 534 (1980); M. Yoshimura, Tohoku preprint 79/143 (1979) and Phys. Lett. $\underline{88B}$, 294 (1979).

40. G. Segrè and M. Turner, Enrico Fermi preprint (1980).

41. See. Ref. 39. For more complete analysis of induced baryon asymmetry in general see E.W. Kolb and S. Wolfram, Phys. Lett. $\underline{91B}$, 217 (1980) and Caltech preprint (1979); J. Fry, K. Olive and M. Turner, Enrico Fermi preprint (1980).

42. See Refs. 9 and 10. An example of partial symmetry nonrestoration at high temperature was designed by Coleman and reported by Weinberg, Ref. 8.

43. R.N. Mohapatra and G. Senjanović, Phys. Rev. $\underline{D21}$, 3470 (1980).

44. R.W. Brown and F. Stecker, Phys. Rev. Lett. $\underline{43}$, 315 (1979); G. Senjanović and F. Stecker, Phys. Lett. $\underline{96B}$, 285 (1980) and references therein. The domain picture becomes a must for the conventional case of symmetry restoration, but with the superheavy Higgs sector responsible for CP breaking at $T \simeq m_X$, as in J.A. Harvey, P. Ramond and D.B. Reiss, Caltech preprint (1980).

45. S. Weinberg, Phys. Rev. Lett. $\underline{37}$, 657 (1976).

46. G.C. Branco, Phys. Rev. Lett. $\underline{44}$, 504 (1980) and Phys. Rev. $\underline{D22}$, 2901 (1980); A.A. Anselm and N.G. Ureltsev, Sov. J. Nucl. Phys. $\underline{30}$, 240 (1979).

47. Recently, K. Sato, Nordita preprint - 80/31 has argued that the problem may be resolved of the phase transition of a vacuum is of the first order. This solution requires some unnatural assumptions and I believe that more work is in order.

COMMENTATOR

BIG BANG BARYOSYNTHESIS AND GRAND UNIFICATION

Michael S. Turner
Astronomy and Astrophysics Center
The University of Chicago, Chicago, Il. 60637
and
Institute for Theoretical Physics
University of California, Santa Barbara, Ca. 93106

ABSTRACT

The universe possesses an overt matter-antimatter asymmetry, quantified as the baryon-to-specific entropy ratio, $kn_B/s \simeq 10^{-11 \pm 1}$. An attractive explanation is that this ratio evolved dynamically due to B, C, and CP violating interactions during a period of non-equilibrium in the GUT epoch ($t \sim 10^{-35}$s). The out-of-equilibrium decay scenario is reviewed and detailed numerical results are discussed. A model for the requisite CP violation within the minimal SU(5) theory (one 5 and one 24 representations of Higgs) is presented. Baryosynthesis is discussed in detail for a generic superheavy boson, and a constraint on low energy unification schemes is obtained.

INTRODUCTION

Although the laws of physics are very nearly matter-antimatter symmetrical (the only observed violation being in the K°-$\bar{K}^\circ$ system, ref. 1), the universe possesses an overt matter-antimatter asymmetry.[2] Within our galaxy antimatter/matter $\lesssim 10^{-4}$ (cosmic ray data), and all evidence indicates that on scales as large as the Virgo cluster ($\sim$ 1000 galaxies) there is negligible antimatter.[2] In addition, baryons account for only a tiny fraction of all the particles in the universe--their ratio to 3K microwave photons being, $n_b/n_\gamma \simeq 10^{-10 \pm 1}$ (ref. 3). These two observations deal a death blow to conventional symmetrical cosmologies. A baryon symmetrical big bang cosmology predicts baryon- and antibaryon-to-photon ratios of,

$n_b/n_\gamma \equiv n_{\bar{b}}/n_\gamma \simeq 10^{-18}$ --owing to the incompleteness of annihiliations (when $kT \lesssim 20$MeV baryons and antibaryons can no longer "find each other" and annihilations cease).

If one wishes to avoid the "annihilation catastrophe," matter and antimatter must be separated when $kT \gtrsim 20$MeV ($t \lesssim 10^{-3}$s). In the standard Friedmann-Robertson-Walker (F-R-W) model the distance over which light signals (and hence causal effects) could have propagated since $t = 0$ is finite, and $\simeq 2ct$. When $kT \simeq 20$MeV, causally connected regions encompassed only $\simeq 10^{-6} M_\odot$ of baryons (the mass of our galaxy is $\simeq 10^{11} M_\odot$); therefore causal processes could

ISSN:0094-243X/81/720224-20$1.50 Copyright 1981 American Institute of Physics

only separate out 10^{-6} $M_\odot$ chunks of matter and antimatter. Alternatively, one might invoke statistical fluctuations to explain the local baryon excess. Consider the comoving volume which today contains our galaxy; within it there are $\gtrsim 10^{68}$ baryons and $\gtrsim 10^{78}$ photons. When the temperature of the universe was $\gtrsim 1$ GeV ($t \lesssim 10^{-6}$s) baryons, antibaryons, and photons were roughly equally abundant $n_b \sim n_{\bar{b}} \sim n_\gamma \sim 10^{78}$ (maintained by $b + \bar{b} \rightleftarrows \gamma + \gamma$, etc.). To avoid the annihilation catastrophe $n_b - n_{\bar{b}}$ must have been $\sim 10^{68}$; however, Poisson fluctuations $\sim n^{1/2} \sim 10^{39}$! There is no simple way to reconcile present observations with a conventional baryon symmetrical cosmology.

The two observations, $n_b \gg n_{\bar{b}}$ and $n_b/n_\gamma \simeq 10^{-10\pm 1}$, can be combined in the baryon number-to-specific entropy ratio, $kn_B/s \simeq 10^{-11\pm 1}$ (s/k $\sim$ number of relativistic particles--today the γ and ν backgrounds). Whenever baryon number is conserved and the expansion is adiabatic this ratio remains constant. This is in contrast to the baryon number-to-photon ratio which changes every time a relativistic species annihilates (e.g., $e^\pm$ pairs when $kT \simeq 1/2$ MeV). kn_B/s could change due to baryon nonconserving interactions or entropy production during a phase transition (e.g., a first-order phase transition associated with spontaneous symmetry breaking). Although the asymmetry today is maximal, $kn_B/s \simeq 10^{-11}$ tells us that when $kT \gtrsim 1$ GeV ($n_b \sim n_{\bar{b}} \sim n_\gamma$ and s/k $\sim Nn_\gamma$ where $N \sim 10$ is the number of relativistic species present), $(n_b - n_{\bar{b}})/n_b \simeq 10^{-10}$--the fractional asymmetry was tiny!

Until a few years ago it was necessary to postulate $kn_B/s \simeq 10^{-11}$ as an initial condition--a rather peculiar one at that. However, as most of you are aware, in the past few years very attractive dynamical explanations involving GUTs have been put forth.[4-6] Baryosynthesis is probably the most exciting development in theoretical cosmology in the past decade--of comparable importance to big bang nucleosynthesis. [The primordial nucleosynthesis calculations done in the 1960s resolved an analogous conundrum--the observed abundances of several light isotopes, especially the large mass fraction of ^{4}He ($\approx 25\%$).[7]] Qualitatively the idea is very simple: At the planck time ($t_p \sim 10^{-43}$ s) the universe begins as a hot, baryon symmetrical "soup" of fundamental particles, and $kn_B/s \equiv 0$. [Earlier than t_p quantum corrections to general relativity are probably important, so that t_p marks the beginning of classical cosmology.] At the end of the GUT epoch ($t \sim 10^{-35}$ s, $kT \sim 10^{15}$ GeV) an asymmetry of $kn_B/s \simeq 10^{-11}$ evolves due to

particle interactions. Finally, when kT falls below $\sim$ 1 GeV
(t $\sim 10^{-6}$ s) baryon-antibaryon pair production is no longer energetically possible, and essentially <u>all</u> the antibaryons are annihilated leaving the $\sim$ one baryon per 10^{10} photons that we see today.

It should be clear that three ingredients are necessary for baryosynthesis: (i) B nonconserving interactions--which of course are predicted by GUTs (for an excellent review see ref. 8). (ii) C and CP violations, an arrow to specify that a matter excess should be produced--such violations are seen in the K°-$\overline{K}^{\circ}$ system, and can be accommodated by GUTs. (iii) A departure from thermal equilibrium; some particle species must have a non-equilibrium distribution. When all particle species have equilibrium distributions CPT and unitarity guarantee that $n_b \equiv n_{\overline{b}}$ (ref. 6). A simple heuristic argument goes as follows: equilibrium distributions are of the form, $n(\underset{\sim}{p})$ = $[\exp(E/kT) \pm 1]^{-1}$, where $E^2 = p^2 + m^2$ (this follows from unitarity and CPT). CPT guarantees that $\tilde{m}_b \equiv m_{\overline{b}}$, and therefore $n_b \equiv n_{\overline{b}}$.

In the remainder of my talk I will review the present status of baryosynthesis. First, I will discuss qualitatively the GUTs scenario which incorporates all three ingredients, the so-called out-of-equilibrium decay scenario of Weinberg and Wilczek.[5] Then I will review the detailed numerical calculations which indicate that the qualitative picture is correct, and in fact is even more efficient than was originally expected.[9-11] Baryosynthesis depends crucially upon the details of C and CP violation, and so I will address this issue and present a model for C and CP violation in the minimal SU(5) theory (one 5 and one 24 of Higgs). This scheme produces an asymmetry of the proper magnitude only if there exists at least one more generation of fermions which are massive (M $\sim$ 30 - 200 GeV). Finally, I will discuss a simplified set of "generic" Boltzmann equations for baryosynthesis in order to: (i) illustrate in detail how the process of baryon generation works, and (ii) put constraints on low energy unification schemes which produce the baryon asymmetry.

THE OUT-OF-EQUILIBRIUM DECAY SCENARIO

Let me begin by introducing the quantities of interest. In the standard Friedmann-Robertson-Walker (F-R-W) model of the universe the expansion scale factor R(t) is governed by ($\hbar = k_B = c = 1$)

$$H^2 \equiv (\dot{R}/R)^2 = 8\pi G\rho/3 = \frac{4\pi^3}{45} g_*(T) T^4/m_p^2 \qquad (1)$$

when the energy density of the universe is dominated by relativistic particles. $g_*(T)$ counts the total effective number of degrees of freedom of the relativistic species ($\equiv \sum_{bosons} g + 7/8 \sum_{fermions} g$, $g_*(10^{15}$ GeV$) \gtrsim 100$), and $m_p = G^{-1/2} \simeq 1.2 \times 10^{19}$ GeV. Since $T \propto R^{-1}$ and $R \propto t^{1/2}$, $|\dot{T}/T| = (\dot{R}/R) = (2t)^{-1}$. Thus the time rate of change of the temperature which sets the time scale for how rapidly particle distributions must change to keep pace with the changing temperatures is,

$$H = |\dot{T}/T| = (2t)^{-1} \simeq 1.66 g_*^{1/2} T^2/m_p. \qquad (2)$$

I will design as S the superheavy ($\gtrsim 10^{14}$ GeV) boson whose interactions violate B conservation--either a Higgs or gauge boson. Let its coupling to fermions and mass be α and M respectively. From dimensional considerations its decay rate ($\Gamma_D \sim \tau_S^{-1}$),

$$\Gamma_D \sim \alpha M. \qquad (3)$$

At the planck time the universe is assumed to be baryon symmetrical ($kn_B/s = 0$) with all the fundamental particle species present in equilibrium distributions. The temperature $T \simeq g_*^{-1/4} m_p \gg M$, so that S and $\bar{S}$ bosons are highly relativistic and as abundant (up to statistical factors) as photons. Nothing of interest occurs until T falls to $\lesssim M$. For $T \lesssim M$, the equilibrium number abundance of S, $\bar{S}$ bosons relative to photons is

$$N_S \simeq (M/T)^{3/2} \exp(-M/T). \qquad (4)$$

[Equivalently, N_S can be thought of as the number of S bosons per comoving volume.] In order for S bosons to maintain an equilibrium abundance, they must be able to diminish in number rapidly compared to $H \sim t^{-1} \sim |\dot{T}/T|$. The most important process in this regard is decay; other processes (e.g., annihilation) are higher order in α. If $\Gamma_D \gg H$ (for $T \simeq M$), then S bosons can adjust their number density rapidly enough so that it "tracks" the equilibrium value. Thermal equilibrium is maintained and thus no asymmetry evolves.

More interesting is the case where $\Gamma_D < H$ (for $T \simeq M$), or equivalently $M > g_*^{-1/2} \alpha\, 10^{19}$ GeV. In this case S, $\bar{S}$ bosons are not decaying on the expansion timescale, and so remain as abundant as photons ($N_S \simeq 1$) for $T \lesssim M$ (i.e., overabundant relative to their equilibrium number). Their overabundance provides the departure from thermal equilibrium. Much later ($T \ll M$) $\Gamma_D \gtrsim H \sim t^{-1}$ (t becomes greater than τ_S), and to a good approximation they decay freely, inverse decays being energetically forbidden since typical fermions have energies $\sim T \ll M$. The time evolution of their abundance is summarized in Figure 1.

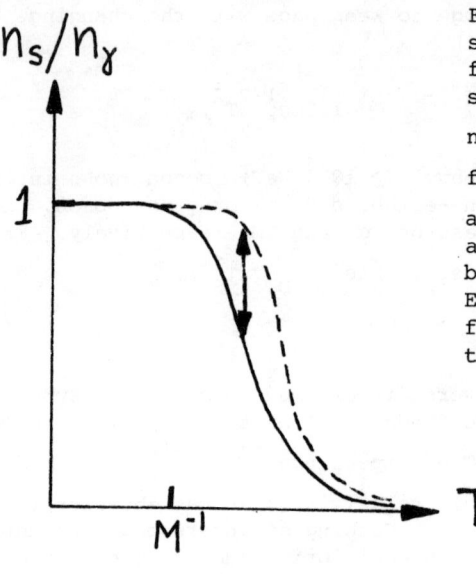

Figure 1.—The abundance of S bosons relative to photons as a function of T. The solid curve shows the equilibrium value of n_s/n_γ, which $\sim (M/T)^{3/2} \exp(-M/T)$ for $T \lesssim M$. If $\Gamma_D < H$ S bosons are overabundant for $T \lesssim M$ -- a departure from thermal equilibrium (indicated by the arrow). Eventually when $\Gamma_D \simeq H$ they decay freely. The broken curve shows the actual abundance of S bosons.

Consider the decay of S and $\bar{S}$ bosons: suppose S decays to channels 1 and 2 with baryon numbers B_1 and B_2, and branching ratios r and $1-r$. The corresponding quantities for the $\bar{S}$ are $-B_1$, $-B_2$, $\bar{r}$, and $1-\bar{r}$.

Then the mean net baryon number of the decay products of the S and $\bar{S}$ are

$$B_S = rB_1 + (1-r)B_2,$$
$$B_{\bar{S}} = -\bar{r}B_1 - (1-\bar{r})B_2, \quad (5)$$

respectively. Thus the decay of an S, $\bar{S}$ pair produces (on average) a baryon number ε,

$$\varepsilon \equiv B_S + B_{\bar{S}} = (r - \bar{r})(B_1 - B_2). \quad (6)$$

If both C and CP are violated, then $r \neq \bar{r}$; $\varepsilon \sim (r-\bar{r})$ measures the C and CP violations.

When the S, $\bar{S}$ bosons finally decay ($T \ll M$, $t \sim \tau_S$) they are still roughly as abundant as photons, so that the total baryon number produced is $n_B \simeq \varepsilon n_\gamma$. The specific entropy $s/k \sim$ number of relativistic particles present, which $\sim g_* n_\gamma$. Thus the baryon asymmetry generated is

$$kn_B/s \simeq \varepsilon g_*^{-1} \sim 10^{-2}\varepsilon \tag{7}$$

since $g_*(10^{15}\text{ GeV}) \gtrsim 100$. Recall that the condition for a departure from equilibrium to occur is: $\Gamma_D < H$ ($T \simeq M$) or $M > g_*^{-1/2} \alpha\, 10^{19}$ GeV. For a gauge boson $\alpha \simeq 1/45$, so that M must be $\gtrsim 10^{16}$ GeV; for a Higgs boson $\alpha \sim 10^{-4} - 10^{-6}$, so that M must be $\gtrsim 10^{12} - 10^{14}$ GeV. If M is $\gtrsim$ the "critical value," only a modest CP violation, $\varepsilon \sim 10^{-9}$, is required to "explain" $kn_B/s \simeq 10^{-11}$. As we shall see later, even when $M \lesssim$ the "critical value," an asymmetry of interesting magnitude evolves. To summarize, the key feature of the out-of-equilibrium decay scenario is the overabundance of S, $\bar{S}$ bosons (for $T \lesssim M$). This occurs when no interaction is rapid enough ($\Gamma > H$) to allow these bosons to diminish in number and "track" their equilibrium abundance as they become non-relativistic.

DETAILED NUMERICAL CALCULATIONS

Two groups have developed numerical codes to follow the evolution of the baryon asymmetry in detail.[9-11] The codes integrate the relevant Boltzmann equations with the following assumptions being made.
- The universe is described by the standard, radiation dominated hot big bang model (F-R-W cosmology).
- B conserving interactions are assumed to be happening rapidly enough ($\Gamma \gg H$) so that the distribution of all particle species in momentum space is thermal, i.e., kinetic equilibrium (for $T \lesssim 10^{16}$ GeV this is a reasonable assumption; see, e.g., ref. 12).
- Interactions included in the network: decays, inverse decays, and annihilations of superheavy bosons; B nonconserving scatterings mediated by superheavy bosons.
- GUTs are used to compute the required matrix elements. Thus far SU(5)[9,11] and SO(10)[11] have been used.

The two groups reach essentially the same conclusions. The results for a specific theory can be fairly complicated and not particularly illustrative. Therefore, I will restrict myself to discussing the effect of a single gauge or Higgs species on the evolution of kn_B/s.

It is very useful to define the quantity K which describes the "effectiveness" of superheavy boson decays (and as I shall discuss later all the interactions involving the superheavy boson),[9]

$$K \simeq (\Gamma_D/H)\bigg|_{T=M} \simeq \frac{3\times 10^{17}\alpha \text{ GeV}}{M}, \tag{7}$$

where the numerical value is for the minimal SU(5) theory. When $K < 1$, decays are not occurring rapidly on the expansion timescale at the critical epoch ($T \simeq M$), and a departure from equilibrium is expected. On the other hand, for $K > 1$, they are occurring rapidly on the expansion timescale, and one expects equilibrium to be maintained to some extent.

In Figures 2a and 2b the final value of kn_B/s produced by a single gauge (2a) or Higgs (2b) species is shown as a function of K. The results scale with ε, the CP violation parameter; note, in these figures $\varepsilon/2$ = mean net baryon number produced by the decay of an S, $\bar{S}$ pair. There are several interesting features to observe. First, for $K \lesssim 1$ the asymmetry which evolves is roughly independent of K and equal to the "saturation" value, $kn_B/s \sim 10^{-2}\varepsilon$--in agreement with the qualitative discussion in the previous section.

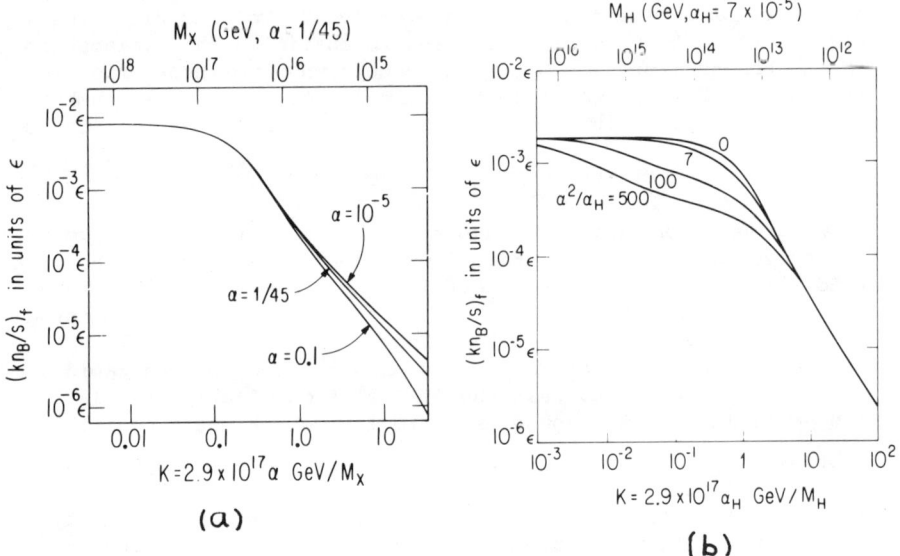

(a)

(b)

Figure 2.—(a) The final value of kn_B/s produced by a gauge boson in units of ε as a function of α and K. For fixed K there is negligible dependence on α. A mass scale is shown for $\alpha = 1/45$. (b) Same as (a) for a Higgs boson. The dependence on α^2/α_H shows the small effect of Higgs annihilations on baryon production (from ref. 9).

For $K > 1$, one might have expected equilibrium to be maintained (i.e., S bosons present in equilibrium numbers), and hence that no asymmetry would evolve--this is certainly not the case (cf. Figures 2a and 2b). For $K \gtrsim 1$, baryosynthesis decreases only as $K^{-1.3}$ (roughly). The reason for this is that equilibrium is difficult to maintain since the temperature of the universe is always changing. This will be discussed in more detail in the section after next. Note that a gauge boson of mass $\sim 5 \times 10^{14}$ GeV (the inferred value for SU(5), ref. 13) can produce the observed $kn_B/s \simeq 10^{-11}$ if $\varepsilon \simeq 3 \times 10^{-5}$. A Higgs boson of mass $\sim 10^{14}$ GeV and coupling $\alpha_H \sim 10^{-4}$ can produce the required asymmetry if $\varepsilon \simeq 10^{-8}$.

Gauge and Higgs bosons can also damp pre-existing asymmetries (i.e., those present prior to $t_{GUT} \sim 10^{-35}$ s). The damping is primarily by the two-step process, e.g., $qq \to \bar{S}$, $\bar{S} \to \bar{q}\,\bar{\ell}$ ($\Delta B = -1$), and the damping factor, (final asymmetry)/(initial asymmetry), $\sim \exp(-AK)$, where $A \sim O(1)$. This is discussed in more detail in ref. 9.

THE CP VIOLATION ε

Since the asymmetry which evolves scales with ε, questions such as: Can ε be calculated?, What is the sign of ε?, and Can ε be related to quantities measured in the K°-$\bar{K}^\circ$ system? naturally arise. The most general statement which can be made is that ε must be $\lesssim O(\alpha)$, where α is the relevant coupling constant.[14] Simply stated, this is because at the tree level, superheavy boson decays are CP conserving (in the Born approximation the Lagrangian is hermetic). Any C and CP violating effects arise from higher order loop diagrams. The lowest order effect results from the interference of a one loop diagram with the tree diagram. If the particle in the loop is a gauge boson, then $\alpha \sim 10^{-2}$ and $\varepsilon \lesssim 10^{-2}$; if it is a Higgs boson, then $\alpha \sim 10^{-4} - 10^{-6}$ and $\varepsilon \lesssim 10^{-4} - 10^{-6}$. One thing is clear, since kn_B/s can be as large as $10^{-2}\varepsilon \lesssim 10^{-4}$ there is plenty of room to explain $kn_B/s \simeq 10^{-11}$.

In order to be more specific let's consider the minimal SU(5) theory (one 5 and one 24 of Higgs) with the usual three generations of fermions. Both the $X\tilde{Y}$ gauge bosons, and a superheavy color triplet of Higgs, H, which is a member of the 5 representation mediate B nonconservation and hence are candidates for producing a baryon asymmetry. For the H, the first non-zero contribution to C and CP violation in its decays is the interference of the tree diagram with a three loop diagram where the particles in the loops are H bosons[15]

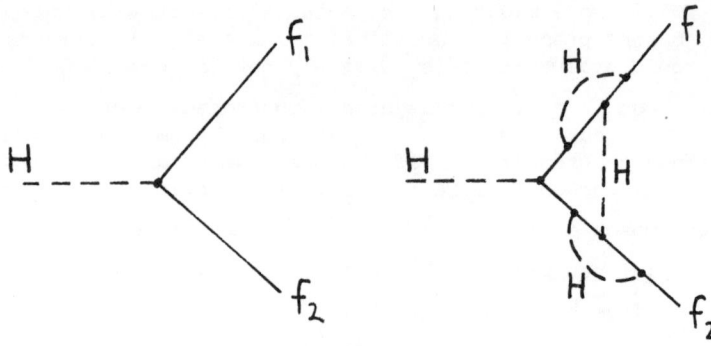

Figure 3.-The tree graph for superheavy Higgs decay and a three-loop graph. In minimal SU(5) their interference is responsible for the lowest order CP violating effects in Higgs decay.

(shown in Figure 3). One expects that $\varepsilon_H \sim 0(\alpha_H^3) \sim (m_f/m_W)^6 \alpha^3$, where m_f^6 is a specific combination of fermion masses, and m_W is the W-boson mass $\sim$ 80 GeV. [Recall that $\alpha_H \sim \alpha(m_f/m_W)^2$.] For the XY gauge bosons, the lowest order contribution results from a four loop diagram involving Higgs bosons in the loops, so that $\varepsilon_{XY} \sim 0(\alpha_H^4) \sim (m_f/m_W)^8 \alpha^4$. With the present three generations of fermions $m_f \lesssim 0$(few GeV), so that

$$\varepsilon_H \lesssim 10^{-18},$$

$$\varepsilon_{XY} \lesssim 10^{-24}$$

(8)

--far too small to be important for baryosynthesis.[15] Yildiz and Cox[16] have shown that the addition of one more 5 representation of Higgs to the minimal model allows non-zero contributions to ε at the one loop level, so that ε can be as large as $0(\alpha_H) \sim 10^{-4} - 10^{-6}$ --making baryosynthesis possible in this extended SU(5) model.

Gino Segrè and I have considered another possible solution which allows the minimal Higgs structure to be retained.[17] We introduce a fourth (and possibly a fifth) generation of heavy fermions (M $\sim$ 30 - 200 GeV). Note, the present cosmological constraints on the number of neutrino flavors permit a fourth and possibly even more generations.[18] In the four generation scheme it can be arranged such that the m_f^6 factor only involves heavy quarks (t, b', and t'), and then $\varepsilon_H \sim 10^{-9}$--just large enough to produce $kn_B/s \simeq 10^{-11}$ if the mass of the H boson $\gtrsim 10^{13}$ GeV. In this model ε_H is a function of quark masses, mixing angles, and one phase which is unmeasurable in low energy experiments (E $\lesssim$ few TeV).

In the five generation scheme if this unmeasurable phase is taken to be zero, then not only can $\varepsilon_H \sim 10^{-9}$, but ε_H is related to quark masses, mixing angles, and the CP violating phase in the reduced KM matrix for the three heaviest generations--quantities, which in principle, are all measurable in low energy experiments (E $\lesssim$ few TeV). The five generation model falls short of relating ε to the $K°-\overline{K°}$ system since the CP violation in that system involves the lightest quarks (u,d,c,s), although in this model both CP violations share a common origin--the KM mechanism. Other models for the CP violation within the SU(5) and SO(10) theories have been investigated by Harvey et al.[11]

The prospects for relating the ε of baryosynthesis to the $K°-\overline{K°}$ system seem dim. In general, the requisite CP violation in both systems involves the exchange of Higgs bosons, whose couplings $\alpha_H \sim \alpha(m_f/m_W)^2$. First and second generation quark masses are O(10 MeV - 1 GeV), so that α_H is $O(10^{-6} - 10^{-10})$. If the CP violation in the superheavy system involved the same light quarks, then even if ε were due to one loop Higgs exchange, it would be difficult to achieve $\varepsilon \sim 10^{-9}$--the minimum value required to explain $kn_B/s \simeq 10^{-11}$.

To summarize, at present the situation with regard to ε is a bit dissatisfying. About the strongest statement that can be made is: $\varepsilon \lesssim O(\alpha) \lesssim 10^{-2}$. Attempts to relate the magnitude or the sign of ε to quantities which can be measured in the laboratory have not met with success. In fact Harvey et al.[11] have discussed an SU(5) model with two 5 representations of Higgs, in which the sign of the baryon asymmetry is not even determined by the sign of ε, it also depends upon the mass ratio of the two superheavy Higgs species.

BARYOSYNTHESIS IN DETAIL

For K < 1 the numerical results[9-11] agreed with the qualitative model, i.e., S bosons remain overabundant when T $\lesssim$ M--a departure from equilibrium, and eventually (T << M) an asymmetry of $kn_B/s \simeq 10^{-2}\varepsilon$ evolves. For K > 1 one naively expects equilibrium to be maintained and no asymmetry to be produced; however the numerical results showed the efficiency of baryosynthesis fell off slowly, roughly as $K^{-1.3}$. In this section I will explain this result by discussing a simplified set of Boltzmann equations for a generic superheavy boson--simplified in the sense that I have taken the exact equations and retained only the terms which are responsible for the basic results.[19] Since K $\simeq 3 \times 10^{17} \alpha$ GeV/M [in SU(5)], as the unification energy scale decreases baryosynthesis become more difficult, and therefore, a discussion of this regime (K >> 1) naturally leads to addressing the question of what constraints baryosynthesis places upon unification at low energies.

I will assume that $K \gg 1$, and $z \equiv M/T \gtrsim 1$; when $z \lesssim 1$ S-bosons are relativistic and are present in equilibrium numbers—all of the "action" occurs for $z \gtrsim 1$. Recall that $K \equiv (\Gamma_{\not B}/H)$ for $T = M$ measures the "effectiveness" of B nonconserving decays of the S-bosons. Suppose that the rate of B nonconserving decays $\Gamma_{\not B} = c_D \alpha M$, c_D = a theory-dependent constant of $O(1)$, then

$$K = \frac{c_D \alpha M}{g_*^{1/2} 1.66 T^2/m_p}\bigg|_{T=M} \simeq \left(\frac{c_D \alpha}{g_*^{1/2} M}\right) 10^{19} \text{ GeV}. \tag{9}$$

From the exact Boltzmann equations we have gleaned the following simplified generic set of equations for baryosynthesis.[19] First, for

$$\Delta \equiv n_S/n_\gamma - N_{EQ}, \tag{10}$$

the departure from equilibrium, the governing equation is

$$\Delta' = -zK[\gamma_D + \gamma_A z^{-3} N_{EQ}]\Delta - N'_{EQ}. \tag{11}$$

Here n_S = number density of S, $\bar{S}$ bosons, n_γ = number density of photons, ' denotes d/dz, and N_{EQ} is the equilibrium value of n_S/n_γ which for $z \gtrsim 1$,

$$N_{EQ} \simeq (\pi/2)^{1/2} z^{3/2} e^{-z}. \tag{12}$$

The γ_D and γ_A terms represent decays--B conserving and nonconserving, and annihiliations. All the rates in (11) have been normalized to $\Gamma_{\not B}$. For example, the total decay rate $\Gamma_D \sim (\alpha_{\not B} + \alpha_B)M \sim \gamma_D \Gamma_{\not B}$, so that $\gamma_D \sim [(\alpha_B/\alpha_{\not B}) + 1]$. Of course, if S has no B conserving decay modes, then $\gamma_D \simeq 1$. The annihilation rate $\Gamma_A \sim n_S \sigma v \sim (n_S/n_\gamma) T^3 \alpha_A^2 M^{-2} \sim N_{EQ} z^{-3} \alpha_A^2 M$ when n_S/n_γ is not too far from its equilibrium value, N_{EQ}. Therefore, $\gamma_A \simeq \alpha_A^2/\alpha_{\not B}$ (α_A is the relevant coupling for annihilation, e.g., if the annihilation channel is electromagnetic $\alpha_A \sim \alpha_{EM}$).

In the absence of the N'_{EQ} term in Equation (11), any departure from equilibrium would relax exponentially, as one might naïvely expect. However, because the universe is cooling, $T \propto z^{-1}$, the equilibrium value, $N_{EQ}(z)$, is changing, and the S bosons are trying to track an equilibrium abundance which is constantly decreasing (see Figure 1). The N'_{EQ} term represents this effect. For $K \gg 1$, Equation (12) relaxes quickly to $\Delta' \simeq 0$, so that to a good approximation the solution to (12) is obtained by solving $\Delta' \simeq 0$,

$$\Delta \simeq \frac{-N'_{EQ}}{zK(\gamma_D + \gamma_A z^{-3} N_{EQ})} \simeq \frac{N_{EQ}}{zK(\gamma_D + \gamma_A z^{-3} N_{EQ})}. \quad (13)$$

Notice that as z increases annihilations become less significant compared to decays--$\gamma_A z^{-3} N_{EQ}/\gamma_D \sim (\gamma_A/\gamma_D) z^{-3/2} e^{-z}$; because they involve two S bosons they tend to "turn themselves off" as S bosons become less abundant.

The equation for

$$B \equiv (n_b - n_{\bar{b}})/n_\gamma, \quad (14)$$

the ratio of baryon number to photon number, is

$$B' = -zK(c_S \alpha_\beta z^{-5} + N_{EQ})B + zK(\varepsilon\Delta). \quad (15)$$

The $-zKN_{EQ}B$ term represents destruction of baryon number by an inverse decay followed by a decay, $q + q \to \bar{S}$, $\bar{S} \to \bar{q} + \bar{\ell}$, and so its rate is limited by the inverse decay rate $\Gamma_{ID} \sim N_{EQ} \Gamma_\beta$ (inverse decay is only possible for fermion pairs more energetic than M). The $-zKc_S \alpha_\beta z^{-5}$ term represents B destruction by scatterings mediated by S bosons, e.g., $q + q \to \bar{q} + \bar{\ell}$; $\Gamma_S \sim n_b \sigma v \sim BT^3 \times c_S \alpha_\beta^2 T^2/M^4 \sim (c_S \alpha_\beta) z^{-5} \Gamma_\beta$ (c_S is a constant of $O(1)$ which depends upon the particular theory). Finally, the $zK\varepsilon\Delta$ term represents B production by the asymmetrical decays of S, $\bar{S}$ bosons. Note that if C or CP is a good symmetry ($\varepsilon = 0$), or if equilibrium is maintained ($\Delta = 0$), then the production term vanishes--in accordance with the theorem mentioned earlier.[5,6] For $K \gg 1$, both the destruction and production terms in Equation (15) are large, so that B' remains $\simeq 0$, and therefore

$$B(z) \simeq \frac{\varepsilon\Delta}{(c_S \alpha_\beta z^{-5} + N_{EQ})} \simeq \frac{\varepsilon N_{EQ}}{zK(\gamma_D + \gamma_A z^{-3} N_{EQ})(c_S \alpha_\beta z^{-5} + N_{EQ})}. \quad (16)$$

Note, that $B(z)$ decreases with time at least as rapidly as z^{-1}. Eventually, the destruction processes become ineffective ($\Gamma_{ID} < H$, $\Gamma_S < H$), and B no longer decreases--it "freezes out."

The condition for freeze out ($\Gamma = H$ or equivalently $|B'_{des} \cdot z| = B$) defines freeze out for the two destructive processes, inverse decays and scatterings:

$$z_{ID}^{7/2} \exp(-z_{ID}) K = 1, \qquad (16a)$$

$$z_S = (c_S \alpha_B K)^{1/3} . \qquad (16b)$$

The final value of B is determined by $B(z_f)$, where $z_f = \max\{z_S, z_{ID}\}$, see Figure 4. Unfortunately (16a) is not solved easily

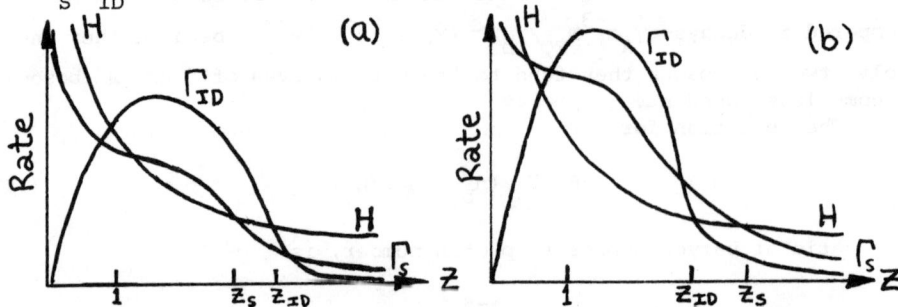

Figure 4.—Relevant rates for baryosynthesis: $H \sim z^{-2}$, $\Gamma_{ID} \sim K e^{-z}$ ($z \gtrsim 1$)--damping by inverse decay, and $\Gamma_S \sim \alpha_B K z^{-5}$ ($z \gtrsim 1$)--damping by scatterings. Freeze out occurs when $\Gamma \simeq H$. (a) For $1 \lesssim K \lesssim K_{crit}$ scatterings freeze out first. (b) For $K \gtrsim K_{crit}$ inverse decays freeze out first. The final value of $k n_B/s$ is determined by the last B destroying process to freeze out.

(asymptotically $z_{ID} \to \ln K$). However, numerically we find that for $K \gtrsim K_{crit}$ $z_S \gtrsim z_{ID}$ and for $K \lesssim K_{crit}$ $z_{ID} \gtrsim z_S$, where

$$K_{crit} \simeq 7 \times 10^3 (c_S \alpha_B)^{-5/4} . \qquad (17)$$

K_{crit} separates baryosynthesis into two qualitatively different regimes: inverse decay dominated and scattering dominated.

(a) <u>Inverse decay dominated</u>, $1 \lesssim K \lesssim K_{crit}$. In this regime inverse decays freeze out after scatterings, so that the final value of B is determined by $B(z_{ID})$ (see Figure 4),

$$B(z \gtrsim z_{ID}) \simeq \frac{\varepsilon}{z_{ID} K[\gamma_D + \gamma_A z_{ID}^{-3} N_{EQ}(z_{ID})]} \simeq \frac{\varepsilon}{z_{ID} K} , \qquad (18)$$

where I have assumed that $\gamma_D > \gamma_A z_{ID}^{-3/2} \exp(-z_{ID})$. [For $K \gtrsim 10$, $z_{ID} \gtrsim 10$ so that this condition is satisfied if $\gamma_D > 10^{-6} \gamma_A$.] A reasonable numerical fit to (18) is

$$B(z \gtrsim z_{ID}) \simeq \gamma_D^{-1} \varepsilon K^{-1.3}, \qquad (19)$$

which is the slow fall off in baryon production discussed previously. Note that for the SU(5) XY gauge bosons $c_S \simeq 1500$ and $\alpha_{\not{B}} \simeq 1/45$ so that $K_{crit} \simeq 100$.

(b) <u>Scattering dominated, $K \gtrsim K_{crit}$</u>. In this regime scatterings freeze out after the inverse decays, so that the final value of B is determined by $B(z_S)$ (see Figure 4),

$$B(z \gtrsim z_S) \simeq \varepsilon \gamma_D^{-1} (c_S \alpha_{\not{B}} K)^{5/6} \exp[-(c_S \alpha_{\not{B}} K)^{1/3}], \qquad (20)$$

where once again $\gamma_D > \gamma_A z_S^{-3/2} \exp(-z_S)$ has been assumed. When baryosynthesis is scattering dominated production falls off very rapidly, decreasing exponentially with $\alpha_{\not{B}} K$.

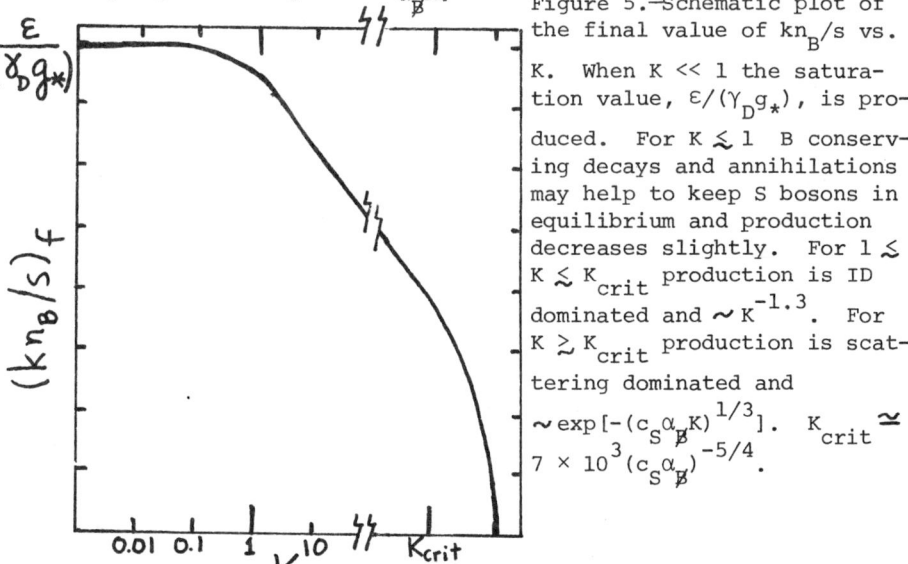

Figure 5.—Schematic plot of the final value of kn_B/s vs. K. When $K \ll 1$ the saturation value, $\varepsilon/(\gamma_D g_*)$, is produced. For $K \lesssim 1$ B conserving decays and annihilations may help to keep S bosons in equilibrium and production decreases slightly. For $1 \lesssim K \lesssim K_{crit}$ production is ID dominated and $\sim K^{-1.3}$. For $K \gtrsim K_{crit}$ production is scattering dominated and $\sim \exp[-(c_S \alpha_{\not{B}} K)^{1/3}]$. $K_{crit} \simeq 7 \times 10^3 (c_S \alpha_{\not{B}})^{-5/4}$.

Figure 5 summarizes schematically baryosynthesis by a single generic superheavy species. For $K \ll 1$, such that Γ_A and Γ_D are both $< H$ for $T \simeq M$, superheavy bosons remain overabundant and the saturation value of kn_B/s is produced,

$$kn_B/s \simeq (g_* \gamma_D)^{-1} \varepsilon, \qquad (21)$$

essentially independent of $\alpha_{\not B}$ and K. If $K \lesssim 1$, but Γ_A is not $< H$ for $T \simeq M$, then annihilations help to maintain an equilibrium abundance of S bosons to some extent, and a slightly lesser asymmetry than that given by (21) results (see ref. 19). For intermediate values of K, $1 \lesssim K \lesssim K_{crit}$, inverse decays are the dominant baryon number destroying process, and the asymmetry produced is,

$$kn_B/s \simeq (g_*\gamma_D)^{-1}\varepsilon K^{-1.3}, \qquad (22)$$

essentially independent of $\alpha_{\not B}$. Finally, for the largest values of K, $K \gtrsim K_{crit}$, S mediated scatterings are the dominant baryon number destroying process, and the asymmetry produced is,

$$kn_B/s \simeq (g_*\gamma_D)^{-1}\varepsilon(c_S\alpha_{\not B}K)^{5/6}\exp[-(c_S\alpha_{\not B}K)^{1/3}], \qquad (23)$$

depending only upon $\alpha_{\not B}K$.

Since $K \propto M^{-1}$, it is the scattering dominated regime which is of interest for constraining low energy unification schemes. For example, suppose $(g_*\gamma_D)^{-1} \sim 10^{-2}$ and $\varepsilon \lesssim 10^{-2}$ then in order to produce $kn_B/s \gtrsim 10^{-12}$, we must have

$$M \gtrsim (c_S c_D g_*^{-1/2})\alpha_{\not B}^2\, 5 \times 10^{14}\text{ GeV}, \qquad (24)$$

which for $c_S c_D g_*^{-1/2} \simeq 10^2$ and $\alpha_{\not B} \simeq 10^{-6}$ implies that $M \gtrsim 5 \times 10^4$ GeV. In general, the constraint on a low energy unification theory (very large K) is

$$10^{-12} \lesssim kn_B/s \simeq 10^{-11\pm 1} \simeq (g_*\gamma_D)^{-1}\varepsilon(c_S\alpha_{\not B}K)^{5/6}\exp[-(c_S\alpha_{\not B}K)^{1/3}]. \qquad (25)$$

SUMMARY

One of the striking predictions of grand unification is proton decay--matter is unstable. Interestingly enough, it is possible that grand unification also provides the explanation for the existence of matter in the universe in the first place, through B, C, and CP violating interactions. The highly attractive idea of baryosynthesis joins together two puzzles--the slight matter-antimatter asymmetry ($kn_B/s \simeq 10^{-11}$) and CP violation. At present we know of only one instance of CP violation--the K°-$\bar{K}^\circ$ system;[1] however, we now have strong reason to suspect (or hope?) that CP is also violated in the S-$\bar{S}$ system. As I have discussed the prospects for relating these two instances of CP violation do not seem good, although the same underlying mechanism might be responsible for both violations (e.g., ref. 17).

At present the statistical mechanics aspects of baryosynthesis are well under control, in fact the out-of-equilibrium decay scenario works better than one had expected. Given a GUT it is possible to write a network code (similar to nucleosynthesis) to follow the evolution of kn_B/s. The biggest uncertainty is in the CP violation ε, which can be anywhere in the range $0 - 10^{-2}$. A good understanding of the origin of kn_B/s (at the same level as primordial nucleosynthesis) would be a spectacular achievement for cosmology and particle physics, and would at the same time add one more fossil to the handful of fossils (e.g., ^4He, D, the μ-wave background, etc.) which remain from the epoch when the universe was the ultimate high energy physics laboratory. As I discussed we can already use $kn_B/s \simeq 10^{-11\pm1}$ to place constraints on low energy unification schemse [cf. Equations (24), (25)].

Of course there are many issues which remain to be resolved. Among these are the possibility that the GUT transition is strongly first order, the details of CP violation, and the rather naïve assumption that there exists a desert between present laboratory energies and the unification scale. As several speakers have pointed out it is perhaps a bit unrealistic to believe that unification is so simple, when our present vantage point is so far from the unification scale. However, the mechanics of baryosynthesis seem equally well suited for prequarks, preons, maons, rishons, or alphons as they are for quarks and leptons. It appears that we are now on the verge of understanding how a symmetrical primordial hot soup of particles gave birth to baryons; perhaps by the time we do understand this we can begin to ask where the primordial hot soup came from.

I would especially like to thank my collaborators, J.N. Fry and K.A. Olive, since most of the work I have discussed was jointly done by the three of us. I also thank R. Mohapatra for several useful conversations during the Blacksburg meeting which stimulated my interest in baryosynthesis at low energies. This work was supported in part by NSF Grant PHY77-27084 at the Institute for Theoretical Physics, UCSB, and by DOE Grant AC02-80ER 10773 at the University of Chicago.

REFERENCES

1. J.H. Christenson, J.W. Cronin, V.L. Fitch, and R. Turlay, Phys. Rev. Lett. 13, 138(1965); Phys. Rev. B140, 74(1965).
2. G. Steigman, Ann. Rev. Astron. Astrophys. 14, 339(1976).
3. K.A. Olive, D.N. Schramm, G. Steigman, M.S. Turner, and J. Yang, Astrophys. J. in press (1981); EFI 80-42.
4. A. Sakharov, JETP Lett. 5, 24(1967); M. Yoshimura, Phys. Rev. Lett. 41, 281(1978); A. Ignatiev, N. Krasnikov, V. Kuzmin, and A. Tankhelidze, Phys. Lett. 76B, 436(1978); J. Ellis, M.K. Gaillard, and D.V. Nanopoulos, Phys. Lett. 80B, 360(1978).
5. S. Weinberg, Phys. Rev. Lett. 42, 850(1979); D. Toussaint, S.B. Treiman, F. Wilczek, and A. Zee, Phys. Rev. D19, 1036(1979).

6. S. Dimopoulos and L. Susskind, Phys. Rev. D18, 4500(1979).
7. R.V. Wagoner, W.A. Fowler, and F. Hoyle, Astrophys. J. 148, 3 (1967); P.J.E. Peebles, Astrophys. J. 146, 542(1966).
8. P. Langacker, "GUTs and Proton Decay," SLAC-PUB-2544 (1980), to appear in Physics Reports.
9. J.N. Fry, K.A. Olive, and M.S. Turner, Phys. Rev. D22, 2953(1980); J.N. Fry, K.A. Olive, and M.S. Turner, Phys. Rev. D22, 2977(1980); J.N. Fry, K.A. Olive, and M.S. Turner, Phys. Rev. Lett. 45, 2074 (1980).
10. E.W. Kolb and S. Wolfram, Phys. Lett. 91B, 217(1980); E.W. Kolb and S. Wolfram, Nucl. Phys. B172, 224(1980).
11. J.A. Harvey, E.W. Kolb, D.B. Reiss, and S. Wolfram, in preparation (1981).
12. J. Ellis and G. Steigman, Phys. Lett. 89B, 186(1980).
13. T.J. Goldman and D.A. Ross, Phys. Lett. 84B, 208(1979).
14. D.V. Nanopoulos and S. Weinberg, Phys. Rev. D20, 2484(1979).
15. S. Barr, G. Segrè, and H.A. Weldon, Phys. Rev. D20, 2494(1979).
16. A. Yildiz and P.H. Cox, Phys. Rev. D21, 906(1980).
17. G. Segrè and M.S. Turner, Phys. Lett. B, in press (1981).
18. K.A. Olive, D.N. Schramm, G. Steigman, M.S. Turner, and J. Yang, Astrophys. J., in press (1981); also, M.S. Turner, "Neutrinos and Cosmology," in this volume (1981).
19. A more detailed discussion of these equations will be given in J.N. Fry and M.S. Turner, in preparation (1981).
20. Freeze out actually occurs for $z < z_f$, so that the final value of B is slightly larger than the estimate given here. For more details see ref. 19.

DISCUSSION

LANGACKER: I just wanted to make a comment on Senjanovic's talk. He argued correctly that if you do not have extra generations of fermions then you need additional Higgs particles in either hard or soft CP violation models. But there is a basic distinction: the soft CP violation models require more than one doublet of light Higgs particles (that you can measure in the laboratory), whereas the hard CP models can get by woth extra Higgs that are superheavy. Now that is very important, not only phenomenologically, but because there is the problem of understanding two different mass scales (the hierarchy problem). Some of the ideas for understanding the two mass scales (the weak interaction scale and the grand unification scale) would suggest that there is likely to be only 1 light Higgs doublet. Other mechanisms might allow more. There is a lot of interest in that question.

SENJANOVIC: I said that if you don't take Michael's scenario with heavy fermions, then at least, in say SU(5), you need an extra light doublet to explain the right baryon number excess.

LANGACKER: You need two 5's, but there is no need for the additional one to have any light particles in it. You see, we don't understand why anything is light. Whatever miracle makes one particle light might not make the other one light.

FRITZSCH: You said that ϵ in your kind of approach was determined by the parameters which also govern the low energy CP violation. But actually there is something else: there is a sign of ϵ, which determines that we are living in a world composed of baryons and not of anti-baryons. Is there a way to connect thin sign with the sign of ϵ or ϵ', observed in the K^0 system?

TURNER: In the 4 generation model there is no hope because ϵ depends upon a phase γ, which is not measurable at laboratory energies ($E \lesssim$ few TeV). It can only be measured at super high energies ($E \sim 10^{14}$ GeV). In the 5 generation model, all the parameters including the sign of ϵ can, in principle, be measured in the laboratory at low energies ($E \lesssim$ few TeV) by determining the "reduced KM matrix" for the three heaviest generations. So in the 4 generation model we cannot determine whether or not we should have matter or antimatter in the universe, while, in principle, in the 5 generation model we can.

SENJANOVIC: I would still say that this is a problem, even with 5 generations. Presumably, at low energies at least part of CP violation comes from the KM matrix; in other words, the complex Cabibbo rotation. Now if the symmetries are restored at some point, which is a picture, then eventually at high temperatures the quarks are massless. There is no KM complex generation, so it is very hard to imagine that there will be a connection, since it can then be only coupling constants which are important. So I would say, independent of the mass of the Higgs today, the mass will be zero eventually.

FRITZSCH: You say it is essentially the opposite of what he was saying?

SENJANOVIC: Yes, I am saying the opposite I think, because eventually I think Cabibbo rotation will become zero exactly since the quark will be massless. For a massless quark of course no rotation makes sense.

TYE: Just one comment on the five generations model. It seems to me there is some internal inconsistency. If you have five generations, you want some neutrinos to have some mass. Is that what you want?

TURNER: That's one of the ways that five generations are permitted. I realized that after I wrote the transparency, but go on.

TYE: With the minimal SU(5) there is no room for neutrino to get a mass. So you have to go beyond the minimal SU(5) anyway.

TURNER: You have to extend the minimal SU(5), but I think that Tony Zee showed that neutrino masses are possible in SU(5) with the addition of a 10 of Higgs. I will discuss the question of cosmological limits on the number of neutrino species tomorrow, but let me briefly explain how it works now. The amount of ^{4}He synthesized in the big bang, Y_p, depends upon the nucleon-to-photon ratio, η, the neutron halflife, $\tau_{\frac{1}{2}}$, and the number of light (<1 MeV) neutrino species, N_ν. An upper bound on Y_p and lower bounds on $\tau_{\frac{1}{2}}$ and η allow one to set an upper limit to N_ν. η is the tricky one; the mass density (m nucleons) of the universe is determined by measuring the mass of a gravitating system by the virial theorem, and dividing by the light energy, i.e. constructing a M/L ratio. Then the average mass density is determined by multiplying the "typical M/L ratio" for the universe by the luminosity density. The photon density is determined by the temperature of the microwave background. Most galaxies are found in binary galaxies and small groups of galaxies (BSG), and if one uses the M/L ratio inferred from BSG the limit $N_\nu \leq 4$ follows. However it might be that the mass which binds BSG is not predominantly nucleons, e.g., massive neutrinos, PBH's, or exotic baryons. In this case the lower bound on η from BSG is not valid, and the limit on N_ν is much worse - $N_\nu \leq 6$ (or perhaps no limit at all). In addition there are uncertainties with regard to $\tau_{\frac{1}{2}}$, Y_p, and the luminosity density of the universe. Although the simplest way around $N_\nu \leq 4$ is neutrino masses, there are other possibilities. I hope I have not given the impression that cosmological neutrino limits are not to be trusted - like most limits they depend upon certain assumptions and input data.

ROSEN: I wonder if there is any way of distinguishing any of these models or classes of models at extremely low energies, by which I mean the energies of accelerators available today?

SENJANOVIC: It wasn't clear to whom the question was addressed, since there are two of us. At least if it has more than one Higgs, there seems to be likely ways of doing so. By requiring perturbation theory to be good, we set the upper limits on the Higgs boson masses. You can argue that the neutron dipole moment is approximately less than $10^{-26} - 10^{-27}$ electron cm, and certainly that would be best way to do this (at least multi-Higgs versus one Higgs) now. But of course this is tied to the former question of whether the extra Higgs is super-heavy or not.

TURNER: If at low energies there were evidence for two Higgs doublets, that would necessitate two five-dimensional Higgs representations in SU(5). Yildiz and Cox have shown that with two $\underline{5}$'s the requisite CP violation is non-zero at the one loop level. The contribution to CP violation that I discussed is at the three loop level, and presumably it would be much smaller than the contribution at the one loop level. The discovery of two Higgs doublets would probably rule out the mechanism I have described as being important.

Lepton Number Non Conservation

R.D. Peccei

Max-Planck-Institut für Physik und Astrophysik, Munich, Fed.Rep.Germany

ABSTRACT

We discuss violation of lepton number, concentrating particularly on the possibility that global lepton number is a spontaneously broken symmetry. We examine the properties of the associated Goldstone boson, the Majoron, which results from the lepton number breakdown, in models with and without right-handed neutrinos. Some of the cosmological consequences of the existence of Majorons are also detailed.

It was assumed until recently that lepton number (and baryon number) were absolute conservation laws. With the advent of grand unified theories, the pendulum seems to have swung the other way. Now everybody seems to take it for granted that baryon and lepton number are violated to some extent [1], although in some special cases some linear combinations of B and L remain as global symmetries [2]. Because quarks carry a charge, it is not possible to introduce directly baryon number violating mass terms. Lepton number violating mass terms can however, be introduced for neutrinos, since they are charge-neutral. Consequently, although it is easy to have lepton number violation at the level of $SU(2) \times U(1)$, baryon number violation necessitates the introduction of lepto-quark interactions.

Because lepton number violation can be introduced without going the further step of considering some (grand) unification scheme, it appears worthwhile to discuss its ramifications even before considering baryon number breakdown. Logically one can conceive of five different possibilitites for lepton number. It can be a

(1) Conserved global symmetry
(2) Conserved local symmetry
(3) Explicitly broken symmetry
(4) Spontaneously broken local symmetry
(5) Spontaneously broken global symmetry

Of these five possibilities only (2) is clearly ruled out by experiment, since it implies the existence of long range interactions coupled to lepton number, which are obviously not present in nature. The last possibility was until now neglected, since spontaneous breakdown of a symmetry implies zero mass Goldstone excitations and these again give rise to long range interactions. However, we shall see that (5) is in fact perfectly viable phenomenologically.

The standard $SU(2) \times U(1)$ model [3] with no right-handed neutrino field and only doublet Higgs fields has lepton number as a global symmetry. The introduction of a right handed neutrino field with an $SU(2) \times U(1)$

invariant Majorana mass, affords an example of an explicit breakdown
of L. If this Majorana mass, however, is generated by the vacuum expectation of some SU(2) x U(1) singlet Higgs field, <u>which carries lepton number itself</u>, then lepton number is a spontaneously broken symmetry.
The symmetry can be locally realized, in which case the spontaneous breakdown causes the relevant gauge field coupled to lepton number to become massive [4]. If, on the other hand, the symmetry is only global then the spontaneous breakdown will always be accompanied by a Goldstone boson.
It is this last novel possibility that I want to describe in detail below.
Explicit breakdown of lepton number has been considered recently in Ref. [5] and we shall not discuss it further here.

The idea that lepton number is a spontaneously broken global symmetry was studied recently by Chikashige, Mohapatra and myself [6] in an SU(2) x U(1) model with right handed neutrinos. A similar analysis was done by Gelmini and Roncadelli [7] in the standard SU(2) x U(1) model, without a ν_R field. In both of these cases the appearance of a Goldstone boson, arising from the breakdown of lepton number, has innocuous phenomenological consequences for any conceivable terrestrial experiments. However, the existence of a Majoron- which is the name we have given to this Goldstone boson - can have important cosmological implications.

I want to emphasize that Majorons are irrelevant physically for laboratory experiments because their couplings to matter are extremely tiny.
These very small couplings arise, however, in a natural way in the models of Ref. [6] and [7], being proportional to the ratios of the two natural scales in the problem. One of these scales is associated with the usual spontaneous breakdown of SU(2) x U(1) which gives rise to the W and Z° masses, while the other scale reflects the presence of almost massless neutrinos in the theory. Although one has a hierarchy problem to explain -
a problem which one has at any rate to explain, if neutrinos have a non vanishing mass - one may use this freedom to essentially suppress any unwanted vestiges of the spontaneous breakdown of lepton number.

Although the explicit details of the models can be found in Refs.
[6] and [7], I want to summarize here some of the results. I will begin by discussing the model in which a right handed neutrino field is present in the theory [6]. The theory, otherwise, is just the standard SU(2) x U(1)

model [3], which I shall take as one generation, for simplicity, for the moment. Let me write the two neutrino states as D_L and S_R, D_L being part of a doublet $\psi_L = \binom{D_L}{e_L}$ and S_R being the singlet field. Their charge conjugate partners are $D_R^c = C(\bar{D}_L)^T$ and $S_L^c = C(\bar{S}_R)^T$, with C the charge conjugation matrix. Further, consider two different types of Yukawa couplings in the theory, involving respectively a doublet, ϕ, and singlet, φ, Higgs fields [8]

$$\mathcal{L}_1 = -h_1 (\bar{S}_L^c \varphi S_R + \bar{S}_R \varphi^+ S_L^c) \tag{1a}$$

$$\mathcal{L}_2 = -h_2 (\bar{\psi}_L \phi S_R + \bar{S}_R \phi^+ \psi_L) \tag{1b}$$

When ϕ acquires non zero vacuum expectation value, a Dirac mass term ensues, while $\langle\varphi\rangle \neq 0$ gives rise to a Majorana mass term. If one wants lepton number to be a good symmetry, one must assume that φ carries lepton number two. Then $\langle\varphi\rangle \neq 0$ signals the spontaneous breakdown of lepton number.

The neutrino mass matrix arising from Eqs. (1) is of the standard type [9]

$$m = \begin{pmatrix} 0 & m \\ m & M \end{pmatrix} \tag{2}$$

where m(M) is the Dirac (Majorana) mass. If $M \gg m$ (hierarchy problem) then the resulting mass eigenstates have widely different masses, M and $m_\nu = \frac{m^2}{M}$. The ordinary neutrino then has a mass much below the Dirac mass value, and this is the presumed reason why it is so light compared to the electron and u and d-quarks. The physical states are, approximately,

$$\nu_L \simeq D_L - \frac{m}{M} S_L^c \tag{3a}$$

with mass m_ν and

$$N_R \simeq S_R + \frac{m}{M} D_R^c \tag{3b}$$

with mass M, along with their charge conjugates. The combinations $\nu_L + \nu_R^c = \nu$ and $\eta_R + \eta_L^c = \eta$ are self conjugate, Majorana fields.

The Higgs field φ, whose vacuum expectation value breaks lepton number spontaneously, is coupled originally to the singlet field S and hence it remains coupled principally to the heavy neutrino field η. Writing

$$\varphi = \langle\varphi\rangle + \frac{1}{\sqrt{2}}(\rho + i\chi) \tag{4}$$

one identifies the Goldstone excitation, the Majoron, with χ. The coupling of the Majoron to ordinary neutrinos is small

$$\mathcal{L}_{\chi-\nu} = -i\frac{h_1}{\sqrt{2}}\left(\frac{m_\nu}{M}\right)\bar{\nu}\gamma_5\nu\,\chi \tag{5}$$

being proportional to the square of the small mixing angle $(\frac{m}{M})$. Because of the mixing, the heavy neutrinos η can decay rapidly into the light neutrinos and the Majoron.

The Majoron has no direct coupling to ordinary matter, but can eventually couple to matter by mixing with the Z^0 through a neutrino loop. (For electrons there is also an induced coupling due to W-exchange). In Ref. [6] we found that the effective interaction of the Majoron with charged fermions was given by

$$\mathcal{L}_{\chi-f} = i g_f h_1 \frac{G_F m_\nu m_f}{16\pi^2} \bar{f}\gamma_5 f\,\chi \tag{6}$$

with $g_f = 1$ for e, u and $g_f = -1$ for d-quarks. Given that the induced coupling is first order weak, one could have guessed the form of (6), since the coupling must be proportional to m_ν - vanishing for zero neutrino masses - and proportional to m_f, as all ordinary Higgs coupling are. The dimensionless coupling constant in Eq. (6) is tiny being of the order of 10^{-20} for typical parameters, $h_1 = 10^{-2}$, $m_\nu = 1$ ev, $m_f = 10$ Mev. This is one of the reasons why Majorons, if they exist, are essentially undetectable in the laboratory. The other main reason is that the Majorons have a γ_5 interaction with matter and this always gives spin dependent effects.

Because Majorons have zero mass, they give rise to a long range force. However, this force is of a dipole-dipole type, with a potential

$$V_{Maj} = \frac{\lambda_f}{r^3} \left\{ 3 (\vec{\sigma}_1 \cdot \hat{r})(\vec{\sigma}_2 \cdot \hat{r}) - \vec{\sigma}_1 \cdot \vec{\sigma}_2 \right\} \tag{7}$$

dropping off as r^{-3}, with

$$\lambda_f = \frac{1}{4\pi} \left(\frac{h_1 G_F m_\nu}{16 \pi^2} \right)^2 \sim \frac{10^{-40}}{m_f^2} \tag{8}$$

Note that this force is only comparable in strength to gravity at typical nuclear distances, dropping off then much more rapidly. There are some bounds, obtained by Feinberg and Sucher [10], on non-magnetic long range residual dipole-dipole interactions. But, as can be expected, these bounds for λ_f are many orders of magnitude too weak

$$(\lambda_f)_{F-S} \lesssim \frac{10^{-8}}{m_\pi^2} \tag{9}$$

Clearly, in the above model, spontaneous breakdown of lepton number is perfectly acceptable phenomenologically, as far as terrestrial experiments go. Indeed, the Majoron's couplings to matter are so small that it is really irrelevant whether Majorons exist or not. Majorons, however, can have non-trivial cosmological implications for the neutrino mass spectrum [6] and we shall see that the existing spectrum of neutrinos can perhaps distinguish whether lepton number breakdown is caused spontaneously or not. The reason why the existence of Majorons may have a bearing on the cosmologically allowed neutrino mass spectrum is simple to understand. Heavy neutrino species, of the light variety, can decay into the lightest neutrinos by Majoron emission. If this process is fast enough on a cosmological scale, the heavy neutrinos are effectively removed and bounds derived for their mass from the observed mass density of the present universe are voided [11].

If there are many fermion generations, the coupling of Majorons to matter remains essentially that of Eq. (6). However, the Majoron's coupling to neutrinos in general need not be flavor diagonal. Its strength, apart from mixing angles, will still be given by m_ν/M. This lack of flavor diagonality allows for the heavier neutrinos among those of the light-neutrino type to decay into the lightest neutrino by Majoron emission. (Presumably $m_{\nu_\tau} > m_{\nu_\mu} > m_{\nu_e}$ and so ν_τ and ν_μ decay into ν_e plus a Majoron). The lifetime for this process is given by the formula

$$\tau(\nu_H \to \nu_L + \chi) = \frac{32\pi}{h_1^2 \sin^2\theta} \left(\frac{M}{m_{\nu_H}}\right)^2 \frac{1}{m_{\nu_H}} \tag{10}$$

where $\sin\theta$ is an intergeneration mixing angle. Clearly this lifetime is long, because presumably M/m_{ν_H} is a large number. However, this lifetime may be short compared to the universe's lifetime. Indeed, taking as typical parameters $h_1 = 10^{-2}$, $\sin\theta = 10^{-1}$, $M = 10^5$ Gev, $m_{\nu_H} = 100$ ev one finds $\tau \sim 2 \times 10^7$ years. Thus one sees that for "sensible" scales M - where one might expect new physics to arise - neutrino decay by Majoron emission can play a role in removing the heavy neutrino component from the mass of the universe.

In Ref. [6] we studied in detail the contribution to the universe's energy density of Majorons coming from the decay of heavy neutrinos. Our analysis was analogous to that done earlier by Dicus, Kolb and Teplitz [12], when they considered radiative neutrino decays. In contrast to radiative neutrino decays, however, we do not have to worry about the effects of Majorons on the future evolution of the universe, since their couplings to matter are so negligible. The energy density of Majorons which are decay by-products, for the case in which $\tau(\nu_H \to \nu_L + \chi) \ll t_u$ - with t_u the universe's lifetime, is easily seen to be [6]

$$\rho = n_H(T_D) \left(\frac{1.9°K}{T_D}\right)^3 m_{\nu_H} \left(\frac{\tau}{t_u}\right)^{1/2}$$

Here $n_H(T_D)$ is the density of heavy neutrinos at the decoupling time, characterized by the temperature T_D. This number is the same as that obtained in the standard analysis [12], since the dominant weak interaction processes for neutrinos are still as in the usual case. The factor $(\frac{1.9°K}{T_D})^3$ accounts for the volume expansion from decoupling to present temperatures. However, because the heavy neutrinos are unstable and the Majorons are massless, the heavy neutrino mass contributions can now be effectively redshifted - by the factor $(\frac{\tau}{t_u})^{1/2}$ - below m_{ν_H}.

The normal bounds for cosmologically allowed neutrino masses arise from the requirement that ρ be less than the critical density $\rho_c \sim 5\times 10^{-3}$ Mev/cm^3 required to close the universe. For stable neutrinos the factor of $(\tau/t_u)^{1/2}$ is missing in Eq. (11) and $\rho \lesssim \rho_c$ implies that neutrinos of mass greater than around 50-100 ev are not allowed cosmologically [11][13]. In our case since τ depends on the mass parameter M (actually really on $\frac{M}{h_1} \sim \langle \varphi \rangle$) one can considerably weaken the cosmological bounds. Taking $h_1 = 10^{-2}$, $\sin\theta = 10^{-1}$ as typical parameters, we found in Ref. [6] that for $M \lesssim 10^6$ Gev, $\rho < \rho_c$ for all values of m_{ν_H}. Thus in this case no cosmological bounds on neutrino masses exist. For large M, however, $M \gtrsim 10^9$-10^{10} Gev the cosmological bounds remain for neutrinos in the interesting mass range, $m_{\nu_H} \lesssim 100$ Mev. The actual determination that the μ or τ neutrinos have masses above 100 ev would constitute the best evidence I know for the existence of Majorons.

The weak coupling of the Majorons to matter in the above model is understandable since the Majoron is an SU(2) x U(1) singlet and therefore it couples essentially only to the heavy right-handed neutrino. Gelmini and Roncadelli [7], recently considered spontaneous breakdown of lepton number in a standard SU(2) x U(1) model with no right-handed neutrinos. There, at first sight, it is not obvious that the existence of a Goldstone excitation can be tolerated. However, even in this case, it turns out that there are no practical phenomenological considerations that rule out this option.

If only left handed neutrinos are present, then the only possible Yukawa interaction which can give rise to spontaneous lepton number breaking involves a triplet of Higgs field $\vec{H}$, with charges 2, 1 and 0. The relevant coupling with the lepton doublet field ψ_L is given by

$$\mathcal{L}_3 = -h_3 \left\{ \bar{\psi}_R^c \tau \cdot \vec{H} \psi_L + \bar{\psi}_L \tau \cdot \vec{H}^\dagger \psi_R^c \right\} \tag{12}$$

Again, H must carry lepton number if one wants lepton number to be a good symmetry. Then $\langle H \rangle \neq 0$ signals the spontaneous breakdown of lepton number. However, now since $m_\nu \sim \langle H \rangle$ this vacuum expectation value must be very small [14]. Clearly in this case, contrary to what happens when a right-handed neutrino field is introduced, it is not possible to gauge lepton number, since the relevant gauge boson would end up with too small a mass, $M \sim m_\nu$, and it could give rise to observable effects in, for example νe scattering. One of course, could always forbid H to carry lepton number [5]. However, if lepton number is to be broken spontaneously, it must necessarily be a global symmetry.

Because H is not an SU(2) singlet, it is no longer totally true that the Goldstone excitation is given by the imaginary part of H°. Rather now there will be mixing between H° and the neutral member of the Higgs doublet field Φ. One linear combination of these fields gets eaten by the Z° and the remaining orthogonal combination is the Majoron. However, because $\langle H \rangle \ll \langle \Phi \rangle$ [14] (again a hierarchy of expectation values) the Majoron will predominantly be proportional to the triplet Higgs field with only a $\frac{\langle H \rangle}{\langle \Phi \rangle}$ admixture of the doublet Higgs field. This circumstance guarantees again that its coupling to matter is weak. In fact Gelmini and Roncadelli [7] show that the coupling of the Majoron to matter is given by

$$\mathcal{L}_{\chi-f} = i \epsilon_f 2\sqrt{2} \; G_F \frac{m_\nu m_f}{h_3} \bar{f} \gamma_5 f \chi \tag{13}$$

with $\epsilon_f = -1$ for e and d and $\epsilon_f = +1$ for u-quarks. The dimensional coupling constant in (13) is not terribly different from that in Eq. (6), except if $h_3 \to 0$ [15]. However, one would naturally expect h_3 to be of $O(10^{-2})$ or so, in which case the Majoron is again totally undetectable.

The model of Gelmini and Roncadelli [7], however, does have some non-trivial differences from that proposed in Ref. [6]. A careful analysis of the potential $V(H, \Phi)$ reveals that besides the Majoron at zero mass, there exists a light neutral scalar field, ρ_L, of mass of $O(\langle H \rangle)$ and hence presumably in the ev range. Such a state, although not unexpected on general grounds, could have disastrous consequences for the model of

Ref. [7], were if not for the fact that it is also extremely weakly coupled to matter, with dimensional coupling constant again of the order of $G_F m_\nu m_f$. Thus ρ_L exchange is not a problem phenomenologically.

The most interesting aspect of the model of Roncadelli and Gelmini is that both Majoron and ρ_L-exchange give rise to substantial interactions among neutrinos, which in principle could be stronger than the regular weak interactions, especially at low energies. It is not clear to me how these interactions could be detected experimentally. However, I should comment that in a multigeneration version of the model of Ref. [7] heavier neutrinos decay very rapidly ($\tau \ll 1$ sec) by Majoron emission into the lighter neutrinos. Thus, no cosmological bounds on heavy neutrinos should ensue.

I have worried in the above about spontaneous breaking of global lepton number, without talking in detail about other manifestation of broken lepton number, like neutrino oscillations and neutrinoless double beta decay. This is because the phenomenology of these processes, when the breakdown arises from a Majorana mass term for the neutrinos, is essentially the same whether the lepton number breakdown is done explicitly or spontaneously, and this phenomenology is already well discussed in Ref. [5]. It should be clear, however, that these processes should be continued to be looked for because they may well be the only "experimental handle" that we have to find out whether lepton number is a symmetry or not.

Acknowledgements

I would like to thank Y. Chikashige, G. Gelmini, R. Mohapatra and M. Roncadelli for some very useful discussions.

References and Footnotes

[1] For a recent review, see for example S. Weinberg, talk presented at the First Workshop on Grand Unification, Durham, New Hampshire April 1980.

[2] For example, in the simplest version of SU(5), B-L remains as a global symmetry.

[3] S. Weinberg, Phys.Rev.Lett. $\underline{19}$ (1967) 1264;
A. Salam, Proceedings of the 9th Nobel Symposium, ed. by N. Svartholm (Almquist and Wicksells, Stockholm 1968);
S.L. Glashow, J. Iliopoulos and L. Maiani, Phys.Rev. $\underline{D2}$ (1970) 1285.

[4] This is the case, for example in SO(10) where B-L is a local symmetry and the right-handed neutrino Majorana masses are generated by spontaneous breakdown of a Higgs transforming according to the 126 representations of SO(10).

[5] R.E. Marshak, R.N. Mohapatra and Riazuddin; VPI preprint 1980;
R. Barbieri, D.V. Nanopoulos, G. Morchio and F. Strocchi, Phys.Lett. $\underline{90B}$ (1980) 91;
T.P. Cheng and L.F. Li, C00-3066-152 preprint, June 1980.

[6] Y. Chikashige, R.N. Mohapatra and R.D. Peccei, MPI-PAE/PTh 36/80 to be published in Phys.Lett. B. and MPI-PAE/PTh 40/80 to be published in Phys.Rev.Lett.

[7] G. Gelmini and M. Roncadelli, MPI-PAE/PTh 50/80.

[8] In principle there can also be a coupling to a triplet Higgs field, but for this model this coupling is ignored since we want to obtain a neutrino matrix of the type given in Eq. (2).

[9] M. Gell-Mann, P. Ramond and R. Slansky, in "Supergravity", ed. by
     P. Van Nieuwenhuizen and D. Freedman (North Holland, 1979).
     T. Yanagida, Proceedings of the Workshop on the Unified Theory and
     the Baryon Number of the Universe (KEK 1979).

[10] G. Feinberg and J. Sucher, Phys.Rev. $\underline{D20}$ (1979) 1717.

[11] R. Cowsik and J. McClelland, Phys.Rev.Lett. $\underline{29}$ (1972) 669;
     For a more recent estimate see D. Schramm and G. Steigman, Gen.Rel. and
     Grav. to be published.

[12] D. Dicus, E. Kolb and V. Teplitz, Phys.Rev.Lett. $\underline{39}$ (1977) 168.

[13] For very heavy neutrinos, $m_{\nu_H} \gtrsim 2$ Gev, the cosmological bounds again
     fail because then $n_H(T_D)$ becomes exponentially small. See B.W. Lee
     and S. Weinberg, Phys.Rev.Lett. $\underline{39}$ (1977) 165.

[14] $\langle H \rangle$ must also be much smaller than $\langle \Phi \rangle$ so as not to cause
     violations to the $\Delta I_w = 1/2$ rule. However, with neutrino masses in the
     ev range unless $h_3 \ll 1$ one is guaranteed that $\langle H \rangle \ll \langle \Phi \rangle$

[15] In this case there is no hierarchy since m_ν can be made small by
     letting $h_3 \to 0$. However, this is unappealing physically.

DISCUSSION

FRITZSCH: Could you have a decay of the ν_e fast enough that, let's say, solar neutrinos would decay on their way to the earth?

PECCEI: No, it is only the heaviest neutrinos that presumably decay.

FRITZSCH: But you don't know which one is light and which one is heavy; ν_e might decay into a ν_μ.

PECCEI: No, the lifetime is short, but it is short in cosmological time; it is not short in real time except in the model of Gelmini-Roncadelli, in which lifetimes are really rather short and you have to worry about whether, in fact, ν_μ beams exist. You have to really tune the parameters carefully.

TSAO: It seems to me, naively, that because of the diagonalization between two states of this psuedoscalar, you would have a very small matter coupling to this particle already at the tree level.

PECCEI: You don't have any coupling at the tree level because this is a singlet.

TSAO: But what if there is a mixing?

PECCEI: No, there is no mixing. I mean if you look at it there is no mixing in this model, at the tree level.

Weak Scalar Boson and Vector Boson Phenomenology

Current Phenomenological Status of Higgs Physics and Technicolor

G.L. Kane[*]

Randall Laboratory of Physics
University of Michigan
Ann Arbor, MI 49109

Although we know that spontaneous symmetry breaking occurs, because $M_{W^\pm}$, $M_Z \neq 0$, we do not yet know the underlying physics or understand what happens. So far, no reasonably attractive theoretical picture has emerged at all, let alone been tested by experiment. Essentially no experiment has put significant constraints on the existence or masses or interactions of scalar or pseudoscalar bosons, and very few experiments are planned for the near future that could change this.

Here I will review the current situation concerning present experimental constraints on Higgs physics [it is useful to use this name for scalar or pseudoscalar bosons of either fundamental or dynamical origin].

First, there are the two major possibilities. I am supposed to concentrate on phenomenological questions, leaving discussion of model building to other speakers.[1]

---

[*] Research supported in part by the U.S.D.O.E.

Fundamental Higgs Bosons	Dynamical Bosons
. This alternative has been introduced in an ad hoc manner. It requires introducing a new fundamental coupling apparently not derivable from a gauge principle.	[Technicolor (TC) + Extended Technicolor (ETC) or Hypercolor + Sideways]
. It is not clear whether a scalar mass on the scale of a few GeV can be maintained in the presence of radiative corrections.	. So far there is no nice model.
	. In principle everything is calculable here.
. No clear experimental predictions.	. Probably some reliable and testable predictions.
. If these scalars were part of a bigger theory, such as a supersymmetric one, the objections would presumably be removed, but there are so far no realistic examples of this.	. Maybe conflicts with experiment already at a detailed level.

Possible experimental predictions will mainly be of two kinds, either new particle states, or scalar or pseudoscalar interactions. First consider the particles.

Expected spectrum

Fundamental Higgs

Minimal SU(2)⊗U(1)	2 or more doublets
. 1 Higgs doublet	. This seems the more likely alternative since it would be required to calculate quark mixing angles, to get CP violation via the Higgs sector, if Higgs arose from supersymmetric theories, etc.
. 1 neutral scalar boson h° expected, with mass $m \gtrsim$ few GeV.	
. Its fermion coupling will be $g_f = g \, m_f/m_W$.	
	. Charged and neutral bosons $h^{\pm}$, h° are expected, with arbitrary masses and arbitrary couplings to fermions and other bosons.
	. Scalar and pseudoscalar neutra occurs.
	. In most supersymmetric approaches $m_h \simeq m_W$.

It could happen that m_h, $\Gamma_h \sim 1$ TeV and no simple experimental signatures occur. Then it would be necessary to find new interactions on the TeV scale, but so far there is no clear understanding of what effects would occur.

Dynamical Theories

In TC and ETC particles arise naturally [2-8] at three mass scales: (i) $\Lambda_T \simeq 1$ TeV, the basic scale of the theory, set to get $m_{W^\pm}$, m_Z correct; (ii) about 250 GeV, set by $m^2 \sim \alpha_s \Lambda_T^2$ for pseudoscalars that would be massless but get mass through color interactions; and (iii) a few GeV, with $m^2 \sim \alpha^2 \Lambda_T^2$, for pseudoscalars that would be massless but get mass from electroweak interactions or the even weaker ETC interactions. The last category can be split into two parts. Charged particles get mass (about 7 GeV) from electroweak interactions -- this should be reliably calculated. Both neutral and charged particles get mass from the ETC interactions. At present the best calculation that can be done in practice with the ETC interactions is to put an upper limit on their contribution, which comes out around 2 GeV, a reasonable result for a very short range force.

In essentially any dynamical theory in which quarks get masses by coupling to new colored fermions, Goldstone bosons which carry color will appear, and will get mass on the scale of 250 GeV; the occurrence of such states will be natural. Similarly, the uncolored states will generally appear and will be in the few GeV region. It would be surprising if the very short range electroweak interaction gave even as much mass as the electroweak interaction.

Thus we have the predictions:

Mass Range	$\Lambda_T \lesssim 1$ TeV	~250 GeV	~8 GeV	~2 GeV
Mass from	Technicolor Scale	QCD	$SU(2) \otimes U(1)$	Extended Technicolor
Particles	$\rho_T, \omega_T \ldots$ baryons	η_T^c, $\pi_T^{\pm c}$, lepton-quark states	$a_T^\pm$	$a_T^\circ, a_T'^\circ$
Observe	FNAL collider	FNAL collider Isabelle	PEP PETRA in $e^+e^- \to a_T^+ a_T^-$ with 1/4 unit of R	$\psi, T \to a_T^\circ \gamma$ $KN \to a_T^\circ X$
Main decays	$\rho_T \to GG, W^+W^-$ $\to q\bar{q}'$	$\eta_T^c \to t\bar{t}, b\bar{b},$ $GG, G\gamma$ GZ°	$a_T^\pm \to c\bar{s}, c\bar{b},$ $\tau\nu$	$a^\circ \to \mu^+\mu^-,$ $s\bar{s}(\phi\pi,$ $KK^*, KK)$ $BR(\mu^+\mu^-)$ large

Present Knowledge

One could learn about spin zero bosons from either seeing scalar or pseudoscalar currents, or finding actual particles. Assume we can write a Lagrangian for the effective coupling to fermions,

$$\mathcal{L} = g \frac{m_f}{m_W} \bar{f}[c_s + c_p \gamma_5] f' \phi$$

where ϕ is a fundamental or dynamical boson. This can be written for charged or neutral bosons. Then the most useful present limits are

(a) For charged Higgs, $m_\phi > m_\tau$ since otherwise the τ would decay semiweakly through the Higgs coupling. Similarly, as soon as it is established clearly (as the reports from CESR at this meeting seemed to demonstrate) that the b quark decays normally, then we know $m_\phi > m_b$.

(b) For neutral Higgs only $(g-2)_\mu$ provides[9] a potentially useful limit. For a scalar boson it gives

m_ϕ	c_s
25 MeV	< 1.2
1 GeV	< 6
10 GeV	< 35

For a pseudoscalar the limits on c_p are less good by about a factor 3. For scalar plus pseudoscalar states cancellations can occur; if they are not excluded there is no restriction. The result is

$$a_\mu = \frac{g^2 m_\mu^2}{8\pi^2 m_W^2} \left[c_s^2 A(m_\phi^2) - (c_s^2 + c_p^2) B(m_\phi^2) \right]$$

where A,B are integrals which depend only on m_ϕ^2/m_μ^2. For a wide range of masses B/A is about 1/3.

Possible conflict

There is one possible conflict of Technicolor physics with experiment. Very crudely, the method of giving masses to quarks leads to a result

$$m_q \sim \frac{g_{ETC}^2}{m_{ETC}^2} \Lambda_T^3$$

In general the theory also has flavor changing neutral currents (FCNC), giving contribution to Δm (K_L-K_S), $K \to \mu e$, etc. Both the quark masses and the FCNC arise from exchange of ETC gauge bosons. The FCNC are of order g_{ETC}^2/m_{ETC}^2 in amplitude. Then the problem is that if g_{ETC}^2/m_{ETC}^2 is large enough to give the heavier lepton and quark masses, it gives too much FCNC (and conversely). Whether this is a basic problem, or just indicates the presence of subtle and interesting mechanisms, is unknown at present.

Conclusion

It is unfortunate that more experimental effort has not gone into establishing useful limits on Higgs physics. In the near future the experiment mentioned for $a_T^\pm$, a_T°, η_T^c should impact greatly on our understanding of spontaneous symmetry breaking -- perhaps a scalar or pseudoscalar boson will be found.

References

1. See other talks at this meeting, such as Beg's, for review of the theory.
2. E. Eichten and K. Lane, Phys. Lett. 90B, 125 (1980).
3. S. Dimopoulos, Nuc. Phys. B168, (1980) 93.
4. M. Peskin, Saclay Preprint, May 1908.
5. J. Preskill, Harvard Preprint, July 1980.
6. V. Baluni, unpublished.
7. J.D. Bjorken, unpublished.
8. S. Dimopoulos, S. Raby, and G.L. Kane, Michigan Preprint UM HE 80-22, to appear in Nuc. Phys. B.
9. J. Leveille, Nuc. Phys. B137, 63 (1978).

COMMENTATOR

A way to observe the Higgs Boson H^0
in Hadronic Interactions

Ling-Lie Chau Wang
Brookhaven National Laboratory, Upton, New York 11973

With its single parameter $\sin^2\theta_W = .23 \pm .02$, the standard $SU(2) \times U(1)$ model has impressively fitted many different types of weak interactions[1]. Though the model has the definite predictions of the intermediate bosons $W^\pm$, Z^0, with Masses $M_W \simeq 80$ GeV, $M_Z \simeq 90$ GeV, the mass-generating mechanism of the theory has been obscure. Since we are at the threshold of checking the existence of the predicted $W^\pm$ and Z^0 with definite signatures in the next generation accelerators[2], it is of great importance to find ways to check the existence of these mass-generating related bosons. Contrasting to the situation with e^+e^- machines, for which signatures and ways to observe such bosons are rather clear, ways of detecting such bosons have been noticably missing in hadronic scatterings.[3]

Here we point out that the bremsstrahlung of the H^0 by a Z^0, produced in hadronic scattering as shown in the insert of Fig. (1), gives a definite signature[4] of a bump in the dilepton mass Q with a sharp fall at $Q \simeq M_Z - M_H$. The production is calculated using the Drell-Yan model.

In Fig. (1) we show the Q distribution for $\sqrt{s} = 800$ GeV, and various Higgs mass $M_H = 5$ GeV, 10 GeV, 15 GeV and 20 GeV. We see that besides the direct Z^0 peak at $Q = M_Z \simeq 90$ GeV, there is a bump peaks about $Q + M_H \simeq M_Z$. Actually the sharp fall around $Q \simeq M_Z - M_H$ comes from the B-W propagator in $\hat{s}$ and the fall on the lower side of the bump is from the B-W propagator in Q. It is interesting to point out that the top shape of the bump is rather insensitive to the Higgs masses and the Z^0 width. The $M_H = 5$ GeV curve indicates that one must be careful in interpreting the lower secondary peak, whether it is a broad second resonance or it is from the effect of such a massive particle bremsstrahlung.

Now we shall discuss briefly the detectability of the Higgs boson through such a mechanism. In Fig. (1) we also show the Q-distribution of $\ell\bar{\ell}$ from Z^0 without the H^0 emission. Agreeing with Ref. 5, the peak at $Q = M_Z$ with the H^0 emission is about three orders of magnitude below the peak at $Q = M_Z$ without H^0 emission. However, the second bump in Q is enticingly not too far below the background of the Z^0 and the virtual photon γ_v production. With the projected luminosity of ISABELLE 10^{33} cm^{-2} sec^{-1}, there will be about 600 events/year (3.15×10^7 sec) from the bump between 66 GeV $\leq Q \leq$ 76 GeV in the case of $\sqrt{s} = 800$ GeV and $M_H = 15$ GeV. As low as it is, however, it is not completely out of the question to accumulate enough events for its detection in case we can eliminate the background above. To get rid of the background, we propose to trigger a third lepton of relatively low momentum besides the original two fast leptons with momentum $\simeq \frac{1}{2}(M_Z - M_H)$, using the property that the H^0 prefers to couple (with a strength $\propto M_F^2 G$) to the most massive fermion pair allowed, i.e., $\tau\bar{\tau}$ and $c\bar{c}$ for $2M_\tau < M_H < 2M_b$, and $b\bar{b}$ for $2M_b < M_H < 2M_t$. The τ, D have $\sim .15\%$ branching ratio decaying into a charged lepton,[5] and the b has even higher portion of its final states having a lepton due to its cascade decay into the charm. Thus by this trigger the bump is reduced less than .30%. However, we anticipate much more reduction for the background from our current understanding of single lepton productions.[6] If the Higgs masses are much bigger than $2M_b$, it may be also triggered by two hadronic jets.

Due to the smallness of the cross section it definitely points to machines with high luminosity besides high energy. Note also that for small M_H, a detector of good mass resolution is needed, in case if the background of the single Z^0 production can be triggered away. In Fig. (2) we show the energy variation of the bump for M_H = 15 GeV at $\sqrt{s}$ = 1000 GeV, 800 GeV, 600 GeV for pp scattering. In the same figure we also show the $p\bar{p}$ production of H at $\sqrt{s}$ = 540 GeV and 2000 GeV. Though the cross section from $\bar{p}p$ is a factor of a few larger than that of pp at the same energy, due to the projected low $\bar{p}$ luminosity ($\simeq 10^{30}$ cm^{-2} sec^{-1}) the Higgs production under the present estimate seems too small to be detected in the next generation $\bar{p}p$ storage rings.

In conclusion the bremsstrahlung of a Higgs boson H^0 by the Z^0 produced in the nucleon-nucleon reactions will produce a bump structure in the dilepton invariant mass Q with a sharp fall around $Q \simeq M_Z - M_H$. To sweep away the background of $\ell\bar{\ell}$ system from other sources ($\gamma_v + Z^0$ without H^0), we suggest a trigger of detecting a third lepton (or jets) from the dominant decay channels of the H^0. Since the mechanism discussed here is a kinematic one, similar effects should exist for any emission of massive particle by a resonance. Thus this phenomena provides a means to discover particles in the cases when the coupling strength of the emission is large enough. Further, the Higgs-boson bremsstrahlung mechanism discussed here provides an explicit example that the detecting of the tri-lepton or dilepton plus jets might be the way for future experiments to discover new particles, in addition to the historically successful method of detecting the dilepton system.

References

1. See review talk by C. Baltay, Proceedings of the XIXth International Conference on High Energy Physics, Tokyo, Aug. 1978. J.K. Kim, P. Langacker, M. Levine and H.H. Williams, University of Pennsylvania preprint, UPR-158T (submitted to Rev. of Mod. Phys.)
2. R.F. Peierls, T.L. Trueman, L.-L. Wang, Phys. Rev. D16, 1397 (1977); F.E. Paige "Updated estimates of W production in pp and $\bar{p}p$ interactions," Proceedings of Workshop on Production of new particles in super high energy collisions, Madison, Wisconsin, 1979; and a review by L.-L. Chau Wang, "Theoretical Implications on ISABELLE Physics," Lecture at the 1980 Erice Summer School.
3. The production of the H^0 and an on-shell Z^0 in hadronic scattering was calculated by S.L. Glashaw, D.V. Nanopoulos, A. Yildiz, Phys. Rev. D18, 1724 (1978).

4. Details see L.-L. Chau Wang "Signature of the Bremsstrahlung of a Higgs Boson by the Z^0 in Hadronic Reactions," Brookhaven preprint #
5. For leptonic decay branching ratios for the τ and the b, see M. Perl, The Tau Lepton, SLAC preprint, SLAC-PUB-2446 (Dec. 1979) (submitted to Ann. Rev. Nucl. Part. Sci.). B. Gittelman's talk at the XXth Int. Conf. on High Energy Physics.
6. For review on the status of prompt lepton production, see talk by H. Wachsmuth, Proceedings of the 1979 Int. Sym. on Lepton and Photon Interactions at High Energies.

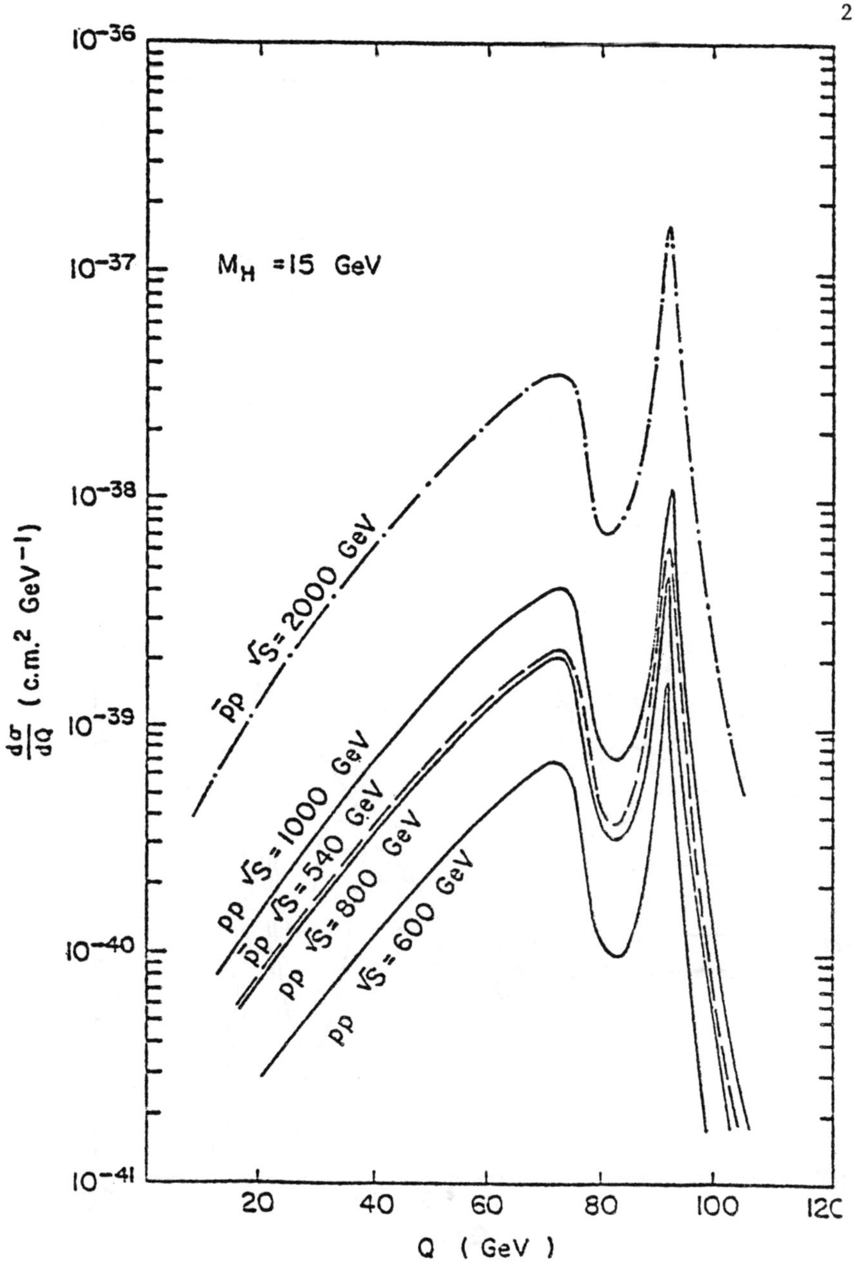

Fig. (1)

Fig. (1). Dilepton mass Q distribution for the bremsstrahlung of a H^0 by a Z^0 produced in $pp \to XZ^0 \to XZ^0H^0 \to \ell\bar{\ell}H^0$ at $\sqrt{s}$ = 800 GeV, and Higgs boson mass m_H = 5, 10, 15, 20 GeV, and for $pp \to X(Z^0 + \gamma_v) \to X\ell\bar{\ell}$ at the same energy.

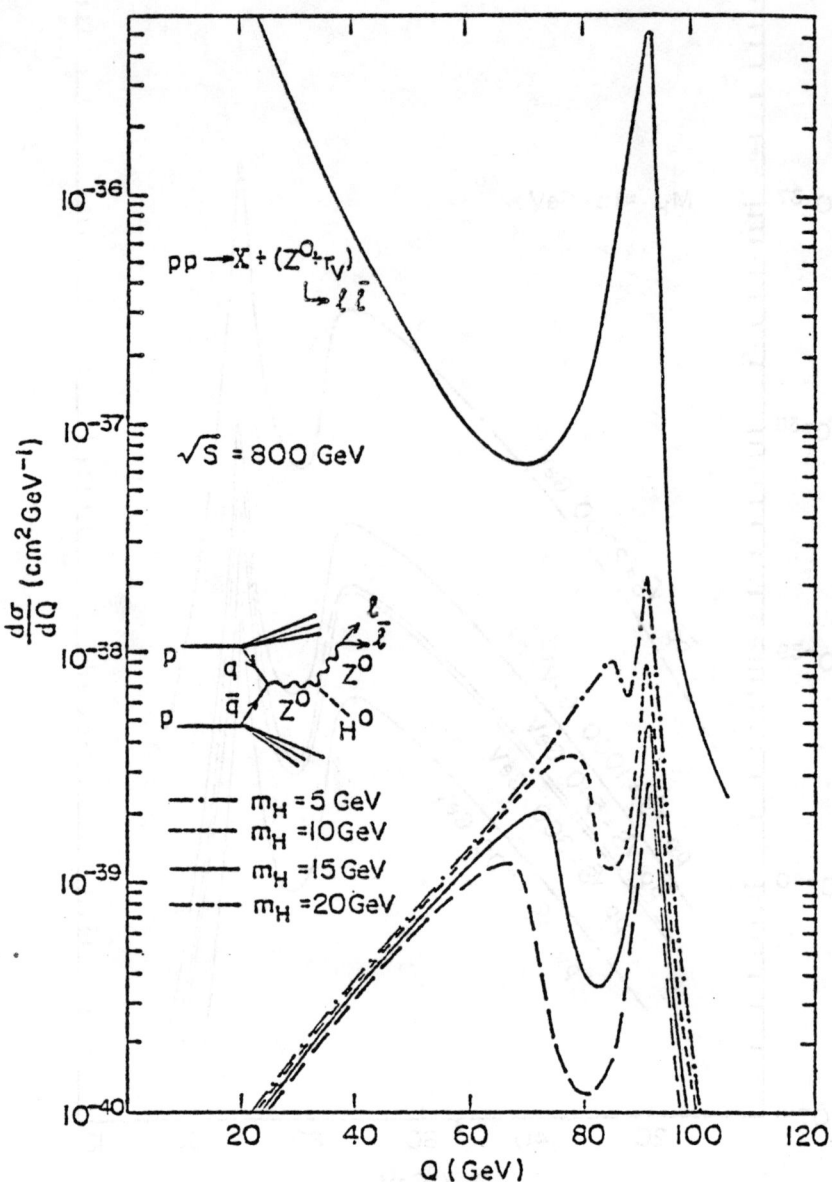

Fig. (2)

Fig. (2). Same as Fig. (1) but for $m_H = 5$ GeV, in pp scattering with $\sqrt{s}$ = 600, 800, 1000 GeV and in p̄p scattering with $\sqrt{s}$ = 540, 2000 GeV.

DISCUSSION

LANDE: Suppose one tried, in the pre-Isabelle and pre-Fermilab collider days, to look at very high energy cosmic ray interactions. What are the signatures one would look for in these things?

KANE: Well, the first comment is that the cross sections are too small to produce them that way. The heavy ones, the 200 GeV sort of things, for example, will have typically two kinds of dominant decay modes. One is pairs of gluons jets each with half the energy. The other is fermion pairs, $t\bar{t}$ for example. Some of the other states will decay into W and a gluon jet or a Z and a gluon jet. So events with multiple W's, multiple Z's, one or two hundred GeV gluon jets or heavy quark jets would be the typical signatures.

The light ones in the dynamical theory have only $s\bar{s}$ and μ pair decay modes, so they decay essentially completely into these, and they have a large branching ratio for μ pairs. All statements about decay branching ratios of any of these things are model-dependent much more than anything else I've said. The charged ones have essentially $c\bar{s}$ and $\tau\nu$ decays, and depending whether they are fundamental or not, the $c\bar{s}$ or the $\tau\nu$ will dominate.

LANDE: I should point out that the muon pairs are a very good, very easily recognized signature of the cosmic ray situation.

KANE: Sure, and at the accelerators, too. For example, an experiment with a K^+ beam is probably the optimum experiment that Fermilab could do to find μ pairs. Since you annihilate strange quarks, which are heavier, the signal-to-noises is within the limits of experimentaly possibility, but not with any other accelerators.

TYE: Just checking a point with the technicolor people. Is it true that in all technicolor models, the processes that L.L. Wang talked about would not occur? If H^0 is a technicolor Higgs, you cannot couple to the Z-pair.

KANE: I believe that's correct. (See Bég's talk).

TYE: So the question is, if you don't see it, can you make a clear statement?

KANE: If you are sure you could have distinguished it from the background, yes. There are a number of ways to distinguish pseudo-scalar Higgs from scalar Higgs. First you should see them and then worry about distinguishing them.

AXION

S. -H. H. Tye
Newman Laboratory of Nuclear Studies
Cornell University, Ithaca, NY 14853

ABSTRACT

Properties of the axion are briefly reviewed. The axion in its original form (as suggested by Peccei, Quinn, Weinberg and Wilczek) has been ruled out by experimental data. Here I describe a new axion. A new super-strong interaction is introduced to enhance the mass of the old axion but leaving its couplings to the quarks and leptons intact. Properties of this new force and the resulting axion are discussed.

The existence of instanton solutions[1] in Yang-Mills gauge theories was discovered by Belavin et al.[1] in 1975. This has far-reaching consequences in Quantum Chromodynamics (QCD). t'Hooft and others[2,3] pointed out that in the presence of instanton solutions labeled by a topological quantum number, the QCD vacuum is no longer the vacuum described by the naive perturbative approach. There are, in fact, a continuous set of possible vacuum labeled by a parameter θ. The Hilbert space of the gauge theory factors into subspaces of states on each θ vacuum and there is no transition between any two states which have different θ values. Clearly, nature must have chosen one particular value of θ.

For any value of $\theta \neq 0$, the effective QCD Lagrangian is the original Lagrangian with an additional term:

$$L_{effective} = L_{QCD} + \frac{i\theta g^2}{32\pi^2} G_\mu{}^a{}_\nu \tilde{G}_\mu{}^a{}_\nu \qquad (1)$$

where $\tilde{G}_\mu{}^a{}_\nu$ is the dual of the non-abelian field stress tensor $G_\mu{}^a{}_\nu$. The presence of the second term induces transitions which violate P and CP invariances. However, we know that nature breaks CP and P only weakly. In fact, the absence of an observed electric dipole moment of the neutron ($D_n \leq 10^{-24}(e)cm$)[4] implies[5] $\theta \leq 10^{-8}$. Clearly, θ must be extremely small if not exactly zero. There are a number of circumstances under which θ (or the effective θ) would be zero:

(1) $\theta = 0$ or $\theta \leq 10^{-8}$ by accident (or by some as yet unknown reason).

(2) the mass of the lightest quark, namely the u-quark mass $m_u = 0$ (ref. 2).
(3) introduction of a left-right symmetry to the Weinberg-Salam model, e.g., $SU(2)_L \times U(1) \to SU(2)_L \times SU(2)_R \times U(1)$ $G \times U_N(1)$ (ref. 6).
(4) introduction of an extra chiral $U(1)$ symmetry so that the effective $\theta = 0$. This results in an axion (ref. 7, 8).

Theoretically, option (1) seems rather arbitrary while option (2) has some apparent contradiction with our present understanding of current algebra. Option (4) has the advantage that it gives predictions which can be tested experimentally.

Here I shall concentrate on the physics of the last option. First, I shall briefly review the properties of the axion and then describe some of the modifications that have been proposed to change the axion properties in order to render it compatible with present experimental bounds. Finally, I shall describe a new suggestion.

The original Peccei-Quinn model is an extension of the Weinberg-Salam model with two Higgs doublets. One Higgs doublet couples only to the right-handed 2/3 charged (p) quarks and the other only to the right-handed -1/3 charged (n) quarks.

$$L = L_{QCD} + L_{W-S} + L_{Higgs}$$

where (I shall follow the notation of ref. 9) L_{Higgs} is the part of the Lagrangian that involves the two Higgs fields

$\chi_1 = \begin{pmatrix} \chi_1^o \\ \chi_1^- \end{pmatrix}$ $\chi_2 = \begin{pmatrix} \chi_2^+ \\ \chi_2^o \end{pmatrix}$ The Higgs potential $V(\chi_1, \chi_2)$ is chosen such that both χ_1 and χ_2 have vacuum expectation values so that it would produce four Goldstone bosons in the absence of gauge field couplings. In the presence of the gauge field couplings, three of them are absorbed by the Higgs mechanism in generating mass for the $W^\pm$ and Z^o bosons. The four Higgs particles that are massive would have typical masses of the order of the W boson mass. Let us consider the remaining "Goldstone" boson, which is the axion. Keeping only the axion field a, χ_1 and χ_2 can be written as

$$\chi_1 = \frac{f_1}{\sqrt{2}} \begin{pmatrix} 1 \\ 0 \end{pmatrix} \exp(ixa/f) \qquad (2a)$$

$$\chi_2 = \frac{f_2}{\sqrt{2}} \begin{pmatrix} 0 \\ 1 \end{pmatrix} \exp(ia/xf) \qquad (2b)$$

where $f^2 = f_1^2 + f_2^2 = (\sqrt{2} G_F)^{-1}$ and $x = f_2/f_1$. The axion Lagrangian becomes

$$L_a = \tfrac{1}{2}(\partial_\mu a)^2 - \sum_j m_{p_j} \{\bar{p}_j p_j \cos(\tfrac{xa}{f}) + i\bar{p}_j \gamma_5 p_j \sin(\tfrac{xa}{f})\}$$

$$- \sum_j m_{n_j} \{\bar{n}_j n_j \cos(\tfrac{a}{xf}) + i\bar{n}_j \gamma_5 n_j \sin(\tfrac{a}{xf})\}$$

$$- \sum_j m_{\ell_j} \{\bar{\ell}_j \ell_j \cos(\tfrac{a}{xf}) + i\bar{\ell}_j \gamma_5 \ell_j \sin(\tfrac{a}{xf})\} \qquad (3)$$

The residual U(1) symmetry is explicit. This allows the θ angle to be rotated away, i.e., $\theta_{effective} = 0$. The current associated with this U(1) axial vector current is

$$J_{5\mu} = f\partial_\mu a + \tfrac{x}{2} \sum_j \bar{p}_j \gamma_\mu \gamma_5 p_j$$

$$+ \tfrac{1}{2x} \sum_j (\bar{n}_j \gamma_\mu \gamma_5 n_j + \bar{\ell}_j \gamma_\mu \gamma_5 \ell_j) \qquad (4)$$

The divergence of this current is zero except for the QCD anomaly[10]

$$\partial^\mu J_{5\mu} = \tfrac{1}{2}(x + \tfrac{1}{x})N \frac{g^2}{16\pi^2} G\mu^a{}_\nu \tilde{G}\mu^a{}_\nu \qquad (5)$$

for 2N flavors.

The presence of the topological index associated with the instanton solutions[2,3] implies that this anomaly actually corresponds to a symmetry breaking in the true vacuum. The symmetry breakdown introduces a mass to the axion. Our experience from QCD tells us that the anomaly breaking ($\simeq m_\eta^2 - m_\pi^2$) is much bigger than the chiral symmetry breaking ($\sim m_\pi^2$). Hence, we expect the axion would acquire a mass smaller than that indicated by Eq.(5). Since nature is close to the chiral symmetry limit, it is convenient to use the current algebra technique to evaluate the coupling and mass of the axion. First, recall that for a massless quark q with the axial current $J_5^q{}_\mu = \bar{q}\gamma_\mu \gamma_5 q$, the divergence is

$$\partial^\mu J_{5\mu}^q = \frac{g^2}{16\pi^2} G\mu^a{}_\nu \tilde{G}\mu^a{}_\nu$$

Hence, we can define a new current associated with the axion:

$$J_{5\mu}^a = J_{5\mu} - \tfrac{1}{2}(x + \tfrac{1}{x})N \{\tfrac{1}{1+Z} \bar{u}\gamma_\mu \gamma_5 u + \tfrac{Z}{1+Z} \bar{d}\gamma_\mu \gamma_5 d\} \qquad (6)$$

(where $Z \simeq m_u/m_d$) whose divergence is

$$\partial^\mu J^a_{5\mu} = -(x + \frac{1}{x}) \frac{N}{1+Z} [m_u\{i\bar{u}\gamma_5 u \cos(\frac{xa}{f}) - \bar{u}u \sin(\frac{xa}{f})$$
$$+ Zm_d\{i\bar{d}\gamma_5 d \cos(\frac{a}{xf}) - \bar{d}d \sin(\frac{a}{xf})\}] \quad (7)$$

which is anomaly-free. It approaches zero as m_u or m_d goes to zero. The current associated with the pion is

$$J^3_{5\mu} = \tfrac{1}{2}(\bar{u}\gamma_\mu\gamma_5 u - \bar{d}\gamma_\mu\gamma_5 d) \quad (8)$$

whose divergence is given by

$$\partial^\mu J^3_{5\mu} = m_u\{i\bar{u}\gamma_5 u \cos(\frac{xa}{f}) - \bar{u}u \sin(\frac{xa}{f})\}$$
$$- m_d\{i\bar{d}\gamma_5 d \cos(\frac{a}{xf}) - \bar{u}u \sin(\frac{a}{xf})\} \quad (9)$$

Standard current algebra methods allow us to determine the axion mass:

$$m_a^2 = f^{-2} f_\pi^2 m_\pi^2 N^2 (x + \frac{1}{x})^2 \frac{Z}{(1+Z)^2} \quad (10)$$

For $Z \sim \tfrac{1}{2}$ and $x \sim 1$, the mass of the axion is roughly 100 KeV. Its coupling to the leptons are given by the original Lagrangian (3). Its strong interaction coupling can be determined by its mixing with the pion, which is given by the overlap of the axion and the pion currents.[8,9]

$$P_{\pi-a} = \tfrac{1}{2}\frac{f_\pi}{f}\{N(x + \frac{1}{x})(\frac{m_d - m_u}{m_d + m_u}) - x + \frac{1}{x}\} \quad (11)$$

However, an axion with the couplings given by eq.(3) and eq.(11) and a mass given by eq.(10) seems to have been ruled out by experimental and astrophysical arguments.[11,12]

In order to salvage the axion idea, something must be modified. For example (referring to Eq.(10) and (11)), we can change the model to modify:

(1) the N factor: this is carried out in Ref. 9 but does not seem to be sufficient.
(2) the x and 1/x factor: this can be achieved by introducing more Higgs fields so that there are more free parameters (ref. 13).

(3) the Z factor: this is done by assuming the u quark mass to be zero or very small. The compatibility of $m_u = 0$ with current algebra may be questioned. In any case, if $m_u = 0$, the whole Peccei-Quinn mechanism is no longer needed.

(4) the f factor: this can be done by introducing new unusual Higgs and quark fields (ref. 14). As a result, the axion mass must be smaller than 10^{-2} eV.[15]

Here I would like to suggest another way to change the axion properties. The present experimental situation is described by L. Mo in this workshop. I just note that the allowed axion mass must be bigger than 50 MeV if its couplings remain unchanged. (To see qualitatively the reason for this bound, we note that if $M_a > 50$ MeV, all the low energy nuclear experiments cannot produce axions; in beam dump experiments, its decay is too fast and hence is hidden by π^0 and other decays.)[12] The new model is going to add a new term to the R.H.S. in eq.(10).

Consider the introduction of a new force, a super strong interaction similar to QCD, but has a gauge group G different from the SU(3) in QCD. Let me consider the simplest version of the idea and refer to this as QC'D. The corresponding QC'D parameter Λ' is much bigger than the QCD parameter Λ ($\Lambda \sim 0.5$ GeV). Its "quarks" Q_j form left-handed isodoublets and right-handed singlets under the Weinberg-Salam SU(2) x U(1) model. For simplicity, let me consider one doublet $(U,D)_L$ only, with masses M_U and M_D. They couple to the two Higgs doublets the same way as the standard u and d quarks. Again for simplicity, (see ref. 16 for a more generalized discussion) let us assume $\Lambda' \gg M_U, M_D$ so that chiral symmetry breaking is much weaker than the anomaly breaking so that current algebra method is again applicable; in this case, the corresponding "pion decay constant" F_π is also of the order of Λ' and the "pion" mass is of the order $M_\pi^2 \sim (M_U + M_D)\Lambda'$; the axion mass is given by

$$f^2 m_a^2 = (x + \frac{1}{x})^2 \{f_\pi^2 m_\pi^2 \frac{Z}{(1+Z)^2} + F_\pi^2 M_\pi^2 \frac{Z'}{(1+Z')^2}\} \quad (12)$$

where it is expected that $Z' = M_U/M_D \sim 1$. The first term is of course negligible. The experimental constraint requires $F_\pi M_\pi / f_\pi m_\pi \geq 10^3$. In general QCD and QC'D have different θ angles, namely θ_c and θ_c' and only one of them can be rotated away. However, we expect these two forces to be unified at a higher energy scale so that $\theta_c = \theta_c'$ (at the tree level). Whatever $\Delta\theta = \theta_c - \theta_c'$ induced by higher order quantum corrections is expected to be small so that the extra U(1) symmetry renders $\theta_{effective} = 0$ for both QCD and QC'D. The small $\Delta\theta$ that may be left over is observable only in a process where QCD, QC'D and weak interaction all come in. Hence, the resulting CP violation is expected to be extremely small. The axion coupling to leptons and QCD quarks remain intact. We note that if $M_U = 0$ and/or $M_D = 0$, the axion would not gain weight from this new force.

Hence, we cannot identify this new force as the technicolor force[17] which requires the lightest doublets to be massless.

To have an idea of the lower bound of the mass scale of this new force, we take $F_\pi \sim M_\pi$ so that $M_\pi > 3$ GeV. This means that the new force, its new "quarks and new "gluons" and its new QC'D singlet bound states may be observable at the next rounds of experiments. (Details of experimental consequences of QC'D will be discussed in ref. 16.)

I thank W. A. Bardeen and J. D. Bjorken for valuable discussions.

REFERENCES

1. A. A. Belavin, A. M. Polyakov, A. S. Schwartz and Yu. S. Tyupkin, Phys. Lett. $\underline{59B}$ 85 (1975).
2. G. t'Hooft, Phys. Rev. Lett. $\underline{37}$, 8 (1976); Phys. Rev. $\underline{D14}$ 3432 (1976).
3. R. Jackiw and C. Rebbi, Phys. Rev. Lett. $\underline{37}$ 172 (1976); C. G. Callan, R. F. Dashen and D. J. Gross, Phys. Lett. $\underline{63B}$ 334 (1976).
4. N. F. Ramsey, Phys. Reports $\underline{43C}$ 409 (1978).
5. V. Baluni, Phys. Rev. $\underline{D19}$ 2227 (1979); R. Crewther, P. DiVecchia, G. Veneziano and E. Witten, Phys. Lett. $\underline{88B}$ 123 (1979).
6. e.g., M. A. B. Beg and H.-S. Tsao, Phys. Rev. Lett. $\underline{41}$ 278 (1978); See also R. N. Mohapatra and G. Senjanovic, Phys. Lett. $\underline{79B}$ 283 (1978); H. Georgi, Hadronic J. $\underline{1}$ 155 (1978); G. Segre and H. A. Weldon, Phys. Rev. Lett. $\underline{42}$ 1191 (1979).
7. R. D. Peccei and H. R. Quinn, Phys. Rev. Lett. $\underline{38}$ 1440 (1977); Phys. Rev. $\underline{D16}$ 1791 (1977).
8. S. Weinberg, Phys. Rev. Lett. $\underline{40}$ 223 (1978); F. Wilczek, Phys. Rev. Lett. $\underline{40}$ 279 (1978).
9. W. A. Bardeen and S.-H. H. Tye, Phys. Lett. $\underline{74B}$ 229 (1978); See also J. Kandaswamy, P. Salomonson and J. Schechter, Phys. Lett. $\underline{74B}$ 377 (1978); Phys. Rev. $\underline{D17}$ 3051 (1978).
10. S. Adler, Phys. Rev. $\underline{177}$ 2426 (1969); J. S. Bell and R. Jackiw, Nuovo Cumento $\underline{60A}$ 49 (1969); W. A. Bardeen, Phys. Rev. $\underline{184}$ 1848 (1969).
11. T. W. Donnelly et al., Phys. Rev. $\underline{D18}$ 1607 (1978); D. A. Dicus, E. W. Kolb, V. L. Teplitz and R. V. Wagoner, Phys. Rev. $\underline{D18}$ 1829 (1978).
12. Recently, an exhaustive analysis on the experimental and astrophysical bounds has been undertaken by J. D. Bjorken (unpublished).
13. P. Ramond and G. G. Ross, Phys. Lett. $\underline{81B}$ 61 (1979).
14. J. E. Kim, Phys. Rev. Lett. $\underline{43}$ 103 (1979).
15. D. A. Dicus, E. W. Kolb, V. L. Teplitz and R. V. Wagoner, Phys. Rev. $\underline{D22}$ 839 (1980).
16. S.-H. H. Tye (to be published).
17. S. Weinberg, Phys. Rev. $\underline{D19}$ 1277 (1979); L. Susskind, Phys. Rev. $\underline{D20}$ 2619 (1979).

DISCUSSION

TEPLITZ: In addition to an upper bound on the light KIM axion of about 1 eV, there is also, from the burning of stars, a lower bound on the Weinberg-Wilczek axion of about 350 KeV.

TYE: Well, the constraints on the mass of the axion have become an experimental issue. Actually that astrophysical bound, you know, is not very stringent; because if you do a survey of the present experimental bounds, it seems that if the axion has a mass ~400 KeV, you can more or less arrange it by taking the X value roughly about 2. Then you find most of the other experimental bounds, taking the uncertainties of the experiments and theory together, are not very stringent. There is a tiny domain of the choice of variables for the axion which may be still viable. And that choice plus, possible instanton contributions to the mass also, essentially, say, the mass would be somewhere above 400KeV. So that is why I say it is not dead, but it is very close to being ruled out.

PECCEI: I found if you took all experiments at face value, then in fact the standard axion wasn't there.

TYE: Yes, if you take all experiments at face value, that's true. However, any time you examine one experiment, one of two things always happens: either the experimental value is shaky by an order of magnitude or the theoretical number is shaky. For example, I would say nuclear reaction experiments are somewhat shaky (by an order of magnitude), as are beam dumps. The theory is shaky on K^+ decay for example, where the experiment is very good. And if you look at the e^- beam dump experiment for example, the bound essentially says X should have a value of ~2 which then tells you that the axion mass is of the order of a few hundred KeV, which is compatible with astrophysical bounds. So it looks very bad, I agree, for the standard model, but the book is not closed.

Search for "Axion-Like" Particles

Luke W. Mo
Virginia Polytechnic Institute
and State University
Blacksburg, Virginia 24061

I will represent our collaboration to give you a very brief report on a newly approved experiment at SLAC (E-137).[1] The experiment is designed to look for low mass, metastable, neutral, and weakly interacting particles. These particles are pseudoscalars. They could be similar to the classical axions[2] or the technicolor π^o's.[3]

The layout of the experiment is shown schematically in Figure 1. The high energy electron beam is dumped at the SLAC Beam Dump East. This beam dump is immediately followed by an earth shielding of approximately 200 m in depth. After the shielding, there is a decay space of about 200 m in length in front of an electromagnetic shower detector of high angular resolution. The possible production and the decay processes are shown in Figure 2. Specifically, the Primakoff production process shown in Fiugre 2a is most pertinent to the experimentation. In this case, the pseudoscalar produced by the photons in the beam dump is very much along the original electron beam direction. The subsequent decay into two photons again gives an electromagnetic shower along the forward direcgon. Because of the low mass under consideration, it is unlikely that the two photons can be separated in the detector. What we would observe are just high energy electromagnetic showers along the forward direction, similar to the case of elastic neutrino-electron scattering. This fact makes the experiment specific and well-defined.

Naturally, people would like to know the angular distribution of the photons, electrons, and positrons in the beam dump. This has been calculated by using the EGS program.[4] The result is shown

SSN:0094-243X/81/277-08$1.50 Copyright 1981 American Institute of Physics in Figure 3 in terms of transverse momenta. It can be seen that the "forwardness" of the processes is preserved. The sensitivity of the planned experiment is shown in Figures 4a and 4b for the case of production by Primakoff process and the decay into two photons. This case can be fairly well calculated. The other cases shown in Figure 2 are more speculative and we will not discuss it here.

The detectors we will use are those used at Fermilab for doing an elastic neutrino-electron scattering,[5] which is shown schematically in Figure 4. These detectors will be stacked into an assembly of 2 m x 3 m x 8 $r.l.$ It has an angular resolution of ±5 mr. By the time we do the experiment, probably the detector will be upgraded both in size and in resolution.

The experiment is approved of running with 30 Coulombs of electrons at an energy of 20 GeV or above. It is planned for the apparatus to be set up in the spring of 1981, and to take data by approximately September of 1981.

References

1. A. Abashian, J. D. Bjorken, S. D. Ecklund, L. W. Mo, W. R. Nelson, and Y. S. Tsai,. Proposal to Search for Low Mass, Metastable, Neutral Particles at SLAC (E-137), 1980.
2. S. Weinberg, Phys. Rev. Lett. $\underline{40}$, 223 (1978); F. Wilczck, Phys. Rev. Lett. $\underline{40}$, 279 (1978).
3. S. Dimopoulos, S. Raby and G. L. Kane, University of Michigan preprint (1980); M.A.B. Beg, invited talk at the XXth International Conference on High Energy Physics, Madison, Wisconsin, July 17-23, 1980.
4. W. R. Nelson, SLAC Report 210 (1978). "EDS Code System: Computer Programs for the Monte Carlo Simulation of Electromagnetic Cascade Showers".
5. R. H. Heisterberg, et al., Phys. Rev. Lett. $\underline{44}$, 635 (1980).

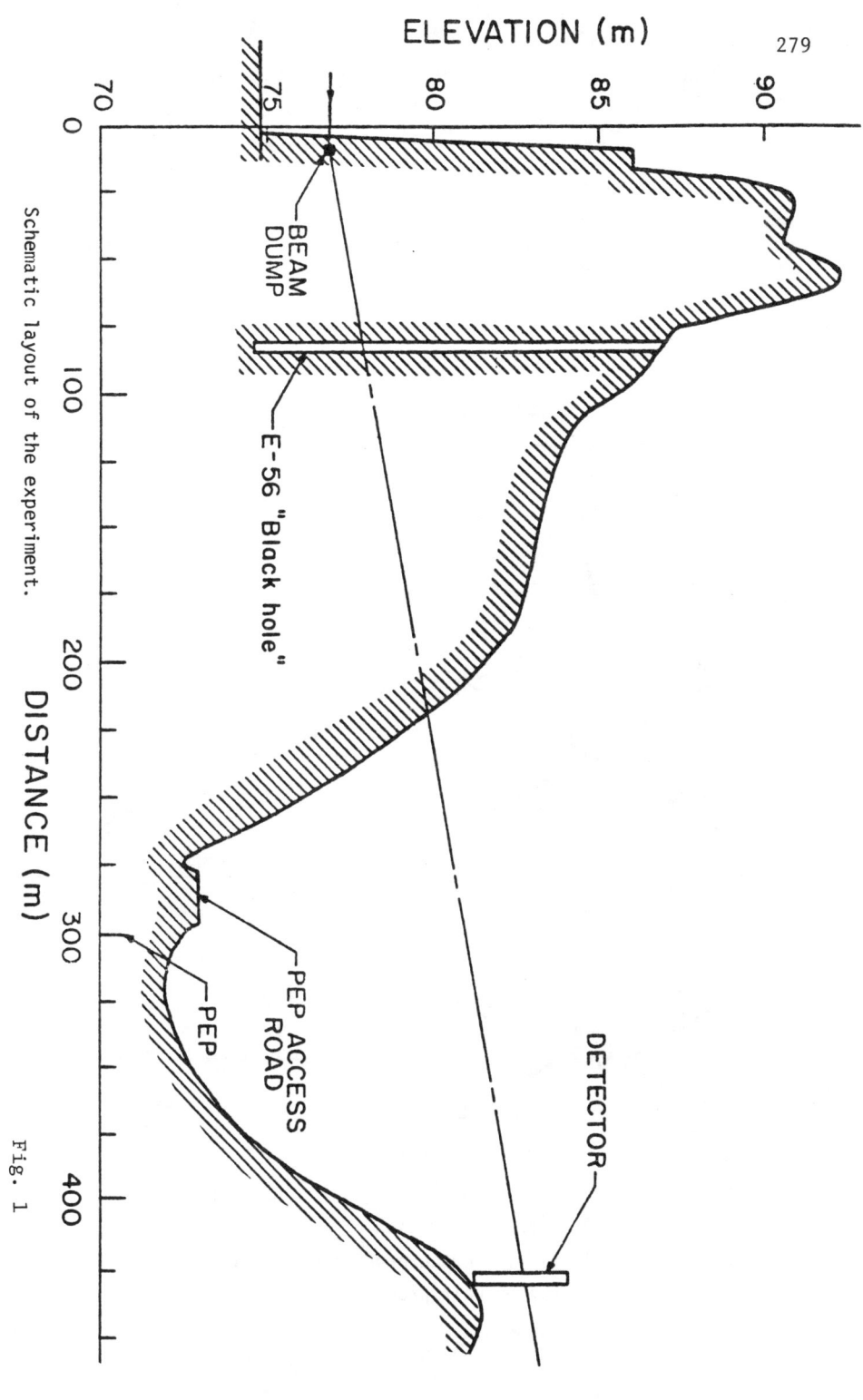

Fig. 1 Schematic layout of the experiment.

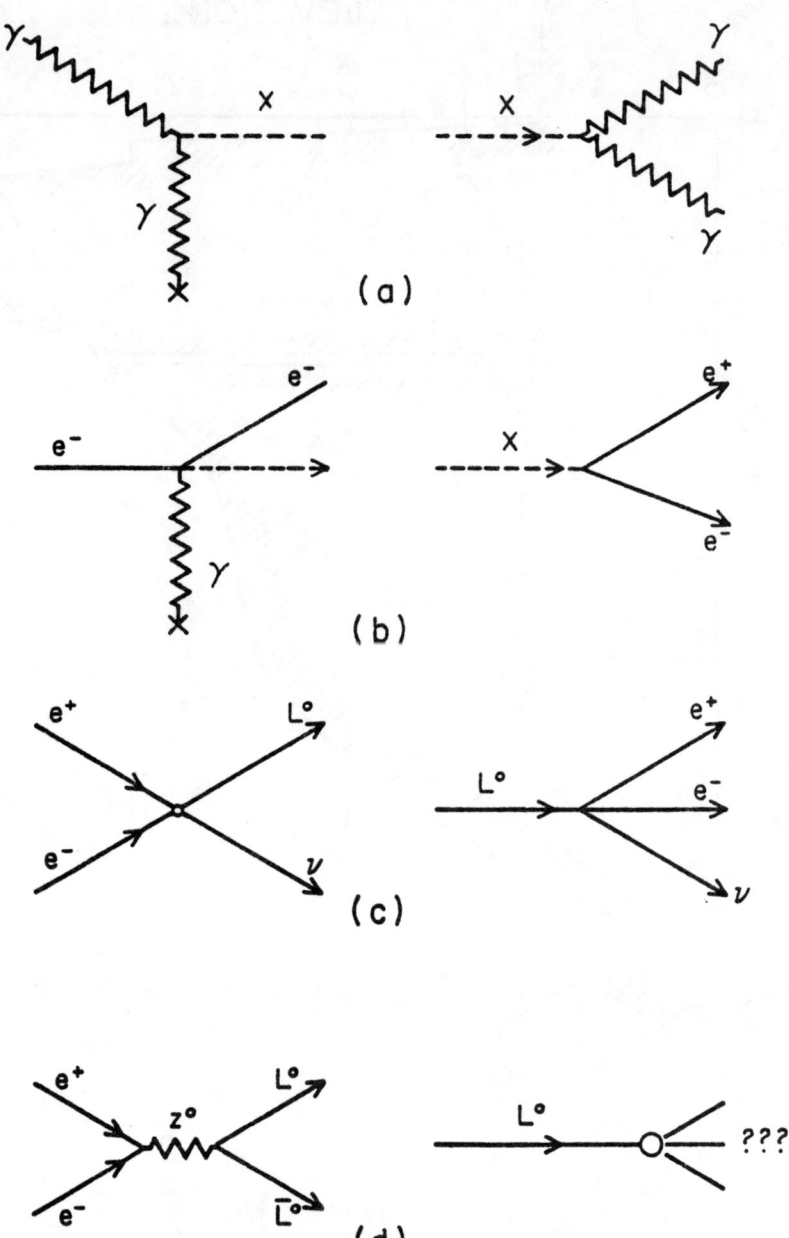

Figure 2   (a) Photoproduction of axion coupled to $\gamma\gamma$;
              (b) Electroproduction of axion coupled to e^+e^-;
              (c) Single lepton production by positron annihilation;
              (d) Lepton pair production by positron annihilation.

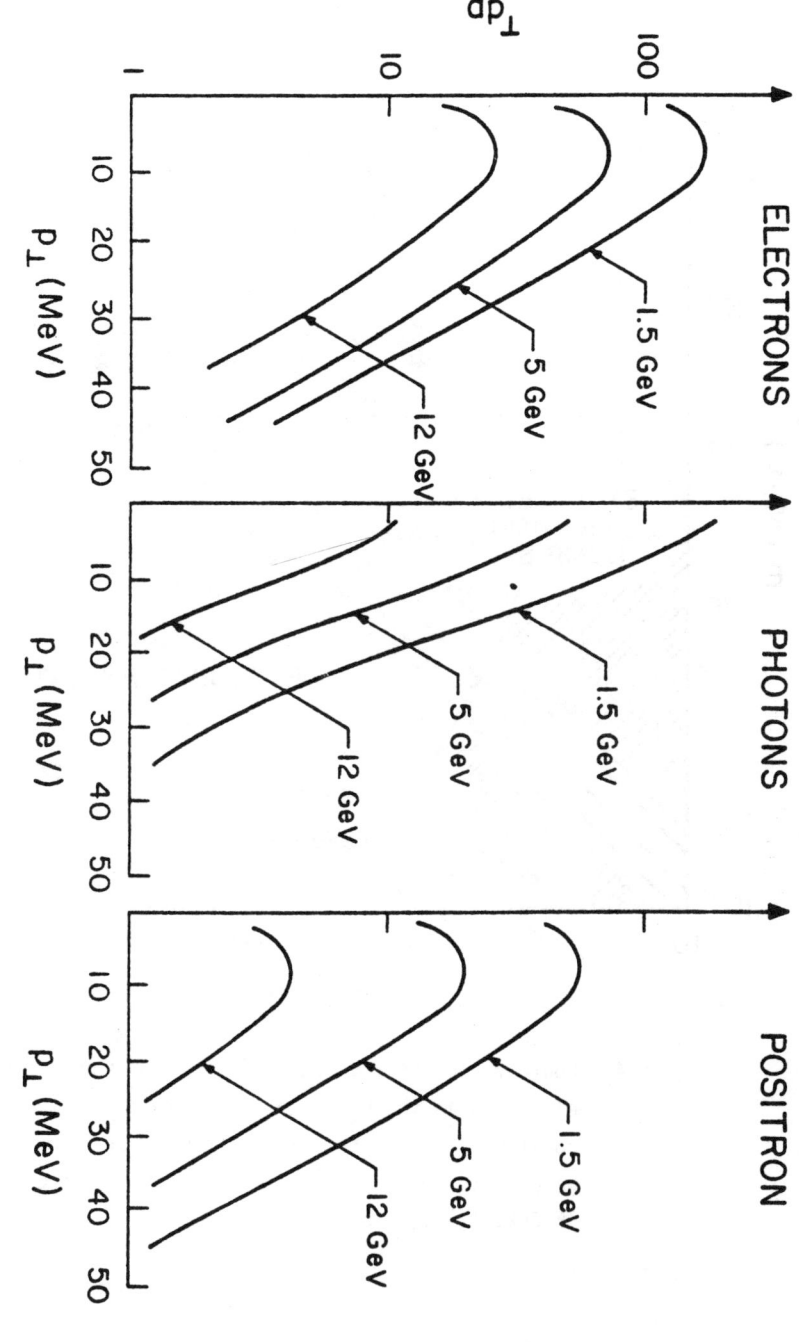

Figure 3 $P_\perp$ distribution of photons at various energies for incident electron energy E_0 = 15 and 30 GeV. The vertical scale is arbitrary.

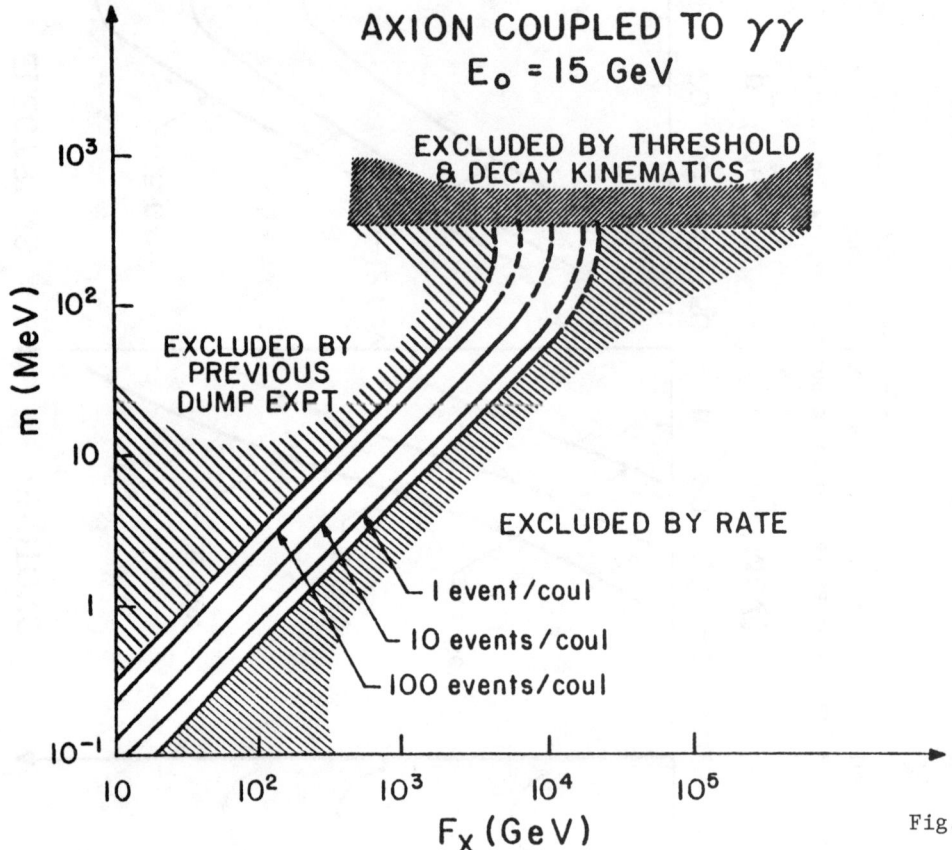

Figure 4 Contour plot of detected axions per Coulomb of incident electrons in the F_X - m plane, where F_X is the coupling constant and m is the mass of the axion. The axions are coupled to $\gamma\gamma$ and the incident electron energy is (a) E_0 = 15 GeV, (b) E_0 = 30 GeV.

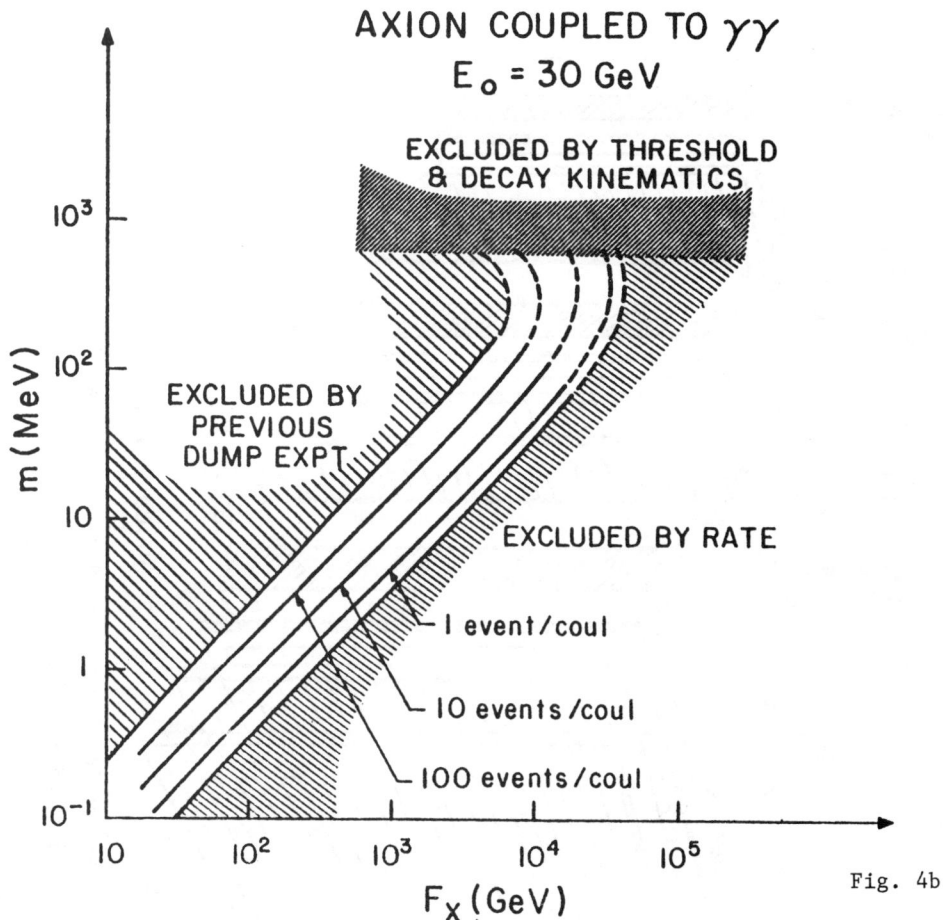

Fig. 4b

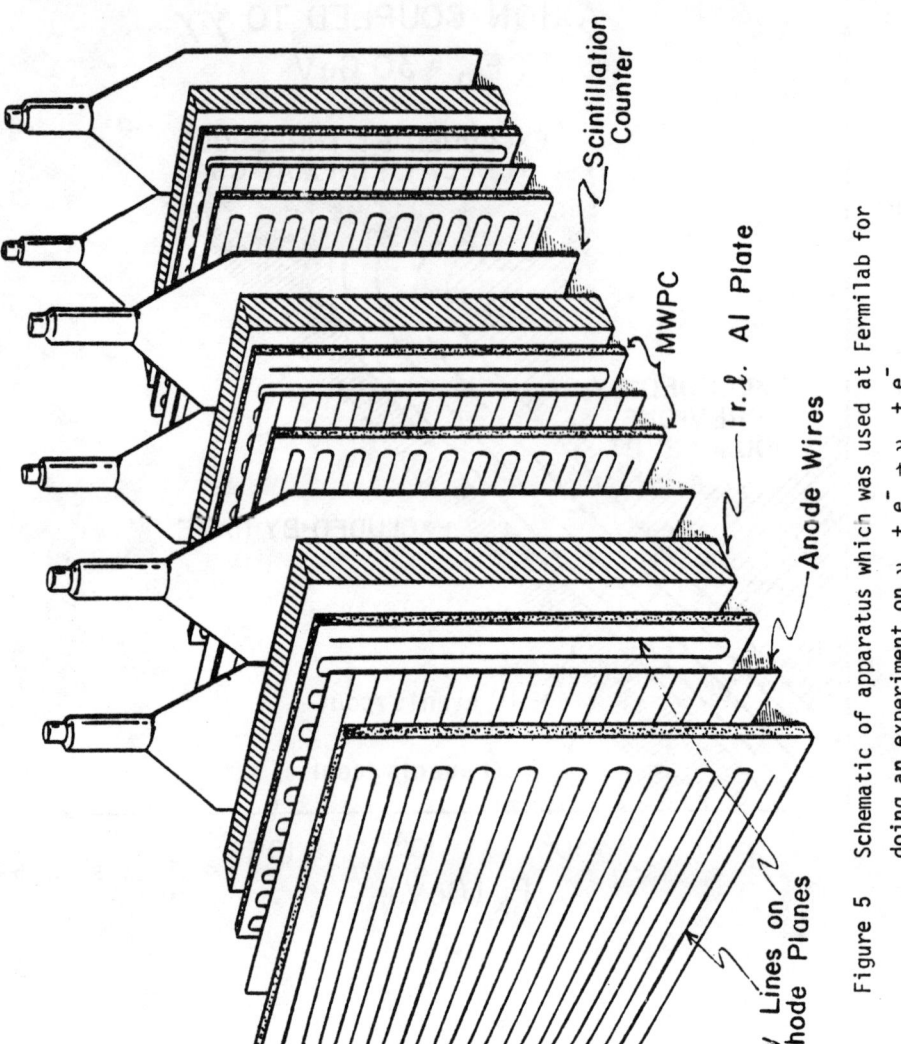

Figure 5 Schematic of apparatus which was used at Fermilab for doing an experiment on $\nu_\mu + e^- \rightarrow \nu_\mu + e^-$.

W's and Z's

T. L. Trueman
Department of Physics
Brookhaven National Laboratory
Upton, New York 11973

There are very few people skeptical enough to doubt that the intermediate vector bosons will be found when machines capable of observing particles of mass of order 100 GeV produced with cross sections of order 1 nb begin operation. In fact, I believe that most people expect that they will be found just where the Glashow[1]-Weinberg-Salam model[2] tells us they should be. Marciano and Sirlin[3] have calculated the masses to four digits in SU(5) to be

$$M_W = 84.36 \text{ GeV},$$
$$M_Z = 94.91 \text{ GeV}.$$

Does anyone have any idea how such precise predictions can be tested experimentally?

One problem is that the Weinberg-Salam model is so attractive and so successful that it takes a real intellectual effort to suggest alternatives against which it can be tested. QCD suffers from the same problem, but here we are better off because there are some consistent alternatives whose predictions at higher energy can be set off against the Weinberg-Salam predictions. For example, there is the $SU(2)_L \times SU(2)_R \times U(1)$ model[4,5] much studied by Mohapatra and Sidhu.[5] Another type is the $SU(2)_L \times U(1) \times U(1)'$ introduced by Georgi and Weinberg[6] and recently generalized to $SU(2)_L \times U(1) \times SU(2)'$ by Barger, Keung and Ma.[7] All of these theories predict additional bosons at high mass, but just what the allowed limits on these are depends on the other assumptions of the model. Thus, the left-right model has $g_L = g_R$ and so gives Z_1^o and $W_L^{\pm}$ near the usual Weinberg-Salam values with lower bounds for their partners:

$$M_{Z_2} \gtrsim 2.5 \, M_{Z_1} \simeq 230 \text{ GeV},$$
$$M_{W_R} \gtrsim 2.4 \, M_{W_L} \simeq 200 \text{ GeV}.$$

In the $SU(2) \times U(1) \times SU(2)'$ model, the couplings are not a priori

related and so there is a range of values with masses and widths
varying all over the place. One example which would be nice
because everything falls in a readily observable range, is

$$M_{W_1} = 43 \text{ GeV}, M_{Z_1} = 45 \text{ GeV}$$
$$M_{W_2} = 91 \text{ GeV}, M_{Z_2} = 100 \text{ GeV}$$

The production and decay of these particles in a $\bar{p}p$ collision at
$\sqrt{s} = 540$ GeV appears in the following single lepton spectrum. Although
the Z's and W's may not be resolvable in this measurement, it's
difference from the Weinberg-Salam prediction which has one peak at
around 40 and the other around 45 is marked. (Fig. 1)

Characteristic of non-abelian gauge theories is the trilinear
boson coupling and one would certainly like to measure the ZWW coupling.
Since it always enters mixed with other processes it is difficult to
isolate it, so one seems forced to work out the predictions of the
theory to compare with the data. Here the alternative theories can
be very useful, too, because they provide self-consistent alternatives
for comparison. Brown and Mikaelian[8] have calculated

$$e^+ + e^- \to W^+ + W^-,$$

amongst other processes, and emphasized it as a means of getting at
the $Z^0 W^+ W^-$ coupling. Figure 2 shows the standard predictions for
various masses. The dashed line shows the result when the trilinear
coupling is set to zero. The effect is dramatic, but perhaps not
very meaningful because such a theory doesn't make sense and will
eventually violate uncertainty. (Note, however, how small the cross
section and how high the energy in this example.) The theory of
Barger, Keung and Ma[7] does not suffer from this defect and so is
perhaps a more useful comparision. They present the following results

for a set of their allowed values--here the heavier Z goes into a pair of the lighter W's (Fig. 3). This is an even more striking difference. It would be nice to see a systematic exploration of the alternatives in this regard.

One very important effect, from a fundamental point of view, surely, but from a practical point of view with regard to W studies, is the transverse momentum with which it is produced. This subject seems on the verge of being cleaned up but I haven't seen the last word yet.

When the W is produced with momentum transverse to the beam direction, as we now expect--and as QCD predicts--the beautiful Jacobian peak in the decay lepton's transverse momentum distribution will clearly be modified. For largish values of this momentum, perturbative QCD makes unambiguous predictions. The following curve from Halzen and Scott[9] shows the signal remains quite strong (Fig. 4).

The behavior right at the peak is a bit more problematic and was handled in a phenomenological way by them. An accurate determination of the mass clearly depends on understanding the behavior right near the peak and there seems to have been a great deal of progress recently in being able to calculate the behavior all the way down to $k_\perp = 0$. Collins & Soper promise a paper in which the somewhat heuristic derivation of Parisi and Petronzio[10] is justified, but I have not seen it yet. Most of what understanding I have is due to discussions with Al Mueller who has his own approach to the problem and which gives the same result as Parisi and Petronzio. Suffice it to say: the distribution of the W near $k_\perp = 0$ is calculable and not dependent on some unknown primordial distribution,

the reason being that it becomes very improbable not to emit large
momentum gluons and so a $k_\perp = 0$ W is obtained by emitting gluons
with compensating momenta, thus washing out the initial distribution.
(This is clearly an approximation, better and better justified as $Q^2 \to \infty$;
Parisi and Petronzio reckon it is pretty good for W production. This
is a bit strange, but that is what they predict.

As a result, the distribution in $k_\perp$ is rather flat for a few GeV
and then switches over to a $1/k_\perp^2$ behavior for the most significant
part of its range, before falling really fast beyond $k_\perp \gtrsim M_W$. For
parameters rather near an 80 GeV W produced at ISABELLE, Parisi and
Petronzio show the following $k_\perp$ distribution. It is rather flat out to about
3 GeV and then plummets, dropping by an order of magnitude by 10 GeV (Fig. 5).

Frank Paige has run a Monte Carlo to determine how this distribution
would be reflected in the lepton $p_\perp$ distribution. (He uses an analytic
form that fits the curve of ref. 10 over the relevant range.) The
histogram[11] in Figure 6 was generated for ISABELLE parameters, between
$45°$ and $135°$. We see quite a sharp drop here at $M_W/2$ (note the linear
scale and true zero). The corresponding calculation has not been done
for the situation considered by Barger, Keung and Ma, but it seems clear
that the two sets of peaks will remain well separated when this is
taken into account.

To first order in $k_\perp$ all that happens, of course, is that the
Jacobian peak is displaced by $k_\perp$ parallel to the electron direction.
Since this $k_\perp$ must be balanced on the other side, presumably as
hadrons, presumably in a jet, it should be possible to correct for
this and shift all the peaks together. Calculations by Kajantie,
Lindfors and Raito,[12] for a similar problem, indicate that a large

fraction of the recoiling jets will occur in an observable region for experiment. However, there will be inevitable errors due to acceptance and resolution, jet identification etc., and the theory of the peak shape clearly needs to be understood so that the best possible mass determination will be achieved. Comparison of theory with experiment for the Z^0 will be very valuable in this regard.

A vexing problem is how to study the hadronic decay modes of W's. The branching ratio of W's into hadrons is supposed to be larger, by color factors, than into leptons and one would very much like to measure the spin structure of this coupling and the flavor dependence of it. There are two related problems here: one is extracting the jets resulting from W decay from the background; the other is determining the flavor of the jets.

One of the most prevalent ideas for extracting W decays from the background is based on the expectation that the flavor distribution in the W decay is quite different from the background jets. The former are expected to be equally $u\bar{d}$, $c\bar{s}$, $t\bar{b}$, etc., while the latter is mainly due to uu, ud, ug, etc. Of course, this is an expectation that one would ultimately want to verify. Various ways of utilizing this difference have been examined;[13] there is no doubt some signal at the quark level. The problem that everyone gets hung up on is that there does not seem to be any very good way of determining the flavor of a jet, either in an event-by-event basis or statistically. This realization goes back to Field and Feynman[14] and is inherited by everyone else who either uses their model or one very much like it.

For example, you might hope to enrich your sample of W's by asking for jets which contain a fast K. A group at BNL looked at this, and there is surely some enrichment but it is not as good as you might expect because u quarks are also a good source of K^+'s. We found, based on Field and Feynman, that the probability of getting a K^+ carrying more than 40% of the parent's momentum was 11% for $\bar{s}$ quarks and 5% for u quarks. So there is an effect but it is not very pronounced, and you must make severe cuts on your data even to achieve this.

Alternatively, you might look at average charges of jets. This was done in Ref. 14 and again recently in a model which includes baryon production, resonance decay and gluons, by Ranft and Ritter.[13] The problem they find is that the expected average charge distribution is too wide to make a safe identification. I show their result for a particular way of averaging charges, which weights fast particles more, suggested by Teper[15] in Figure 7. They don't give the corresponding result for d quarks but it is clear that there is substantial overlap and distinction of +2/3 and -1/3 quark jets would not be very certain.

We clearly have a great deal yet to learn about jet fragmentation both experimentally and theoretically and I hope that as we learn more, some reliable method for studying this interesting question will be developed.

In the Drell-Yan model of W production, the W's are produced predominantly with helicity -1. This comes about because they are made by left-handed quarks, right-handed antiquarks and tend to move in the direction of the quarks, which carry a larger fraction of the proton momentum. For the same reason, the decay lepton (e^-, μ^-, in particular) will move in the same direction as the boson while the decay anti-lepton (e^+, μ^+) will move in the opposite direction. This leads to very striking differences in the angular distrubution.

The next figure shows some old results of Peierls, Wang and myself[16] which shows this clearly. (When $p_\perp = M_W/2$ the lepton is coming out at 90° in the W rest frame and so simply maps out the longitudinal distribution of the parent W and is basically the same for both charges.)

This effect obviously results from <u>two</u> parity violating steps; in the production leading to longitudinally polarized W's and in the decay giving a forward backward asymmetry, or $\vec{J}\cdot\vec{k}_\ell$ term. However, viewed overall it is not manifestly parity violating: it is not a pseudoscalar; in fact, there is no pseudoscalar in the problem in this approximation because there are only two directions; the beam and the lepton momenta. Clearly, one would have to be a hardened skeptic, if angular distributions of this type are observed, not to associate them with parity violation; still, one would like to see clearly this basic attribute of the W.

As soon as one thinks about QCD corrections to W production, one realizes that now a place is formed by the incident momenta $\vec{p}$, the recoil jet momenta (be it quark or gluon) $\vec{q}$ and so a pseudoscalar can be formed with the lepton momenta $\vec{k}$. So we are in the classical situation of looking for an up-down asymmetry with regard to the production plane. Sad to say, it is quite difficult to predict the size of this effect.

Define the W density matrix with regard to any of the various convenient axes in the production plane, such as the beam direction, and measure the azimuthal angle of about that direction from the production plane. The asymmetry $A(\theta,\phi)$ is the normalized rate for leptons coming out at angle (θ,ϕ) minus those arising out at angle $(\theta,2\pi-\phi)$. For some generality, which allows us to consider Z decay or right handed W decays as well, let the decay amplitude into left handed leptons be a_- and into right handed leptons be a_+ then

$$A(\theta,\phi) = (|a_+|^2 + |a_-|^2)(\text{Im}\rho_{-1,1} \sin^2\theta \sin 2\phi$$
$$+ \sqrt{2}\,[\text{Im}\rho_{1,0} - \text{Im}\rho_{0,-1}]\cos\theta \sin\theta \sin\phi)$$
$$+ (|a_+|^2 - |a_-|^2)(\sqrt{2}\,[\text{Im}\rho_{1,0} + \text{Im}\rho_{0,-1}]\sin\theta \sin\phi)\,.$$

Parity conservation in the decay, as it nearly is for the standard Z^0, gives $|a_+|^2 = |a_-|^2$ and the last term would be absent. The first two terms instead depend on parity violations in the production and so are proportional to the (appropriately weighted) vector-axial vector interference of the quark couplings. Since the angular dependence of all these terms is different, it is clear that such measurement would help disentangle the boson vector and axial vector couplings to quarks and leptons.

The reason the calculation is difficult can be seen in the fact that every term in $A(\theta,\phi)$ contains the imaginary part of a density matrix element and the Born graphs (Fig. 9a) are real. Thus, in order to get a non-vanishing result one must go to next order in QCD and calculate graphs such as those in the next Fig. 9b. This is obviously a very extensive calculation and I haven't done it. Let me just make some comments: I think it is sensible in that soft gluons will not contribute to the phase of ρ_{ij}. (I haven't proved this by any means but have seen it to be true for certain kinds of graphs). The graphs are similar to those calculated by Devoto, Pumplin, Repka and Kane[17] for $q + g \to q + \gamma$, but are more complicated because of V-A interference and the fact that the W has mass. They found photon linear polarization of a few percent and I would expect that in this case the result would be even smaller. First of all this is the same factor of α_s in these graphs; in addition, I expect that $\text{Im}[\rho_{1,0} - \rho_{0,-1}] \propto k_\perp/M_W$ and $\text{Im}\rho_{-1,1} \propto (k_\perp/M_W)^2$, and we expect from Parisi and Petronzio[10] that the cross section for $k_\perp/M_W > 0.1$ is very small. Finally, the calculation is interesting because it is crucially dependent on higher order QCD calculations (as is the case in Ref. 17) and so can be the source of very valuable information about

QCD in addition to the weak interactions.

However, interesting as it might be, it clearly will be very difficult and will require very precise and extensive studies of W decays.

Finally, I would like to emphasize the virtue that polarized proton (or anti-proton) beams would have in studying the intermediate bosons. This was pointed out by Okun[18] some years ago and some detailed calculations were carried out by Haber and Kane[19] and by Paige, Tudron and myself.[20] The point is very simple: left-handed quarks make W's, right handed ones don't. To the extent that the quark helicity is correlated to its parent proton's helicity--and this correlation appears to be quite strong[21]--left handed protons will be a better source of W's than right handed ones. In any case if one subtracts the results with right handed protons from that with left handed, only parity violating signals will remain. The important point is that the signal from W production expected from these calculations is very large; very little is lost by the subtraction because of the high correlation of quark spin and proton spin. The next two figures show the results of Paige, Tudron and myself (updated by Paige to include non-scaling structure function and weak-strong interference[22]). These figures show $\sigma_{--} - \sigma_{++}$ compared with the parity conserving background for jets at 90° and at 20°, $\sqrt{s}$ = 800 GeV. Evidently, it requires quite precise experiments to insure that the parity conserving background is properly subtracted, but it does seem feasible.

At a much cruder level, the single lepton signal from W decay has very little background and this manifestly parity violating asymmetry should hit you right in the eye.

In conclusion then, I think there are several interesting issues regarding W and Z^0 production and decay that require further theoretical consideration. There seems to be much that can be learned from experiments which study their properties, but many of them will have to be of high precision and require extensive data.

References

1. S. Glashow, Nucl. Phys. $\underline{22}$, 579 (1961).
2. S. Weinberg, Phys. Rev. Lett. $\underline{19}$, 1264 (1967).
   A. Salam, Nobel Symposium No. 8, ed. N. Svartholm (Almquist and Wiksell, Stockholm, 1968).
3. W. Marciano and A. Sirlin, preprint (1980).
4. R. Mohapatra and D. Sidhu, Phys. Rev. Lett. $\underline{38}$, 665 (1977) and Phys. Rev. $\underline{D16}$, 2843 (1976).
5. D. Sidhu, BNL 27614 (1979).
6. H. Georgi and S. Weinberg, Phys. Rev. $\underline{D17}$, 275 (1978), E.H. deGroot, G.J. Gounaris, and D. Schildknecht, Phys. Lett. $\underline{85B}$, 399 (1979).
7. V. Barger, W.Y. Keung, E. Ma, Phys. Rev. (to be published).
8. R.W. Brown and K.O. Mikaelian, Phys. Rev. $\underline{D19}$, 922 (1979).
9. F. Halzen and D.M. Scott, Phys. Lett. $\underline{78B}$, 318 (1978).
10. G. Parisi and R. Petronzio, Nucl. Phys. $\underline{B154}$, 427 (1979).
11. F. Paige, private communication.
12. K. Kajantie, J. Lindfors, and Risto Raitio, Nucl. Phys. $\underline{B144}$, 422 (1978).
13. P.K. Williams, S.U. Chung, V. Flaminio, F.E. Paige, E.A. Paschos and T.L. Trueman, Proceedings of 1977 Summer Workshop on ISABELLE, p. 224; M. Abud, R. Gatto, and C.A. Savoy, Ann. Phys. (NY) $\underline{122}$, 219 (1979); J. Ranft and S. Ritter, KMU-HEP-80-10 (1980).
14. R.D. Field and R.P. Feynman, Nucl. Phys. $\underline{B136}$, 1 (1978).
15. M.J. Teper, RL-79-096 (1979).
16. R.F. Peierls, T.L. Trueman and L.L. Wang, Phys. Rev. $\underline{D16}$, 1397 (1977).
17. A. Devoto, J. Pumplin, W. Repko and G.L. Kane, Phys. Rev. Lett. $\underline{43}$, 1064 (1980).

18. L.B. Okun, Preprint ITEP-66 (1976).

19. H.E. Haber and G.L. Kane, Nucl. Phys. <u>B146</u>, 109 (1978).

20. F.E. Paige, T.L. Trueman, and T.N. Tudron, Phys. Rev. <u>D19</u>, 935 (1979).

21. M.J. Alguard, et al., Phys. Rev. Lett. <u>37</u>, 1258 (1976); <u>37</u>, 1261 (1976); <u>41</u>, 70 (1978) and references cited therein.

22. F.E. Paige, Invited talk at Topical Workshop on Production of New Particles in Super High Energy Collisions, Madison, Wisconsin (1979). Explicit QCD corrected calculations of these asymmetries have recently been carried out by K. Hidaka, Westfield College preprint (1980) and substantial asymmetries are found.

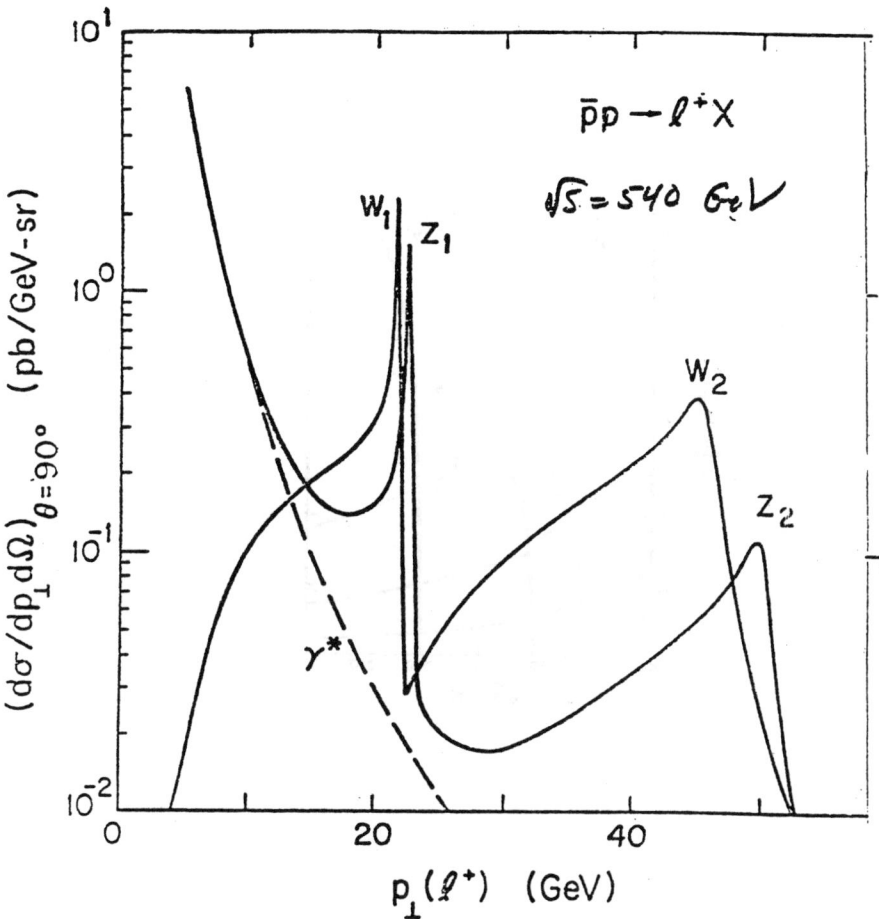

Fig. 1 Single lepton spectrum for the model of Barger, Keung and Ma, taken from ref. 7.

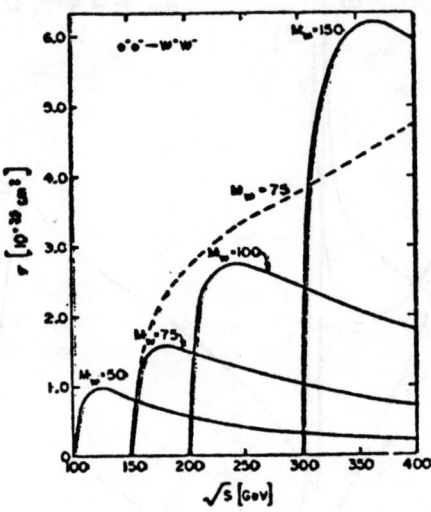

Fig. 2 Total cross sections, in units of 10^{-36} cm^2, for $e^+e^- \rightarrow W^+W^-$ in the Weinberg-Salam gauge theory, from ref. 8.

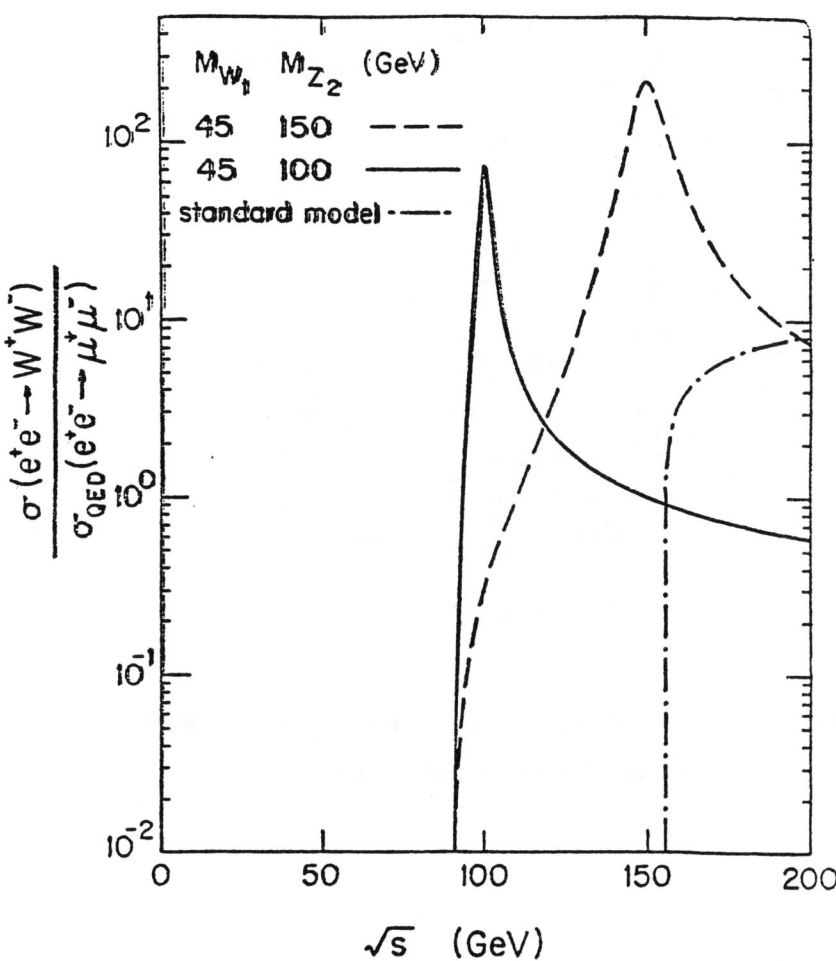

Fig. 3 The ratio of $\sigma(e^+e^- \to W^+W^-)/\sigma(e^+e^- \to \mu^+\mu^-)$ vs. $\sqrt{s}$, from ref. 7.

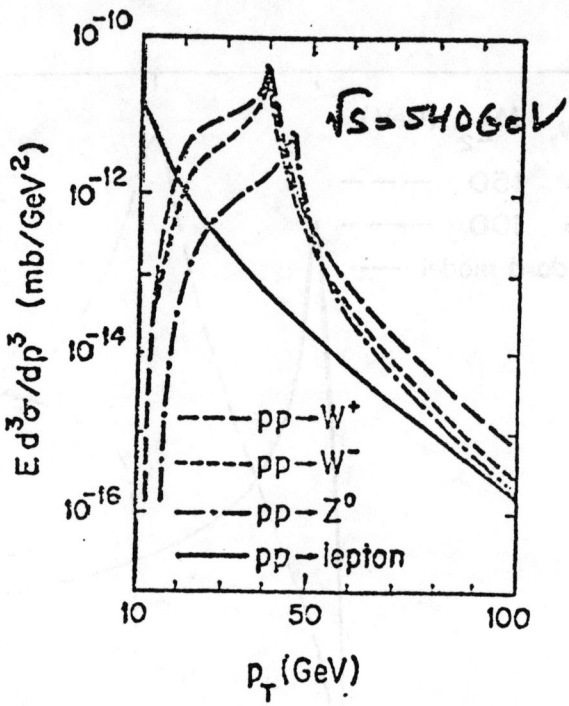

Fig. 4 Single lepton spectrum with some transvere momentum smearing folded in, from Halzen and Scott ref. 9.

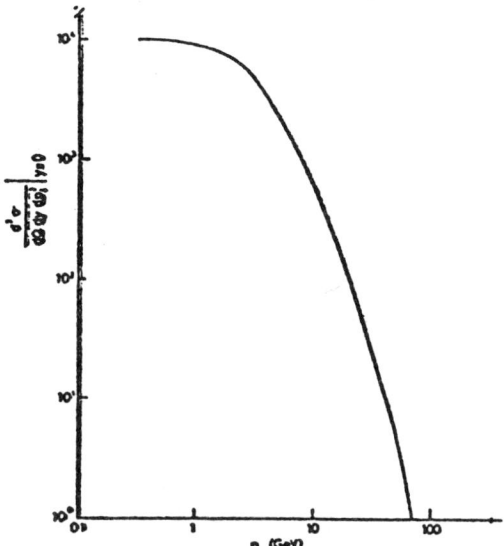

Fig. 5 $p_\perp$ distribution of lepton pair for $s = 7.5 \times 10^5$ GeV^2 and $Q^2 = 7500$ GeV^2 from Parisi and Petronzio, ref. 10.

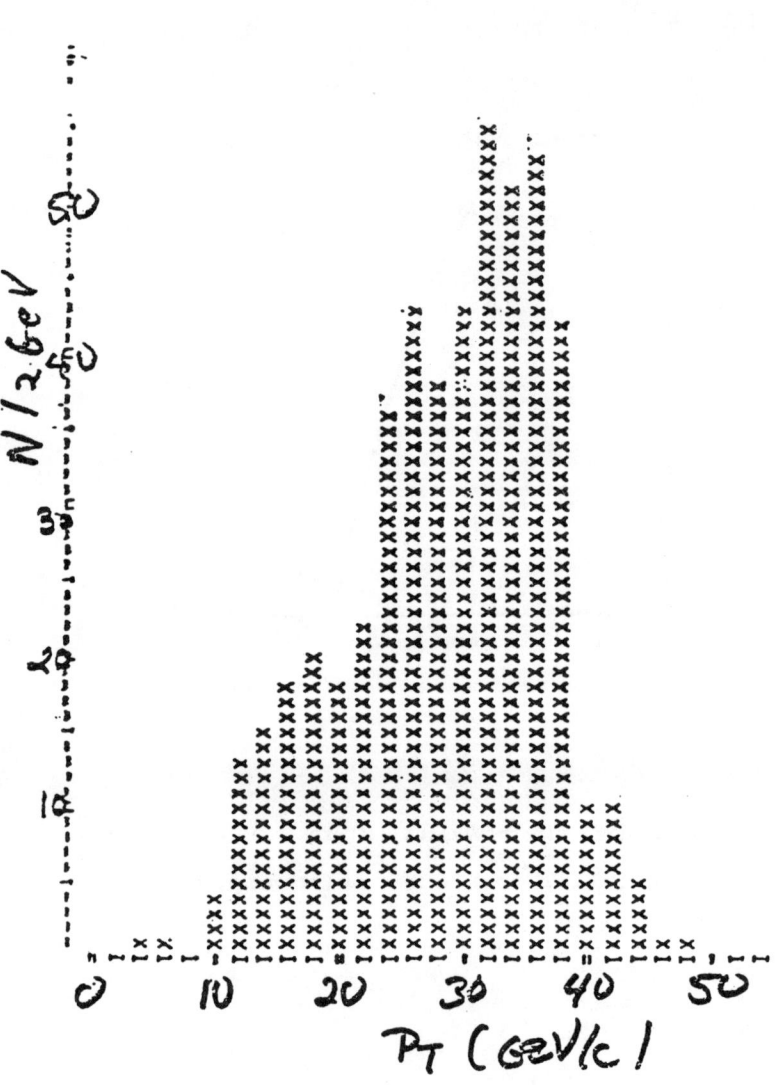

Fig. 6   Monte Carlo prediction of single lepton spectrum due to F. Paige and S. Protopopescu (private communication) M_W = 77.8 GeV.

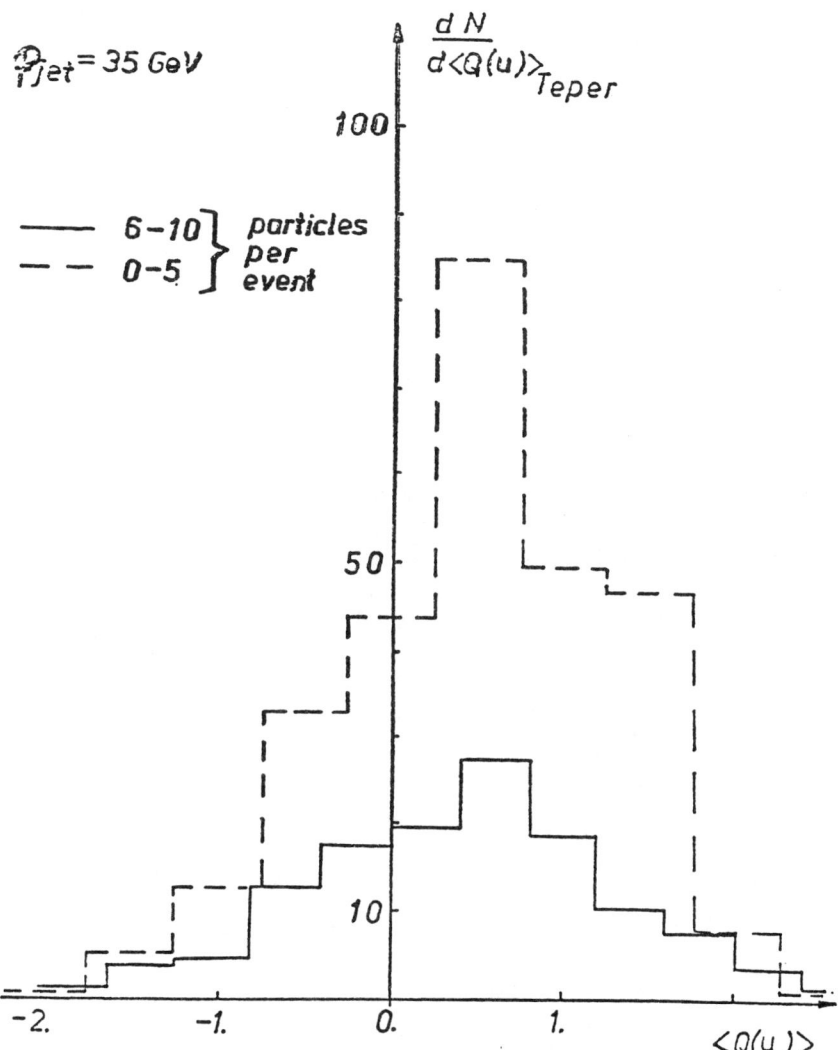

Fig. 7 Average charge distribution according to Ranft and Ritter, ref. 15.

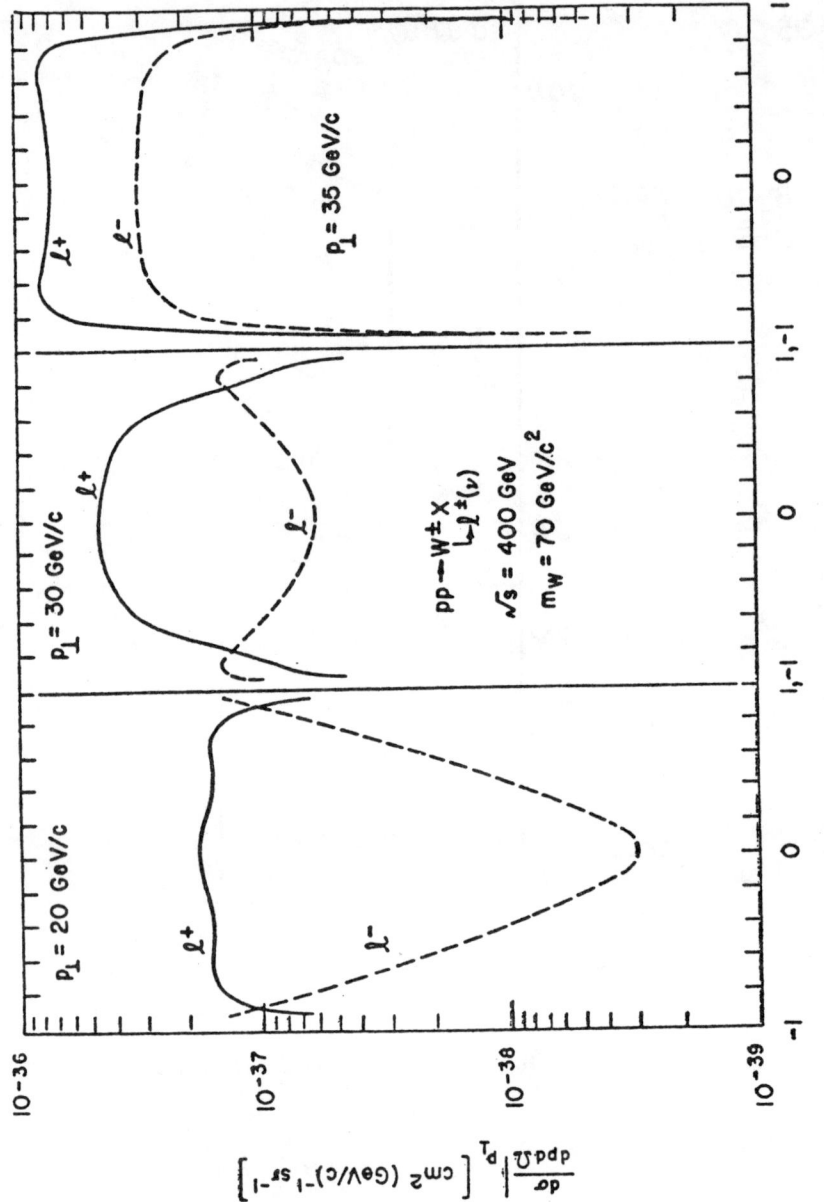

Fig. 8 Single lepton angular distribution from Peierls, Trueman and Wang, ref. 16.

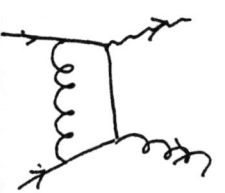
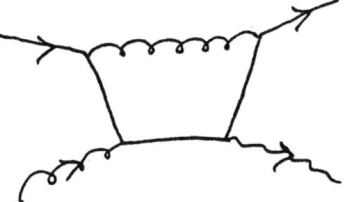

Figure 9a

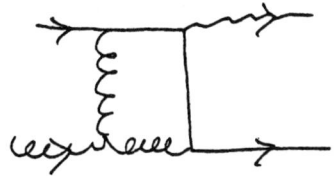

Figure 9b

Fig. 9 (a) Two of the lowest order QCD corrections yielding large $k_\perp$ W's.
(b) Some typical second order corrections.

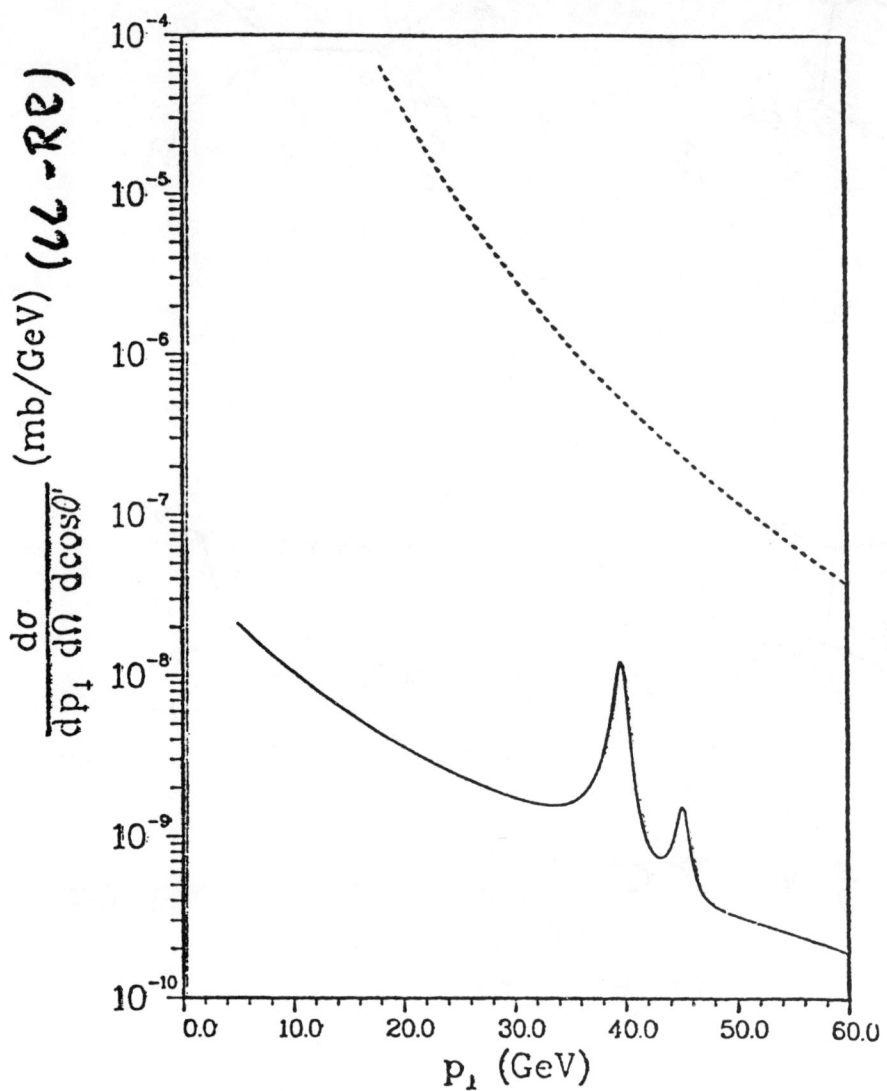

Fig. 10 Asymmetry in production of hadronic jets from W and Z decay at $\sqrt{s}$ = 800 GeV, at 90°. The dashed curve is the unpolarized backgrou

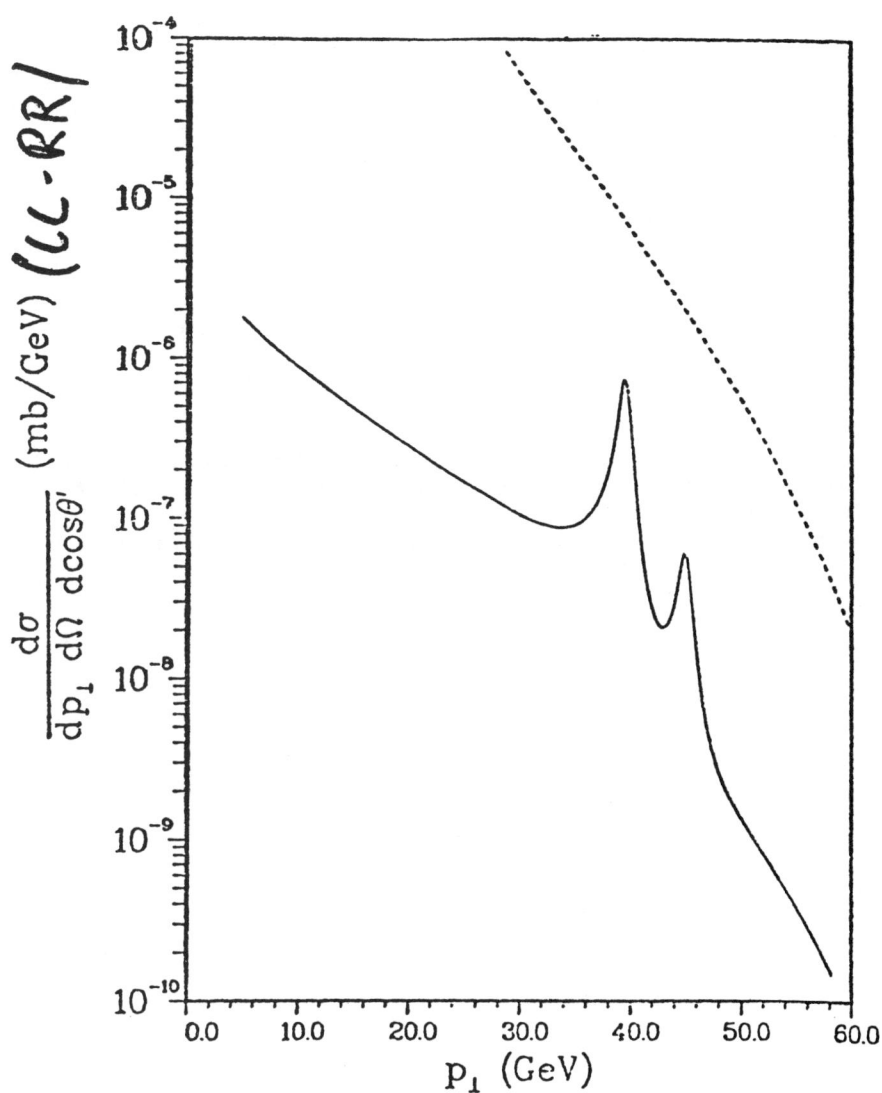

Fig. 11 Same as Fig. 10 but for jets at 20°.

DISCUSSION

STROVINK: I wanted to comment that Chris Day of LBL using a QCD-based Monte Carlo has demonstrated, at least on a statistical basis, that it should be quite possible to distinguish t jets from b and light quark jets, using the existing lower limit on the t quark mass and assuming that it is light enough that W's or Z's decay into it. The distinction is made using the "fatness" or shape of the jet.

SENJANOVIC: I just want to use the opportunity to make a comment on left-right symmetric theories here, which deals with the lower bound on the right-handed charged boson. It turns out that the best way to understand the smallness of the neutrino mass is to have a model in which the right-handed neutrino is a heavy Majorana lepton. This is the work of Mohapatra and myself about a year ago. In that case the world is V-A at low energies, not because the W_R is heavy, but because the light neutrino is left-handed. So you don't have any lower limit on the W_R mass but what the neutral current analysis puts on it. It is amusing that **this allows** a low W_R mass (this is the work of Tom Rizzo and myself) about 150 GeV and also the W_L mass could be as low as 70 GeV, which would be clearly different even if the radiative corrections are included.

MULTI-GENERATION PROBLEMS

Leptons

NEUTRINO OSCILLATIONS

V. Barger

Physics Department, University of Wisconsin, Madison, WI 53706

ABSTRACT

Aspects of neutrino oscillation phenomenology are discussed, including correlations of oscillation averages, a reexamination of solar neutrino oscillations, possible tests for CP violation in the neutrino sector, the results of reactor measurements, future prospects with deep mine experiments, and the possibility of doublet-singlet neutrino oscillations.

INTRODUCTION

If neutrinos have non-degenerate masses and mix, then oscillations will occur in neutrino beams with interchanges of neutrino types. Three classes of neutrino oscillations have been entertained: (i) <u>Neutrino-antineutrino oscillations</u>($\nu \leftrightarrow \bar{\nu}$) were suggested[1] by Pontecorvo in 1958, but these are suppressed by $(m_\nu/E_\nu)^2$ for a V-A coupling. (ii) <u>Flavor oscillations</u> ($\nu_e \leftrightarrow \nu_\mu \leftrightarrow \nu_\tau$) were first introduced[2] by Maki, Nakagawa, and Sakata in 1962 (and independently by Pontecorvo in 1967) on the basis of a quark-lepton analogy. (iii) <u>Doublet-singlet oscillations</u> ($\nu \leftrightarrow \eta$) may also exist in which the ordinary weak doublet members ν oscillate into essentially sterile (non-interacting) singlets η. This possibility has been widely discussed in the past year.[3] In this report we focus primarily on flavor oscillations, which have been the subject of most phenomenology. At the end we will briefly address the doublet-singlet oscillation possibility.

The usual charged weak current neutrino eigenstates ν_α (with n flavors $\alpha = e, \mu, \tau, \ldots, f_n$) are related to mass eigenstates ν_i (mass m_i with $i = 1, 2, 3, \ldots, n$) by a unitary transformation

$$|\nu_\alpha\rangle = \sum_{i=1}^{n} U_{\alpha i} |\nu_i\rangle . \qquad (1)$$

The $\nu_\alpha \to \nu_\beta$ transition probability at a distance L from a relativistic ν_α source is

$$P(\alpha \to \beta) = \left| \sum_i e^{-i\Delta in} U_{\alpha i} U_{\beta i}^* \right|^2 \qquad (2)$$

where

$$\tfrac{1}{2}\Delta_{in} = \frac{(m_i^2 - m_n^2)L}{4E} = \frac{1.27 \delta m_{in}^2 (eV^2) L(m)}{E(MeV)} . \qquad (3)$$

CORRELATIONS OF OSCILLATION AVERAGES

If all $\delta m^2 (eV^2) \gg E(MeV)/L(m)$, a complete oscillation average occurs with

$$\langle P(\alpha \to \beta) \rangle = \sum_{i=1}^{n} |U_{\alpha i}|^2 |U_{\beta i}|^2 \geq \frac{1}{n}. \quad (4)$$

Oscillation averages may also result in other situations due to ΔL or ΔE resolution [i.e., $\Delta L \gg E/\delta m^2$ or $\Delta E \gg E^2/(\delta m^2 L)$]. The complete oscillation average has long been of interest in connection with the solar neutrino measurements, where $L/E \sim 10^{10}$ m/MeV. The solar minimizing solution for three neutrinos

$$\langle P(e \to e) \rangle = \frac{1}{3} \quad (5)$$

is obtained with $|U_{ei}|^2 = 1/3$ for $i = 1,2,3$. Two extreme variants of the solar minimizing solution are:
  I. The CP conserving case
  II. The maximal CP violating case.[4,5]
The average $\nu_\mu \to \nu_\mu$ transition probability is given in these two cases by

$$\text{I.} \quad \langle P(\mu \to \mu) \rangle = \frac{1}{2} \qquad \text{II.} \quad \langle P(\mu \to \mu) \rangle = \frac{1}{3} \quad (6)$$

Only with CP violation (CPV) can $\langle P(\mu \to \mu) \rangle$ be less than 1/2 for $\langle P(e \to e) \rangle = 1/3$. Solution II is a fully symmetric minimizing solution, with all $\langle P(\alpha \to \beta) \rangle = 1/3$ and all $|U_{\alpha i}|^2 = 1/3$.

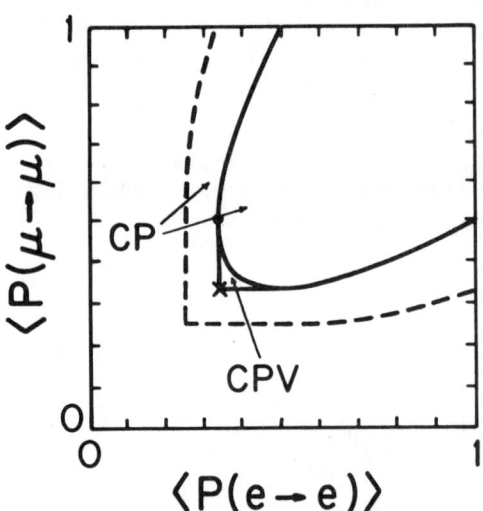

Fig. 1. Correlated ranges of the average transition probabilities $\langle P(\mu \to \mu) \rangle$ and $\langle P(e \to e) \rangle$ in the CP-invariant and CPV cases (from Ref. 6.)

More generally, for correlated pairs of average probabilities the allowed ranges depend on the number of neutrinos n and whether CP is violated.[6] This is illustrated in Fig. 1 for the case of $\langle P(\mu \to \mu) \rangle$ versus $\langle P(e \to e) \rangle$ for n = 3 (solid curves) and n = 4 (dashed curves). CPV expands the region of possible values for the correlated pairs of average probabilities. In this figure the solar minimizing solutions are denoted by the dot (I) and cross (II).

Any experiments which average over all oscillations can be used in correlation tests of CPV or n. If all $\delta m^2 > 10^{-5}$ eV^2, deep mine measurements of average

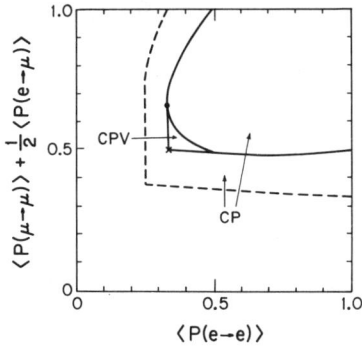
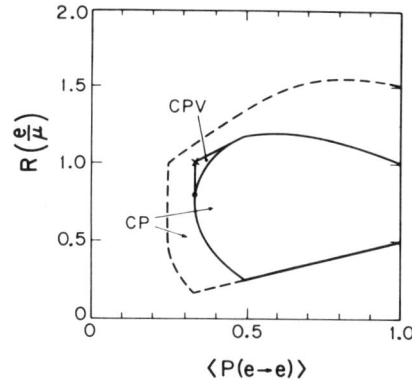

Fig. 2. Correlations of average probabilities measurable in deep mine and solar ν experiments.

Fig. 3. The e/μ ratio for deep mine experiments versus the solar $<P(e \to e)>$ in the CP and CPV cases.

transition probabilities may be used in conjunction with solar neutrino measurements of $<P(e \to e)>$. The atmospheric neutrino flux source[7] of these experiments is approximately $2(\nu_\mu + \bar{\nu}_\mu) + (\nu_e + \bar{\nu}_e)$. In experiments that detect muons from $\nu_\mu N \to \mu X$ interactions in the surrounding rock, the combination $<P(\mu \to \mu)> + \frac{1}{2}<P(e \to \mu)>$ can be determined from the ratio of observed to expected muon events. Figure 2 shows the correlations of this combination with $<P(e \to e)>$. In the proton decay experiments electron and muon signals from neutrino interactions in a large water detector will be detected[8] and the electron to muon ratio

$$R(e/\mu) = \frac{N(e^\pm)}{N(\mu^\pm)} = \frac{<P(e \to e)> + 2<P(\mu \to e)>}{2<P(\mu \to \mu)> + <P(e \to \mu)>} \qquad (7)$$

can be measured for upper or lower hemisphere $\nu_e, \bar{\nu}_e$ and $\nu_\mu, \bar{\nu}_\mu$ events. Figure 3 shows the correlation of $R(e/\mu)$ with $<P(e \to e)>$. Note that $R(e/\mu) = 0.5$ for no oscillations, 0.75 for case I, and 1.0 for case II.

SOLAR NEUTRINO OSCILLATIONS

In an experiment which spans the past decade Davis et al.[9] have determined that the solar neutrino capture rate $\nu_e + {}^{37}Cl \to e^- + {}^{37}Ar$ is s = 2.1 ± 0.3 SNU (captures/sec/10^{36} atoms). The expected capture rate calculated by Bahcall et al.[10] in the standard solar model is s = 7.5 ± 1.5 SNU. If the discrepancy is due to neutrino oscillations,[11] then

$$<P(e \to e)>_s = 0.3 \pm 0.1 \qquad (8)$$

where the subscript s denotes an average at least over the energy spectrum of the solar neutrino flux.

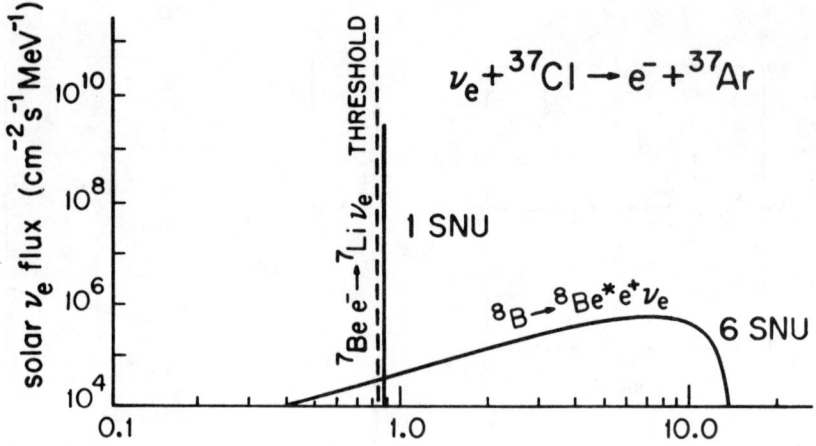

Fig. 4. Dominant components of the solar emission spectrum for the ^{37}Cl experiment, based on Ref. 10.

The dominant components of the solar emission spectrum in the standard model[10] are shown in Fig. 4. The contributions to the ^{37}Cl capture rate are 6 SNU from the ^{8}B continuum and 1 SNU from the ^{7}Be line at 0.862 MeV.

For three neutrino oscillations with symmetrical mixing $|U_{ei}|^2 = 1/3$, the transition probability is

$$<P(e \to e)>_s = 1 - \frac{4}{9} <\sin^2 \tfrac{1}{2}\Delta_{12} + \sin^2 \tfrac{1}{2}\Delta_{23} + \sin^2 \tfrac{1}{2}\Delta_{31}>_s . \quad (9)$$

The value of $<P>_s$ depends on averages of the factors $\sin^2 \tfrac{1}{2}\Delta_{ij}$ over the spectrum weighted by the capture cross section. The dependence of $<\sin^2 \tfrac{1}{2}\Delta>_s$ on δm^2 is given[12] in Fig. 5. The major contribution is due to the ^{8}B continuum, with the ^{7}Be line giving superimposed rapid oscillations. From this figure we can conclude that (i) <u>for all $\delta m^2 \gg 10^{-10}$ eV^2</u>, the complete oscillation average sets in, <u>$<P>_s = 1/3$</u>; (ii) <u>for one $\delta m^2 \sim 10^{-10}$ eV^2</u>, values of $<\sin^2 \tfrac{1}{2}\Delta>_s$ as high as 0.9 are possible, allowing the probability to fall to $<P>_s \simeq 0.1 - 0.2$; (iii) <u>for two $\delta m^2 \sim 10^{-10}$ eV^2</u>, the transition probability can be as low as <u>$<P>_s = 0.05$</u>.

If one or more δm^2 is of order 2×10^{-10} eV^2, an annual variation[13] of the solar rate could occur as the earth-sun distance varies between perihelion and aphelion ($\Delta L \simeq \bar{L}/30$). The time variation comes mainly from the ^{7}Be spectral line at 0.862 MeV. Figure 6 illustrates possible maximal variations of order 1 SNU that could occur in the ^{37}Cl experiment.

CP VIOLATION

If there is CP violation in the lepton sector, neutrino oscillations offer one of the few possibilities for detecting it. The difference $P(\overline{\mu \to e}) - P(\mu \to e)$ of $\bar{\nu}_\mu \to \bar{\nu}_e$ and $\nu_\mu \to \nu_e$ transition

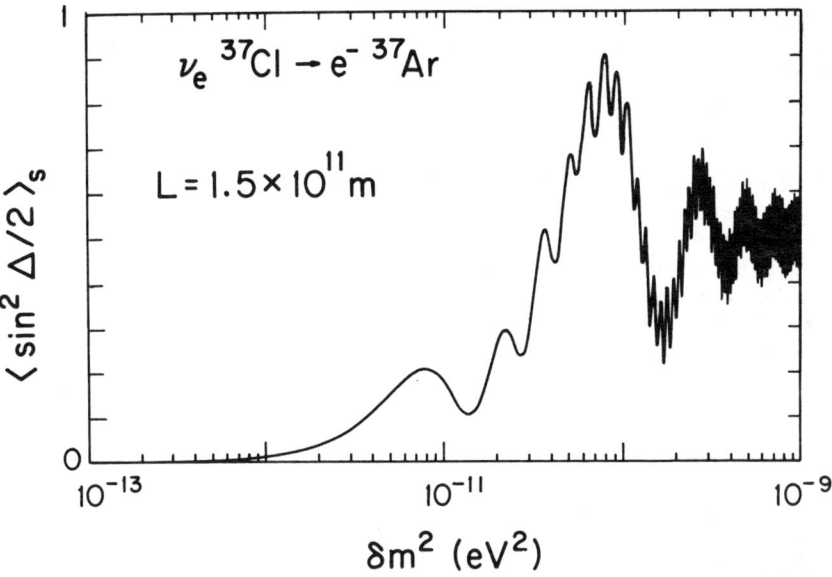

Fig. 5. Spectrum averaged values of $\sin^2(\tfrac{1}{2}\Delta)$ versus δm^2 at $L = 1.5 \times 10^{11}$ m for the ^{37}Cl detector.

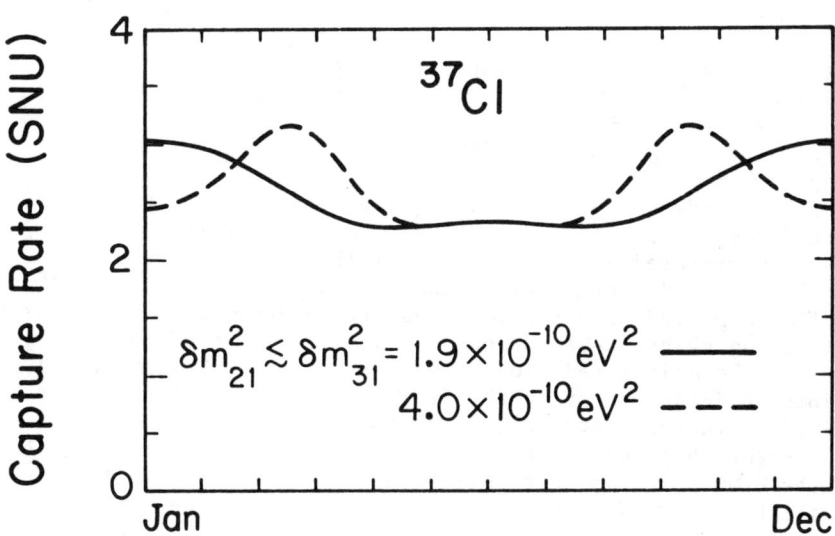

Fig. 6. Possible maximal annual variations of the solar neutrino flux in the ^{37}Cl experiment.

probabilities is a direct measure of CP violation.[4] The equality $P(\bar{\mu} \to e) = P(e \to \mu)$ follows from CPT. In oscillations of three neutrinos the size of CPV is equal in all three channels ($\nu_e \leftrightarrow \nu_\mu$, $\nu_\mu \leftrightarrow \nu_\tau$, $\nu_e \leftrightarrow \nu_\tau$).

CP-violating effects can be detected only over an L/E range where more than one δm^2 plays a significant role. For two comparable δm^2, it is possible that the bound $|P(\bar{\mu} \to e) - P(\mu \to e)| \leq 1$ will be saturated at certain L/E values. If one δm^2 is much bigger than another, the high frequency oscillations essentially average out in the region where CPV becomes appreciable and the bound on the average CPV effect is $|P(\bar{\mu} \to e) - P(\mu \to e)| \leq 0.38$. Figure 7 illustrates possible maximal CP effects[6] based on the symmetric solar-minimizing solution ($|U_{\alpha i}| = 1/\sqrt{3}$). An experiment is proposed[14] at LAMPF to search for CPV in $\bar{\nu}_\mu \to \bar{\nu}_e$ and $\nu_\mu \to \nu_e$ oscillations.

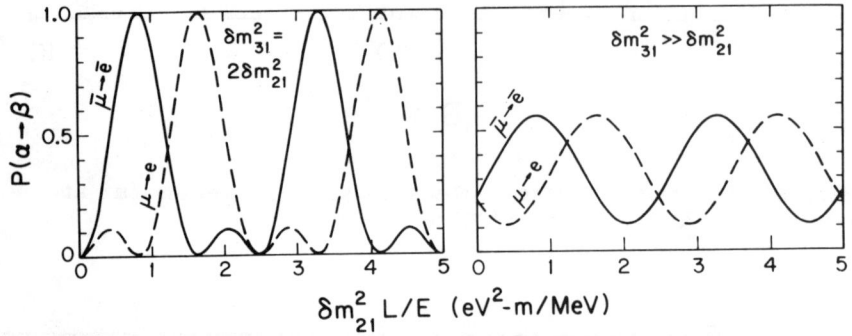

Fig. 7. Maximal CP-violating splittings of the transition probabilities $P(\bar{\mu} \to e)$ and $P(\mu \to e)$ for three-neutrino oscillations.

REACTOR RESULTS

The evidence on neutrino oscillations from reactor experiments is controversial. In proton target experiments the reactor $\bar{\nu}_e$ flux has been measured at a distance from the reactor core center of $L = 11.2$ m by Reines et al.[15] and $L = 8.7$ m by Boehm et al.[16] The $\bar{\nu}_e$ flux expected from fissions has been calculated by Davis et al.[17] (DVMS) and Avignone-Greenwood[18] (AG). The ratio of the observed flux to the calculated flux determines $P(e \to e)$. The principal uncertainty is disagreement in the flux calculations, which can be resolved by measurements of the reactor e^- fission spectrum.

Figure 8 shows $P(\bar{e} \to e)$ deduced from the two experiments, based on the DVMS spectrum which gives the smaller values of $1 - P(\bar{e} \to e)$. Oscillation curves with $\delta m^2 = 0.9$ eV^2 and amplitude $\sin^2 2\alpha = 0.3$ or 0.5 are shown for comparison. The $L = 11.2$ m data suggest an oscillation effect of this scale. It is not evident whether a similar effect is present in the $L = 8.7$ m data.

The Irvine group has reported[19] indications of oscillations based on simultaneous measurements of charged-current and neutral-

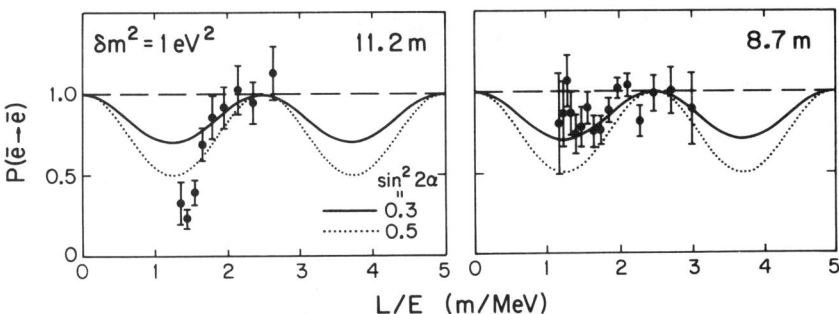

Fig. 8. Reactor results for P(e→e) based on the proton target experiments and the DVMS spectrum.

current deuteron breakup reactions, $\bar{\nu}_e d \to nne^+$ and $\bar{\nu} d \to np\bar{\nu}$. Since the neutral current process is the same for all flavors of neutrino, it is immune to oscillations and provides a monitor of the initial $\bar{\nu}_e$ flux. The average probability for $\bar{\nu}_e$ oscillations can be extracted from

$$\overline{P}(\overline{e \to e}) = \frac{[\overline{\sigma}(CC)/\overline{\sigma}(NC)] \text{experiment}}{[\overline{\sigma}(CC)/\overline{\sigma}(NC)] \text{theory at L=0}} \quad (10)$$

where $\overline{\sigma}$ denotes a spectrum averaged cross section. Variations in the theoretical ratio $\overline{\sigma}(CC)/\overline{\sigma}(NC)$ of up to 20% result from different final state interaction parameters.[20] The quoted experimental result of $\overline{P}(\overline{e \to e}) = 0.40 \pm 0.22$ for the DVMS spectrum is based[21] on ^1S scattering lengths a(nn) = a(np) = -23.7 F and effective ranges r_s(np) = 2.72 F, r_s(nn) = 2.8 F. For the experimentally favored values of a(nn) = -16.6 to -18.5 F, the transition probability becomes $\overline{P}(\overline{e \to e}) = 0.43 \pm 0.24$. Figure 9 shows a range of $\overline{P}(\overline{e \to e})$ predictions for various oscillation parameters. The observed effect in the deuteron experiment is compatible with oscillation parameters that can describe the L = 11.2 m data.[22]

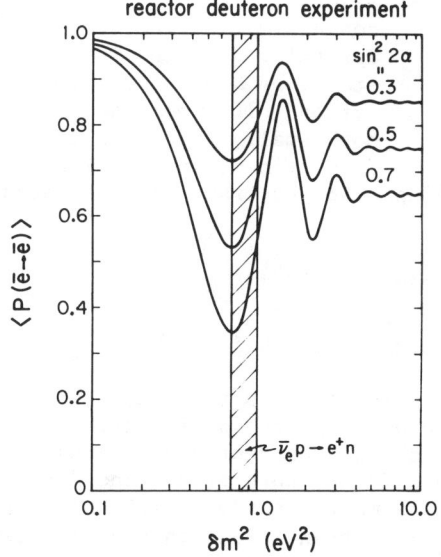

Fig. 9. Oscillation predictions for the reactor deuteron experiment.

OSCILLATIONS IN MATTER

For neutrinos which pass through more than 20% of the earth's radius, coherent forward ν_e charge current scattering from electrons in matter can appreciably modify vacuum oscillations.[23] The ν_e part of the neutrino state is changed relative to ν_μ, ν_τ parts by a phase factor $e^{ik(n-1)x}$. The index of refraction n is given by $k(n-1) = \sqrt{2} G_F N_e$ where N_e is the electron number density. Matter corrections are significant for energies $E(MeV) \gtrsim 5 \times 10^5 \, \delta m^2 (eV^2)$, where δm^2 is the smallest vacuum mass-squared difference.[24] Figure 10 illustrates $R(e/\mu)$ results for vacuum oscillations and matter corrections associated with $\delta m^2 = 10^{-2} \, eV^2$.

DOUBLET-SINGLET OSCILLATIONS

The assignments in SU(2) × U(1) of a single generation are (ν_L, ℓ_L^-) as a doublet and ℓ_L^+, η_L as singlets. The neutral state η_L is needed in the construction of the most general neutral lepton mass matrix [the usual right-hand singlet is $\eta^c_R = C(\overline{\eta_L})^T$ where C is the charge conjugation matrix.] Dirac mass terms are of the form $\overline{\nu_L} \eta^c_R$ while Majorana mass terms are of the forms $\overline{\nu_L} \nu^c_R$ and $\overline{\eta_L} \eta^c_R$, which violate lepton conservation by two units. The mass eigenstates are Majorana fields (i.e., self-conjugate).[25] The states ν_L, η_L are related to the mass eigenstates by a unitary transformation. If the masses are both small, then doublet-singlet neutrino oscillations

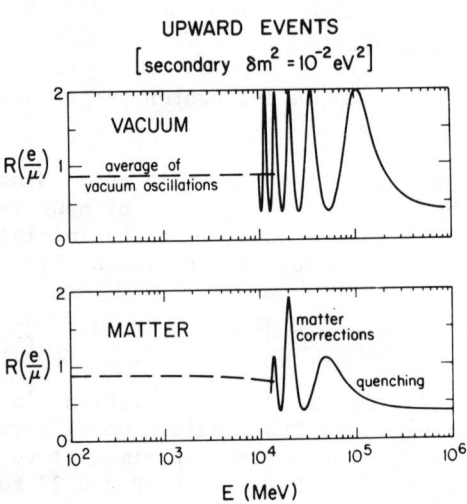

Fig. 10. Example of electron to muon event ratio in deep mine experiments for neutrinos through earth's core.

can occur.[3] In the single generation case the transition probabilities are

$$P(\nu_L \to \nu_L) = 1 - \sin^2 2\alpha' \sin^2 \tfrac{1}{2}\Delta'$$
$$P(\nu_L \to \eta_L) = \sin^2 2\alpha' \sin^2 \tfrac{1}{2}\Delta' . \qquad (11)$$

The singlet η_L is effectively non-interacting so the transition probability into this sterile component appears to be lost. In doublet-singlet oscillations, both charged current and neutral current cross sections oscillate. In general, both flavor and doublet-singlet oscillations can be present simultaneously. With both, the minimum average probability for $\nu_e \to \nu_e$ transitions is $P(e \to e) = 1/6$.

ACKNOWLEDGMENTS

I am indebted to Kerry Whisnant for valuable assistance in the preparation of this review and to Roger Phillips and Sandip Pakvasa for discussions. The contributions of my collaborators are gratefully acknowledged.

This research was supported in part by the University of Wisconsin Research Committee with funds granted by the Wisconsin Alumni Research Foundation, and in part by the Department of Energy under contract DE-AC02 76ER00881-187.

REFERENCES

1. B. Pontecorvo, Zh. Eksp. Theor. Fiz. $\underline{34}$, 247 (1958) [Sov. Phys. J.E.T.P. $\underline{7}$, 172 (1958)]; J. N. Bahcall and H. Primakoff, Phys. Rev. $\underline{D15}$, 3463 (1978).
2. Z. Maki, M. Nakagawa and S. Sakata, Prog. Theor. Phys. $\underline{28}$, 870 (1962); M. Nakagawa et al., ibid. $\underline{30}$, 258 (1963); B. Pontecorvo, Zh. Eksp. Theor. Fiz. $\underline{53}$, 1717 (1967) [Sov. Phys. JETP $\underline{26}$, 984 (1968).
3. V. Barger, P. Langacker, J. Leveille and S. Pakvasa, Phys. Rev. Lett. $\underline{45}$, 692 (1980); Dan-di Wu, Phys. Lett. $\underline{96B}$, 311 (1980); J. Schechter and J. P. Valle, Phys. Rev. $\underline{D22}$, 2227 (1980); S. M. Bilenky, J. Hosek and S. Petcov, Dubna preprint (1980); T. Yanagida and M. Yoshimura, KEK preprint TH-14 (1980); T. P. Cheng and L. F. Li, Carnegie-Mellon preprint 152 (1980). M. Gell-Mann, R. Slansky and G. Stephenson, unpublished (1980); I. Yu. Kobzarev, B. V. Martemyanov, L. B. Okun and M. G. Schepkin, ITEP report 90 (1980).
4. N. Cabibbo, Phys. Lett. $\underline{72B}$, 333 (1978).
5. L. Wolfenstein, Phys. Rev. $\underline{D18}$, 958 (1978); S. Nussinov, Phys. Lett. $\underline{63B}$, 201 (1976).
6. V. Barger, R.J.N. Phillips and K. Whisnant, Phys. Rev. Lett. $\underline{45}$, 2084 (1980) and paper in preparation.
7. J. L. Osborne, S. S. Said and A. W. Wolfendale, Proc. Phys. Soc. (London) $\underline{86}$, 93 (1965); E.C.M. Young, Cosmic Rays at Ground Level (Inst. of Phys. Press, London and Bristol, 1973), edited by A. W. Wolfendale.
8. See, e.g., B. Cortez and L. R. Sulak in Proc. Erice Workshop on Grand Unification (1980); J. Losecco in Neutrino Mass 1980, edited by V. Barger and D. Cline, Univ. Wisc. report 186.
9. R. Davis, Jr., in these proceedings and Neutrino Mass 1980, edited by V. Barger and D. Cline, Univ. Wisc. report 186.
10. J. N. Bahcall et al., Phys. Rev. Lett. $\underline{45}$, 945 (1980); J. N. Bahcall in Neutrino Mass 1980, edited by V. Barger and D. Cline, Univ. of Wisc. report 186.
11. V. N. Gribov and B. Pontecorvo, Phys. Lett. $\underline{28B}$, 493 (1979); J. N. Bahcall and S. C. Frautschi, Phys. Lett. $\underline{29B}$, 623 (1979).
12. V. Barger, R.J.N. Phillips and K. Whisnant, Univ. Wisc. report DOE-ER/00881-179 (1980).

13. R. Ehrlich, Phys. Rev. D18, 2323 (1978); R. Barbieri, J. Ellis and M. K. Gaillard, Phys. Lett. 90B, 249 (1980).
14. T. Bowles, T. Dombeck and R. Burman, LAMPF proposal 638 (1980).
15. H. W. Sobel, F. Reines and E. Pasierb, Univ. Cal.-Irvine report 48 (1980).
16. F. Boehm et al., Phys. Lett. 97B, 310 (1980).
17. B. R. Davis et al., Phys. Rev. C19, 2259 (1979); P. Vogel in Neutrino Mass 1980, edited by V. Barger and D. Cline, Univ. Wisc. report 186.
18. F. T. Avignone and Z. D. Greenwood, Univ. of South Carolina report (1980).
19. F. Reines, H. W. Sobel and E. Pasierb, Phys. Rev. Lett. 45, 1307 (1980).
20. A. Dar, Technion preprint (1980).
21. A. Soni, private communication.
22. D. Silverman and A. Soni, UCLA/80/TEP/25.
23. L. Wolfenstein, Phys. Rev. D17, 2369 (1978).
24. V. Barger, K. Whisnant, S. Pakvasa and R.J.N. Phillips, Phys. Rev. D22, 2718 (1980).
25. S. M. Bilenky and B. Pontecorvo, Lett. Nuovo Cimento 17, 569 (1976).

COMMENTATORS

T. Goldman
Los Alamos Scientific Laboratory

I would just like to make a couple of comments on something that Jerry Stephenson and I have come on at Los Alamos recently about the mixing angles for these neutrino oscillations. You've seen this neutrino mass matrix in the Majorana basis here before: the off-diagonal submatrix m is the standard quark mass matrix before it's been diagonalized; if the rest vanished, there would be Dirac neutrino masses. Our idea is to study the diagonalization of the full matrix, where following Gell-Mann, Ramond and Slansky, one assumes that the right-handed neutrino mass submatrix M has large entries, 10^9eV or larger. Then this matrix, with blocks V on the diagonal and $mM^{-1}V$ off-diagonal, will approximately block diagonalize the mass matrix where V is just the KM matrix. What you can see immediately is that if M is proportional to a power of m, then the whole mass matrix is completely diagonalized. We went on to ask what further diagonalization is necessary to complete the diagonalization for arbitrary M, and found an interesting result. When we tried various matrices for M, we found they didn't make much difference from the obvious case. You can begin to get the idea of why this is so from the block diagonal form. If you absorb in the KM rotation, you have some random M^{-1} with diagonal quark mass matrices multiplying it on both sides. So we are finding that the extra rotation necessary to diagonalize this matrix is small, and our results show it is small because the quark mass ratio is small. In the 2 x 2 case you can see this analytically: the extra rotation needed for $mM^{-1}m$ is reduced by the quark mass ratio from that needed for M, or by roughly the square of the Cabibbo angle. Rather than analyze n x n cases analytically, we did a numerical study. One thing we also did was to ask what if the number of right-handed neutrinos does not parallel the number of left-handed neutrinos and again we didn't see any big effect.

The result of a typical statistical study shows that if one puts in the Cabibbo angle and virtually any angle you want that is allowed by the Lee-Shrock analysis or the Shrock-Treiman-Wang analysis in the quark sector, that these outputs are the angles produced in the neutrino sector. The Cabibbo angle turns out to be almost exactly the same for the electron-muon neutrino mixing. The first and third generation mixing we've assumed small so the corrections are larger. This effect that I said depends on the separation of the quark masses shows up by a narrowing of the distribution of μ and τ neutrino mixing so that if, for example, the mixing correspond to a sine of 0.4, that may be slightly different in the neutrino sector; but the larger the quark mass is, the smaller the difference will be. I think we can say that within this class of theories it would be very hard to understand why the neutrino mixing angle for e-μ mixing would be less than the Cabibbo angle. It would be larger if the neutrinos are made degenerate.

P. Rosen
Purdue University

I would like to make a comment on the question of neutrino oscillations as they apply to neutrino electron scattering because some interesting issues arise in that process. One of the main interests in neutrino electron scattering in general is the interference between the charged current diagram and the neutral current diagram. Now the simplest Weinberg-Salam model has a very definite prediction for this interference term, namely that it will be destructive. Without oscillations, the scattering of electron type neutrinos or antineutrinos will have a much larger cross section than the purely neutral current cross section associated with the scattering of muon-type neutrinos because the charged current diagram is significantly larger than the neutral current one.

Now take an experiment in which you start out with electron-type neutrinos or antineutrinos and consider what happens if there are oscillations. Part of the time the electron-neutrino becomes a muon type neutrino when it scatters off the electrons, and it will therefore scatter with a reduced cross section. This means that the oscillation can, to some extent, mimic the kind of effect you get from the destructive interference between charged and neutral current diagrams. Thus, the oscillation is going to reduce the effective cross section, and you could interpret this reduction conceivably coming from a destructive interference even though in reality, there might not be a destructive interference actually taking place.

That is one qualitative effect, and it means that when you go to analyze the one experiment that does exist at the present time, the one with electron-type antineutrinos from the reactor, you have a hard time trying to untangle the possibility of oscillations from the existence of a destructive interference here. Now the sign of that interference is also important theoretically because, if it does not turn out to be negative, as you expect from the simplest model, I would suggest that in fact you would have to have a model not with one Z^o, but with two or more Z^o's.

If you start out with a beam of accelerator neutrinos, which generally are born as muon-type neutrinos, and if they have enough time to oscillate before they actually scatter, then you are going to have an enhanced cross section from the components of the electron-type neutrino that begin to appear in the beam. The enhancement from a $\nu_\mu \rightarrow \nu_e$ oscillation can actually be as large as a factor of 3 or 4. Such an enhancement could be used as a test of the second class oscillations that Barger just described: if accelerator neutrinos oscillate into "inert" neutrinos, you will never get an enhancement, but always a depletion of the cross section. For details see S. P. Rosen and B. Kayser, Phys. Rev. D23, 669 (1981).

DISCUSSION

AVIGNONE: I would like to make a point that I think should be made concerning the Grenoble experiment versus the Irvine experiment. It's true that our spectrum is based on some old electron measurements, as Barger pointed out, from the fission products of uranium. And they agreed pretty well but, of course, if those experiments are wrong we are obviously wrong, as Peter Vogal thinks we are. If we are wrong, that means that we would be in total disagreement with the Grenoble results without oscillations and that would throw the Reines results into serious conflict. If, however, we are correct and Vogal and company are wrong, then we agree exactly with the Grenoble experiments and with the Reines experiments if there are oscillations. And the second point was about the Ahrman-Dar (1980) paper. There are lots of things in that paper that appear wrong. He starts out with one formalism and ends up with an expression right out of Ahrens et al 1979 paper. Stevenson and Gibbs at Los Alamos have calculated the influence of using the correct scattering length and get less than 10% effect. I have calculated it and get ($6 \pm 1\%$), which is essentially the same thing as Stevenson. So we are talking about new ratios of 0.40 for our spectrum, 0.42 for the Davis et al spectrum and so you are still near three standard deviations from unity for the Reines experiment. That's it.

WOLFENSTEIN: I have also done things similar to Goldman and you find that the mixing angles are like the quark mixing angles and also the masses have to have a hierarchy like the quark mass generation masses. That means that if you have the ν_τ limited by, say, cosmological consideration, if you believe in the 50 eV then the ν_e would have to be very, very light. Because ν_τ is the heaviest and ν_e is the lightest neutrino when you fit that kind of matrix. The other comment that I want to elaborate is that there is an alternative model, due to Zee, that I have looked at, that does allow the possibility of having the electron neutrino and the τ neutrino both heavy and of similar mass, which then permits a lot of $\nu_e - \nu_\tau$ mixing and, at the same time, allows ν_e to be heavy.

SOLAR NEUTRINO EXPERIMENTS AND NEUTRINO OSCILLATIONS

Bruce T. Cleveland, Raymond Davis Jr., and J. K. Rowley
Brookhaven National Laboratory, Upton, New York 11973

ABSTRACT

This report will give the results of the Brookhaven solar neutrino experiment that is based upon the neutrino capture reaction, $^{37}Cl(\nu,e^-)^{37}Ar$. The experiment was built in 1967 to test the theory of solar energy production, and it is well known that the neutrino capture rate in the detector is lower than that expected from theoretical models of the sun. The results will be compared to the current solar model calculations. One possible explanation of the low solar neutrino capture rate is that the neutrinos oscillate between two or more neutrino states, a topic of particular interest to this conference. We will discuss this question in relation to the ^{37}Cl experiment, and to other solar neutrino detectors that are capable of observing the lower energy neutrinos from the sun. A radiochemical solar neutrino detector located deep underground has a very low background and is capable of detecting the monoenergetic neutrinos from megacurie sources of radioisotopes that decay by electron capture. Experiments of this nature will be described that are capable of testing for neutrino oscillations with a δm^2 as low as 0.2 eV^2 if there is maximum mixing between two neutrino states.

THE ^{37}Cl EXPERIMENT

The operation and results of this experiment have been reported many times[1]. The ^{37}Ar produced in 615 metric tons of C_2Cl_4 after exposing for 1 to 4 months is removed by a helium purge, purified, placed in a small proportional counter and counted for 100 to 300 days. The pulse height and pulse rise-time are measured to differentiate ^{37}Ar decay events from most events produced by background processes. The events having the correct pulse rise-time and pulse height for ^{37}Ar decays are resolved into a decaying component and a uniform background component (approximately 0.5 to 1 counts per month) by a maximum likelihood statistical method. The ^{37}Ar production rate observed in the detector over the period 1972-1979 is given in figure 1.

Backgrounds. To obtain the residual rate that could be attributed to neutrinos from the sun, the rate of ^{37}Ar production by cosmic-ray muons and by cosmic ray associated neutrinos must be determined and subtracted. This cosmic ray rate was estimated by exposing smaller tanks of C_2Cl_4 (11.2 m tons) at various levels in the mine from 25 to 1080 hg/cm^2 and extrapolating the observed rates to the full depth of the solar neutrino detector tank at 4400 hg/cm^2.[2] This extrapolation of the measured values with depth is guided by the known variations of total muon intensity and average muon energy with depth, and the variation of the photonuclear cross-section with muon energy. The background ^{37}Ar production

ISSN:0094-243X/81/720322-13$1.50 Copyright 1981 American Institute of Phys.

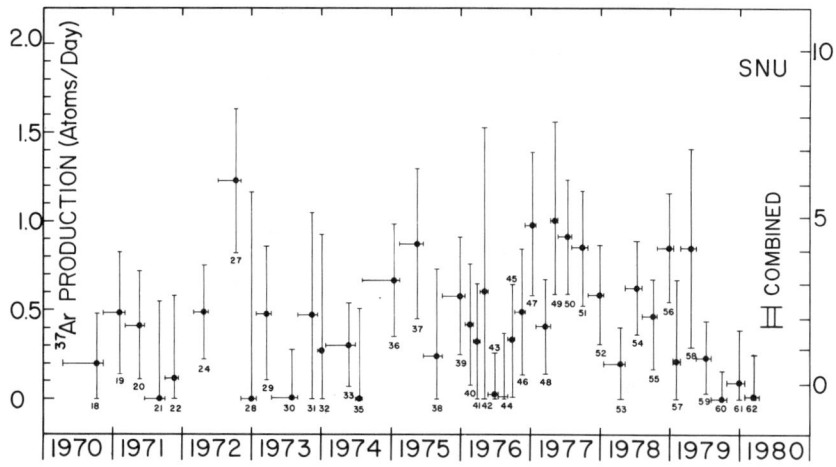

Figure 1

rate so derived in 615 m tons of C_2Cl_4 is 0.08 ± 0.03 atoms per day. Upon subtracting this background from the average ^{37}Ar production rate of 0.47 ± 0.05 atoms per day observed, one obtains the ^{37}Ar production rate that could be ascribed to solar neutrinos of 0.39 ± 0.06 per day. This rate corresponds to a solar neutrino rate of 2.1 ± 0.3 SNU (where SNU = solar neutrino unit 10^{-36} ν-captures sec^{-1} target atom^{-1}). The cosmic ray background is presently the major uncertainty in determining the solar neutrino signal. E. L. Fireman (Smithsonian Astrophysical Observatory) is currently studying the photonuclear interactions of energetic muons as a function of the depth by a radiochemical method based upon the process $^{39}K(\mu^{\pm},\mu^{\pm}np)^{37}Ar$. His current results scaled to the ^{37}Ar production rate in C_2Cl_4 yield a background rate of 0.09 ^{37}Ar atoms per day for 615 m tons of C_2Cl_4.[3] Backgrounds from internal alpha contamination, and fast neutrons from the rock wall are negligibly small. The ^{37}Ar produced by cosmic ray produced neutrinos is calculated to be small, 0.0065 SNU[2].

Comparison with solar theory. The observed solar neutrino rate of 2.1 ± 0.3 can be compared to the rate of 7.8 ± 1.5 expected from the most recent standard model calculations.[4] These results differ from earlier standard solar model calculations that gave an expected rate of about 5 SNU[5] because improved opacities, solar composition and nuclear reaction cross-sections were used. Table 1 lists the neutrino capture cross-sections and the neutrino fluxes prediced by the most recent standard model calculation. Note that

Table 1
SOLAR NEUTRINO FLUXES* AND CROSS SECTIONS**

$$\nu + {}^{37}Cl \rightarrow e^- + {}^{37}Ar^* \rightarrow {}^{37}Ar$$

Neutrino Sources & Energies in Mev	Flux on Earth ϕ in cm^{-2} sec^{-1}	Cross Section σ in cm^2	Capture Rate ^{37}Ar $\phi\sigma \times 10^{36}$ sec^{-1} SNU
$H + H \rightarrow D + e^+ + \nu$ (0-0.42)	6.1×10^{10}	0	0
$H + H + e^- \rightarrow D + \nu$ (1.44)	1.5×10^8	1.56×10^{-45}	0.23
^{7}Be decay (0.861)	4.1×10^9	2.38×10^{-46}	0.98
^{8}B decay (0–14)	5.85×10^6	1.08×10^{-42}	6.31
^{15}O decay (0–1.73)	3.7×10^8	6.61×10^{-46}	0.24
^{13}N decay (0–1.19)	2.6×10^8	1.66×10^{-46}	0.04

$\Sigma\phi\sigma = 7.8$ SNU

* J. N. Bahcall, W. F. Heubner, S. H. Lubow, N. H. Magee, A. L. Mertz, P. D. Parker, B. Rozsnyai and R. K. Ulrich

** J. N. Bahcall, Rev. Mod. Phys. 50, 881 (1978)

6.3 SNU out of a total of 7.8 SNU arise from the low flux of energetic neutrinos from ^{8}B decay. This high capture rate results from the fact that ^{8}B decay neutrinos have sufficient energy (0 to 14 Mev) to feed the analog state in ^{37}Ar at 5.2 Mev, a super allowed transition.

The production of ^{8}B in the sun is very sensitive to the internal temperatures. To account for the low neutrino capture rate observed in the chlorine experiment a number of solar models have been proposed in which various mechanisms are invoked to reduce the internal temperatures. These mechanisms include internal convection, periodic mixing, a fast rotating core, high internal magnetic fields, a helium rich core, and a depleted heavy element composition. The resulting models give a total neutrino capture rate in the range of 1-2 SNU[6]. However, none of these models has gained general acceptance because the mechanisms are considered unreasonable. Some have features that conflict with observations of the sun, or of other main sequence stars. One of the more attractive models with a low ^{8}B flux presumes that the proto-sun had a heavy element content (mostly C, N and O) a factor of ten lower than that presently observed in the sun's photosphere. This model gives a ^{8}B flux diminished by a factor of 10, and a total neutrino capture rate of 1.5 SNU.[7]

Neutrino oscillations. Of direct interest to this conference is the possibility that the low solar neutrino capture rate observed by the chlorine experiment is a result of neutrino oscillations as first suggested by Pontecorvo.[8] The rate reported here is a factor of 3.9 ± 0.9 below the rate expected from the most recent standard solar model. This reduction factor could conceivably be explained by maximum mixing between three neutrino types. Because of the large distances involved a solar neutrino experiment is sensitive to very small neutrino mass differences ($\delta m^2 \sim 10^{-10}$ eV^2). A recent analysis by Barger, Whisnant and Phillips[9] suggests that reduction factors of as much as 0.1 are possible. A recent analysis of the Zee model involving extra Higgs bosons leads to reduction in solar neutrino fluxes by a factor of 1/2 to 1/3.[10] However, in view of the sensitivity of the ^{37}Cl experiment to the flux of ^{8}B neutrinos and of the strong dependence of the flux of these neutrinos on various solar processes, one should be very cautious in attributing the low observed rate to neutrino oscillations. On the other hand, the flux of the low energy neutrinos from the chain initiating proton-proton reaction, is predicted with great certainty, and a measurement of the flux of these neutrinos could permit a valid test for neutrino mixing or oscillations.

NEW SOLAR NEUTRINO EXPERIMENTS AND NEUTRINO OSCILLATIONS

The proton-proton chain of reactions is almost certainly the primary source of the sun's energy. The indicative neutrino

(1) ^{1}H + ^{1}H $\rightarrow$ ^{2}D + e$^+$ + ν (99.75%) and ^{1}H + ^{1}H + e $\rightarrow$ ^{2}D + ν (0.25%)

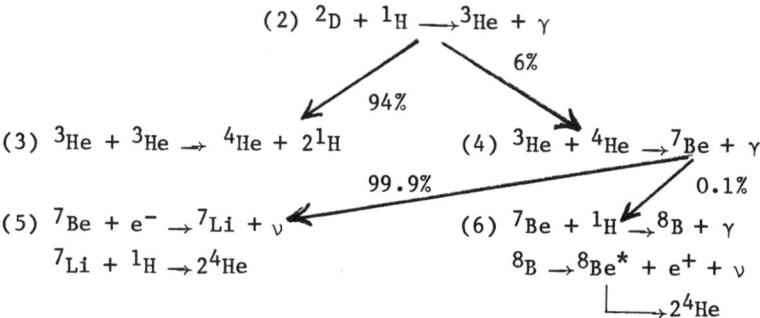

sources in this chain of reactions are the H + H $\rightarrow$ D + e$^+$ + ν reaction, ^{7}Be decay and ^{8}B decay. The fluxes of these neutrinos are listed in Table 1. A direct measurement of the solar neutrino spectrum would verify our present understanding of these processes.

Unfortunately, observing low fluxes of low energy neutrinos is beyond our present capability. A method has been proposed based upon the ^{115}In$(\nu,e^-)^{115}$Sn$^* \longrightarrow {}^{115}$Sn reaction that is, in principle, capable of giving this information. A pilot indium detector with a ~ 35% resolution is being developed.[11] Another approach to resolving the solar neutrino spectrum is a set of measurements with radiochemical detectors each having a dominant sensitivity to a particular neutrino source. Table 2 lists four radiochemical detectors that could be used. The table gives the total neutrino capture rates in SNU, the tons of the element needed to obtain a capture rate of 1 per day according to the standard model and the percent of the total signal that is produced by each neutrino source in the sun. Detectors based upon three of these inverse-beta processes would be capable of resolving the solar neutrino spectrum: ^{71}Ga$(\nu,e^-)^{71}$Ge for the very low energy neutrinos from the H-H reaction, ^{71}Br$(\nu,e^-)^{81}$Kr$^* \longrightarrow {}^{81}$Kr for the monoenergetic ^{7}Be decay neutrinos, and ^{37}Cℓ$(\nu,e^-)^{37}$Ar for the energetic ^{8}B decay neutrinos.

Table 2

COMPARISON OF RADIOCHEMICAL SOLAR NEUTRINO DETECTORS

Percent of signal from various solar neutrino sources, standard model

Neutrino Soruce	^{37}Cl-^{37}Ar	^{7}Li-^{7}Be	^{71}Ga-^{71}Ge	^{81}Br-^{81}Kr
H + H $\rightarrow$ D + e$^+$ + ν	0	0	65	0
H + H + e$^-$ $\rightarrow$ D + ν	3	22	2	8
^{7}Be decay	12	10	27	69
^{13}N decay	1	3	1	4
^{15}O decay	3	21	3	10
^{8}B decay	81	44	2	9
$\sum \phi \sigma$	7.8	41	100	6.3
Tons Element for 1 ν-capture/day	356	3.5	34	495

Gallium. A solar neutrino detector based upon the ^{71}Ga isotope is the best choice to test the question of neutrino oscillations. The flux of the H-H reaction neutrinos is essentially independent of many of the parameters used in the model calculations such as opacities and solar composition, and within broad limits, internal structure. The rate of the H-H reaction is closely related to the solar luminosity, and the neutrinos from this reaction provide 67 SNU. If neutrinos oscillate between 2 or 3 states and the mixing is near to the maximum allowed, a rate below 30 SNU would result, and this could be observed with a 35-50 ton gallium detector. A pilot experiment is now in operation that uses 1.3 tons of gallium. The gallium is in the form of a water solution of gallium chloride and hydrochloric acid. The ^{71}Ge is removed from this solution as $GeCl_4$ by purging with gas. The extracted ^{71}Ge is converted to germane, GeH_4, and counted in proportional counter.[19] The next stage of development will be to build a similar system using 10 tons of gallium. With this 10 ton system a measurment will be made of the neutrino capture cross-reaction with a 2.5 M curie source of ^{51}Cr. This measurement will also serve as a test for neutrino oscillations over short distances. The project is a joint effort of the Max Planck Institute for Nuclear Physics, Heidelberg, the Weizmann Institute, the Institute of Advanced Study, Oak Ridge National Laboratory, the University of Pennsylvania and Brookhaven National Laboratory.

Bromine. The ^{81}Br-^{81}Kr experiment was originally proposed as a test for variations in the solar neutrino luminosity over the last 10^5 years by measuring the ^{81}Kr (half-life 2.1 x 10^5 years) that would accumulate in a natural salt deposit.[12] Some recent developments have led us to consider using this reaction to observe the present ^{7}Be neutrino flux from the sun. Techniques have been developed at Oak Ridge National Laboratory that make it possible to measure a few hundred atoms of ^{81}Kr.[13] It now appears feasible to fill the Homestake chlorine detector tank with a bromine compound (CH_2BrCH_2Br or $C_2Cl_4Br_2$) and to measure the ^{81}Kr atoms that accumulate in a year (~300). Another important development is that Bennett and his colleagues have measured the electron-capture partial lifetime of the 13 second isomeric state in ^{81}Kr that is produced by neutrino capture.[14] Their measurement allows the calculation of the neutrino capture cross-section, which in the past had to be estimated.[15] A ^{81}Br solar neutrino detector has a high sensitivity to ^{7}Be decay neutrinos and would serve to measure this important neutrino flux. The ^{7}Be neutrinos are monoenergetic, an important consideration for studying neutrino oscillations.[9] Ehrlich has already considered the possibility that the signal presently being observed by the ^{37}Cl experiment is perhaps mainly from ^{7}Be decay neutrinos and it is possible that a change in the rate has been observed as an annual variation in earth-sun distance.[16]

Summary. The possibility of neutrino oscillations is obviously of great importance in the general field of neutrino astronomy and, more particularly, in the field of solar neutrino astronomy. Although the question raised by the results of the chlorine experiments can be answered by electron neutrino oscillations, these experimental results alone do not constitute conclusive evidence for neutrino oscillations. Many other explanations are possible. A logical step on the way to a full-scale gallium solar neutrino experiment is to calibrate a smaller gallium detector using a source of neutrinos with a known flux and energy. A similar source calibration of the present chlorine experiment is highly desirable. Although our principal objective is to study solar neutrinos, either of these calibrations could also easily serve as a neutrino oscillation experiment. A detailed consideration of such experiments is the topic of the next section.

OBSERVING NEUTRINO OSCILLATIONS WITH MEGACURIE RADIOACTIVE SOURCES

This section deals with possible radiochemical experiments that can test for the existence of short range neutrino oscillations. This type of experiment follows the original suggestion of L. W. Alvarez to use a ^{65}Zn source to measure the neutrino absorption cross-section of ^{37}Cl.[17]

The general principles of a radiochemical oscillation experiment are quite straightforward: a strong source of a radioactive isotope that decays primarily by electron capture is prepared. The shielded source is placed deep underground within or near to a target material that is able to capture the emitted neutrinos by an inverse beta decay process that leads to a moderately long-lived product isotope. This product is extracted from the target by chemical means and then counted by detecting the beta decay back to the original target isotope. Possible sources include ^{51}Cr and ^{65}Zn; possible target isotopes are ^{7}Li, ^{37}Cl, ^{71}Ga, ^{81}Br, etc.

Some of the advantages of radiochemical neutrino oscillation experiments are:

1. The neutrinos are monochromatic
2. The neutrinos are of the electron type for which oscillations have been reported by Reines et al.[18]
3. The source can be made quite compact.
4. The source strength can be accurately calibrated
5. The neutrino absorption cross-section is well-known.
6. Background production rates due to α-particles, neutrons, and muons can be made negligibly small. There is only a minor residual background effect from solar neutrinos.
7. Different neutrino oscillation lengths can be sampled by placing concentric detectors about the source.

The principal difficulty in experiments of this type is that very large source strengths are required. With the strongest available reactor flux (the HFIR reactor at Oak Ridge) an irradiation time of approximately one year is needed. Even then, large targets are required (1-10 m in diameter) and the neutrino capture rates are low (no more than 5 per day).

We shall now outline the procedures used in the calculation of the expected capture rate and then present two radiochemical experiments that are feasible at the present time.

Consider a point source that emits s neutrinos per second and an extended absorber that contains n atoms/cm^3 of a neutrino absorbing isotope. From the definition of the cross-section σ, the neutrino capture rate R is given by R = A G where A = nsσ depends on the physical properties of the source and absorber and

$$G = \frac{1}{4\pi} \int_{\text{absorber}} \frac{dV}{r^2} \qquad (1)$$

depends only on the source and absorber geometry (r here is the distance from the source to an arbitrary point in the absorber). The factor A is given in Table 3 for cross-sections calculated by Bahcall.[15] The geometrical factor G can be expressed in closed form for simple geometries such as a sphere or a cylinder, and can be easily evaluated for more complex geometries or for an extended source by Monte Carlo techniques.

Table 3

Absorber	Source	Detected Isotope	A (cm^{-1} Mc^{-1})
C_2Cl_4	^{65}Zn	^{37}Ar	1.12
7.1 M GaCl$_3$	^{51}Cr	^{71}Ge	2.84
7.1 M GaCl$_3$	^{65}Zn	^{71}Ge	3.67

We suppose that the electron type neutrinos emitted by the source oscillate into <u>one</u> other type and that the probability of no oscillation at the distance r (m) from the source is given by the standard formula,

$$P(r) = 1 - \sin^2(2\theta_m) \sin^2(1.27\, \delta m^2 r/E) , \qquad (2)$$

where θ_m is the mixing angle, δ_m^2 is the neutrino mass difference in eV^2, and E is the neutrino energy in MeV. The neutrino capture rate in the presence of oscillations is then easily calculated by including P(r) from eq (2) into the integrand of G in eq (1).

<u>Chlorine.</u> We have two radiochemical experiments to search for neutrino oscillations under active consideration. The first involves placing a 1 Mc source of ^{65}Zn inside a tube that extends to the center of the 100,000 gallon C_2Cl_4 detector in the Homestake mine. Two obvious advantages of this experiment are that the detector presently exists and that the background has been well determined. The dependence of the capture rate on δm^2 for the case of maximal mixing is given in Figure 2. Without oscillations the capture rate of neutrinos from the source is 4.7 per day and there is also present a background production rate of 0.5 per day. The error bars indicated on this figure are at the 68% confidence level and are due only to random effects. These errors have been

calculated by a simulation of the entire process of ^{37}Ar
production, extraction, and counting for a set of experimental
conditions that are believed to be realizable in practice. The
production rate due to the neutrino source was extracted from the
simulated counting times by a maximum likelihood method. It is
evident from Figure 2 that for $\delta m^2 > 0.2$ eV^2 the production rate at
the 90% confidence level falls below the predicted rate in the
absence of oscillations. The lower limit on δm^2 that is obtained
for other choices of the mixing angle is presented in Figure 4.

<u>Gallium</u>. Another attractive experiment employs a target of a
water solution of GaCl$_3$.[19] All of the procedures for extraction,
purification, and counting of the ^{71}Ge that is produced by neutrino
capture have been well developed. Background effects due to
neutrons and α-particles are known and can be controlled by
suitable choice of location and materials of construction. The
muon produced background is much less severe than for C$_2$Cl$_4$ and is
less than 0.1 per day for depths greater than 1500 hg/cm^2. The
background effect due to solar neutrinos is believed to be 0.2 to
0.3 per day for the approximately 10 tons of Ga that is needed for a
neutrino source experiment. In this case since an experiment can
be designed specifically to search for neutrino oscillations, it
can be made in two different zones. An experimental configuration
that could be built consists of a central tank that forms the inner
zone surrounded by six close-packed tanks as the outer zone. With
a 2.5 Mc source of ^{51}Cr in a central annulus the capture rates in
the inner and outer zones of a 10.6 ton Ga detector are given as a
function of δm^2 in Figure 3. This again is for maximal mixing and
the error bars are at the 68% confidence level for a reasonable
schedule of experimental operations. A zero background has been
assumed in the inner zone and the background in the outer zone has
been assumed to have been measured to an accuracy of ±0.25
atoms/day. This rate given for the outer zone had this known
background subtracted with the errors combined in quadrature. If
we compare these predicted rates in the two zones, there are two
regions of δm^2 over which the rates differ by more than two
standard deviations. These regions are plotted for this and for
other mixing angles with a dashed line in Figure 5. These regions
happen to overlap to a considerable extent with the regions
suggested by the reactor ν_e experiment of Reines <u>et al</u>.[18] We can
also look at only the rate in the outer zone, and, in a manner
similar to that for the Cl experiment, obtain the lower bound for
δm^2 given by the solid line in Figure 5.

It is apparent from these figures that interesting neutrino
oscillation experiments can be performed with a strong radioactive
source and either a gallium or chlorine target. Such experiments,
sensitive to oscillation lengths of the order of several meters,
are logical steps to be taken on the way to a full-scale gallium
solar neutrino experiment. They would settle the question of the
possible oscillations suggested by the Reines experiment. In
addition, if it is shown that such oscillations do not exist, the
calculated neutrino capture cross sections of these targets can be

confirmed. Then a full scale gallium solar neutrino experiment could be confidently undertaken which would yield critical information about the reason for the present discrepancy between the results of the chlorine experiment and the predictions made by the standard solar model.

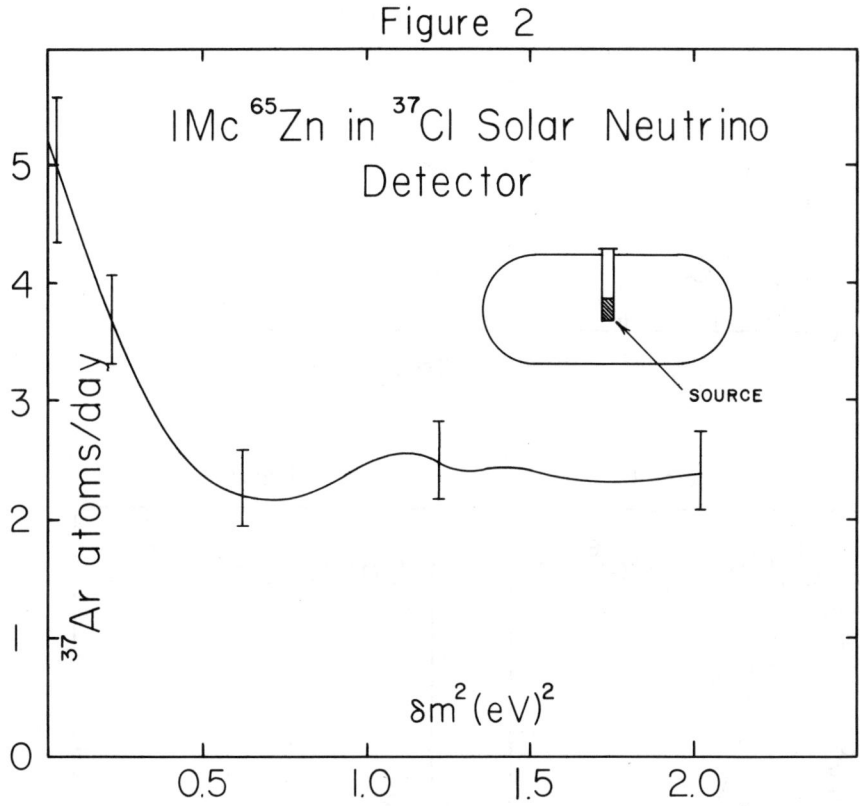

Figure 2

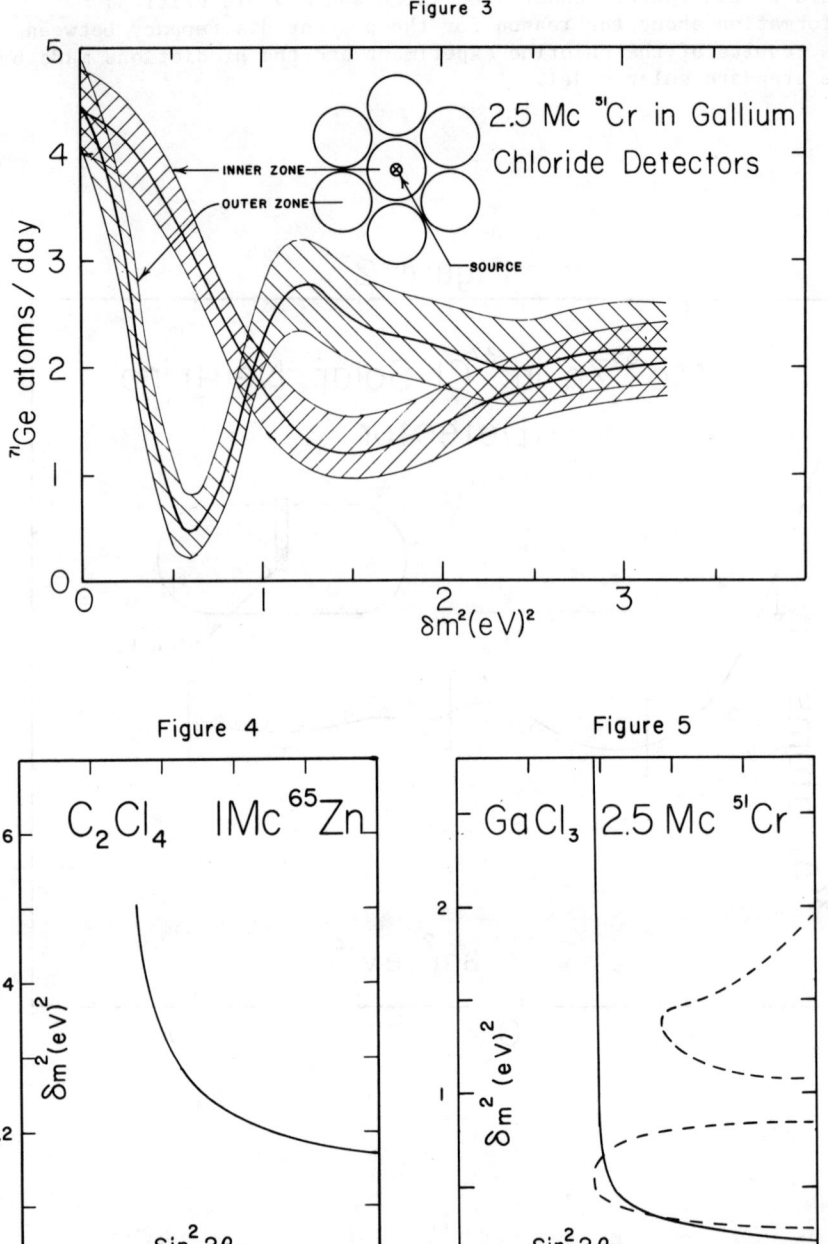

Figure 3

Figure 4

Figure 5

REFERENCES

1. R. Davis, Jr., D. S. Harmer, and K. C. Hoffman, Phys. Rev. Letters 20, 1205 (1968); R. Davis, Jr., Proc. Informal Conf. on Status and Future of Solar Neutrino Research (BNL 50879) 1, 1 (1978) G. Friedlander, Ed.; J. K. Rowley, B. T. Cleveland, R. Davis, Jr., W. Hampel and T. Kirsten, Proc. Conf. Ancient Sun (1980) R. O. Pepin, J. A. Eddy and R. B. Merrill Eds; B. T. Cleveland, R. Davis, Jr., and J. K. Rowley, Proc. Mini Conference on Neutrino Masses, Oct. 2-5, 1980, V. Barger and D. Cline, Eds. API Conf. Proc. to be published; J. N. Bahcall, and R. Davis, Jr., Science 191, 264 (1976).

2. A. W. Wolfendale, E. C. M. Young, and R. Davis, Jr., Nature Physical Sciences 238, 130 (1972); G. L. Cassiday, Proc. Int. Cosmic Ray Conf. 13th Conf. (Denver) Papers 3, 1958 (1973); Cosmic ray neutrino background, see M. A. Rudzskii and Z. F. Seidov, Soviet J. Nucl. Phys. 30, 653 (1979).

3. E. L. Fireman, Proc. Int. Cosmic Ray Conf. 16th Conf. (Kyoto) Papers, Supplement 13, 389 (1979).

4. J. N. Bahcall, S. H. Lubow, W. F. Huebner, N. H. Magee, Jr., A. L. Merts, M. F. Argo, P. D. Parker, B. Rozsnyai, and R. K. Ulrich, Phys. Rev. Letters, 45, 945 (1980).

5. J. N. Bahcall, W. F. Huebner, N. H. Magee, Jr., A. L. Merts, and R. K. Ulrich, Astrophys. J. 184, 1 (1973); J. Christensen-Dalsgaard, D. O. Gough, and J. G. Morgan, Astron. and Astrophys. 73, 121 (1979); J. N. Bahcall, Space Sci. Rev. 24, 227 (1979).

6. G. Friedlander (Editor). Proc. Informal Conf. on Status and Future of Solar Neutrino Research (1978) BNL 50879, Volume I, see reports by J. N. Bahcall p 55, P. D. Parker p 77, W. F. Huebner p 107, S. H. Lubow and R. K. Ulrich p 157, R. T. Rood p 175, and I. W. Roxburgh p 207; J. N. Bahcall, Proc. Int. Conf. on Nucl. Physics, Vo. 2, p 681, Invited Papers, J. de Boer and H. J. Mang, Editors (1973).

7. J. N. Bahcall and R. K. Ulrich, Ap. J. 170, 593 (1971); Z. Abraham and I. Iben, Jr. Ap. J. 170, 157 (1971); J. Christensen-Dalsgaard, D. O. Gough and J. G. Morgan Astron. Astrophpys. 73, 121 (1979).

8. B. Pontecorvo, JETP 53, 1717 (1967); S. M. Bilenky and B. Pontecorvo Physics Reports (Sect. C of Physics Letters) 41, 225 (1978).

9. V. Barger, K. Whisnant and R. J. N. Phillips DOE-ER/00881-179, U. Wisconsin-Madison, Madison, WI 53706.

10. L. Wolfenstein, Nucl. Phys. B17, 93 (1980).

11. R. Raghavan, Univ. Wisconsin Mini conference on the Neutrino Mass, Oct. 2-5, 1980, to be published in the AIP Conf. Proc. Series; R. Raghavan, Phys. Rev. Letters 37, 259 (1976).

12. R. D. Scott, Nature 264, 729 (1976); T. Kirsten, Proc. Int. Conf. on Status and Future of Solar Neutrino Research, BNL Report 50879, Vol. 1, p 305, G. Friedlander, Editor (1978); J. K. Rowley, B. T. Cleveland, R. Davis, Jr., W. Hampel, and T. Kirsten, Proc. Conf. Ancient Sun p 45, R. O. Pepin, J. A. Eddy and R. B. Merril, Editors (1980).

13. G. S. Hurst, M. G. Payne, S. D. Kramer, and C. H. Chen, Physics Today, September 1980.

14. C. L. Bennett, M. M. Lowry, R. A. Naumann, F. Loeser and W. H. Moore, Phys. Rev. 22C, 2245 (1980).

15. J. N. Bahcall, Rev. Mod. Phys. 50, 881 (1978).

16. R. Ehrlich Phys. Rev. 18D, 2323 (1978).

17. L. W. Alvarez, Memo no. 767, Lawrence Radiation Laboratory, Univ. of California, March 23, 1973.

18. F. Reines, H. W. Sobel and E. Pasierb, Phys. Rev. Letters 45, 1307 (1980).

19. J. N. Bahcall, B. T. Cleveland, R. Davis, Jr., I. Dostrovsky, J. C. Evans, Jr., W. Frati, G. Friedlander, K. Lande, J. K. Rowley, R. W. Stoenner, and J. Weneser, Phys. Rev. Letters 40, 1351 (1978).

NEUTRINOS AND COSMOLOGY

Michael S. Turner

Astronomy and Astrophysics Center
The University of Chicago
Chicago, IL 60637
and
Institute for Theoretical Physics
University of California
Santa Barbara, CA 93106

ABSTRACT

Neutrinos perhaps best illustrate the close relationship that exists between cosmology and elementary particle physics. Cosmological and astrophysical observations can be used to constrain neutrino properties. The abundance of ^{4}He probably limits the number of light (< 1 MeV) neutrino species to be less than or equal to 4. An unstable neutrino with lifetime α m_ν^{-5} is restricted to be less massive than $\sim$ 200 eV or more massive than $\sim$ 10-100 MeV. Neutrinos more massive than a few eV have important cosmological consequences: if the sum of neutrino masses $\gtrsim$ 3.5 eV the universe is neutrino-dominated, and if it is $\gtrsim$ 100 eV the universe is closed. The role that neutrinos play in the development of structure in the universe (galaxies, clusters, etc.) is also discussed.

INTRODUCTION

The evidence that the universe began from a hot big bang is very convincing and includes the following: the universal expansion ("the Hubble flow"), the 3K cosmic microwave background, the abundance of D, ^{3}He, ^{4}He, and ^{7}Li, and on the theoretical side general relativity, in particular the singularity theorems of Hawking and Penrose[1]. In the standard hot big bang model, the Friedmann-Robertson-Walker (F-R-W) cosmology[2], the temperature of the universe is related to the age by,

$$t = (45/16\pi^3)^{\frac{1}{2}} g_*^{-\frac{1}{2}}(T) m_p/T^2 ,$$

$$\cong 2.41 \times 10^{-6} g_*^{-\frac{1}{2}}(T) T_{GeV}^{-2} \text{ sec},$$

(1)

when the energy density of the universe is dominated by relativistic particles. Here $\hbar = k_B = c = 1$, $m_p = G^{-\frac{1}{2}} = 1.22 \times 10^{19}$ GeV, and $g_*(T)$ counts the total number of effective degrees of freedom of all the relativistic species ($\equiv \sum_{\text{bosons}} g + 7/8 \sum_{\text{fermions}} g$). Classical cosmology (as described by general relativity) should be valid for $t \gtrsim t_p \sim 10^{-43}$s; earlier, quantum corrections to general relativity are probably important. At the planck time the temperature of the universe was,

$$T_p = 0.55 \, g_*^{-\frac{1}{4}} \, m_p \qquad (2)$$
$$\simeq g_*^{-\frac{1}{4}} \, 10^{19} \text{ GeV} \, .$$

For this reason it is often said, "The early universe <u>was</u> the ultimate high energy physics laboratory."

The relationship between cosmology and particle physics is one that works both ways: cosmology can be used to constrain particle theory, and particle theory can have important consequences for cosmology. There is perhaps no better example of this symbiotic relationship than that of neutrinos and cosmology. In this talk I will begin by discussing how cosmology can put constraints on the properties of neutrinos. Through the abundance of ^{4}He an upper limit to the number of light (< 1 MeV) neutrino species can be obtained[3,4]: if baryons dominate the mass of binary galaxies and small groups of galaxies, then at most four flavors of neutrinos are permitted[4]. Consideration of the microwave and other photon backgrounds, the present density of the universe, and supernova energetics restrict unstable neutrinos with lifetimes $\alpha \, m_\nu^{-5}$ to be lighter than ~ 200 eV or heavier than ~ 10-100 MeV. Next I will discuss the important cosmological consequences of neutrinos more massive than a few eV. If the sum of the neutrino masses $\gtrsim$ 3.5 eV, then neutrinos contribute more to the present mass density than baryons do, and may resolve the mass density hierarchy problem in astrophysics. If the sum of neutrino masses $\gtrsim$ 100 eV, then neutrinos provide enough mass density to close the universe. If neutrinos do dominate the mass density of the universe, then the standard picture of galaxy formation must be modified significantly. I will briefly review the standard gravitational instability scenario, and how it must be changed if the universe is neutrino-dominated. Although I will not discuss neutrinos less massive than a few eV, a neutrino with mass as small as 10^{-6} eV can have important astrophysical implications; for example, its existence might lead to a resolution of the solar neutrino problem.[5]

THE ABUNDANCE OF ^{4}He AND THE NUMBER OF NEUTRINO FLAVORS[6]

The universe is by mass 77%-68% H and 23%-30% ^{4}He, with all the other elements accounting for the remaining 0-2%. In some sense the hydrogen is primordial (although the baryons themselves may have been synthesized ~ 10^{-35}s after the big bang[7]), and it is generally believed that the elements heavier than ^{4}He were synthesized in stars. However, based on energetics alone, the types of stars we observe today could not have produced the 23%-30% ^{4}He which is observed. It was originally suggested by Gamow[8] and later firmly established by detailed computer calculations[9] that ~ 25% of the mass of the universe should have been "cooked" into ^{4}He during a period of nucleosynthesis from ~ 0.01s to ~ 200s after the big bang. [Lesser amounts of D, ^{3}He, and ^{7}Li should also have been produced; however, significant nucleosynthesis beyond ^{4}He is prevented by the lack of stable nuclei with A = 5 or 8 and by coulomb barriers.] The remarkable concordance of these calculations with the observations makes primordial nucleosynthesis one of the crowning jewels of the hot big bang model.

Because of the large amount of ^{4}He produced primordially ($\sim 25\%$) and the probable small contamination by subsequent stellar production ($\lesssim 7\%$), ^{4}He is an important relic from an earlier epoch when the temperature of the universe was $\sim 0.1\text{-}10$ MeV. At temperatures $\gtrsim 1$ MeV light neutrinos (< 1 MeV) were in thermal equilibrium with the rest of the universe, with "thermal contact" maintained by the usual neutral and charged current weak interactions. Below 1 MeV they ceased to interact, and thereafter expanded freely, with their temperature being redshifted $\alpha\ R(t)^{-1}$. During the epoch of nucleosynthesis the energy density of the universe was dominated by relativistic particles: γ, e^-, $\nu_e\ \bar{\nu}_e$, $\nu_\mu\ \bar{\nu}_\mu$, $\nu_\tau\ \bar{\nu}_\tau$,... Since $H^2 \equiv (\dot{R}/R)^2 = 8\pi G \rho/3$, the number of neutrino species affects the timescale of the expansion, $t \sim H^{-1}$, and hence the production of light elements -- in particular ^{4}He. Essentially all the neutrons in the universe eventually end up in ^{4}He. As long as the weak interactions which tranform $n \leftrightarrow p$ ($n + e^+ \leftrightarrow p + \bar{\nu}_e$, $n + \nu_e \leftrightarrow p + e^-$, $n \rightarrow p + e^- + \bar{\nu}_e$) are occurring rapidly on the expansion timescale ($\Gamma > H$) the n/p ratio "tracks" its equilibrium value, $\exp(-\Delta m/T)$. When these reactions cease to be rapid ($\Gamma < H$), the n/p ratio "freezes out" ($T \sim 10^{10}$K), and determines the amount of ^{4}He produced (which occurs at $T \sim 10^9$K). More neutrino species speed up the expansion, and so freeze out occurs at a higher temperature and value of the n/p ratio. Therefore, additional neutrino species increase ^{4}He production.

The mass fraction of the universe synthesized into ^{4}He, Y_p, is only a function of three parameters: N_ν, the number of light (< 1 MeV), two-component neutrino species, η, the present baryon-to-photon ratio, and $\tau_{\frac{1}{2}}$, the neutron halflife which sets the scale of the weak interactions which transform $n \leftrightarrow p$. Y_p increases with increasing values of all three of these parameters (with an important exception to be discussed later). Therefore, if one places an upper limit on Y_p and lower limits on $\tau_{\frac{1}{2}}$ and η, an upper limit on N_ν can be obtained.

(a) Primordial abundance of ^{4}He

Since ^{4}He is also produced by stars, the primordial mass fraction is non-trivial to determine. For normal metal abundance ($Z \sim 2\%$) HII regions $Y \cong 0.28\text{-}0.32$. [An HII region is a region of hot ionized hydrogen gas.] For HII regions which have low metal abundance (and presumably a lesser amount of stellar contamination since stars synthesize both ^{4}He and metals) $Y \cong 0.23\text{-}0.25$, with typical stated errors of ± 0.01. It is possible to try to extrapolate to a truly primordial value, e.g., by $Y \cong Y_p + 3Z$; however, it is not at all clear that such a relationship is valid, or even if any such relationship exists. Such extrapolations lead to $Y_p \cong 0.22\text{-}0.25$. We have taken $Y_p = 0.25$ as an upper limit on the primordial mass fraction of ^{4}He (ref. 4) synthesized.

(b) The neutron halflife

Until recently, the accepted value of the neutron halflife was $\tau_{1/2} = 10.61 \pm 0.16$ min (ref. 10). There are two recent determinations which are not totally in agreement with this value: $\tau_{1/2} = 10.13 \pm 0.09$ min (ref. 11) and $\tau_{1/2} = 10.82 \pm 0.2$ min (ref. 12). We have adopted as a lower limit on the neutron halflife $\tau_{1/2} = 10.13$ min.

(c) The baryon-to-photon ratio

Finally, there is the most elusive of the three parameters which need to be known, η. The baryon-to-photon ratio is related to quantities which can be directly measured by,

$$\eta = 2.83 \times 10^{-8} \, \Omega_N \, h_o^2 \, (2.7K/T_\gamma)^3. \qquad (3)$$

Where Ω_N is the ratio of the mass density in nucleons to the critical density, $\rho_c = 3 H_o^2/8\pi G$, $H_o = 100 \, h_o$ kms^{-1} Mpc^{-1} is the Hubble parameter, and T_γ is the present temperature of the microwave background. Observations restrict T_γ to the range (2.7K, 3.0K), and h_o to the range (½, 1). So we obtain

$$\eta \geq 5.16 \times 10^{-9} \, \Omega_N . \qquad (4)$$

The standard dynamical approach to determining Ω_N (see, e.g., ref. 13) involves measuring the gravitational mass of a region (e.g., by analyzing rotational curves of galaxies, or by applying the virial theorem to groups of galaxies) and comparing this to the light emitted by that region -- constructing a mass-to-light ratio. Then the mass density of the universe can be found by multiplying the mass-to-light ratio times the luminosity density of the universe. There are at least two inherent problems with this procedure: (i) the mass inferred does not distinguish between baryons and other forms of gravitating material (e.g., massive neutrinos) and (ii) one is not sure to what extent a given mass-to-light ratio is "typical" of most of the luminous mass in the universe. The mass-to-light ratios and Ω's which are inferred from them are compiled in Table 1 for various scales.

Table 1

Scale	M/L (Solar Units)	Ω
Solar Neighborhood Material	2 ± 1	$(0.0014 \pm 0.0007)/h_o$
Central Region of Galaxies	$(8-20) \, h_o$	$0.006-0.014$
Binaries and Small Groups of Galaxies (BSG)	$(60-180) \, h_o$	$0.04-0.13$
Rich Clusters	$(300-1000) \, h_o$	$0.2-0.7$
Hot Gas in Clusters	---	$\gtrsim 0.007 \, h_o^{-3/2}$

The material in the solar neighborhood is most certainly baryons, and from equation (4) we obtain the rather weak lower bound $\eta \geq 0.14 \times 10^{-10}$. There is every reason to believe that the luminous material in the central portions of galaxies is predominantly baryonic; the range in the M/L's does not reflect observational uncertanties but rather real variations from spiral (closer to 8 h_o) to elliptical galaxies (closer to 20 h_o) due to the different stellar populations present. Using $\Omega_N = 0.006$ we obtain the lower bound $\eta \geq 0.29 \times 10^{-10}$.

Most galaxies in the universe are found in binaries or small groups, so it seems reasonable that the M/L inferred from BSG is characteristic of the luminous mass in the universe. Using $\Omega_N = 0.04$, we obtain $\eta \geq 2 \times 10^{-10}$. There is, however, the uncertainty as to whether or not the mass which is observed is baryonic. Schramm and Steigman[14] have pointed out that neutrinos of mass $\gtrsim O(10eV)$ may cluster on these scales and dominate the mass of BSG.

From rich clusters we can deduce a lower bound of $\eta \geq 10^{-9}$; however, since most galaxies are not in rich clusters, there is reason to believe that the M/L ratio inferred for these objects is not characteristic of most of the luminous matter in the universe. Finally, the x-ray emitting hot gas found in clusters is most certainly baryons and implies a lower bound on η of $\geq 1 \times 10^{-10}$. Again, there is the uncertainty as to whether or not all galaxies have this much gas associated with them; in BSG or in isolated galaxies this gas, even if it were present, would not be hot enough to emit detectable amounts of x-rays as the gravitational potential wells of these objects are not nearly as deep as those in rich clusters. I should also mention that for $\eta < 1 \times 10^{-10}$ a large abundance of deuterium is produced primordially: $X_D \gtrsim 3 \times 10^{-4}$, more than an order of magnitude greater than the abundance observed. Unless more than 90% of all baryons have been cycled through stars, this also suggests that $\eta \gtrsim 1 \times 10^{-10}$ (note: deuterium production decreases with increasing η).

In Table 2, I have summarized the lower bounds on η, from most reliable (M/L's for the solar neighborhood) to least reliable (BSG). Because most of the luminous matter in the universe is not in rich clusters I have not included that entry.

Table 2

Method of Determination	Lower Bound
Solar neighborhood material	$\eta \geq 0.14 \times 10^{-10}$
Central regions of galaxies	$\eta \geq 0.29 \times 10^{-10}$
Hot gas/Deuterium	$\eta \geq 1.0 \times 10^{-10}$
BSG	$\eta \geq 2.0 \times 10^{-10}$

(d) What limits can be placed upon N_ν?

First, recall that N_ν is the number of light (< 1 MeV), two-component neutrino species. Massive neutrinos can be of two varieties: (i) 2-component Majorana neutrinos ($\nu \equiv \bar{\nu}$), in which case each light species contributes one unit to N_ν, or (ii) 4-component Dirac neutrinos. In this case, if the right-handed components have interactions strong enough so that they remain in thermal contact with the rest of the universe at least until $T \cong 10^{12}K \cong 100$ MeV (this is when π's and μ's decoupled heating all those species still in thermal equilibrium), then they will affect nucleosynthesis just as their left-handed counterparts do, and each light species will contribute two units to N_ν. However, if they interact much more weakly than left-handed neutrinos, then they will decouple earlier, will have a lower temperature during nucleosynthesis, and will contribute much less to the energy density than a left-handed neutrino. If the right-handed components couple sufficiently weakly ("G_R" < $G_F/500$), then each light species effectively contributes only $\cong$ one unit to N_ν. Right-left interactions are not sufficiently strong to keep right-handed neutrinos in thermal contact late enough (i.e., until $T \cong 10^{12}K$) so that they contribute significantly to the energy density during nucleosynthesis. Unless there exist purely right-handed interactions of sufficient strength, right-handed neutrinos will have no significant effect on nucleosynthesis and each new species will change N_ν by only one unit. To summarize, for Majorana neutrinos each species changes N_ν by one unit, and for Dirac neutrinos each species changes N_ν by between one unit ("G_R" << G_F), and two units ("G_R" $\cong G_F$).

Case 1: $m_\nu \lesssim$ few eV (for all species). In this case neutrinos cannot cluster in BSG (this point will be discussed later) and there is no reason to believe the mass which binds BSG is anything but baryons, so that the lower bound on η from BSG, $\eta \geq 2 \times 10^{-10}$, is applicable. The number of allowed neutrino species, N_ν, as a function of $\tau_{\frac{1}{2}}$ for $Y_p \leq 0.25$ is shown below in Figure 1. For $\tau_{\frac{1}{2}} \geq 10.13$ min and $\eta \geq 2 \times 10^{-10}$, N_ν must be ≤ 4.

Case 2: $m_\nu \gtrsim 0(10$ eV$)$ for at least one species. If at least one neutrino species is more massive than $0(10$ eV$)$, then it may cluster in BSG and dominate the mass; or other non-baryonic matter (e.g., monopoles, PBHs, heavy stable neutral leptons, etc.) may dominate the mass of BSG, and in either case $\eta \geq 2 \times 10^{-10}$ cannot be used as a lower bound. Therefore, a more reliable lower bound must be used. If the lower bound based on hot gas/deuterium[15] is used (recall the uncertainties), $\eta \geq 1 \times 10^{-10}$, then for $\tau_{\frac{1}{2}} \geq 10.13$ min and $Y_p \leq 0.25$ N_ν must be ≤ 6 (also see Figure 1).

A very reliable lower limit to η is that derived from the inner luminous parts of galaxies, $\eta \geq 0.29 \times 10^{-10}$. Surprisingly, unless Y_p can be constrained to be ≤ 0.21 this bound results in no upper limit to N_ν. Let me briefly explain the new twist. Increasing N_ν

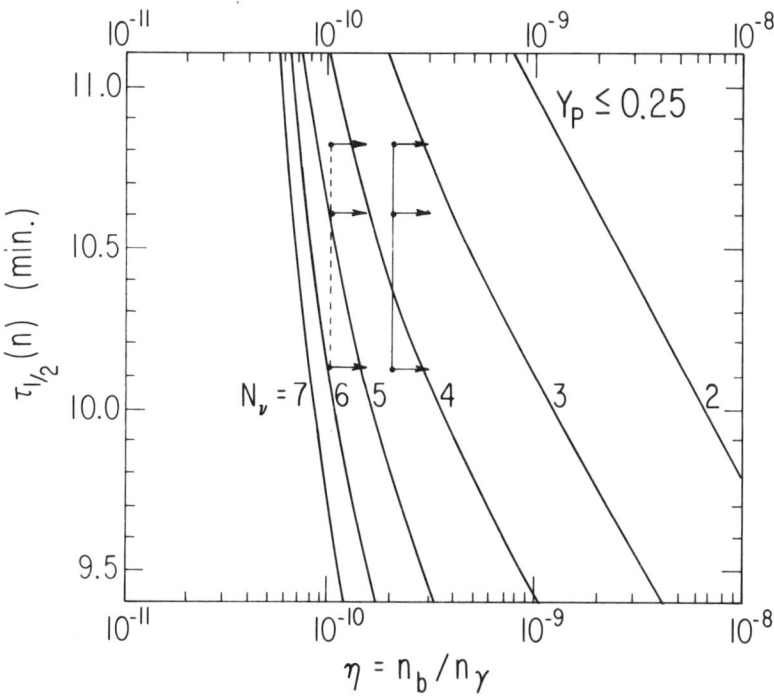

FIGURE 1 - The number of allowed neutrino species as a function of η and $\tau_{1/2}$ for $Y_p \leq 0.25$.

speeds up the expansion and leads to an earlier freeze out for the n/p ratio. This in turn results in more ^{4}He production. However, if the expansion is sped up too much, the reactions which produce ^{4}He [e.g., ^{3}H(^{2}H,n)^{4}He] are not occurring rapidly on the expansion timescale, and so there is not enough time to make ^{4}He. Hence, for large enough values of N_ν Y_p decreases with increasing N_ν. This is exacerbated for small values of η since these rates (per baryon) are α η. For this reason, Y_p as a function of N_ν achieves a maximum value (see Figures 2a and 2b). For $\eta = 0.29 \times 10^{-10}$, this maximum value is only 0.21, and so no limit to N_ν can be derived from the upper bound $Y_p \leq 0.25$.

To summarize, if baryons provide most of the mass observed in BSG, then N_ν must be ≤ 4. If non-baryonic matter provides most of the mass which binds BSG, then using the hot gas/deuterium lower bound on η, N_ν must be ≤ 6. If only the bound derived from the luminous inner parts of galaxies is used, then at present there is no upper limit on N_ν.

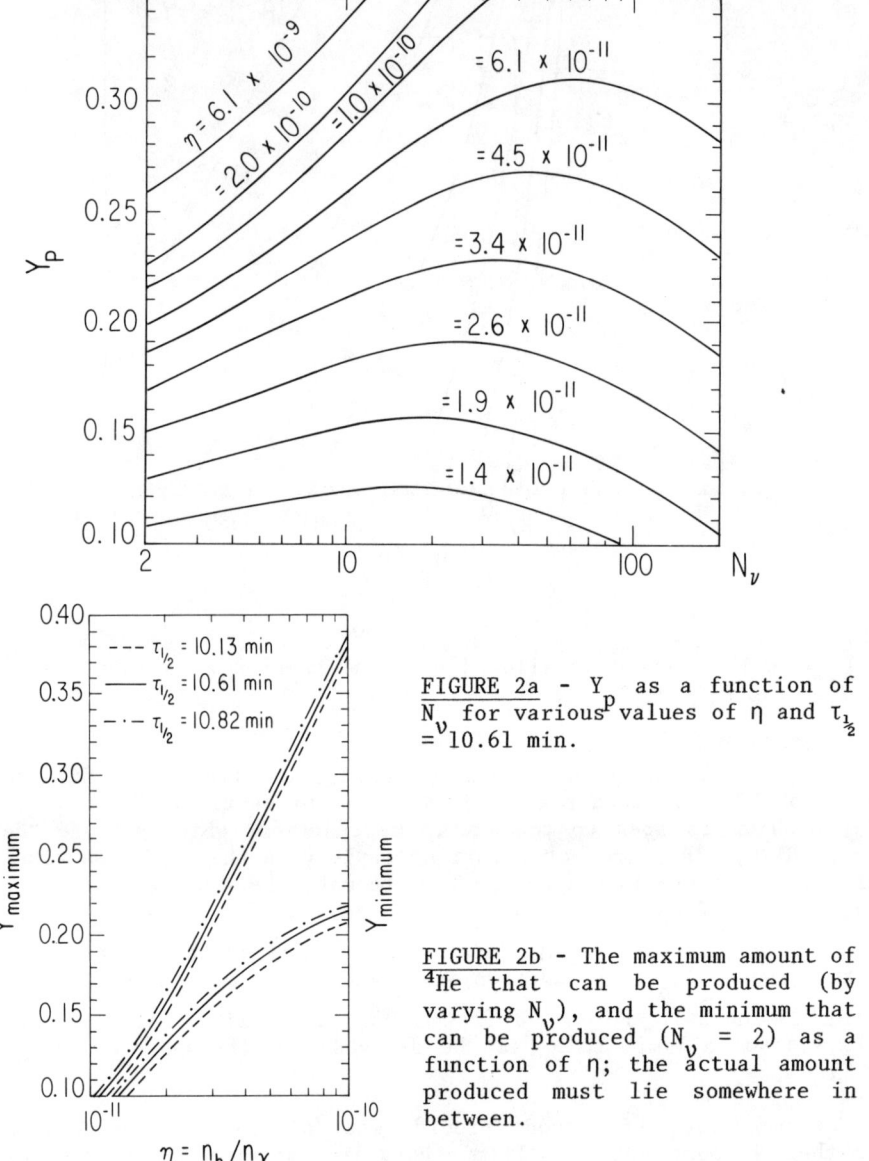

FIGURE 2a - Y_p as a function of N_ν for various values of η and $\tau_{1/2} = 10.61$ min.

FIGURE 2b - The maximum amount of ^{4}He that can be produced (by varying N_ν), and the minimum that can be produced ($N_\nu = 2$) as a function of η; the actual amount produced must lie somewhere in between.

Table 3

^{4}He Derived Limits on N_ν ($Y_p \leq 0.25$ and $\tau_{\frac{1}{2}} \geq 10.13$ min)

$\eta \geq 2 \times 10^{-10}$ (BSG)	$N_\nu \leq 4$
$\eta \geq 1 \times 10^{-10}$ (Hot Gas/D)	$N_\nu \leq 6$
$\eta \geq 0.29 \times 10^{-10}$ (Galaxies)	No Limit

ASTROPHYSICAL AND COSMOLOGICAL CONSTRAINTS ON NEUTRINO PROPERTIES

The astrophysical and cosmological constraints on neutrino mass and lifetime are summarized in Figure 3, as compiled from the work of various groups.[16-24] The assumptions made in deriving these limits are (in brief):

- standard hot big bang model (F-R-W universe).
- the temperature of the universe was at one time greater than max $\{m_\nu c^2, 1 \text{ MeV}\}$ so that at one time neutrinos were present in thermal equilibrium numbers.
- roughly half of the decay energy goes into photons of energy $\sim \frac{1}{2} m_\nu$. The decay modes of the heavy neutrino might include:

$$\nu_H \rightarrow \nu_L + \gamma , \tag{5a}$$

$$\nu_H \rightarrow \nu_L + e^\pm , \tag{5b}$$

$$\nu_H \rightarrow \nu_L + \mu^\pm , \tag{5c}$$

$$\nu_H \rightarrow \nu_L + \pi^\pm , \tag{5d}$$

$$\nu_H \rightarrow \nu_L + \nu \bar{\nu} , \tag{5e}$$

$$\nu_H \rightarrow \nu_L + \emptyset , \tag{5f}$$

where $\emptyset$ is an axion. For modes (5a - 5d) $\sim \frac{1}{2}$ of the decay energy will end up in photons since the annihilation of the pairs which are produced should be very rapid ($\Gamma \gg H$).

Let me briefly explain the origin of the various constraints. A "stable" neutrino ($\tau \gtrsim 10^{24}$s) with mass in the range (200 eV, 2 GeV) would contribute too much mass density to the universe (i.e., $\Omega \gtrsim 2$) causing a larger deceleration of the expansion than is observed.[17,18,22] If the hypothetical neutrino possesses a lepton asymmetry, $(n_L - n_{\bar{L}})/n_\gamma$, of the same order as the baryon asymmetry, $\eta \simeq 10^{-10 \pm 1}$ (ref.4), then the mass range $m_L \gtrsim 60$ GeV is also forbidden for the same reason.[19]

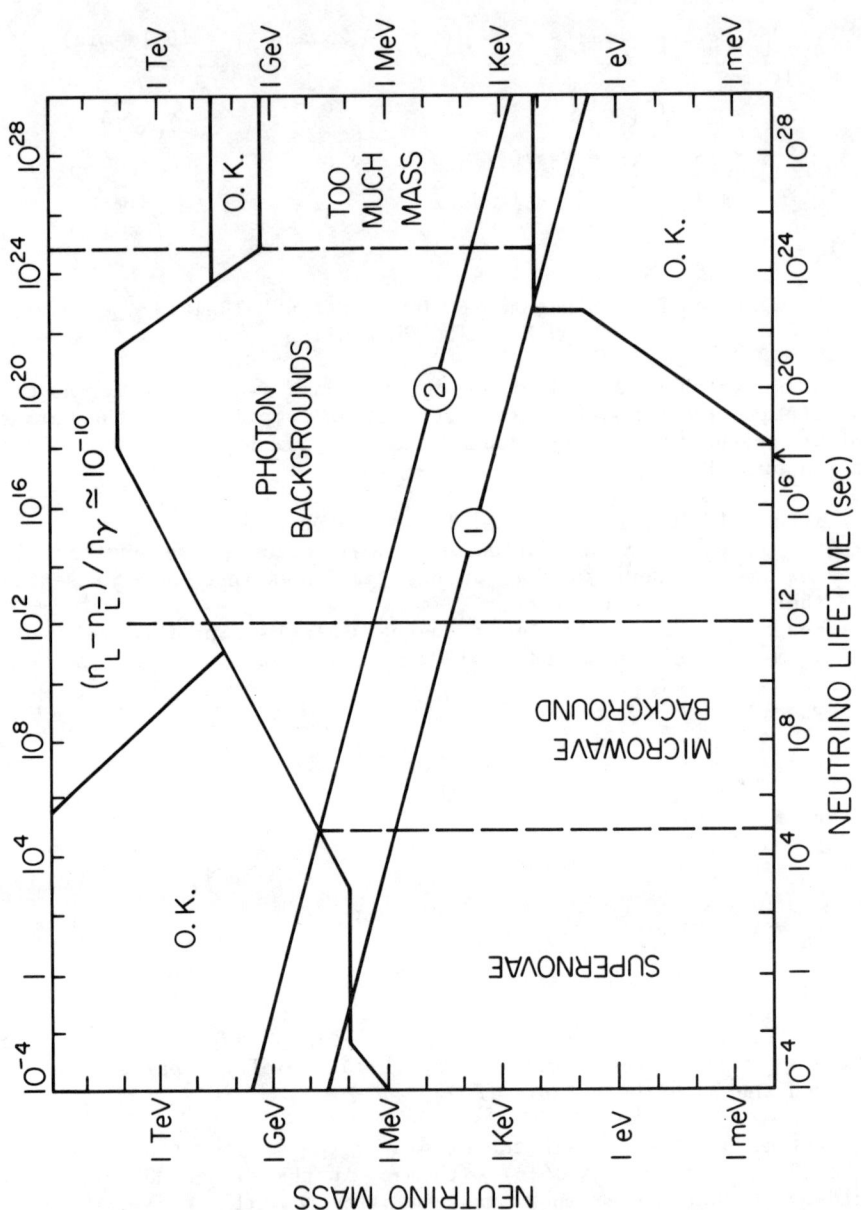

FIGURE 3 - Approximate astrophysical and cosmological restrictions on the mass and lifetime of a hypothetical neutrino. The arrow indicates the present age of the universe. The assumptions and origin of the constraints are explained in the text. The allowed regions are labeled "O.K."

Unstable neutrinos with lifetimes in the range (10^{12}s, 10^{24}s) decay after the universe becomes transparent to photons, and therefore their decay photons remain unthermalized. These photons contribute to the present UV, X-, and γ-ray backgrounds and the restrictions follow from insisting that they do not produce photon fluxes greater than those observed.[16,22,23] The constraint in the upper mass region is only applicable if the hypothetical neutrino possesses a lepton asymmetry $\sim 0(10^{-10\pm1})$.[24]

The decay photons from unstable neutrinos with lifetimes in the range (10^4s, 10^{12}s) and masses in the restricted mass range would distort the present microwave background. Although the decay photons would be degraded in energy and thermalized so that they would not distort the shape of the blackbody spectrum, they would result in there being too many photons. Put simply, the temperature derived from the shape of spectrum and the temperature derived from the normalization would be different.[16,22] Again, the restriction on the triangular region is only valid if this species possesses a lepton asymmetry.[24]

Finally, neutrinos with lifetimes in the range (10^{-3}s, 10^4s) and masses $\lesssim$ 10 MeV will affect supernova energetics. When a massive star collapses to form a neutron star (or perhaps also a black hole), the bulk of the gravitational binding energy ($\sim 10^{53}$ ergs) is released in neutrinos. Only $\sim 10^{51}$ ergs is released as visible energy, in the KE of the expanding shell and the optical supernova "fireworks." Neutrinos with lifetimes in this range would, via their decay photons, deposit more than $\sim 10^{51}$ ergs in the expanding shell or release more than $\sim 10^{51}$ ergs in visible light.[21]

For neutrinos less massive than 1 MeV the dominant decay mode is probably $\nu_H \to \nu_L + \gamma$ (shown in figure 4a), which in models where neutrinos and charged leptons are paired in isodoublets is GIM suppressed. Scaling the neutrino lifetime as the muon lifetime with additional factors for the photon, GIM supression, and mixing between ν_L and ν_H,

$$\tau \cong 10^4 \text{sec } (m_{\nu_H}/1 \text{ MeV})^{-5} \alpha^{-1} (M_W/M_\ell)^4 \sin^{-2} 2\theta , \qquad (6)$$

where $M_\ell \sim$ mass of the heaviest lepton in the intermediate state (Figure 4a). For neutrinos more massive than 1 MeV, the dominant mode is probably $\nu_H \to \nu_L + e^{\pm}$ which is not GIM surpressed (shown in Figure 4b). Again, scaling with the muon lifetime,

$$\tau \cong 10^4 \text{sec } (m_{\nu_H}/1 \text{ MeV})^{-5} \sin^{-2} 2\theta . \qquad (7)$$

In Figure 3 two mass-lifetime lines are shown:
(1) $\tau = 10^4 \text{sec } (m_{\nu_H}/1 \text{ MeV})^{-5}$, and (2) $\tau = 10^{14} \text{sec } (m_{\nu_H}/1 \text{ MeV})^{-5}$

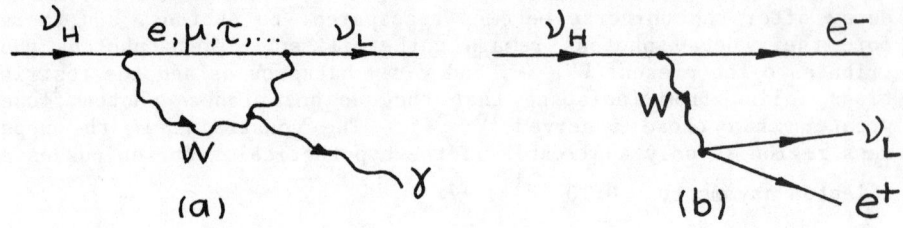

FIGURE 4a,b - Feynman diagrams for $\nu_H \to \nu_L + \gamma$ (a), and for $\nu_H \to \nu_L + e^{\pm}$ (b).

which should bracket the physically interesting cases. From Figure 3 it can be seen that neutrinos whose decays ultimately produce photons, and which lie between the two mass-lifetime lines are forbidden unless they are less massive than ~ 200 eV or more massive than $\sim 10\text{-}100$ MeV. The boundary for low mass neutrinos is very firm, as will be discussed in the next section. However, the boundary near 10-100 MeV is less secure, and needs to be investigated more thoroughly.[24]

MASSIVE NEUTRINOS AND COSMOLOGY

(a) Mass density of neutrinos

The temperature of the microwave background is in the range (2.7K, 3.0K), which corresponds to a photon number density

$$n_\gamma = 400 \cdot (T_\gamma/2.7K)^3 \text{ cm}^{-3} \ . \tag{8}$$

Because neutrinos ceased to be thermal contact with the rest of the universe for $T \lesssim 1$ MeV ($\Gamma < H$), they did not share in the energy released when $e^{\pm}$ annihilations occurred ($T \sim 0.5$ MeV). The temperature today of any light neutrino species ($m_\nu < 1$ MeV) is related to T_γ by[2]

$$T_\nu = (4/11)^{1/3} T_\gamma \ . \tag{9}$$

[Note for massive neutrinos this is not the kinetic temperature, i.e., KE $\not\propto T_\nu$, rather it is the temperature which specifies the distribution of the neutrinos in momentum space.] Each two-component species has a present number density (counting both νs and $\bar\nu$s) given by

$$\begin{aligned} n_\nu &= (4/11)(3/4)\ 400 \cdot (T_\gamma/2.7K)^3 \text{ cm}^{-3} \\ &\cong 109 \cdot (T_\gamma/2.7K)^3 \text{ cm}^{-3} \ , \end{aligned} \tag{10}$$

the factor of 4/11 coming from T_ν^3 and the factor of 3/4 coming from the difference between Fermi-Dirac and Bose-Einsten statistics. From the abundance of ^{4}He and the mass density deduced from the luminous inner parts of galaxies, the baryon-to-photon ratio, η, is restricted to the range[4]

$$\eta = 10^{-9.9\pm0.9} \tag{11}$$

Comparing the mass density contributed by each species to the critical mass density ($\equiv 3H_o^2/8\pi G$), we find that

$$\Omega_\gamma = 3.26 \times 10^{-5} (T_\gamma/2.7K)^3 h_o^{-2}, \tag{12a}$$

$$\Omega_\nu = (\sum m_\nu/100 \text{ eV})(T_\gamma/2.7K)^3 h_o^{-2}, \tag{12b}$$

$$\Omega_b = 10^{-2.4\pm0.9} (T_\gamma/2.7K)^3 h_o^{-2} \tag{12c}$$

where $\sum m_\nu$ is the sum of the masses of the light (< 1 MeV), two-component neutrino species. Taking the ratio of (12b) to the maximum value of (12c), we find that if

$$\sum m_\nu \gtrsim 3.5 \text{ eV} \tag{13}$$

the universe is neutrino-dominated. Since $T_\gamma \geq 2.7K$ and $h_o \leq 1$ if

$$\sum m_\nu \geq 100 \text{ eV} \tag{14}$$

neutrinos close the universe. Finally, from the observation that the universe is not too closed ($\Omega \leq 2$) and $h_o \leq 1$, we obtain:[17]

$$\sum m_\nu \leq 200 \text{ eV}. \tag{15}$$

(b) The hierarchy problem

As I discussed earlier, when the mass density of the universe is sampled on different scales, one obtains different values of Ω (see Table 1). There appears to be an increasing trend in M/L ratio and Ω, going from small scale to large scale, which perhaps indicates that there exists dark matter that preferentially clusters on large scales.[14] In addition, it is known that for galaxies the mass density (as determined by rotation curves) and light density do not decrease at the same rate moving outward from the center of the galaxy -- the light decreases much more rapidly, also indicating the presence of dark matter in galactic halos. Various candidates have been suggested for the dark matter: black holes, "jupiters", very massive stable neutrinos, and light neutrinos.[14,17,25,26]

I will now describe the role a light [$m_\nu \sim O(10 \text{ eV})$] neutrino might play in explaining the dark matter in the universe. Since luminous matter appears to contribute little of the observed mass in BSG, rich clusters, and probably even in galaxies, consider a self-gravitating sphere of neutrinos, characteristic size r and velocity dispersion $\sigma \sim \langle v^2 \rangle$. Since neutrinos are fermions, their phase space number density must always be $\lesssim 1$. [If we assume that their initial momentum distribution was given by an equilibrium distribution and that they are non-interacting, this statement follows from Louisville's theorem and applies to bosons as well.] The phase space volume available in the sphere is

$$\mathcal{V}_p \sim \int d^3p\, d^3x \sim m_\nu^3 \int d^3v\, d^3x$$
$$\mathcal{V}_p \sim m_\nu^3\, r^3\, \sigma^3 ,\qquad(16)$$

so that the mass of such a configuration is

$$\mathcal{M} \lesssim m_\nu \mathcal{V}_p \sim m_\nu^4\, r^3\, \sigma^3 .\qquad(17)$$

If the sphere is to be bound gravitationally, then by the virial theorem $\sigma^2 \sim G\mathcal{M}/r$, so that the self-consistency requirement for this configuration is

$$m_\nu \gtrsim G^{-\frac{1}{4}}\, \sigma^{-\frac{1}{4}}\, r^{-\frac{1}{2}} .\qquad(18)$$

That is, for a given self-gravitating system, defined by σ and r, there is a critical neutrino mass. Below this critical mass, neutrinos <u>cannot</u> be the dominant source of mass for such a system, while above this critical mass they <u>can</u> be. [This argument is derived rigorously in ref. 26.] The critical masses for the astrophysical systems of interest are:

Galactic Halo	$m_\nu \geq 20$ eV
BSG	$m_\nu \geq 14$ eV $h_o^{\frac{1}{2}}$
Rich Clusters	$m_\nu \geq 5$ eV $h_o^{\frac{1}{2}}$

To explain the discrepancy between the Ω inferred for rich clusters and BSG one could postulate a neutrino of mass $\sim O(5\text{-}10\text{ eV})$ -- which could cluster in rich clusters but not in BSG.[14] Alternatively, one might try to explain <u>all</u> the dark matter by a neutrino species of mass ≥ 20 eV which <u>could</u> cluster in galactic halos, BSG, and rich clusters.[27] The discussion above has been one of self-consistency -- could a neutrino of mass m_ν cluster on such a scale? The question, where would a massive neutrino actually end up?, is much more subtle. In order to determine whether or not neutrinos actually cluster on a given scale, one must consider the details of the formation of structure in the universe.

(c) Galaxy formation without massive neutrinos

One of the conundrums of modern cosmology involves the following paradoxical observations: the universe is remarkably homogeneous (on very large scales, $\gtrsim 1$ Gpc), and at the same time very inhomogeneous (on smaller scales -- stars, galaxies, clusters, superclusters). In the gravitational instability theory of galaxy formation (for a complete review, see ref. 28), the present structure in the universe evolved from small density inhomogeneities which were present from a much earlier epoch, and whose origin is a very fundamental question which remains unresolved. Once matter and radiation decouple, i.e., cease to interact, at $t \sim 10^{12}$s density fluctuations on scales smaller than the horizon ($\sim 10^{18} M_\odot$ at $t \sim 10^{12}$s) can grow due to the classical gravitational (or Jeans) instability. In the linear regime ($\delta\rho/\rho \ll 1$) they grow $\sim t^{2/3}$. Earlier than decoupling, the matter-radiation fluid is "too stiff" ($p \sim 1/3\rho$) to collapse.

There are two generic types of primordial inhomogeneities: (i) adiabatic, in which both the matter and radiation are perturbed, such that the baryon number-to-specific entropy ratio is constant, $\delta(kn_B/s) = 0$, and (ii) isothermal, which are just fluctations in kn_B/s, and which during the radiation dominated epoch correspond to spatial variations in the baryon number, $\delta\rho/\rho$ being equal to zero. Adiabatic or curvature fluctuations (on scales larger than the horizon) grow during the radiation epoch $\sim t$. Physically, they can be thought of as spatial variations in the curvature of the universe. Isothermal fluctuations on the other hand do not grow during the radiation epoch -- they are just fluctuations in baryon number.

There are two basic scenarios of galaxy formation and they correspond to the two types of density perturbations. The so-called "pancake" or fragmentation scenario is based upon adiabatic fluctuations.[29] During decoupling adiabatic perturbations on scales less than the Silk mass,

$$M_S = 1.2 \times 10^{12} M_\odot (\Omega h_0^2)^{-2}/(1 + 0.04/(\Omega h_0^2))^3 \quad . \tag{19}$$

are damped by photon diffusion. Just after decoupling only fluctuations on mass scales $\gtrsim 10^{12}$-$10^{15} M_\odot$ remain. As these density perturbations grow due to the Jeans instability, their collapse becomes asymmetric (one axis collapsing more rapidly than the other two), and pancake-like structures form. Eventually the pancakes hydrodynamically fragment, hopefully producing the structure observed today [Note: $10^{12} M_\odot$ is a typical galactic mass, while $10^{15} M_\odot$ is a typical supercluster mass]. Density peturbations of $\delta\rho/\rho \sim 10^{-3}$-$10^{-4}$ at decoupling are required for this scenario. This model predicts fluctuations in the microwave background of $\delta T/T \sim \delta\rho/\rho \sim 10^{-3}$-$10^{-4}$ on scales $\gtrsim 0(30")$. Such fluctuations have not yet been detected, and these negative observations may already be in conflict with the pancake theory.[30]

The isothermal or aggregation scenario[28] is based upon isothermal fluctuations. During decoupling isothermal perturbations on mass scales $\lesssim 10^5 M_\odot$ are damped due to photon diffusion. Just after decoupling perturbations on scales $\gtrsim 10^5 M_\odot$ begin to grow due to the Jeans instability. Assuming that the initial spectrum is given by a power law, $\delta\rho/\rho \sim M^{-n}$, then perturbations on the smaller scales will go nonlinear ($\delta\rho/\rho \gtrsim 1$) first, and the present structure develops by successively building up larger and larger structures. Numerical studies[28] indicate that the structure observed today could have resulted from such a process. In this scenario much smaller fluctuations in the microwave background are predicted, and at present there is no conflict between the predictions of the theory and the observations.[30]

(d) Galaxy formation with massive neutrinos

How is the picture described above altered by the existence of massive neutrinos? This issue has recently been addressed by Bond, Efstathiou, and Silk.[31] Let me briefly review how the two scenarios described in (c) change. For this discussion I will assume that the universe is neutrino-dominated, i.e., $\sum m_\nu \gtrsim 3.5$ eV. Since for $T \lesssim 1$ MeV neutrinos effectively cease to interact, they are collisionless particles. For this reason, they can freely stream and all density perturbations in the neutrino sea on scales $\lesssim$ horizon diffuse away, more precisely on scales[31]

$$M \lesssim 4 \times 10^{15} M_\odot \, (m_\nu/30 \text{ eV})^{-2} \, . \tag{20}$$

When the neutrinos become non-relativistic, perturbations in the neutrinos can grow via the Jeans instability. In the adiabatic (or pancake) picture things proceed very much like they did in the baryon-dominated universe. Fluctuations on scales $\gtrsim 10^{15} M_\odot$ collapse and "take" the baryons with them, so that $\delta\rho_b/\rho_b \sim \delta\rho_\nu/\rho_\nu$. Eventually, the baryons fragment hydrodynamically as before. Since the neutrinos are essentially non-interacting particles they have no microscopic means to dissipate their energy, and therefore do not relax as much as the baryons. They form an extended halo around the supercluster. In this scenario neutrinos do not cluster in galactic halos or in BSG, they only cluster on the largest scales.

In the isothermal picture the scenario changes dramatically. Since perturbations in the neutrinos on scales $\lesssim 10^{15} M_\odot$ diffuse away, the only fluctuations on smaller scales are those in the baryons, so that on scales $\lesssim 10^{15} M_\odot$

$$\delta\rho_T/\rho_T \cong \delta\rho_b/(\rho_b + \rho_\nu) \cong (\rho_b/\rho_\nu) \, \delta\rho_b/\rho_b \, . \tag{21}$$

Therefore, the primordial inhomogeneities on small scales are effectively reduced by a factor of (ρ_b/ρ_ν). In addition Bond et al.[31] have shown that the growth rate of the perturbations in the neutrino-dominated universe is $\sim t^p$, where

$$p = \frac{1}{6} \left[(1 + 24(\rho_b/\rho_T))^{1/2} - 1 \right] \, . \tag{22}$$

For $\rho_b = \rho_T$ this reduces to $p = 2/3$, however, for $\rho_b/\rho_T \cong 0.1$, $p \cong 1/7$ --the growth rate is significantly slower. For both these reasons, much larger initial isothermal perturbations are needed, probably $\delta\rho_b/\rho_b \gg 1$. In this scenario the isothermal perturbations "take" the neutrinos along with them as they grow, so that if the initial perturbations are large enough, neutrinos can cluster on all scales -- from galactic halos to rich clusters.

After this brief review of galaxy formation, the question which begs for an explanation is the origin of the primordial inhomogeneities. Several authors have addressed this question[32-34]; however, at present it is fair to say that this question is still one of the fundamental puzzles of cosmology. For the isothermal perturbations, the question is even more preplexing. In the standard F-R-W model it has been shown[35] that if one adopts the GUTs scenario for baryosynthesis[7], it is not possible to produce isothermal perturbations. Though recently, John Barrow and I have shown[36] that it is possible to produce isothermal perturbations if the universe was initially shear-dominated (also see ref. 37), or if it underwent a de Sitter phase due to a strongly-first order phase transition.[38]

ACKNOWLEDGEMENTS

It is a pleasure to thank Craig Hogan for several very helpful discussions about galaxy formation. This work was supported by the Institute for Theoretical Physics (NSF PHY77-27084) at UCSB and DOE AC02-80ER10773 at the University of Chicago.

REFERENCES

1. S. Hawking and G.F.R. Ellis, The Large Scale Structure of Space-Time (Cambridge University Press: Cambridge, England, 1973).
2. S. Weinberg, Gravitation and Cosmology (John Wiley: NY, 1972).
3. V.F. Shvartsman, JETP Lett. 9, 184 (1969); G. Steigman, D.N. Schramm, and J.E. Gunn, Phys. Lett. 66B, 202 (1977); J. Yang, D.N. Schramm, G. Steigman, and R.T. Rood, Astrophys. J. 227, 697 (1979).
4. K.A. Olive, D.N. Schramm, G. Steigman, M.S. Turner, and J. Yang, Astrophys. J. in press (1981).
5. J.N. Bahcall, S.H. Lubow, W.F. Huebner, N.H. Magee, Jr., A.L. Merts, M.F. Argo, P.D. Parker, B. Rozsnyai, and R.K. Ulrich, Phys. Rev. Lett. 45, 945 (1980).
6. This section is a summary of the more detailed discussions in ref. 4 and in M.S. Turner, "What limits (if any) does big bang nucleosynthesis place on the number of neutrino flavors," to appear in The Proceedings of the Wisconsin Neutrino-Mass Mini Conference (1981).
7. M.S. Turner, "Big bang baryosynthesis and grand unification," in these proceedings (1981).
8. G. Gamow, Phys. Rev. 70, 572 (1946).
9. R.V. Wagoner, W.A. Fowler, and F. Hoyle, Astrophys. J. 148, 3 (1967); P.J.E. Peebles, Astrophys. J. 146, 542 (1966).
10. C.J. Christensen, A. Nielsen, A. Bahnsen, W.K. Brown, and B.M. Rustad, Phys. Rev. D5, 1628 (1972).
11. L.N. Bondarenko, V.V. Kurguzov, Yu.A. Prokof'ev, E.V. Rogov, and P.E. Spivak, JETP Lett. 28, 303 (1978).
12. J. Byrne, J. Morse, K.F. Smith, F. Shaikh, K. Green, and G.L. Greene, Phys. Lett. 92B, 274 (1980).
13. S.M. Faber and J.S. Gallagher, Ann. Rev. Astron. Astrophys. 17, 135 (1979).

14. D.N. Schramm and G. Steigman, Astrophys. J. 243, 1 (1981).
15. At present we are attempting to use the abundance of D and ^{3}He to obtain a reliable lower limit to η, J. Yang, M.S. Turner, G. Steigman, D.N. Schramm, and K.A. Olive, work in preparation (1981).
16. D.A. Dicus, E.N. Kolb, and V.L. Teplitz, Phys. Rev. Lett 39, 168 (1977); D.A. Dicus, E.N. Kolb, and V.L. Teplitz, Astrophys. J. 221, 327 (1978); D.A. Dicus E.N. Kolb, V.L. Teplitz, and R.V. Wagoner, Phys. Rev. D17, 1529 (1978).
17. R. Cowsik and J. McClelland, Phys. Rev. Lett. 29, 669 (1972).
18. B.W. Lee and S. Weinberg, Phys. Rev. Lett. 39, 165 (1977).
19. P. Hut and K.A. Olive, Phys. Lett. 87B, 144 (1979).
20. A.D. Dolgov and Ya.B. Zel'dovich, Rev. Mod. Phys. 53, 1 (1981) and references therein.
21. S.W. Falk and D.N. Schramm, Phys. Lett. 79B, 511 (1978).
22. J.E. Gunn, B.W. Lee, I. Lerche, D.N. Schramm, and G. Steigman, Astrophys. J. 223, 1015 (1978).
23. R. Kimble, S. Bowyer, and P. Jakobsen, University of California, Berkeley, preprint (1981).
24. E.N. Kolb, M.S. Turner, and D.N. Schramm, in preparation (1981).
25. A.S. Szalay and G. Marx, Acta Phys. Sci. Hung. 35, 113 (1974).
26. S. Tremaine and J.E. Gunn, Phys. Rev. Lett. 42, 407 (1979).
27. E. Witten, Proceedings of the First Workshop on Grand Unification (1981); M. Davis, M. Lecar, C. Pryor, and E. Witten, "The formation of galaxies from massive neutrinos," Princeton University preprint (1981).
28. P.J.E. Peebles, The Large-Scale Structure of the Universe (Princeton University Press: Princeton, NJ, 1980).
29. Ya.B. Zel'dovich, Astron. Astrophys. 5, 84 (1970); Ya.B. Zel'dovich, in IAU Symposium No. 79, 409 (1978).
30. M.L. Wilson and J. Silk, University of California, Berkeley, preprint (1981).
31. J.R. Bond, G. Efstathiou, and J. Silk, Phys. Rev. Lett. 45, 1980 (1980).
32. C.J. Hogan, Nature 286, 360 (1980).
33. W.H. Press, Phys. Scripta 21 (1980).
34. Ya.B. Zel'dovich, Mon. Not. R. Astron. Soc. 192, 663 (1980).
35. M.S. Turner and D.N. Schramm, Nature 279, 303 (1979).
36. J.D. Barrow and M.S. Turner, "Baryosynthesis and the origin of galaxies," ITP preprint 80-03, to appear in Nature (1981).
37. J.R. Bond, E.N. Kolb, and J. Silk, in preparation (1981).
38. A. Guth, Phys. Rev. D., in press (1981).

DISCUSSION

WEINBERG: I'd like to make a comment which adds on to this talk about the possibility of detecting these galactic neutrinos experimentally in the laboratory. Assuming that the Russian estimate of the mass of the neutrino is correct, and that it really is in the neighborhood of 10 or 20 eV, which the astrophysicists would also like, the neutrinos really represent quite a concentration. There would be about 10^7 neutrinos per cm^3 in our galaxy as compared with the ambient background from the big bang of about 100 neutrinos per cm^3, so it represents a concentration enhancement by about a factor of 10^5.

Now you can think of one way of detecting them. If you do a beta decay experiment, like say the tritium experiment, which is done to measure the neutrino mass, these neutrinos would produce, in addition to very tiny effects right near the end point, a line beyond the end point. That is if you plot the number of events per unit energy versus electron energy the normal spectrum just decreases to zero at the nominal end point of the tritium spectrum, 18 keV. At an energy 2 m_ν beyond the end point, there would be an extremely sharp line caused by the fact that in addition to the normal event which is $A \rightarrow B + e + \bar{\nu}$ you can have $A + \nu \rightarrow B + e$ using galactic neutrinos. So instead of having to produce $m_\nu c^2$ you get $m_\nu c^2$ from the galactic neutrinos.

Now this line is, of course, very weak and, naturally, even though you have a concentration enhancement by a factor of 10^5, there are still not that many around. In fact, if you make an estimate for a typical, allowed beta decay, it turns out to be reasonably insensitive to the parameters of the beta decay. The number of events in the line is a few hundred per day per mole of source. Now a mole of a beta decay source of (even if it's a long-life beta decay like tritium) is about 100,000 curies. I notice Ray Davis was talking megacurie sources, so I suppose I am allowed to also. As one can easily imagine, the background problems are ferrocious. If you take a source with a few thousand curies so that you just get something of the order of an event per day, you have a huge background problem.

I think the only conceivable hope for this kind of experiment lies, first of all, in the fact that you are looking for a line, so that even if you have a background that is much larger than the number of events in the line, you can hope to pull the line out of the background. The line is very narrow because the neutrinos in our galaxy are very cold. Their typical velocity is a few hundred km/s, so the line has a fractional width of about 10^{-5} or 10^{-6}. So it is quite a remarkable line.

I think what is needed to do this kind of experiment, though, given the terrible background problems, is to try to design some kind of beta spectrometer, perhaps using electrostatic as well as magnetic fields, in which you have an absolute veto against registering any electron with an energy below the energy for which you are setting the spectrometer. Normally though, the problem with spectrometers is straggling, which is at the high energy end, because high energy electrons tend to look lower than they really are. So this may not be so impossible, and I hope Wu Chien-Shung, or any other beta decay physicist here, could give some attention to the possibility of designing beta spectrometers that really have absolutely square resolution shapes, at least at one end of the spectrum.

TURNER: I would like to add a comment. We have very few relics from the big bang; the 3K background tells us that the big bang was hot, at least as hot as ~3000 K. If we were able to detect these relic neutrinos, then we would have one more important fossil - we only have a handful altogether. The existence of the cosmic neutrino sea would tell us that the universe was once hotter than ~10^{11}K (the temperature necessary to produce and thermalize neutrinos). However, the technique you suggest to detect them depends upon the assumption that they cluster in galactic halos, which if they do, enhances their local number density by 10^5 - from ~100 cm^{-3} to ~10^7 cm^{-3}. Although neutrinos more massive than ~ 20 eV <u>can</u> cluster in galactic halos, whether or not they do depends upon the details of galaxy formation (as I discussed in my talk). In fact, it appears unlikely that they would unless there were very large isothermal perturbations at decoupling. It seems more likely that they would cluster on larger scales only.

WEINBERG: The number 10^7 can be derived in several ways, but the simplest way is to just take estimates of the mass in the galactic halo, as made for example, by Ostreiker and Caldwell, and divide by 20 eV. That doesn't rule out the possibility that there is more clustering on larger scales. It just takes the (supposedly) observed mass in the galactic halo and interprets it as neutrinos, leaving as an open question what goes on at larger scales.

TURNER: The observed mass in galactic halos might be something other than massive neutrinos, e.g., black holes or Jupiter-sized objects.

WEINBERG: Well, it's dependent on the model of the galactic halo, but this is actually a low estimate of the halo mass density that Ostreiker and Caldwell give. It is also, of course, dependent on the assumption that neutrinos have a mass of 20 eV which I wouldn't recommend anyone believe at this time until the Russian experiments are really confirmed. In fact the kind of thing I am talking about is absolutely pointless if the Russian measurements of the neutrino mass are not confirmed. I wasn't here yesterday so I didn't hear the discussion of the neutrino mass experiments and I'm not sure if I have the latest information on the reliability of these experiments. But eventually this will settle down, and if the neutrino mass is in the neighborhood of 20 eV, we will know it. Then this kind of experiment will be worth thinking about, but not until then.

RARE MUON PROCESSES

R. E. Mischke
Los Alamos Scientific Laboratory
Los Alamos, New Mexico 87545

Rare muon processes provide important tests of the weak interaction and its role in schemes for unification. It is appropriate to begin with a status report of the experimental limits, all but one of which have results which are less than two years old (see Fig. 1). It is interesting to note the progress which has been made in the last 30 years (Fig. 2).

I do not have time to review the relevant theoretical calculations, but I would like to emphasize the points that predictions as large as 10^{-8} have been published, the results are very model-dependent, and no single decay mode or process is certain to dominate. This means that experimenters must continue to examine all processes and that each improvement further constrains models even if it is just a better upper limit.

There are new efforts underway at each of the meson factories as indicated in Fig. 1. Each will be discussed briefly starting with the Crystal Box program; I know most about this effort since I am a member of the collaboration consisting of physicists from LASL, Stanford, and University of Chicago. A schematic of the detector is shown in Fig. 3.

Muons from a surface muon beam will stop in a thin extended target at the center of a cylindrical drift chamber. The drift chamber has eight concentric layers of cells and is surrounded by 396 NaI(Tℓ) crystals. Between the drift chamber and the crystals, a layer of 36 scintillation counters serves both as veto counters for the crystals and a trigger for electrons from muon decays.

The status of the various components is shown in Fig. 4. It is expected that testing of major components of the system will take place in February-March 1981, and the complete detector will be assembled by July-August 1981. The branching ratio limits which can be attained depend on the stopping rate, running time, and detector resolution functions. Most of the resolution parameters have been verified experimentally and are listed in Fig. 5. Using the more conservative estimates and assuming an average stopping rate of 5×10^5/s, the results to be expected are shown in Fig. 6.

The other major program to search for lepton number violating muon decays is at SIN where SINDRUM has been proposed. A schematic of the detector is shown in Fig. 7. It is a copy of a Russian detector which was designed to be optimum for $\mu \to 3e$ and consists of several concentric cylindrical MWPCs inside a superconducting solenoid. A thin Pb sheet can be placed at an intermediate radius to convert photons. The outer layers then serve as a pair spectrometer.

Work by groups from SIN, ETH Zürich, and U. Zürich is underway to make a prototype detector to check how the contained leptons behave. The real detector and a sufficiently intense beam are not expected until 1985. The goal of the experiment is to achieve 10^{-12} for $\mu \to 3e$ and $\mu \to e\gamma$. They also anticipate studying many other reactions with the same apparatus.

Further in the future, another attack on $\mu \to e\gamma$ is planned at LAMPF using the detector arrangement shown in Fig. 8. The NaI crystals from the Crystal Box will be rearranged into a wall, and if funds are available, a second wall will be added. This experiment is expected to reach 10^{-12}.

The other class of experiments which I will discuss is $\mu \to e$ conversion in a nucleus. At LAMPF, there are plans to search for $\mu^- \to e^+$ on ^{88}Sr. This experiment by a group from U. Chicago, Lakehead, and LASL plans to reach 10^{-12} by observing a very specific decay chain ^{88}Kr $\to$ ^{89}Rb. At present, they are checking the detection scheme by looking for Δ^{++} production in the reaction n^{88}Sr $\to$ ^{89}Kr + Δ^{++}. Later, they would run online in a stopped muon beam.

Another LAMPF experiment by a group from Yale, Penn, and SIN involves a superconducting solenoid with a drift chamber detector (Fig. 9). They will use a could μ^- beam with a stopping rate of order $5 \times 10^5 \mu^-$/s. The target may be Argon and sufficiently energetic electrons will be detected on a TPC-like end cap shown in Fig. 10. This experiment aims for a limit of 10^{-12} and is one to two years away.

The new experiment which is farthest along is the effort at TRIUMF by a group from Chicago, VPI, Victoria, UBC, Montreal, Carleton, and TRIUMF. They have built a large time projection chamber, and this year it has been tested (Fig. 11). Next year, they plan to begin data-taking on the process $\mu^- \to e^-$ with a goal of 10^{-12}.

All of these efforts which have been outlined are very difficult experiments involving many people and large resources. I am sure the motivation is based on a real expectation of significant results bearing on unification.

DISCUSSION

MARSHAK: I would like to reiterate the importance of the $\mu^- + A(Z) \to e^+ + A(Z-2)$ experiment. If you don't have a spin zero nucleus, the reaction is greatly inhibited. I hope your ingenuity will be applied to increasing the detection sensitivity with a spin zero nucleus. You need a heavy Majorana neutrino for that process.

Rare Muon Processes

current limits

$\mu^- \to e^- \gamma\gamma$ $\quad < 5 \times 10^{-8}$

$ \to e^- e e$ $\quad < 1.9 \times 10^{-9}$

$ \to e^- \gamma$ $\quad < 1.9 \times 10^{-10}$

$ \to e^- \nu_e \bar{\nu}_\mu$ $\quad < .098$

$\mu^- \to e^-$ $\quad < 7 \times 10^{-11}$

$\mu^- \to e^+$ $\quad < 3 \times 10^{-10}$

new experiments

	LAMPF	SIN	TRIUM
$\mu^+ \to e\gamma\gamma$	} Crystal Box	SINDRUM	
eee			
$e\gamma$			
$\mu^- \to e^-$	Solenoid		TPC
$\mu^- \to e^+$	Radiochem		

Fig. 1

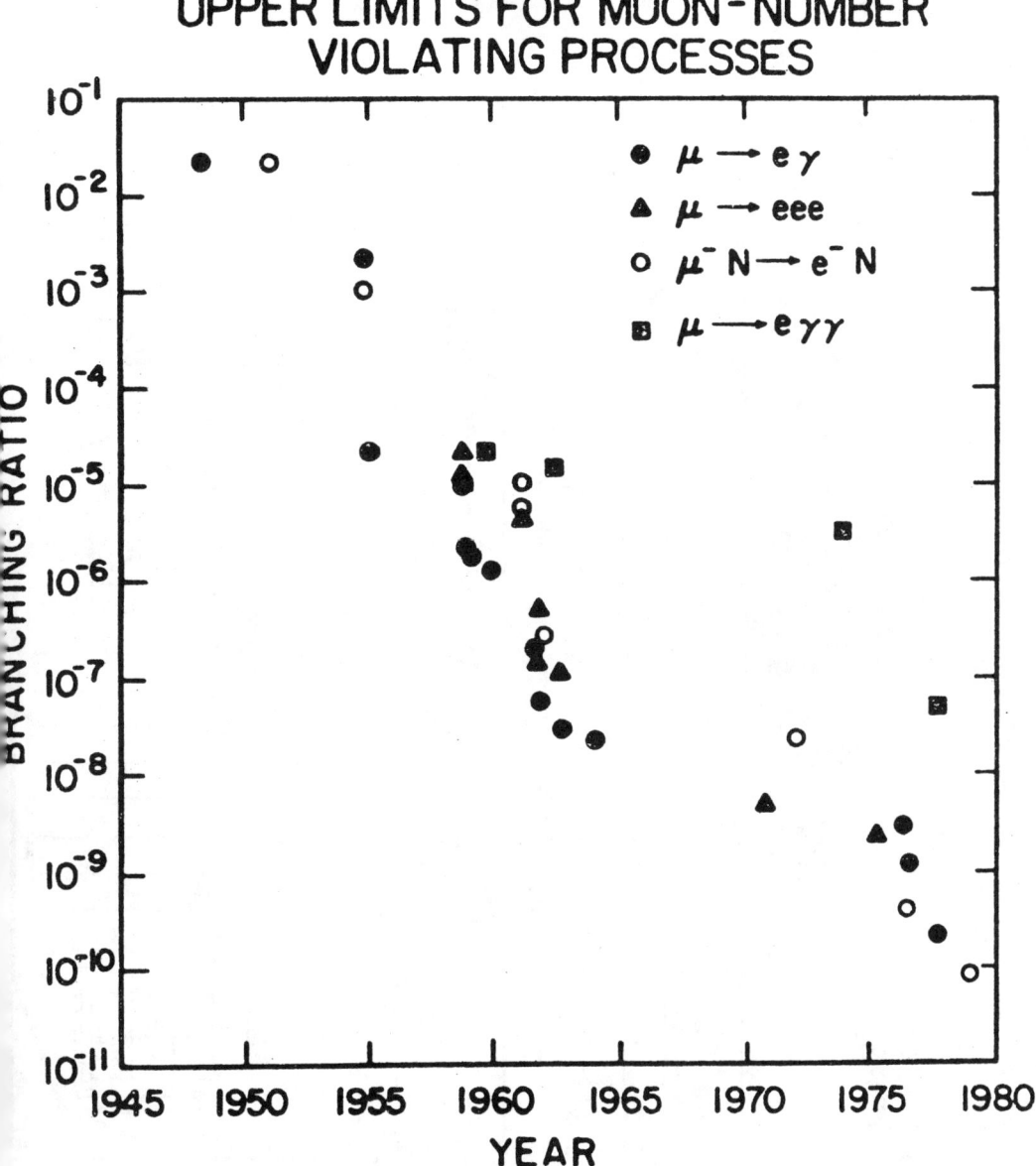

Fig. 2

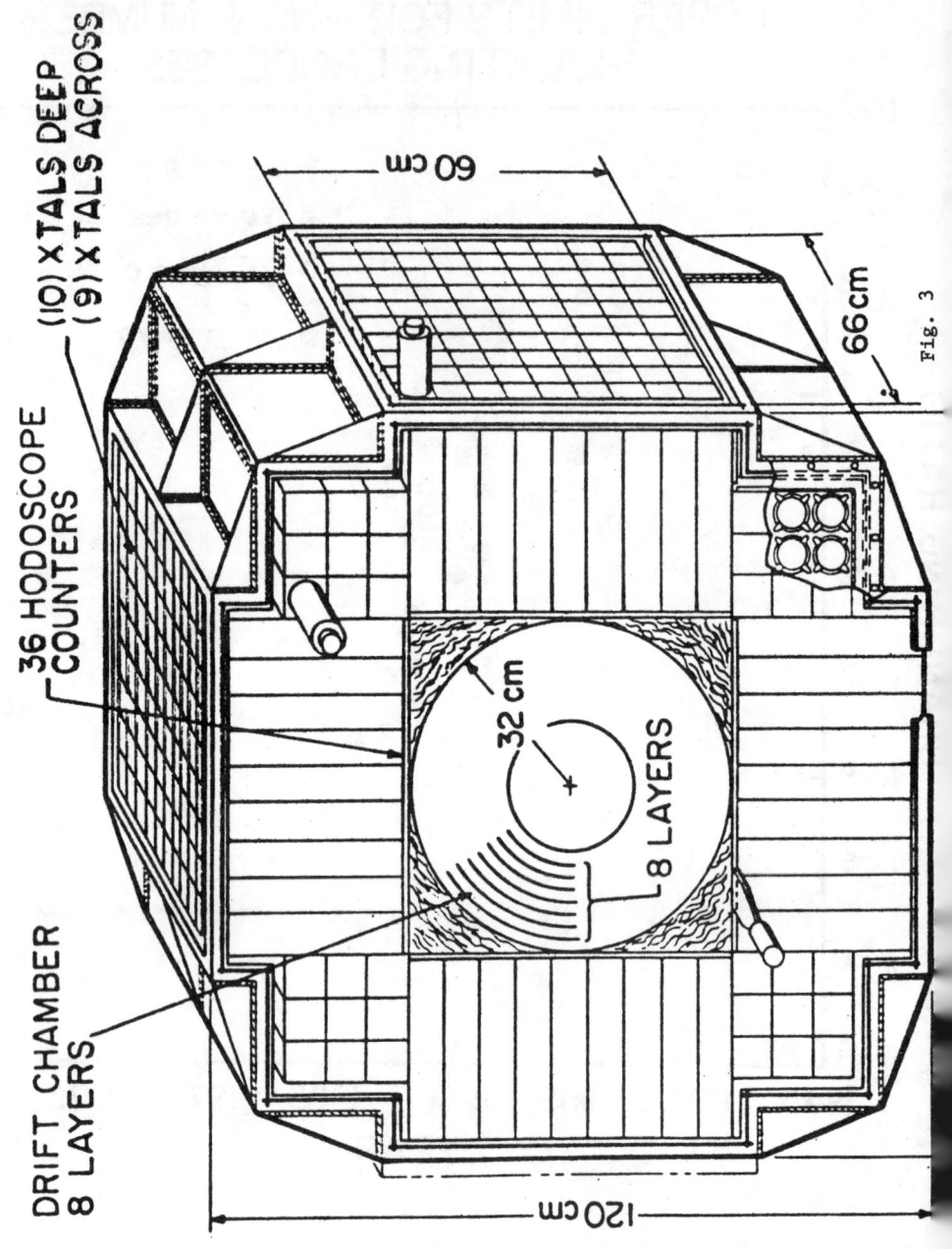

Fig. 3

Crystal Box Status

NaI	enclosure	1 Dec '80
	delivery from Harshow	1 April '81
	fiber optics added	
Scintillators		
	construction	1 Jan '81
Drift chamber		
	end plates	1 Feb '81
	completion	1 Jul '81
Electronics		
	ready for production or procurement	
	designing multiplexing of NaI via microprocessors	
Software		
	designed and coding underway	

Fig. 4

Crystal Box Resolutions

	Addendum (6/79)	Achieved (7/80)
Drift Chamber Position :	$\sigma = 200$ µm	$\sigma \leq 150$ µm
Scintillator Timing:	$\Delta t = 1$ ns (FWHM)	$\Delta t = 0.35$ ns (FWHM)
NaI Energy:	$\frac{\Delta E}{E} = 0.06$ @50 MeV (FWHM)	$\frac{\Delta E}{E} \simeq 0.05$ @ 50 MeV (FWHM)
NaI Timing:	$\Delta t = 1$ ns (FWHM)	$\Delta t = 0.40$ ns (FWHM) @50 MeV

Fig. 5

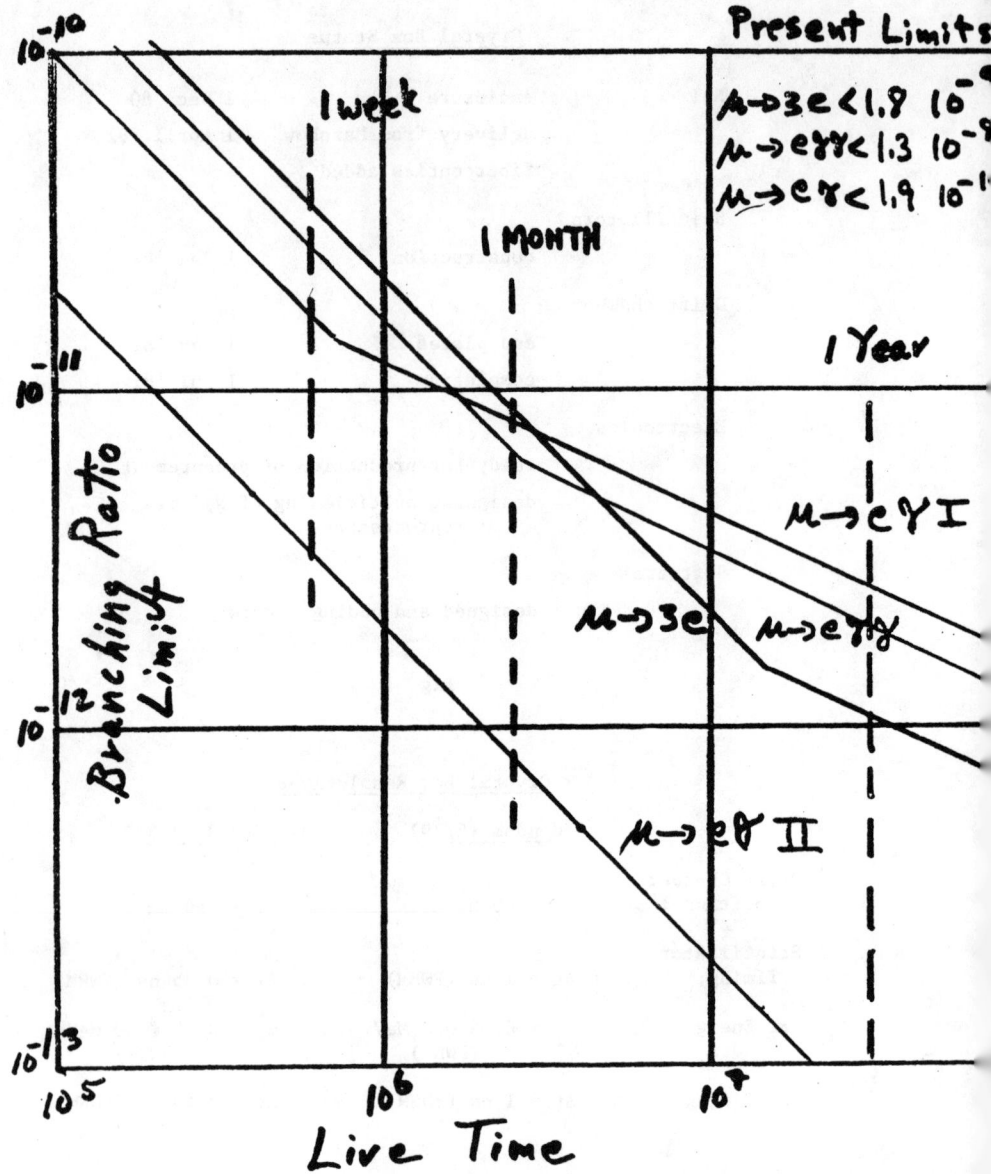

Fig. 6

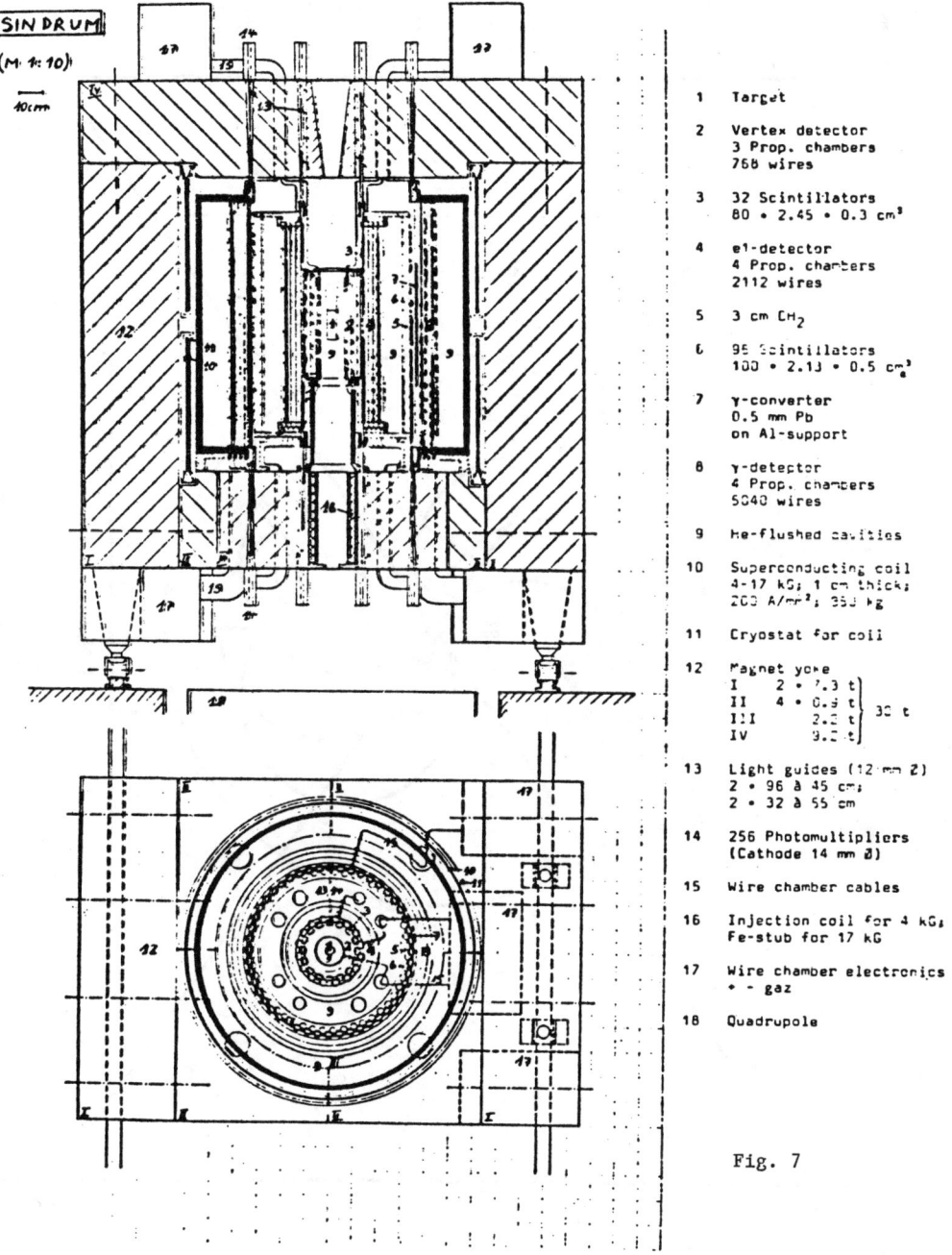

Fig. 7

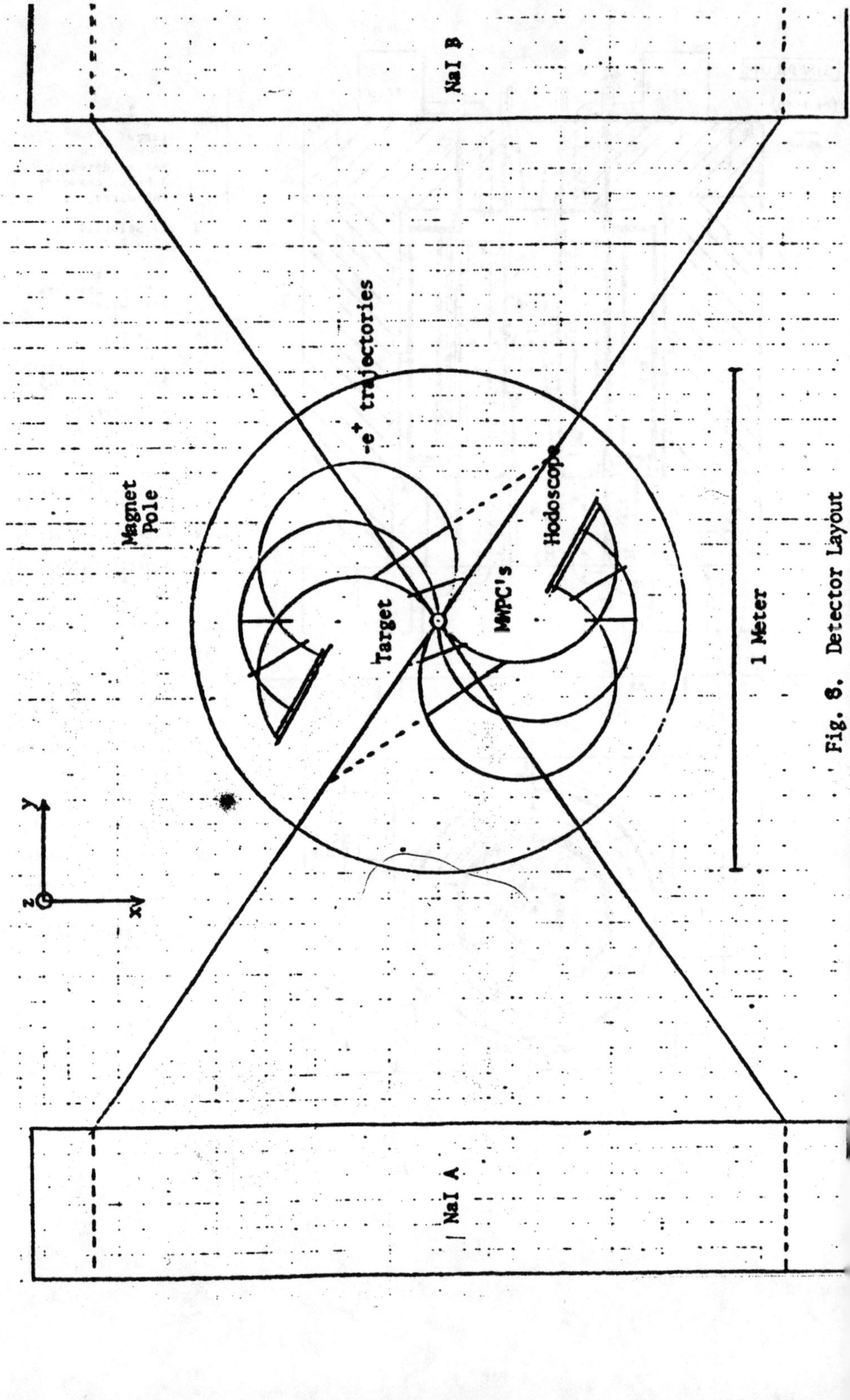

Fig. 8. Detector Layout

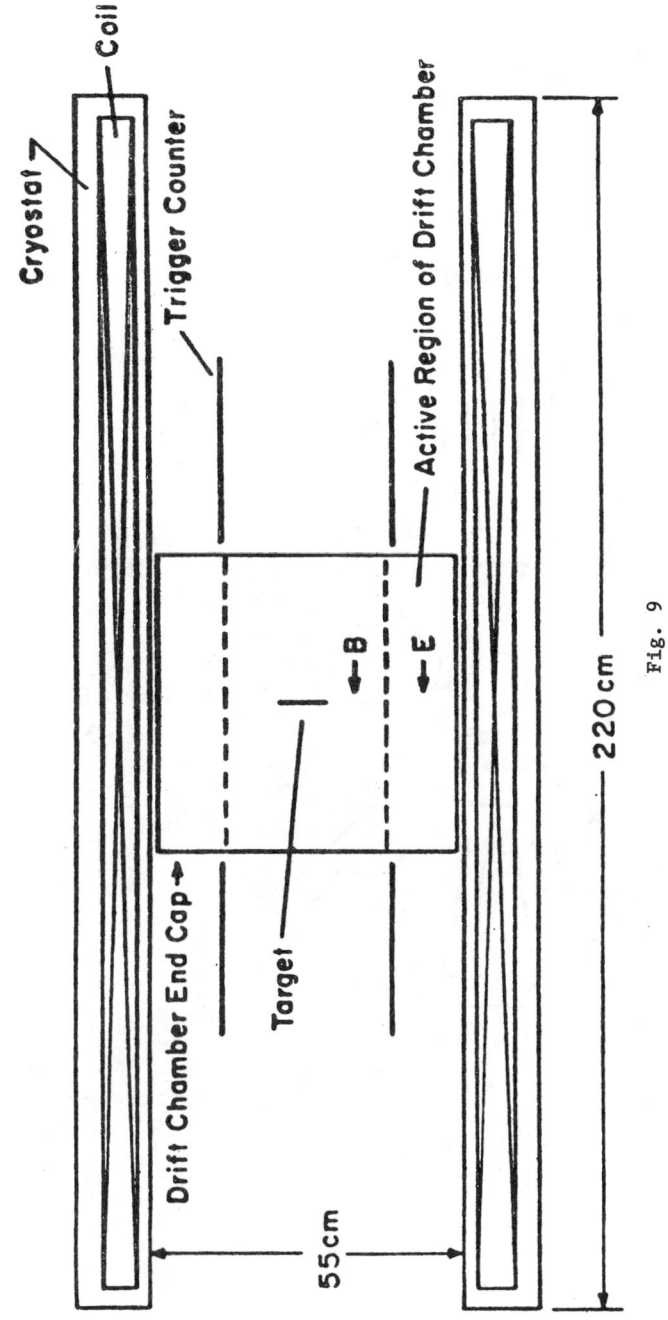

Fig. 9

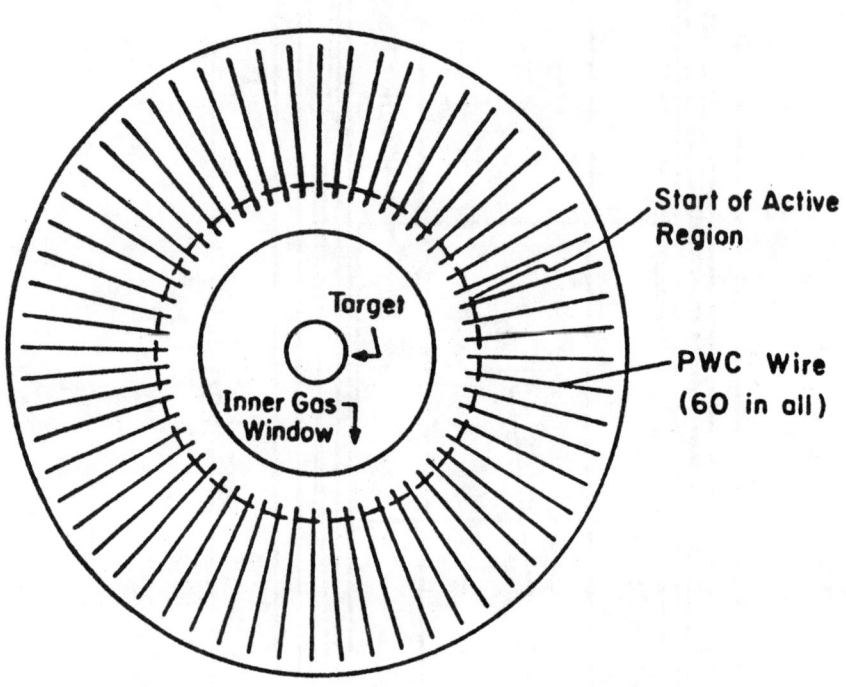

Fig. 10

Search for $\mu^- Z \rightarrow e^- Z$
TRIUMF <u>T</u>ime <u>P</u>rojection <u>C</u>hamber

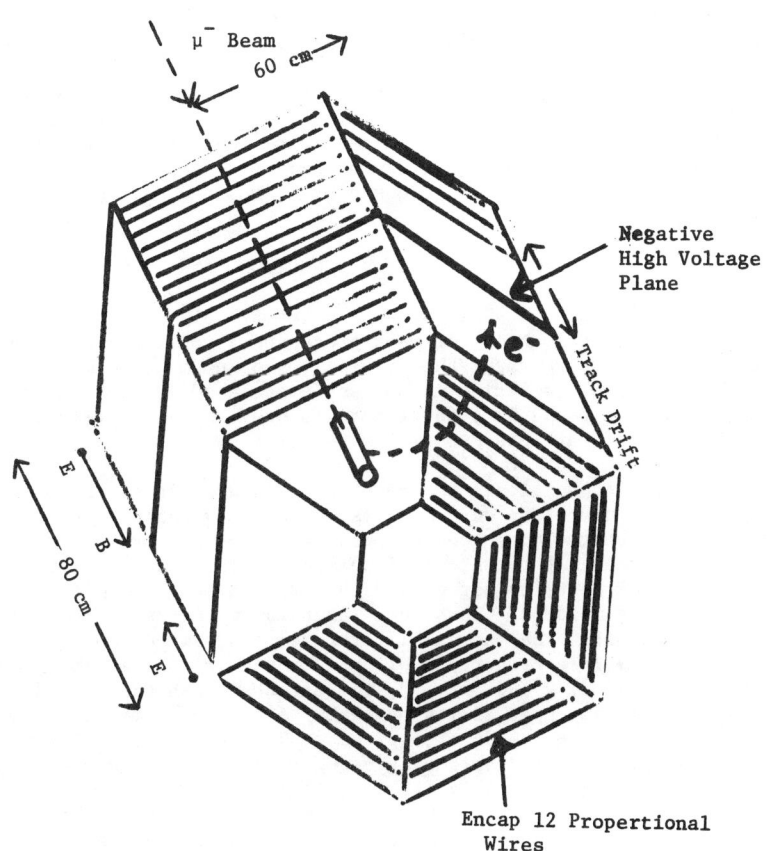

Fig. 11

IMPLICATIONS OF NEUTRINO MASSES AND MIXING FOR WEAK PROCESSES

Robert E. Shrock
ITP - State University of New York, Stony Brook, NY 11794

ABSTRACT

We present a general theory of weak processes involving neutrinos which consistently incorporates the possibility of nonzero neutrino masses and associated lepton mixing. The theory leads to new tests for and bounds on such masses and mixing. These tests make use of $(\pi,K)_{\ell 2}$ decay, nuclear β decay, and μ and τ decays, among others. New experiments at SIN and KEK to apply our tests are mentioned. We further discuss some implications for: (1) the analysis of the spectral parameters in leptonic decays to determine the Lorentz structure of the weak leptonic couplings; (2) fundamental weak interaction constants such as G_μ, $G_V{'}$, f_π, f_K, V_{uq}, $q = d$ or s, m_W, and m_Z; and (3) neutrino propagation.

I. INTRODUCTION

In this talk we shall present a summary of some of the main features of a general theory of weak processes involving neutrinos which consistently incorporates the possibility of nonzero (and, in particular, substantial) neutrino masses and associated lepton mixing. Further details may be found in a short note on neutrino mass limits in the 1980 Review of Particle Properties[1], a brief report[2], and two longer papers.[3,4]

In the conventional theory of weak interactions, it was generally assumed that neutrinos were massless. This assumption was not always used, but when the more general possibility of nonzero neutrino masses was considered in the context of weak decays, it was implicitly assumed that the weak neutrino eigenstates ν_e, ν_μ and more recently, ν_τ, were also mass eigenstates. Thus, one reads in textbooks of nuclear and particle physics, the journal literature, and past editions of the Review of Particle Properties of the "masses" of the electron and muon neutrinos, "$m(\nu_e)$", and "$m(\nu_\mu)$", the effects which they would have if nonzero, and the upper limits on them. A similar comment applies for "$m(\nu_\tau)$". One also reads of a given weak leptonic or semileptonic decay being considered as a <u>single</u> decay, e.g. $\pi^+ \to \mu^+ \nu_\mu$, $\mu \to \nu_\mu e \bar{\nu}_e$, or nuclear β decay, $(Z,A) \to (Z, + 1, A) + e^- + \bar{\nu}_e$, not only in the case of zero neutrino mass, but also in the case where these masses are allowed to be nonzero. However, in precisely the case where the above assumption that there existed well-defined entities $m(\nu_{\ell_a})$, $\ell_a = e,\mu,\tau$, was made and was nontrivial, viz. the case where these were allowed to be nonzero, this assumption was in general false. The reason is very basic: if neutrinos are massive (and nondegenerate), then the weak neutrino eigenstates ν_{ℓ_a}, defined as the states which couple with unit strength to the corresponding charged leptons ℓ_a (where $\{\ell_a\} = \{\ell_1 \equiv e, \ell_2 \equiv \mu, \ell_3 \equiv \tau, \ldots \ell_n\}$) are not themselves mass eigenstates, but rather linear combinations of the neutrino mass eigenstates, (denoted ν_i, $i = 1, \ldots n$), as specified by the unitary transformation

$$\nu_{\ell_a} = \sum_{i=1}^{n} U_{ai} \nu_i \tag{1}$$

We work within the context of the standard $SU(2)_L \times U(1)$ electroweak gauge theory,[5] appropriately generalized to allow for neutrino masses and mixing,[6] and including n generations of fermions. The lepton and/or Higgs sectors of the theory are thus assumed to be expanded as necessary to allow for Dirac and/or Majorana neutrino masses as phenomenological possibilities.

If the lack of validity of the assumption mentioned above were just a minor matter of notational pedanticism, it would have little interest. But it is not; it has led to very real consequences in the history of experiments which sought to observe or set upper limits on neutrino masses. For example, nuclear β decay experiments attempting to detect or place an upper bound on "$m(\nu_e)$" have always assumed that there was only one such limit to be obtained in a given experiment, have chosen the best measured β decay with the smallest Q value, $^3H \to {}^3He + e^- + \nu_e$, to study, and have always searched only for the early falloff in the Kurie plot near the maximum electron energy which a nonzero ν_e "mass" would allegedly cause. The most recent experiments on this decay reported in 1980 still follow this practice![7] But, in fact as we will show[1-3], (1) contrary to what is stated in textbooks, this early endpoint falloff in the Kurie plot is not the most general signature of nonzero masses for one or more of the mass eigenstates ν_i contained in the weak eigenstate ν_e, and even if "$m(\nu_e)$" were nonzero there would not necessarily be any early endpoint falloff, but rather one or more kinks at lower E_e in the Kurie plot; (2) the Kurie plot from any well measured β decay (obviously also the recoil energy spectrum in electron capture and μ capture) can serve as a test for neutrino masses; the Q value does not have to be small; (3) in view of item (1), it was perfectly possible that decades of nuclear β decay experiments which had purported to set upper limits on "$m(\nu_e)$" had, because of their tacit neglect of lepton mixing, missed a positive signal of nonzero neutrino masses as large as several MeV. Similarly, the muon decay experiment[8] which set an upper limit on "$m(\nu_\mu)$" searched only for an early endpoint falloff in the positron energy spectrum. However, again, this is not the most general manifestation of "$m(\nu_\mu)$" $\neq 0$ and indeed could be absent although the gauge group eigenstate ν_μ contained one or more mass eigenstates with masses of ~ 50 MeV, for example! Finally, the experiment at the Swiss Institute for Nuclear Research (SIN)[9] which yielded the best upper limit on "$m(\nu_\mu)$" assumed that there was only one line in the muon momentum spectrum and searched for the shift in this line which would be caused by a nonzero value of "$m(\nu_\mu)$". In fact, contrary to this assumption, we shall show that in general the ℓ momentum spectrum in $(\pi$ or $K)_{\ell 2}$ decay would consist of $k \leq n$ monochromatic lines, not just one. Further, even if quite heavy neutrinos should be emitted in a $(\pi$ or $K)_{\ell 2}$ decay, the main line in

$|\vec{p}_\ell|$ might not be shifted by a measurable amount, so that experiments such as that at SIN[9] might well have missed a massive neutrino signal. This is especially true since, as a consequence of their tacit assumption, they set momentum or energy cuts which could have excluded such events.

It should be noted that existing experiments do not rule out a substantial mass $\lesssim 250$ MeV, for at least one neutrino, ν_3. Moreover, since the number of lepton generations, n, is not known and could be larger than 3, there remains the very real possibility of several neutrinos with considerable masses which could occur, subdominantly coupled, in the decays of light leptons and hadrons, as well as (dominantly coupled) in the decays of heavier particles such as τ and F. In passing, we note that astrophysical arguments do not preclude neutrino masses in the multi-MeV region which is of greatest interest for our new tests.

II(A) THE GENERAL STRUCTURE OF WEAK DECAYS AND THE MEANING OF NEUTRINO MASS LIMITS

The observation underlying the general theory for decays is that, if one allows the possibility of nonzero neutrino masses at all, then one must realize that a decay of the form $X \to Y + \ell_a^\pm + (\bar{\nu})_\ell$, where X denotes the parent particle, and Y denotes a possibly null set of final state particles, would actually consist of an incoherent sum of the separate modes $X \to Y + \ell_a^\pm + (\bar{\nu})_i$, where i runs over the subset of the n neutrino mass eigenstates allowed by phase space. The branching ratio for the ith mode is modulated by the mixing matrix factor $|U_{ai}|^2$ as well as a kinematic factor depending on $m(\nu_i)$. Examples include leptonic π, K, or F decay (for which Y = {ϕ}), $\pi_{\ell 3}$ decay, $K_{\ell 3}$ decay, nuclear beta decay, and hyperon decay. A similar observation applies to two other types of weak decays: a nuclear transition which proceeds via electron or muon capture, $(Z,A) + \ell_a^- \to (Z-1,A) + \nu_{\ell_a}$ (a = 1 or 2), and the semi-hadronic decay of a (heavy) lepton, $\ell_a \to \nu_a + Y$, where $a \geq 3$. An obvious generalization of this observation applies to the pure leptonic decay of the form $\ell_a \to \nu_\ell \ell_b \bar{\nu}_\ell$, which in general consists of all of the modes $\ell_a \to \nu_i \ell_b \bar{\nu}_j$ kinematically allowed, each with U-dependence $|U_{ai}^* U_{bj}|^2$. These remarks also apply, with straightforward changes to the decays of massive neutrinos.

Let us introduce a convenient qualitative classification of the ν_i. The purpose of this system is to divide the ν_i into a light set i ε $\{i_L\}$ which would appear effectively massless in the weak decays under consideration, and a heavy set i ε $\{i_H\}$ with masses sufficiently large that they could be detected by their kinematic effects in at least one of these decays. For a given decay such as $X \to Y + \ell_a + \bar{\nu}_i$ we define the light dominantly and subdominantly coupled (LDC and LSC) modes as those for which i ε $\{i_L\}$ and $|U_{ai}|^2$ is of order unity or $|U_{ai}|^2 \ll 1$, respectively. The heavy dominantly and subdominantly coupled (HDC and HSC) modes are similarly defined but with i ε $\{i_H\}$, as allowed by phase space. For i ε $\{i_H\}$ the various constraints imply that in a light particle decay involving ℓ_a, a ≠ i, $|U_{ai}|^2 \ll 1$.

Given the observation above, it is then clear how to reinterpret the neutrino mass limits obtained in past experiments. The true meaning of the limits "$m(\nu_{\ell_a})$" < c_a quoted in these experiments is that $m(\nu_i) < c_a$ for a dominantly coupled mode i in the relevant decay. In particular, "$m(\nu_e)$" < 60 eV (90% CL),[10] "$m(\nu_\mu)$" < 0.57 MeV (90% CL),[9] and "$m(\nu_\tau)$" < 245 MeV (2σ level)[11] really apply to ν_1, ν_2, and ν_3, respectively. A similar comment applies to the recent result quoted as 14 eV < "$m(\nu_e)$" < 46 eV.[7] These limits do not significantly constrain the masses of subdominantly coupled ν_i in the respective decays. It is the HSC modes that our new class of tests probe.

II(B) THE $M_{\ell 2}$ TEST

The first new test for neutrino masses and mixing which follows from our basic observation of section II(A) makes use of the leptonic decays $M^+ \to \ell_a^+ \nu_{\ell_a}$, where M = π or K. If neutrinos are massive, the ℓ_a momentum spectrum in this decay would in general consist not just of a single shifted line, but rather of k (monochromatic)[12] lines at

$$|\vec{p}_a| = \frac{m_M}{2} \lambda^{\frac{1}{2}}(1, \delta_a^M, \delta_i^M) \tag{2}$$

where

$$\delta_a^n \equiv \frac{m_{\ell_a}^2}{m_M^2} \tag{3}$$

$$\delta_i^n \equiv \frac{m(\nu_i)^2}{m_M^2} \tag{4}$$

and

$$\lambda(x,y,z) \equiv x^2 + y^2 + z^2 - 2(xy + yz + zx) \tag{5}$$

When this was pointed out in Ref. 2, <u>no experiment had searched for this clear signature of massive neutrinos, and accordingly, such a search for additional spectral lines was proposed</u>. In Fig. 1 we show an illustrative spectrum that might be observed in K_{e2} decay for the case n = 4.

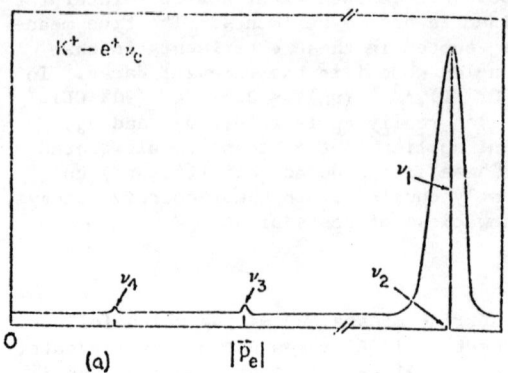

Fig. 1 Illustrative spectrum in K_{e2} decay

Let us describe this test further. First, it has the advantage of being purely kinematic, independent of whether other exotic effects, such as very weakly coupled currents with different Lorentz structure than V-A, or flavor changing Higgs bosons are present. Secondly, if a signal is observed, one can immediately determine independently and unambiguously for each line, $|\vec{p}_a^{(i)}|$ the corresponding $m(\nu_i)$. A very important merit of the test is that this mass determination is independent of mixing angles, except in the minimal sense that $|U_{ai}|$ must be large enough for a signal to be detected in a given experiment. This should be contrasted with another test for HSC modes, viz. the ratio $B(M^+ \to e^+\nu_e)/B(M^+ \to \mu^+\nu_\mu)$, and with the ν oscillation test. The former only measures an integrated effect due to all HSC modes present and is incapable of determining $m(\nu_i)$ or $|U_{ai}|$ for any particular mode. The latter can only yield correlated information on the quantities $m(\nu_i)^2 - m(\nu_j)^2$ and U_{ai}, never the set of $m(\nu_i)$ in isolation. On the other hand, it is sensitive to values of $m(\nu_i)$ below the level detectable by our tests. Next, knowing the mass $m(\nu_i)$ and, at this point, assuming that the relevant couplings are V-A, (which assumption can be tested - see below), one can use simple kinematics to determine

$$R_{ai} \equiv \frac{|U_{ai}|^2}{\sum_{j \in \{j_L\}} |U_{aj}|^2}$$

$$\simeq |U_{ai}|^2 \qquad (6)$$

for each of the HSC lines i . The rate for the mode $M^+ \to \ell_a^+ \nu_i$ relative to that for the mode $M^+ \to \ell_a^+ \nu_a$ with $m(\nu_a) = 0$, is given by

$$\frac{\Gamma(M^+ \to \ell_a^+ \nu_i)}{\Gamma(M^+ \to \ell_a^+ \nu_a)_0} = \frac{|U_{ai}|^2 \rho(\delta_a^m, \delta_i^m)}{\delta_a^m (1-\delta_a^m)^2} \quad (7)$$

where

$$\rho(x,y) = f_m(x,y) \lambda^{\frac{1}{2}}(1,x,y) \quad (8)$$

and

$$f_m(x,y) = x + y - (x-y)^2 \quad (9)$$

In Eq. (8) we have displayed the kinematic rate factor ρ in terms of a part of $f_m \propto |M(M^+ \to \ell_a^+ \nu_i)|^2$, and another part, $\lambda^{\frac{1}{2}}(1, \delta_a^M, \delta_i^M) \propto$ to the two-body phase space factor. A helicity effect acts to enhance f_m as δ_i^M increases; for fixed δ_a^M, f_m increases from a minimum at $\delta_i^M = 0$ to a maximum at $\delta_i^M = \frac{1}{2} + \delta_a^M$.[13] The magnitude of this increase is indicated by the ratio $(f_m)_{max}/f_m(\delta_a^M, 0) = (2\delta_a^M + \frac{1}{2})/[\delta_a^M(1 - \delta_a)]$. For $K^+ \to \mu^+ \nu_i$ this increase is a factor of eight; for $M^+ \to e^+ \nu_i$ it is $\simeq (4\delta_a)^{-1}$ and hence quite dramatic; 2×10^4 for $\pi^+ \to e^+ \nu_i$ and 2×10^5 in the case of $K^+ \to e^+ \nu_i$. In the $M^+ \to \mu^+ \nu_i$ decays this effect offsets, and in the $M^+ \to e^+ \nu_i$ decays it completely overwhelms, the monotonic decrease of the phase space factor $\lambda^{\frac{1}{2}}(1,\delta_a^M,\delta_i^M)$ until $m(\nu_i)$ reaches values rather near to the kinematic limits. Thus, for example, in the decays $\pi^+ \to \mu^+ \nu_i, \pi^+ \to e^+ \nu_i, K^+ \to \mu^+ \nu_i,$ and $K^+ \to e^+ \nu_i$; the relative rate factors

$$\bar{\rho}(\delta_a^m, \delta_i^m) \equiv \frac{\rho(\delta_a^m, \delta_i^m)}{\rho(\delta_a^m, 0)} \quad (10)$$

reach maximal values of 1.00004, 1.11×10^4, 4.13 and 1.38×10^5 at $m(\nu_i) = 3.4, 80.6, 263,$ and 285 MeV, respectively. In Figs. 2 and 3 we show plots of the functions f_m, $\lambda^{\frac{1}{2}}$, and ρ, divided by their values at $m(\nu_i) = 0$, as indicated by the bars (in analogy to Eq. (10)) for $K_{\mu 2}$ and K_{e2} decays. Similar plots for $\pi_{\ell 2}$ decay are given in Ref. 3.

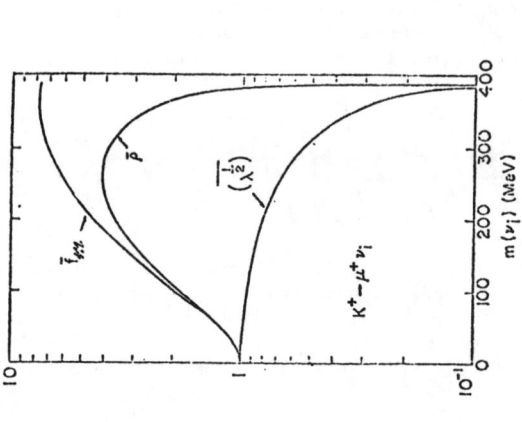

Fig. 2 The functions $\bar{F}_m$, $\overline{\frac{1}{\lambda^2}}$, and $\bar{\rho}$ in $K_{\mu 2}$ decay.

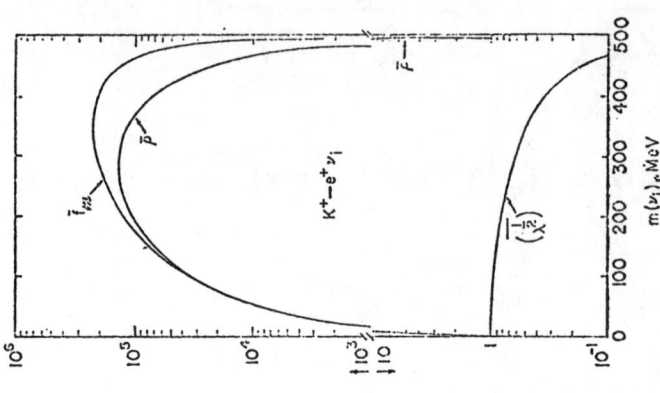

Fig. 3 Same as Fig. 2 but for K_{e2} decay.

Two important points follow from this analysis. First, a search for additional spectral lines in $M^+ \to \ell_a^+ \nu_i$ is not inhibited by kinematic suppression until $m(\nu_i)$ reaches almost its phase space limit, as a consequence of the helicity enhancement effect together with the slow falloff of the two-body phase space factor; indeed, in the $M^+ \to e^+\nu_i$ and $K^+ \to \mu^+\nu_i$ decays there is actually very strong net kinematic enhancement up to quite large values of $m(\nu_i)$. This is to be contrasted with the case for three-body decays involving massive neutrinos; for example, in the decay $\mu \to \nu_i e \bar{\nu}_j$ (HSC i, LDC j) for the value $m(\nu_i)/m_\mu = 0.5$, the kinematic rate factor is only ≈ 0.16 of its value at $m(\nu_i) = 0$. Thus the above test is quite sensitive to neutrino masses throughout most of the kinematically allowed ranges for the various specific decays. Secondly, because of the monochromatic nature of the signal, the test can be applied on an event-by-event basis, in contrast to tests using three-body decays, in which one must search for deviations from a continuous momentum or energy distribution. Hence the test is capable of finding a very small signal or, alternately, of setting a commensurately stringent correlated upper bound of the form "If $m(\nu_i) \in (m_{min}, m_{cut}, \text{ or } m_M - m_a)$, where m_{min} is the minimum value of $m(\nu_i)$ such that the HSC line $|\vec{p}_a^{(i)}|$ is experimentally resolvable from the LDC line(s) $|\vec{p}_a^{(j)}|_{LDC\,j} \approx |\vec{p}_a|_o \equiv |\vec{p}_a|(m(\nu_a) = 0)$ and $m_{cut} \leq m_M - m_a$ is a mass determined by experimental cuts, then

$$R_{ai} < \epsilon / \bar{\rho}(\delta_a^m, \delta_i^m)$$

where ϵ is determined by the statistics and sensitivity of the particular experiment.

The ultimate sensitivity of this test is limited by such factors as the e or μ energy loss in the target and by soft bremsstrahlung, and the resolution in $|\vec{p}_a|$ or E_a of the detector. Two possible background processes include $M^+ \to \ell_a^+ \nu_j \gamma$ and, for M = K, also $K^+ \to \pi^0 \ell_a^+ \nu_j$ (LDC j). However, since their $|\vec{p}_a|$ distributions are continuous and can be calculated, and since such events can be vetoed in an experiment with good photon detection capability, they should not constitute important backgrounds. Concerning the question of whether the test would be more advantageously applied to $M^+ \to e^+\nu_e$ or $M^+ \to \mu^+\nu_\mu$ decay, several comments are relevant. Since the helicity enhancement of the ith HSC mode is much greater, relative to the LDC mode(s), in the former case, the electron modes offer potentially greater sensitivity to small $|U_{ai}|^2$. Also, of course, if $m(\nu_i)$ is sufficiently large, the decay of an $M^+ = \pi^+$ or K^+ into $\mu^+\nu_i$ may be kinematically forbidden while that into $e^+\nu_i$ is still allowed. On the other hand, to the extent that lepton mixing is, like quark mixing, hierarchical, i.e. $|U_{ai}| < |U_{bj}|$ if $|a - i| > |b - j|$, one would expect that a given HSC mode i would be more weakly coupled in $M^+ \to e^+\nu_i$ than in $M^+ \to \mu^+\nu_i$ decay. Perhaps most important, the rate for $M^+ \to e^+\nu_{LDC\,j}$ is much smaller than that for $M^+ \to \mu^+\nu_{LDC\,j}$; hence a search for HSC lines in $M^+ \to e^+\nu_e$ decay is hindered by the fact that μ-e misidentification would produce spurious HSC $M^+ \to e^+\nu_i$ events, since $|\vec{p}_\mu(LDC\,j)| < |\vec{p}_e|_o$.

We have applied our $M_{\ell 2}$ spectral test to existing data on $(\pi$ and $K)_{\ell 2}$ decay. This includes data on $\pi_{\mu 2}$ decay from experiments at SIN [9] and ANL[14] and on $K_{\ell 2}$ decay from experiments by a CERN-Heidelberg collaboration[15] and others. As an example of the sensitivity of the technique, the bound resulting from our analysis of the CERN-Heidelberg K_{e2} data is that for

$$240 \text{ MeV} > |\vec{P}_e| > 220 \text{ MeV} \tag{11}$$

i.e.,
$$82 \text{ MeV} < m(\nu_i) < 163 \text{ MeV} \tag{12}$$

$$R_{1i} \lesssim \frac{0.25}{\bar{\rho}(\delta_e^k, \delta_i^k)} \tag{13}$$

The right-hand side of (13) varies from 1.03×10^{-5} to 3.0×10^{-6} as $m(\nu_i)$ varies over the indicated range. We have also analyzed emulsion data on $\pi \to \mu + \ldots$ decay.[16,17] This data was considered to represent the decay $\pi^+ \to \mu^+ \nu_\mu \gamma$ by the experimenters of Ref. 17, but the operational signature was only a shorter-than-normal muon track from the stopped pion and did not include detection of any photon. Starting from the tacit assumption that $\pi_{\mu 2}$ decay consisted of a decay into two particles, each of definite mass, and that the $\pi_{\mu 2}$ spectrum thus consisted of just a single peak, these authors inferred that the short muon range implied the presence of a photon. However, in the case of massive neutrinos and lepton mixing this assumption and the inference based on it are not in general valid. Hence, it is necessary to reinterpret the results of these experiments. One must say that in general the events observed are due to the set of radiative decays $\pi^\pm \to \mu^\pm (\bar{\nu})_i \gamma$, which must certainly occur, and, in addition, to possible decay modes of the form $\pi^\pm \to \mu^\pm (\bar{\nu})_i$, HSC i. Leaving open the possibility that some of these events do represent HSC $\pi^+ \to \mu^+ \nu_i$ decays, we have taken the conservative course of using this emulsion data to set quite stringent upper bounds on such HSC decays.

Combining the upper limits on $R_{2i} = |U_{2i}|^2$ from all of these sources, we obtain the results shown in Fig. 4.

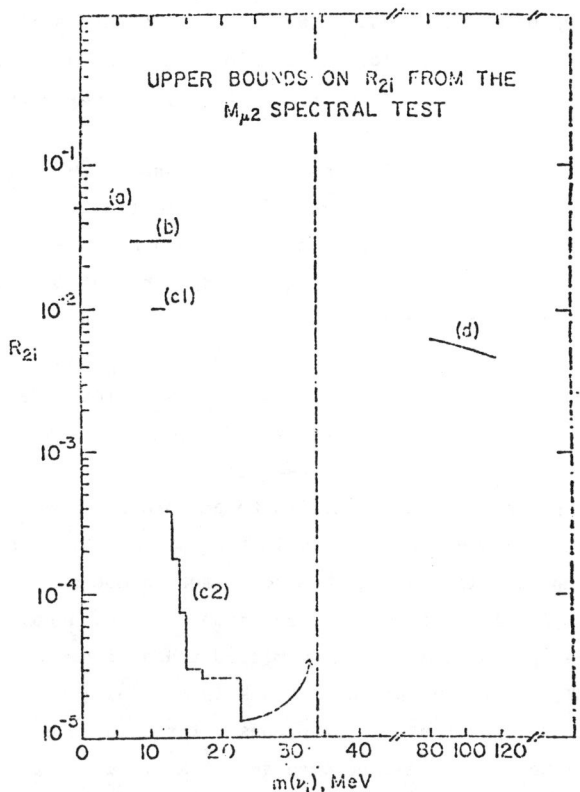

Fig. 4. Upper bounds on R_{2i} from the $M_{\mu 2}$ spectral test

An additional upper bound on R_{2i} can be extracted from an experiment which searched for the decay $K^+ \to \mu^+ \nu_\mu \nu\bar{\nu}$.[18] Re-analyzing this experiment, we find the limit $R_{2i} < 2.0 \times 10^{-6}/[D(T_\mu^{(i)})\rho(\delta_\mu^K, \delta_i^K)]$ for 100 MeV > $T_\mu^{(i)}$ > 60 MeV, i.e., 230 MeV $\lesssim m(\nu_i) \lesssim$ 300 MeV, where $D(T_\mu)$ is the experimental detection efficiency (which varies rapidly over the above interval of T_μ, reaching a maximum of ~ 1.2).

A very powerful feature of the spectral test is that it serves as a probe of the number, n, of lepton generations. If, for example, one observed more than two peaks in addition to the main one,[19] or, alternatively, a peak corresponding to a value $m(\nu_i) > m(\nu_3)_{max}$, then one would have proved that n > 3. Thus, by using old and supposedly well measured decays of π and K mesons (as well as the decays of heavier 0^- mesons), one has the <u>potential</u> of being able to gain information that may not be attainable with an upgraded PEP or PETRA, and maybe not even with LEP itself!

The current status of experiments designed to apply the $M_{\ell 2}$ spectral test proposed last spring in Ref. 2 is as follows. We include here only experiments which have run or at least been approved to run. First, at SIN a group has used the magnetic spectrometer of Ref. 9 to conduct a preliminary search for a second shoulder in the μ^+ momentum spectrum in $\pi_{\mu 2}$ decay, down to $|\vec{p}_\mu| \simeq 27$ MeV $\simeq 0.91 |\vec{p}_\mu|_o$ and has reported the tentative upper limits $B(\pi^+ \to \mu^+ \nu_{HSC\ i}) < 3\%$ for 4 MeV $< m(\nu_i) < 9$ MeV and $B(\pi^+ \to \mu^+ \nu_{HSC\ i}) < 2\%$ for 6 MeV $< m(\nu_i) < 14$ MeV.[20] Second, a University of Virginia group has submitted a proposal which has received favorable consideration at SIN to perform a Ge counter experiment to search for additional peaks in the $|\vec{p}_\mu|$ spectrum in $\pi_{\mu 2}$ decay.[21] This group envisions being able to detect an HSC mode with a branching ratio below 10^{-7} of the dominant mode. Third, a University of Tokyo-KEK-Tsukuba proposal has been submitted and approved, as Experiment E89 at KEK, to search for an HSC peak in the $|\vec{p}_{\ell_a}|$ spectrum in $K^+_{\ell_a 2}$ decay, where $\ell_a = e$ or μ.[22] The proposal for this experiment states that it will be able to detect such an HSC $K^+ \to (e^+$ or $\mu^+)\nu_i$ peak if $|U_{ai}|^2 > 10^{-7}$, a = 1 or 2, respectively, and if $m(\nu_i) \in$ (50 MeV, 350 MeV). In Fig. 5 we show a diagram taken from Ref. 22, of the apparatus for this experiment, which utilizes a wide-acceptance precision magnetic spectrometer with NaI(Tl) crystals to detect and veto background events from $K^+ \to \ell_a^+ \nu \gamma$ and $K^+ \to \pi^o \ell_a^+ \nu_{\ell_a}$ decays.

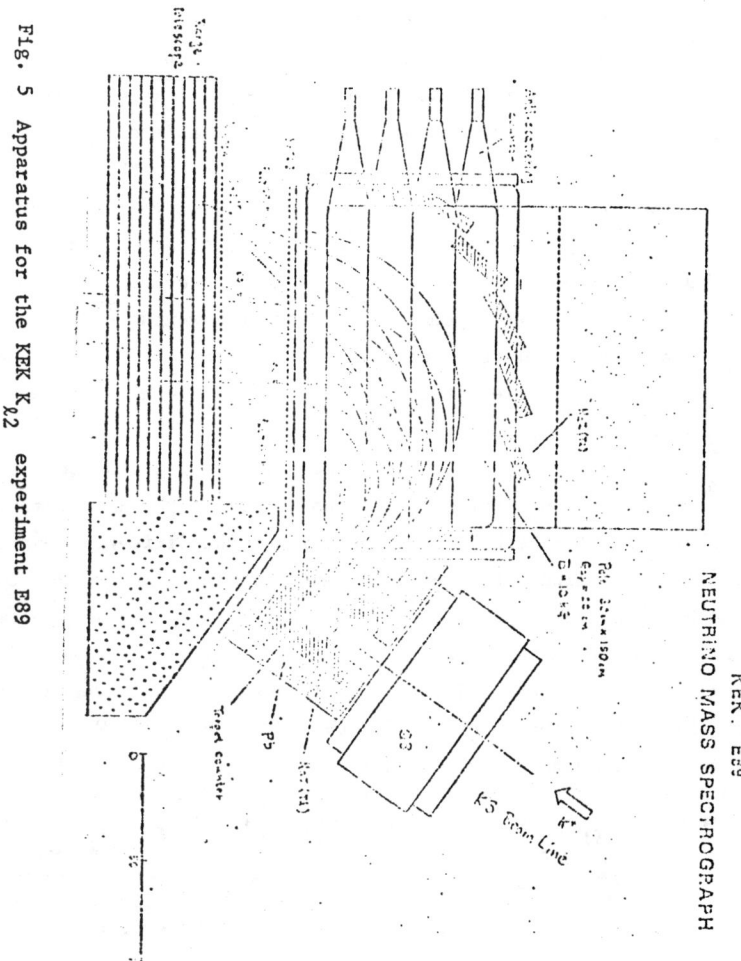

Fig. 5 Apparatus for the KEK $K_{\ell 2}$ experiment E89

All of these experiments are scheduled to have data-taking runs in 1981. Armed with a new understanding of the effects of massive neutrinos (viz., not just a slightly shifted main peak, but instead a multitude of peaks, with the main one not necessarily shifted at all), they will search over regions of $|\vec{p}_e|$ and $|\vec{p}_\mu|$ never before studied as $M_{\ell 2}$ data. One awaits their findings with great interest.

The full $M_{\ell 2}$ test proposed in Ref. 2 consists not only of a careful scan of the $|\vec{p}_a|$ or E_a spectrum to search for the discrete lines from possible HSC modes, but also a measurement of the polarization of the $\ell_a^\pm$. For the general case of $M^\pm \to \ell_a^\pm \stackrel{(-)}{\nu}_i$ decay we denote this (longitudinal) polarization by $P_{\ell_a^\pm}^{(i)}(M_{\ell_a 2})$. The test would thus proceed as follows. If an HSC ν_i peak is observed at momentum $|\vec{p}_a^{(i)}|$ or energy $E_a^{(i)}$ one can immediately determine the mass $m(\nu_i)$. Next, assuming that this peak is sufficiently removed from the LDC peak, one can, by momentum selection, pick out on an event-by-event basis, the $\ell_a^\pm$'s from $M^\pm \to \ell_a^\pm \stackrel{(-)}{\nu}_{\text{HSC } i}$ decay and channel them to the apparatus to measure their polarization. In principle, one can determine the $\ell_a^\pm$ polarization individually for each HSC ν_i line which is sufficiently separated from the main LDC line. Knowing $m(\nu_i)$ and assuming standard V-A couplings, one has a definite prediction for this polarization (see below), which can thus be checked to verify the consistency of the assumption or, if the V-A prediction is violated, to investigate the Lorentz structure of the couplings responsible for the HSC ν_i events. With standard V-A charged current couplings

$$P_{\ell_a^\pm}^{(i)}(M_{\ell 2}) = \frac{\mp(\delta_a^m - \delta_i^n)\lambda^{\frac{1}{2}}(1, \delta_a^m, \delta_i^n)}{f_m(\delta_a^m, \delta_i^n)} \quad (14)$$

Note the antisymmetry of the ℓ_a^+ (or ℓ_a^-) polarization as a function of δ_a^M and δ_i^M, or equivalently m_{ℓ_a} and $m(\nu_i)$; in somewhat figurative language, the lighter lepton has its "preferred" helicity, and hence the heavier one is forced to have a "dispreferred" helicity. The polarization $P_{\ell_a^+}^{(i)}$ is plotted as a function of $m(\nu_i)$ (as the dashed curve) in Figs. 6 and 7, for the decays $\pi^+ \to \mu^+ \nu_i$, $K \to \mu^+ \nu_i$, respectively.

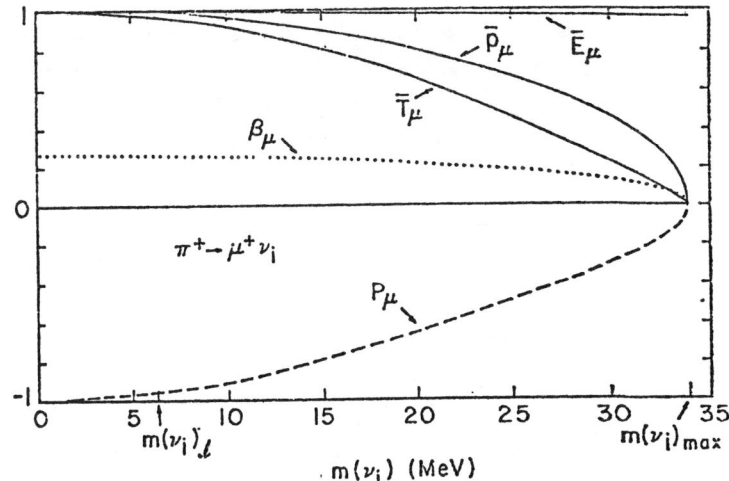

Fig. 6. P_{μ^+} in $\pi^+ \to \mu^+ \nu_i$ decay (dashed curve).

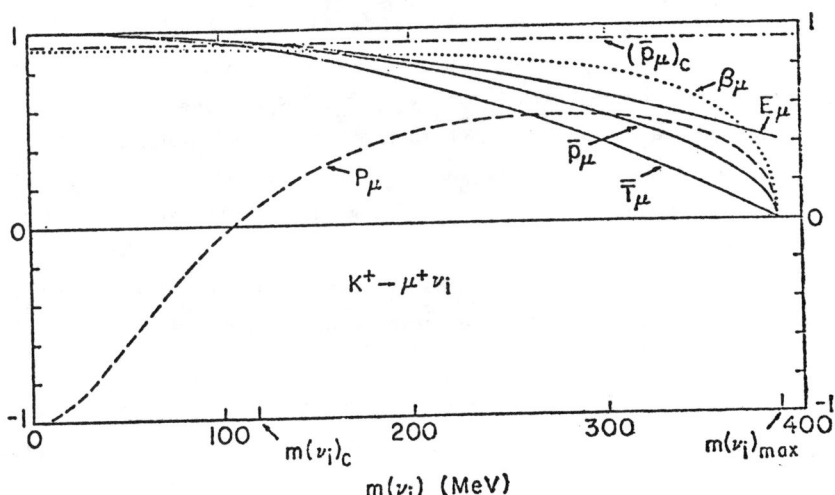

Fig. 7. P_{μ^+} in $K^+ \to \mu^+ \nu_i$ decay (dashed curve).

As $m(\nu_i)$ increases from zero, $P^{(i)}_{\ell_a+}$ rises, at first monotonically. It will cross through zero and assume positive values as $m(\nu_i)$ equals and then exceeds m_{ℓ_a}; this is possible iff $m_{\ell_a} \le m_M/2$, respectively. The inequality $m_{\ell_a} < m_M/2$ is true of M_{e2} and $K_{\mu 2}$, but not $\pi_{\mu 2}$ decay. Assuming that this condition is satisfied, $P^{(i)}_{\ell_a+}$ ascends through positive values and reaches a maximum at

$$m(\nu_i)_{pol. crit.} = \frac{M_M}{\sqrt{3}}(1-\delta_a^m)^{\frac{1}{2}}$$

(15)

where it is equal to

$$\left(P^{(i)}_{\ell_a^\pm}\right)_{crit.} = \frac{\pm(1-4\delta_a^m)^{3/2}}{(1+8\delta_a^m)(1-\delta_a^m)^{1/2}}$$

(16)

In $M^+ \to e^+ \nu_i$ decay, with $M = \pi$ or K, P_{e^+} ascends sharply from -1 and is very close to $+1$ throughout most of the allowed ranges of $m(\nu_i)$. Thus, for almost the entire range of $m(\nu_i)$ in each of the $M_{\ell 2}$ decays, the e or μ polarization would have a grossly different value than the $m(\nu_i) = 0$, V-A one.

A constraint on HSC $M_{\ell 2}$ decays with sufficiently small $m(\nu_i)$ can also be obtained from the ratio of branching ratios

$$R_n = \frac{B(M^+ \to e^+ "\nu_e")}{B(M^+ \to \mu^+ "\nu_\mu")}$$

(17)

where $M = \pi$ or K, and the notation "ν_{ℓ_a}" expresses the fact that the experiments do not observe the gauge group eigenstates ν_{ℓ_a} but instead, indirectly, the mass eigenstates ν_i which, in the case of interest, are different. Thus, the quantity $B(M^+ \to \ell_a^+ "\nu_{\ell_a}")$ is not fully defined without a precise specification of the cuts that were used to select the $\ell_a^+ "\nu_{\ell_a}"$ events. Indeed, if an HSC ν_i had sufficiently large mass that the corresponding $|\vec{p}_a^{(i)}|$ or $E_a^{(i)}$ fell below their respective cuts, this mode would not even have been counted as part of the $\ell_a^+ \nu_{\ell_a}$ event sample! Moreover, for sufficiently large $m(\nu_i)$ either or both of the decays $M^+ \to \mu^+ \nu_i$ and $M^+ \to e^+ \nu_i$ might be kinematically forbidden. The latter case was discussed in Ref. 6. The constraints from the measurement of R_M complement those derived above from a search for HSC peaks in the charged lepton spectra in $M^+ \to \ell_a^+ \nu_{\ell_a}$ decays. The latter apply with greatest sensitivity for $m(\nu_i)$ such that the corresponding $|\vec{p}_a^{(i)}|$ or $E_a^{(i)}$ are outside the main light neutrino peak, whereas, in contrast, the measurement of R_M only places stringent bounds on the couplings $|U_{ai}|$, $a = 1, 2$, of these ν_i with $m(\nu_i)$ such that ν_i events are included in at least one of the "$\ell_a^+ \nu_{\ell_a}$" samples. In Fig. 8 we show the upper limits on $|U_{1i}|^2$ and $|U_{2i}|^2$ from the R_π constraint, as a function of $m(\nu_i)$. We have carried out a similar analysis for the R_K constraint.

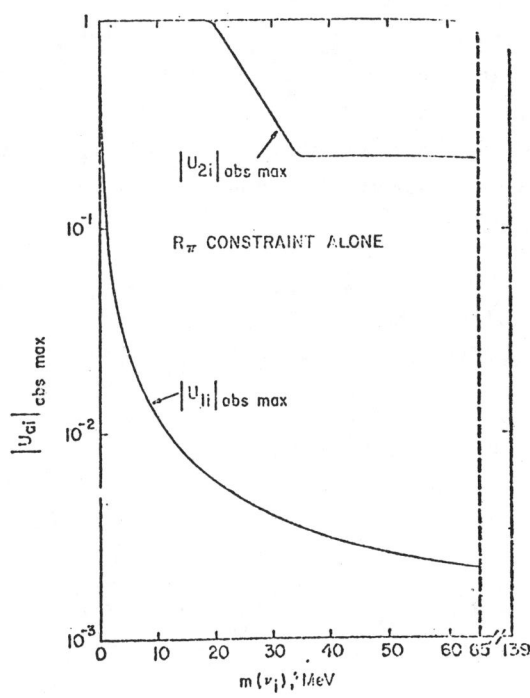

Fig. 8. Upper bounds on $|U_{1i}|^2$ and $|U_{2i}|^2$ from the R_π constraint.

The $M_{\ell 2}$ test can also be applied in F decay. Details are given in Ref. 3.

III. NUCLEAR β DECAYS

Our basic observation can also be applied to nuclear β decays and to nuclear transitions involving e- or μ-capture. Since regular β decay yields a three-body final state, it does not provide the monochromatic signal and corresponding sensitivity to small $|U_{ai}|$ of the first test. However, it has the advantages of high statistics and sensitivity to $m(\nu_i)$ in the range from ≈ 0.1 keV to several MeV; the $M^+ \to \ell^+ \nu_i$ test could not detect an $m(\nu_i) \ll m_a$. As was noted before, past searches for "$m(\nu_e)$" ≠ 0 did not recognize that the early falloff near $(E_e)_{max}$ in a Kurie plot is not the only signature of massive neutrinos. Rather a Kurie plot would in general consist of k' components due to the separate decays $(Z_1, A_1) \to (Z_1 \pm 1, A_1) + e^{\pm} + \overset{(-)}{\nu_i}$. The characteristic signature of the i'th HSC mode is a kink in the Kurie plot at its endpoint energy $(E_e^{(i)})_{max} = \{M_1^2 + m_e^2 - [M_2 + m(\nu_i)]^2\}/(2M_1)$, where $M_{1,2}$ are the initial and final nuclear masses, together with the small incremental addition which it contributes for $E_e < (E_e^{(i)})_{max}$. From the position of the i'th kink one can determine $m(\nu_i)$. Next, one can determine R_{1i} by measuring the increment due to this ith HSC mode and taking into account that

$$\frac{dN}{dE_e} \propto |\vec{p}_e| E_e E_{\nu_i} \lambda^{\frac{1}{2}}(M_2^2, q_{i e}^2, m(\nu_i)^2) / q_{i e}^2 \qquad (18)$$

where

$$q_{ie}^2 = M_1^2 + m_e^2 - 2 M_1 E_e \qquad (19)$$

rather than

$$\frac{dN}{dE_e} \propto |\vec{p}_e| E_e E_{LDC\,\nu} [(E_e)_{MAX}^{(0)} - E_e] / q_{ie}^2 \qquad (20)$$

$$= M_1^{-2} |\vec{p}_e| E_e [(E_e)_{MAX}^{(0)} - E_e]^2 [1 + \mathcal{O}(\frac{E_e}{M_1})] \qquad (21)$$

where $(E_e)_{max}^{(0)} = [M_1^2 + m_e^2 - M_2^2]/(2M_1)$. We have applied this test to the Kurie plots for a large number of β decays, including all superallowed decays and a number of forbidden decays for which the shape correction factors are well known. We obtained the rough upper limit that for $m(\nu_i) \in (0.1 \text{ KeV}, 3 \text{ MeV})$, $|U_{1i}|^2 \lesssim 0.1$.[2]

Using the corrected $\mathcal{F}t$ values for the set of best-measured superallowed β decays, one can also test indirectly for HSC ν_i modes[3]; if $m(\nu_i)$ were such that ν_i could occur as a product in some of these decays but not in others, then $\mathcal{F}t$ would exhibit a decrease as a function of the Q-value of the decay. From the _relative_ agreement of the $\mathcal{F}t$ values for these superallowed decays, we then infer that $|U_{1i}|^2 \lesssim 10^{-3}$ for $m(\nu_i)$ in the relevant range of $\lesssim 1$ MeV to ~ 7 MeV. It should be cautioned, however, that this limit is sensitively dependent upon the validity of the Coulomb corrections that are applied to the raw ft measurements. A possible experimental application of our proposed kink search to various β spectra is being considered by Calaprice, who has also applied our test to existing data on certain well-measured β decays.[2,3]

The signature of neutrino masses and mixing in nuclear transitions involving e- or μ-capture is similar to that in $M_{\ell 2}$ decays, viz. additional peaks in the nuclear recoil spectra. We have examined available data on such nuclear transitions and have obtained bounds on $|U_{ai}|^2$ in agreement with those from the $M_{\ell 2}$ test (although over a much more restricted range of $m(\nu_i)$).

IV. LEPTON DECAYS

(A) The Decay $\ell_a \to \nu_{\ell_a} \ell_b \bar{\nu}_{\ell_b}$

In a general theory of weak processes "the" decay $\ell_a \to \nu_{\ell_a} \ell_b \bar{\nu}_{\ell_b}$ consists of an incoherent sum of the individual decays $\ell_a \to \nu_i \ell_b \bar{\nu}_j$ into all the mass eigenstates ν_i and $\bar{\nu}_j$ allowed by phase space. Given this basic observation, a number of interesting questions arise. For example, what are the energy and angular distributions for an individual (i,j) mode? How, then, is the total observed differential distribution changed as a result of the presence of massive neutrino decay modes and concomitant lepton mixing? Which kinematic quantities and which regions of phase space are most sensitive to these heavy neutrino decays? What is the comparative kinematic suppression of different kinds of massive neutrino modes? In order to answer these questions, we have calculated the differential decay distribution $d^2\Gamma/dE_b d\cos\theta$ for an individual (i,j) mode and for the resulting sum over all such allowed modes. We can write

$$\frac{d^2\Gamma^{(-)}}{dE_b d\cos\theta}(\ell_a \to \nu_i \ell_b \bar{\nu}_j) = \Gamma_o(m_a) |U_{ai}^* U_{bj}|^2 \frac{d^2\bar{\Gamma}^{(-)}}{dE_b d\cos\theta} \tag{22}$$

where

$$\Gamma_o(m_a) \equiv \frac{G_o^2 m_a^5}{192\pi^3} \tag{23}$$

$$\frac{G_o}{\sqrt{2}} = \frac{g^2}{8 m_W^2} \tag{24}$$

and θ denotes the angle between $\hat{p}_b$ and $\langle \hat{s}_a \rangle$. We then find (in units where $m_a \equiv 1$)

$$\frac{d^2\bar{\Gamma}^{(-)}}{dE_b d\cos\theta} = f_1 + |\vec{P}_a| \cos\theta \, f_s \tag{25}$$

where

$$f_1(E_b; m(\nu_i), m(\nu_j), m_b) =$$

$$\lambda^{1/2}(1, Q^2, m_b^2) \left[2Q^2(1 + m_b^2 - Q^2) A_{ij} + ((1-m_b^2)^2 - Q^4) B_{ij} \right] \tag{26}$$

$$f_s\left(E_b; m(\nu_i), m(\nu_j), m_b\right) =$$

$$\lambda(1, Q^2, m_i^2)\left[2Q^2 A_{ij} - (1 - m_i^2 - Q^2) B_{ij}\right]$$

(27)

and

$$A_{ij} = \lambda^{\frac{3}{2}}\left(1, \frac{m(\nu_i)^2}{Q^2}, \frac{m(\nu_j)^2}{Q^2}\right)$$

(28)

$$B_{ij} = 2\lambda^{1/2}\left(1, \frac{m(\nu_i)^2}{Q^2}, \frac{m(\nu_j)^2}{Q^2}\right)\left[1 + \left(\frac{m(\nu_i)^2 + m(\nu_j)^2}{Q^2}\right)\right.$$

$$\left. - 2\left(\frac{m(\nu_i)^2 - m(\nu_j)^2}{Q^2}\right)^2\right]$$

(29)

The distribution for ℓ_a^+ decay is obtained from that for ℓ_a^- decay by the replacement $\cos\theta \to -\cos\theta$. Figs. 9 and 10 show the functions f_1 and f_s for an (i,j) mode with one neutrino (taken here to be ν_i) of non-negligible mass and, to reflect the actual conditions in μ or τ decay, $m_b^2 \ll m_a^2 \equiv 1$. Note that $d^2\Gamma/dE_b d\cos\theta$ is symmetric under interchange of $m(\nu_i)$ and $m(\nu_j)$.

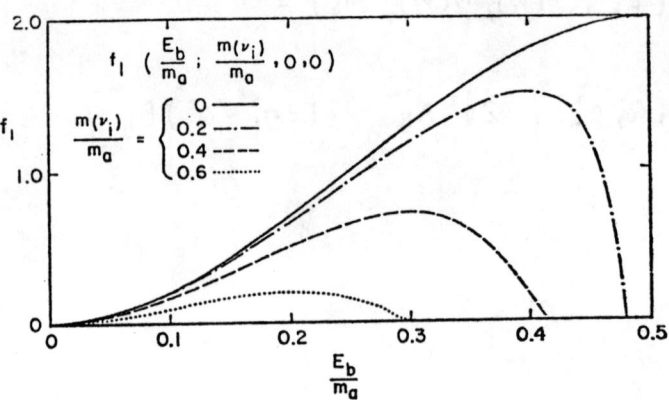

Fig. 9 The function $f_1(E_b; m(\nu_i), 0, 0)$

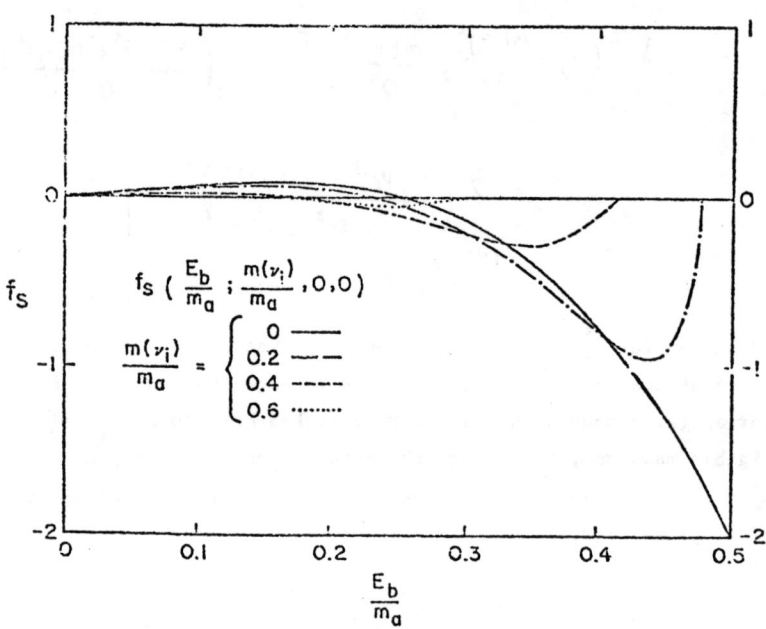

Fig. 10 The function $f_s(E_b; m(\nu_i), 0, 0)$

The total differential decay distribution for "the" decay $\ell_a \to \nu_{\ell_a} \ell_b \bar{\nu}_{\ell_b}$, is then

$$\frac{d^2\Gamma^{(-)}}{dE_b\, d\cos\theta}(E_b, \theta; \ell_a \to \nu_{\ell_a} \ell_b \bar{\nu}_{\ell_b}) = \Gamma_o(m_a) \sum_{i,j} |U^*_{ai} U_{bj}|^2 \times$$

$$\times \frac{d^2\bar{\Gamma}^{(-)}}{dE_b\, d\cos\theta}(E_b, \theta; \ell_a \to \nu_i \ell_b \bar{\nu}_j)$$

(30)

where the sum runs over all (i,j) modes allowed by phase space (henceforth, this will be implicit), and similarly for the charge-conjugate decay. Hence,

$$\Gamma(\ell_a \to \nu_{\ell_a} \ell_b \bar{\nu}_{\ell_b}) = \Gamma_o(m_a) \sum_{i,j} |U^*_{ai} U_{bj}|^2\, \bar{\Gamma}\left(\frac{m(\nu_i)}{m_a}, \frac{m(\nu_j)}{m_a}, \frac{m_b}{m_a}\right)$$

(31)

The reduced decay rate $\bar{\Gamma}(m(\nu_i), m(\nu_j), m_b)$ for the individual mode $\ell_a \to \nu_i \ell_b \nu_j$ is a symmetric function of its three variables. In the physically relevant case at least one of the final state masses is negligible. With no loss of generality we can take this mass to be m_3; then defining the variables

$$a_i \equiv m_i^2, \quad i = 1, 2 \tag{32}$$

$$s \equiv a_1 + a_2 \tag{33}$$

$$D \equiv (a_1 - a_2)^2 \tag{34}$$

and the functions

$$r(U, s, D) \equiv (U^2 - 2Us + D)^{1/2} \tag{35}$$

$$t(s, D) \equiv \tfrac{1}{2}(s^2 - D)^{1/2} = m_1 m_2 \tag{36}$$

$$L_1(U,s,D) \equiv \log\left(\frac{U - s + r(U,s,D)}{2t(s,D)}\right)$$

and

(37)

$$L_2(U,s,D) \equiv \log\left(\frac{sU - D - D^{1/2} r(U,s,D)}{2Ut(s,D)}\right)$$

(38)

We calculate

$$\bar{\Gamma}(m_1, m_2, 0) =$$

$$\frac{1}{2}\left(2 - 3s^3 - s^2 + (5D-14)s - 13D\right) r(1,s,D)$$

$$-\frac{3}{2}\left(s^4 - 2(D+2)s^2 + D^2 - 4D\right) L_1(1,s,D)$$

$$+ 12 s D^{1/2} L_2(1,s,D)$$

(39)

Numerical evaluation of this expression shows that (for obvious reasons) in $\ell_a \to \nu_i \ell_b \bar{\nu}_j$ decay, by contrast with $M_{\ell 2}$ decay, massive neutrino decay modes are strongly kinematically suppressed.

Fig. 11 shows a plot of the reduced double differential decay distribution $d^2\bar{\Gamma}^{(-)}/dE_b d\cos\theta$ for an $\ell_a \to \nu_i \ell_b \bar{\nu}_j$ decay in which $m(\nu_j) \ll 1$, $m_b \ll 1$, and $m(\nu_i) = 0.2$.

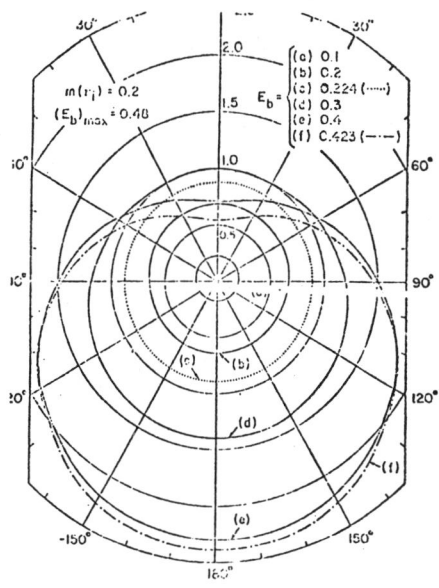

Fig. 11. $d^2\bar{\Gamma}/dE_b d\cos\theta$ $(E_b; \theta; m(\nu_i) = 0.2, m(\nu_j) \ll 1, m_b \ll 1)$

Since (for V-A couplings) the ℓ_a^- decay distribution vanishes at $\theta = 0°$ and $E_b = (E_b)_{max} = (1 + m_b^2)/2$, if $m(\nu_i) = m(\nu_j) = 0$, the $\theta = 0°$, high E_b region of phase space has the potential of being a useful place to search for the effects of HSC neutrino modes. However, a quantitative analysis[4] indicates that these effects will probably be quite small.

A general signature of neutrino masses and mixing in the isotropic decay distribution $d\Gamma/dE_b$ is the presence of kinks, occurring at the endpoints of the various (i,j) modes. Past experiments on μ and τ decay have searched only for the early endpoint falloff in $d\Gamma/dE_b$ which would result if a dominantly coupled neutrino had a sufficiently large mass. We have carried out a search for kinks at intermediate values of E_b corresponding to decay modes involving HSC neutrinos in μ and τ decay data and set resulting upper limits on their relative branching ratios.

The effects of neutrino masses are also apparent in the average quantities such as $\langle\cos\theta\rangle^{(\mp)}(E_b)$, the overall average $\langle\cos\theta\rangle^{(\mp)}$, and

$\langle E_b \rangle$, where the superscript denotes $Q(\ell_a)$. We have calculated these analytically. For example, Fig. 12 shows $\langle \cos\theta \rangle^{(-)}(E_b)$ for the case of a single (i,j) mode with $(m(\nu_i), m(\nu_j), m_b) = (m, 0, 0)$ or $(0, m, 0)$. (Clearly, any of the (mass) = 0 entries may be replaced by (mass) << 1).) The resulting overall average, $\langle \cos\theta \rangle^{(-)}$ is shown as one of the curves in Fig. 13.

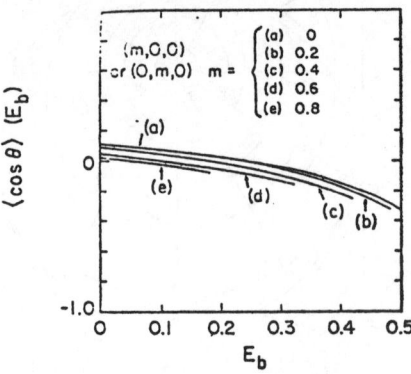

Fig. 12 $\langle \cos\theta \rangle^{(-)}(E_b)$

We have also calculated the general ℓ_b polarization and have found that the presence of neutrino masses and mixing has very little effect on this quantity.

Historically, the analysis of the differential distribution and related spectral parameters in μ and later, τ decay played a very important role in establishing the V-A nature of the respective weak couplings. However, this conventional determination is not applicable in a general theory which admits neutrino masses and mixing. Contrary to the conventional view, the measurement of the spectral parameters (with radiative corrections taken into account to the requisite level of accuracy) does not test the Lorentz structure of the relevant weak couplings in isolation! Rather, these parameters depend not just on the Lorentz structure of these couplings, but also on the masses and mixing angles of the (anti)neutrinos that occur in the various decay modes. The measurement of ρ, η, ξ, and δ would provide a direct test of the Lorentz structure of the weak couplings only if $m(\nu_i) = 0$ for all i, so that $U \equiv 1$. Operationally, of course, one can never verify this condition exactly, so that, in practice, in order to use the

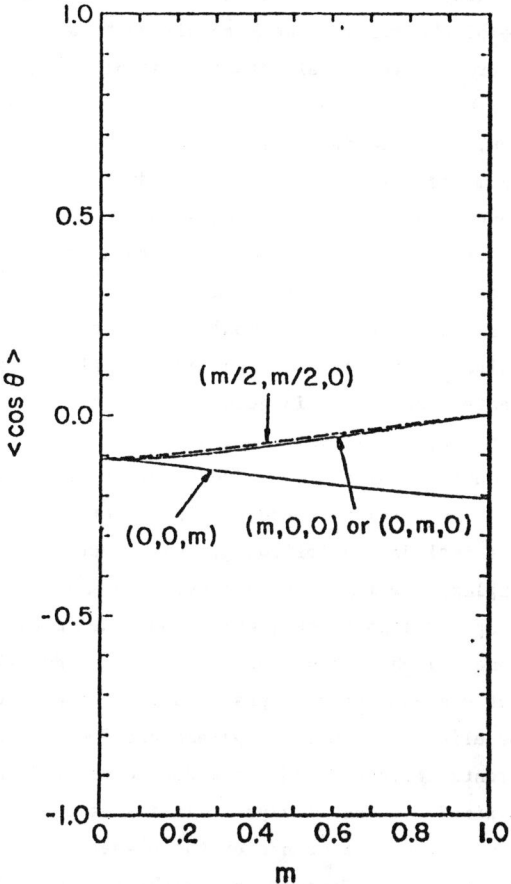

Fig. 13 $<\cos\theta>^{(-)}$

measured values of these spectral parameters to determine the Lorentz
structure of the relevant weak couplings to a given degree of precision,
one must prove that the effects of possible neutrino masses and mixing
are negligible to this order of accuracy. No such proof has previously
been given, and indeed, at present, deviations of the spectral parameters from their V-A values (after radiative corrections have been
divided out) cannot be attributed alone to a difference in the Lorentz
structure from V-A, but must be regarded as possibly being due in part
to massive neutrinos and lepton mixing. Thus, specifically, _even_ if
the relevant weak couplings should be exactly of the V-A type the observed values of the spectral parameters, after radiative corrections
are extracted, would _not_ in general have their conventional V-A values,
$\rho = 3/4$, $\eta = 0$, $\xi = 1$, and $\delta = 3/4$.

We have calculated the effects of neutrino masses and mixing on
each of the spectral parameters. Interesting questions which present
themselves include the following. Given our constraints[2,3] on lepton
mixing angles, how important are neutrino mass and mixing effects for
the four spectral parameters which have been measured in μ decay and
for the one (ρ) which has been measured in τ decay? Are there any
tests that one can apply to past and future data to distinguish between the effects on spectral parameters due to possible non V-A structure, Lorentz structure and those due to possible neutrino masses and
lepton mixing? Can one analyze existing data on μ and τ decay to derive useful correlated bounds on the masses and coupling coefficients
of heavy neutrinos? Finally, how might the forthcoming τ and very
high precision μ decay experiments best search for the manifestations
of heavy neutrinos and, given their projected measurement accuracies,
how large an effect might they be able to detect?

Let us consider the case of ℓ_a decay involving one (anti)neutrino,
which we shall denote as $\overset{(-)}{\nu_k}$, of non-negligible mass. It is straightforward to generalize this to the case of several such (anti)neutrinos.
In Figs. 14 and 15 we show the effective values (with radiative corrections divided out) of the spectral parameters ρ and η which would be
observed in an experiment which made a fit to the complete ℓ_b spectrum.
The curves for these and the other spectral parameters are computed

assuming V-A weak couplings. As in the actual experiments, our values of ρ and η are determined from a simultaneous two-parameter χ^2 fit to $d\Gamma/dE_b$. The curves are plotted for two values of $|U_{2k}|^2$, viz. 10^{-2} and 10^{-3}. As should be clear from the preceeding discussion concerning other constraints on $|U_{rk}|^2$ as a function of $m(\nu_k)$, we do not mean to imply that values this large are allowed for all $m(\nu_k)$ in μ and τ decay. One is referred to Refs. 2 and 3 for an analysis of precisely how large $|U_{rk}|^2$ is allowed to be, as a function of $m(\nu_k)$, by other relevant constraints. Figs. 16 and 17 give similar curves for the spectral parameters ξ and δ. Since η, ξ, and δ have only been measured in μ decay, the units of the $m(\nu_k)$ axes in Figs. 15-17 are chosen accordingly. The short-dashed horizontal lines drawn, by convention, around the $m(\nu_i) = 0$, $U \equiv 1$, V-A values of these parameters represent the projected statistical measurement accuracy of forthcoming μ decay experiments.[24]

From our analysis we propose that these future μ and τ decay experiments try to search for the effects of possible HSC as well as HDC neutrino modes and concomitant lepton mixing on the spectral parameters, especially ρ. We have devised a test which can be used to distinguish between these effects and those due to possible non-V-A Lorentz structure in the weak couplings. This test would involve comparison of the values of a given spectral parameter obtained by fits using different intervals of E_b. If (after appropriate radiative corrections are extracted), a spectral parameter is established to be different from its V-A value and the test proposed above is applied, then (1) if the test yields the same non V-A value of the spectral parameter for all of the ranges of E_b that were used, then one can conclude that the effect is due to non V-A couplings and not to massive neutrinos, to the requisite level of accuracy; (2) if different ranges do yield different results, then one can conclude that at least part of the effect is due to massive neutrinos; and (3) if the deviation disappears as one uses progressively higher ranges of E_b (given an appropriate definition of the area-normalized fitting function), then one can conclude that the effect is due to massive neutrinos rather than non V-A Lorentz structure.

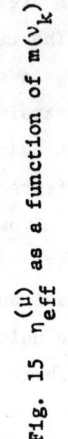
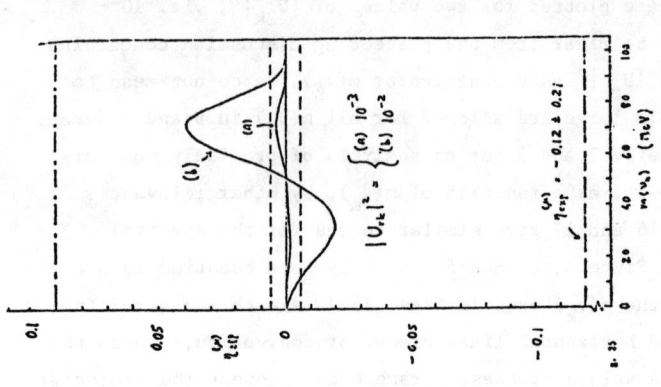

Fig. 15 $\eta_{eff}^{(\mu)}$ as a function of $m(\nu_k)$

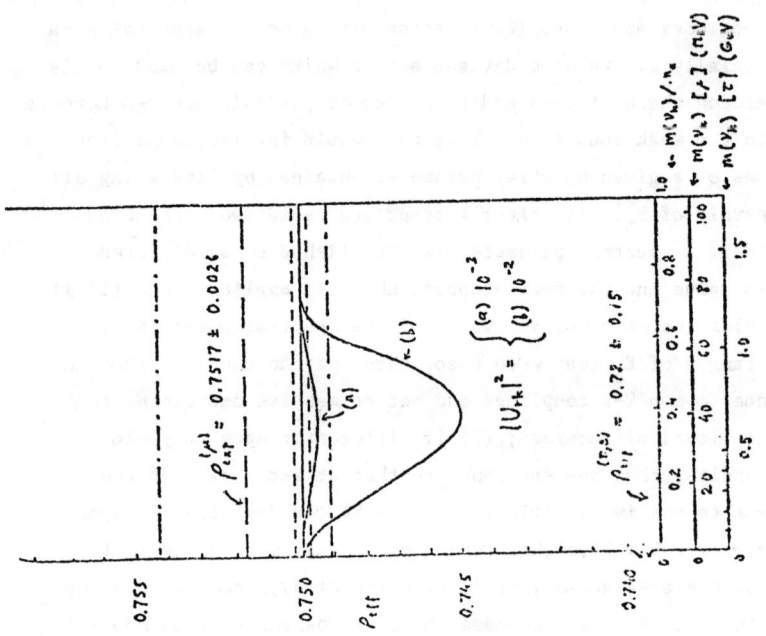

Fig. 14 ρ_{eff} as a function of $m(\nu_k)$

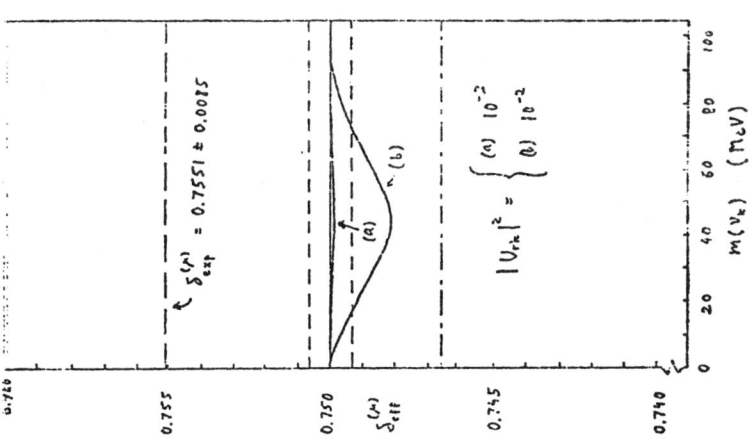

Fig. 17 $\delta^{(\mu)}_{eff}$ as a function of $m(\nu_k)$

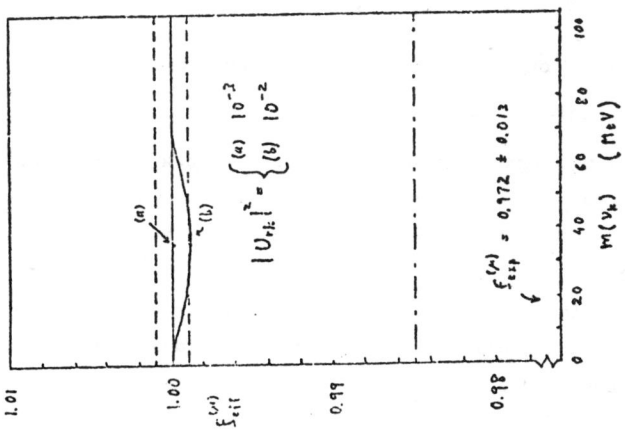

Fig. 16 $\xi^{(\mu)}_{eff}$ as a function of $m(\nu_k)$

Applying our analysis of neutrino mass effects on ρ to existing μ decay data, we can obtain an upper bound on the effective coupling coefficient, $|U_{rk}|^2$, where r = 1 or 2. This bound is shown in Fig. 18 and is useful for r = 2 but weaker than our previous bounds for r = 1.

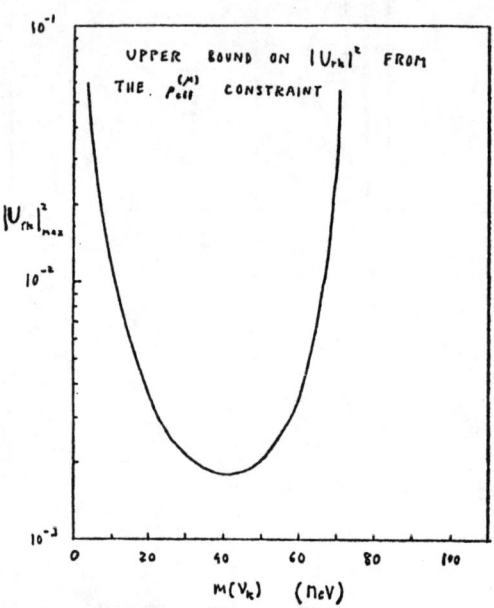

Fig. 18 Upper bound on $|U_{rk}|^2$ from the $\rho^{(\mu)}$ constraint

(B) Semihadronic Decays of Heavy Leptons

The analysis which we carried out for the spectra and branching ratios of $M_{\ell 2}$ decays can also be applied in an obvious way to two-body semihadronic decays of heavy leptons such as τ. For example, the rest-frame spectrum of $|\vec{p}_h|$ in the decay $\tau^- \to \nu_\tau h^-$, where h = π, ρ, etc., would consist of a discrete number of lines. In contrast to leptonic decays of pseudoscalar mesons, however, there is no enhancement of a massive neutrino decay mode, relative to a massless one, in these semihadronic decays. Further, there would be changes in the ratio of branching ratios for various decays, due to both the possible massive HDC ν_3 and to any HSC neutrino modes.

(C) Neutrino Decays

If sufficiently heavy neutrinos are emitted in various leptonic or semileptonic decays, these neutrinos may themselves decay with short enough lifetimes to be experimentally observable. If $m(\nu_i) \lesssim 2m_e$ then the dominant decay is in general the one-loop process $\nu_i \to \nu_j \gamma$. In the standard electroweak theory the lifetime for this decay is many orders too long for it to be detectable. For $m(\nu_i) \gtrsim 2m_e$ the tree-level leptonic decay $\nu_i \to e^- \bar{\nu}_j e^+$ will dominate, unless the relevant mixing angles are exceedingly small. For $m(\nu_i) > m_e + m_{\pi^+}$, the semihadronic $\nu_i \to e^- \pi^+$ will occur and, given the value of $m(\nu_3)_{max}$, will dominate in the case $i = 3$. The nontrivial part of studying the rates and branching ratios for these various modes is to insert properly for a given $m(\nu_i)$ the resulting upper bounds on the relevant coupling coefficient which arise from our earlier analysis.[2,3]

One of the most promising places to look for neutrino decays is in data on τ decay (or F decay, when it becomes available with sufficient statistics). The reason for this is that in these decays, in contrast to decays of π, K, μ, and other light particles, ν_3 can occur as a dominantly coupled neutrino. Given the current upper limit of ~ 245 MeV on $m(\nu_3)$, there is then a reasonable chance that τ_{ν_3} would be short enough for the ν_3 decay to be detected in the sensitive volume of an experimental apparatus. We propose in particular that one should look for $\tau^+ \tau^-$ events, tagged, say, by the $e^{\pm} \mu^{\mp}$ signature, with wide-angle $e^{\pm} e^{\mp}$ or $e^{\pm} \pi^{\mp}$ pairs. The vertices of these pairs would in general be separated by a macroscopic distance from the $e^+ e^-$ interaction point. The invariant mass of the $e^{\pm} \pi^{\mp}$ pairs would have a fixed value, to within experimental uncertainties, while that of any $e^+ e^-$ pairs that are observed would vary but exhibit a fixed maximum. These pairs would arise from the leptonic or semihadronic decay of the DC $\overset{(-)}{\nu_3}$ emitted in leptonic $\tau^{\pm}$ decay, and $m(e^{\pm} \pi^{\mp}) = m(e^+ e^-)_{max} = m(\nu_3)$. Similar comments apply for data on F decay.

(D) Rare μ Decays

As has been discussed in detail elsewhere,[2,5] neutrino masses and mixing give rise to decays which violate e- and μ-type lepton family number such as $\mu \to e\gamma$, $\mu \to e e \bar{e}$, and the conversion process $\mu N \to e N'$. However, the upper limits on these processes do not provide very useful constraints on these masses or mixing angles, at least for moderate neutrino masses. For example, in the necessary generalization of the standard electroweak theory to include neutrino masses,

$$B(\mu \to e\gamma) = \frac{3\alpha}{32\pi} \left| \sum_{i=1}^{n} U_{2i}^* U_{1i} \frac{m(\nu_i)^2}{m_W^2} \right|^2$$

(40)

For $n = 3$ and various choices of the $m(\nu_i)$, taking into account the resultant upper bounds on the $|U_{1i}^* U_{2i}|$ factor, one obtains numbers around 10^{-20} for the maximum of this branching ratio, which are far smaller than the present upper limit of 1.9×10^{-10} (90% CL)[26] and indeed also the eventual limit of $\sim 10^{-12}$ which the LAMPF $\mu \to e\gamma$ experiment hopes perhaps to achieve. (Of course, for $n > 3$ and $m(\nu_{i \geq 4}) \gg m(\nu_3)_{max}$ one could obtain a commensurately larger $\mu \to e\gamma$ branching ratio.) Similar comments apply to the processes $\mu \to e e \bar{e}$ and $\mu N \to e N'$.

V. OTHER IMPLICATIONS OF NEUTRINO MASSES AND MIXING

(A) Other Decays

The general observations of Refs. 2 and 3 and resulting tests for HSC modes can of course also be applied to other decays such as $K_{\ell 3}$ decay, semileptonic baryon decays, and neutrinoless double β decay. For example, we have examined data on $K_{\ell 3}$ decay to search for effects of possible HSC decay modes. However, first, this decay is not nearly as sensitive as leptonic pseudoscalar meson decays, and second, such a search is hindered by important backgrounds, so that the resulting limits that one can derive on masses and mixing are not as restrictive as those from our other analyses.

(B) Implications for Fundamental Weak Interaction Constants

As was stressed in Ref. 3, neutrino masses and mixing have profound implications for the meaning of weak interaction constants such as the μ decay constant G_μ, the Fermi β decay G_V', the pseudoscalar meson decay constants f_π and f_K, quark mixing matrix coefficients such as V_{ud} and V_{us}, $\sin^2\theta_W$, and the predicted, versus true, values of m_W and m_Z. In the conventional theory, μ decay, being purely leptonic, was thought to provide a direct measure of the weak decay constant G_o, defined at tree-level by Eq. 24. However, in general, G_μ is <u>not equal</u> to G_o; instead

$$G_\mu = \chi^{\frac{1}{2}} G_o \tag{41}$$

where

$$\chi = \frac{\sum_{i,j=1}^{n} |U_{2i}^* U_{1j}|^2 \, \overline{\Gamma}\left(\frac{m(\nu_i)}{m_\mu}, \frac{m(\nu_j)}{m_\mu}, \frac{m_e}{m_\mu}\right)_{[c]} \theta(m_\mu - m_e - m(\nu_i) - m(\nu_j))}{\overline{\Gamma}\left(0, 0, \frac{m_e}{m_\mu}\right)_{[c]}} \tag{42}$$

where the subscript [c] indicates that, in accord with convention, the pure electromagnetic corrections are included where necessary. Note that $\kappa \leq 1$. It follows that the true value of m_W would in general lie lower than the conventional prediction:

$$m_W = \chi^{1/4} (m_W)_{conv.\ pred.} \tag{43}$$

where

$$(m_W)_{conv.\ pred.} = \left(\frac{\pi\alpha}{\sqrt{2}\, G_\mu}\right)^{\frac{1}{2}} \frac{1}{\sin\theta_W} \left(1 + \frac{\Delta r}{2}\right) \tag{44}$$

and the one-loop electroweak correction $\Delta r/2$ has been computed by Sirlin[27] (and others, in different renormalization frameworks[28]). Note that one can measure the true value of $\sin^2\theta_W$, even in the presence of neutrino masses and mixing, by using only polarized electron scattering data. Thus, one has the very exciting prospect of gaining crucial information on the neutrino **mass** spectrum from a precise measurement of the W boson mass! In the renormalization framework of Ref. 27 the tree-level relation for m_Z, viz., $m_Z = m_W \sec\theta_W$, remains exact in higher orders. The resulting conventional prediction for m_Z would thus also in general be higher than the true value:

$$m_Z = \mathcal{H}^{1/4} (m_Z)_{conv.\ pred.} \tag{45}$$

Proceeding on, we note that $G_V'^{(apparent)} \neq G_V'$ and, consequently, the value of $|V_{ud}|^{(apparent)}$ obtained by the usual method[29] would differ from the true value. Similarly, from the standard Cabibbo fit one would derive a $|V_{us}|^{(apparent)} \neq |V_{us}|$. However, we stress that, unless one obtained an obviously pathological result such as $|V_{u,d\ or\ s}| > 1$, one would have no way of knowing from these results alone, that the derived values for the quark mixing matrix coefficients were not the true ones. The so-called "constraint" of quark-lepton universality is not, in general, a constraint at all, since one does not know what the true value of

$$(|V_{ud}|^2 + |V_{us}|^2)^{1/2} = \left(1 - \sum_{j=3}^{n} |V_{1j}|^2\right)^{1/2}$$

should be. Following a different route, one could check the value of $|V_{ud}|^{(apparent)}$ determined by the usual method with the value determined from the decay $\pi^+ \to \pi^0 e^+ \nu_e$, but the accuracy with which the rate for the latter decay is known is not sufficiently good at the present time for this to be a stringent consistency check.

Both because $G \neq G_0$ and because $\sum_{i=1}^{n} |U_{2i}|^2 \rho(\delta_\mu^n, \delta_i^n) \neq \rho(\delta_\mu^n, 0)$, the values of $f_M^{(apparent)}$ derived from $M_{\ell 2}$ decay, where $M = \pi$ or K, would similarly differ from their true values. A corresponding statement holds for g_A, as determined with greatest precision from neutron

decay. We have examined quantitatively how large the effects of these
changes would be on the apparent versus true forms of the Goldberger-
Trieman and Adler-Weisberger relations and have found that the effects
of neutrino masses and mixing are small compared to the uncertainties
in the theoretical derivations and/or numerical evaluations of these
relations.

Furthermore, the apparent value of $\sin^2\theta_W$ extracted from a fit
to neutral and charged current neutrino reaction data would differ from
the true value of this parameter, for several reasons. The effect of
neutrino oscillations (in the light neutrino case) is well recognized.
However, for heavy neutrinos there are also other effects due to the
altered polarization of the neutrinos and to the incoherent propagation,
at significantly different velocities, of different neutrino mass
eigenstates.

(C) Effects on the Propagation and Reactions of Neutrinos

Neutrino masses and mixing would, of course, change the propagation and reactions of neutrinos. Indeed, the positive effect reported
by one recent experiment[29] has been a subject of considerable interest
recently. If confirmed, however, this effect would involve neutrinos
with masses much lighter than those which could be detected in particle
(although not nuclear) decays. In the usual formalism[30] for describing
the propagation and subsequent interaction of massive neutrinos it is
assumed that the $m(\nu_i)$ are sufficiently small, relative to the average
neutrino energy, E, that, in accordance with the general rules of
quantum mechanics, the mass eigenstates propagate effectively as a
coherent state,

$$|\nu_\mu(t)\rangle = \sum_{j=1}^{n} U_{2j} |\nu_j\rangle e^{-iE_j t} \simeq e^{-iE} \sum_{j=1}^{n} U_{2j} |\nu_j\rangle e^{-im(\nu_j)^2 t/(2|\vec{p}|)}$$

There follows the well known phenomenon of neutrino "oscillations".
The assumption underlying this quasi-coherent formalism is not in general valid in the general theory where some $m(\nu_i)$ could be substantial,
so that different ν_i's would travel at significantly different velocities and momenta, and indeed, a small number might decay before even
reaching the detector. In this case, for example, one might actually
use precise timing techniques to detect the delayed arrival of

individual massive neutrinos (relative to the light neutrinos or a reference RF signal), which would constitute a clear manifestation of the incoherent nature of the propagation of these heavy ν_i's. Both the propagation differences and the fact that $U \neq 1$ would alter the charged current cross sections.[31] For heavy neutrinos this alteration would not be characterized by spatial oscillations.

Moreover, for heavy neutrinos there are further interesting effects (in addition to the possible decays already noted). <u>Both</u> the charged <u>and</u> neutral current cross sections would be altered by the fact that the polarization of heavy neutrinos would differ from that of light or massless neutrinos. This could strongly increase the yield of anomalously charged leptons from Majorana neutrinos. Another effect of $m(\nu_i) \neq 0$ would be to alter the cross section expressions themselves.

Finally, a massless chiral neutrino cannot have a magnetic moment. The same is true for a massless or massive Majorana neutrino. However, a massive chiral Dirac neutrino does have such a moment, which is given in the standard electroweak theory by[32]

$$\mu_{\nu_i} = \frac{3 e G_0 \, m(\nu_i)}{8 \pi^2 \sqrt{2}} \qquad (46)$$

This would cause the neutrino spin to rotate as it passed through a magnetic field. Such a phenomenon might affect supernova cooling.

ACKNOWLEDGMENT

This work supported in part by NSF Grant # PHY 790637A01.

REFERENCES

[1] R.E. Shrock, p. S63 in Particle Data Group, Rev. of Particle Properties, Rev. Mod. Phys. 52, S1 (1980).

[2] R.E. Shrock, Phys. Lett. 96B, 159 (1980).

[3] R.E. Shrock, SUNY-Stony Brook preprint ITP-SB-80-56.

[4] R.E. Shrock, SUNY-Stony Brook preprint.

[5] S. Weinberg, Phys. Rev. Lett. 19, 1264 (1967); S. Salam, in Elementary Particle Theory: Relativistic Groups and Analyticity, edited by N. Svartholm (Almqvist and Wiksell, Stockholm, 1968), p. 367; S. Glashow, Nucl. Phys. B22, 579 (1961).

[6] B.W. Lee and R.E. Shrock, Phys. Rev. D16, 1444 (1977).

[7] V. Lubimov et al., Phys. Lett 14B, 266 (1980).

[8] M. Bardon et al., Phys. Lett. 14, 449 (1965).

[9] M. Daum et al., Phys. Lett. 74B, 126 (1978); Phys. Rev. D20, 2692, (1979).

[10] K. Bergkvist, Nucl. Phys. B39, 317 (1972); J. Simpson, in Proc. of the Neutrino 79 Conf., eds. A. Haatuft and C. Jarlskog (Bergen, 1979) Vol. 2, p. 208.

[11] C. Blocker, Ph.D. thesis, LBL Report 10801 (1980); see also W. Bacino et al., Phys. Rev. Lett. 42, 749 (1979).

[12] More precisely, the lines are monochromatic in the limit of no e or μ energy loss in the target or by soft bremsstrahlung, and a perfect spectrometer resolution.

[13] This maximum in f_m is in the physical region iff $m_a < m_M/4$, which condition is satisfied in $K_{\mu 2}$, K_{e2} and π_{e2} decay but not in $\pi_{\mu 2}$ decay.

[14] L. Hyman et al., Phys. Lett. 25B, 376 (1967).

[15] K. Heard et al., Phys. Lett. 55B, 327 (1975); J. Heinze et al., Phys. Lett. 60B, 302 (1976); Nucl. Phys. B149, 365 (1979).

[16] W.F. Fry, Phys. Rev. 91, 130 (1953).

[17] C. Castagnoli and M. Muchnik, Phys. Rev. 112, 1779 (1958).

[18] C.Y. Pang et al., Phys. Rev. D8, 1989 (1973).

[19] Given upper limits on $m(\nu_2)$ it would be difficult to resolve the ν_2 peak from the ν_1 peak in $M_{\ell 2}$ decays.

[20] R. Abele et al., SIN Newsletter, No. 13, Nov. 1980, p. 11.

[21] R. Minehart and K. Ziock, spokesmen, SIN proposal.

[22] T. Yamazaki, spokesman, KEK proposal 89.

[23] F. Calaprice, private communication.

[24] H.L. Anderson and J. Bowman, spokesmen, LAMPF proposal. Similar

experiments are planned at TRIUMF.

[25] B. W. Lee and R. E. Shrock, op. cit., Ref. 6. See also W. Marciano and A. I. Sanda, Phys. Lett. 67B, 303 (1977). A number of authors studied a different model which involved right-handed currents; see, e.g., T.P. Cheng and L.F. Li, Phys. Rev. Lett. 38, 381 (1977); Phys. Rev. D16, 1425 (1977); S. B. Treiman et al., Phys. Rev. D16, 152 (1977); J. Bjorken, K. Lane, and S. Weinberg, Phys. Rev. D16, 1474 (1977).

[26] J. Bowman et al., Phys. Rev. Lett. 42, 556 (1979); J. Bowman et al., LAMPF proposal.

[27] A. Sirlin, Phys. Rev. D22, 971 (1980).

[28] M. Veltman, Phys. Lett. 91B, M. Green and M. Veltman, Nucl. Phys. B169, 137 (1980); F. Antonelli et al., Phys. Lett. 91B, 90 (1980).

[29] F. Reines et al., Phys. Rev. Lett. 45, 1307 (1980). See, also R.P. Feynman and P. Vogel, Caltech note; F. Boehm et al., to be published; and D. Silverman and A. Soni, UCLA preprint.

[30] Z. Maki, M. Nakagawa, and S. Sakata, Prog. Theor. Phys. 28, 870 (1962); V. Gribov and B. Pontecorvo, Phys. Lett. 28B, 495 (1969).

[31] R. Shrock, Phys. Rev. Lett. 40, 1688 (1978).

[32] K. Fujikawa and R. Shrock, Phys. Rev. Lett. 45, 963 (1980).

DISCUSSION

WEINBERG: I understand from Calaprice that he has looked for your kinks in old measurements of β decay spectra and is able, in fact, to rule out mixing angles as large as the Cabibbo angle over a certain range of neutrino masses. You probably know better than I do, but this range is, I think, something like a few hundred KeV to a few MeV.

SHROCK: Yes; in a careful study of Kurie plots and ft values he has refined the rough bounds which I originally gave. I should also mention that Calaprice has just informed me that he has proposed an experiment at Indiana to apply our peak search test to $\pi_{\mu 2}$ decay. It will be very interesting to see what resutls he obtains.

RARE KAON PROCESSES

P. Rosen
Purdue University
West Lafayette, IN

I would like to give a brief review of rare K decays which violate both quark and lepton flavor simultaneously. The most frequently discussed modes are $K_L \to \mu^{\pm} e^{\mp}$ with an experimental limit on the branching ratio of $\leq 2 \times 10^{-9}$, and $K^o \to \pi^o + \mu^{\pm} + e^{\pm}$ with a limit of $\leq 5 \times 10^{-9}$.

In the present framework of theoretical ideas, detection of these decays will be taken as a very strong signal that there exist new heavy particles of various kinds. The interesting question will then be to try to decide on the basis of the observations whether the new particles are new heavy leptons, new types of gauge bosons or possibly Higgs mesons. I shall consider these three cases one by one.

In the case of rare K decays and heavy leptons, there was much discussion just a few years ago of theories based upon the idea that these decays would be suppressed by a GIM-type mechanism operating both in the quark sector and also in the lepton sector[1]. The heart of the mechanism is shown in Figure 1; the suppression factor associated with it involves various mixing angles and the mass difference of the two neutral leptons N_1 and N_2 divided by the W mass:

$$\delta = |U_{13} U_{24}| (m_1^2 - m_2^2)/m_W^2.$$

If we assume that the mixing angle factor is somewhere in the ballpark of 0.1 to 0.01, we can draw the following conclusions.

If the masses of N_1 and N_2 are in the eV range, then δ will be so small that we would never be able to observe it. If they are in the MeV range, then the value for the parameter δ will be well below the expected range, which was thought in the past to be 10^{-5} to 10^{-6}. In order to achieve a range of that magnitude, we would need a mass difference between the leptons in the ballpark of several GeV. The typical branching ratios for these processes (as given in the analysis of Lee and Shrock[2], for example) are then 10^{-13} to 10^{-15} for $K_L \to \mu e$ and $10^{-11} - 10^{-13}$ for $K_L \to \pi \mu e$. These are our expectations if there are heavy leptons in the 1-10 GeV mass range.

Another possibility that has been considered, and here I am going to rely very much on the work of Peter Herczeg[3] at Los Alamos, is that these rare K decays are engendered by horizontal gauge bosons. The typical coupling involves a horizontal gauge boson X_λ coupled to the lepton combination $(\bar{e}\gamma_\lambda\gamma_5\mu)$ on the one hand, and to a corresponding hadron flavor-violating axial current J_λ^A on the other hand. The effective interaction for the processes of interest is:

$$\mathcal{L}_{eff} = (g'g''/m_X^2)(\bar{e}\gamma_\lambda\gamma_5\mu)J_\lambda^A + \text{h.c.}$$

where g' and g'' are the coupling constants and m_X is the mass of the horizontal boson.

Herczeg estimates the branching ratio for $K_L \to \mu e$ to be

$$\text{B.R.}(K_L \to \mu e) = 3.7 \times 10^3 \left(\frac{g'g''}{g^2}\right)^2 \left(\frac{m_W}{m_X}\right)^4,$$

where the numerical factor 3.7×10^3 comes from standard estimates of the matrix element of the hadronic current. He then goes on to observe that the horizontal boson X also contributes to the mass difference $K_L - K_S$, and he uses this property in order to extract some limits on the branching ratio. Calculating the mass difference in the standard way, and making a vacuum insertion between the product of two currents, he finds that $\Delta m_X = 2.5 \times 10^3 (g''/g)^2 \cdot (m_W/m_X)^2$. This mass difference certainly has to be smaller than the experimental mass difference of 3.5×10^{-6} eV. (We assume that most of the mass difference comes from the quark sector and the charmed quark.) The resulting bound on the branching ratio for K_L is $\leq 7 \times 10^{-15} (\frac{g'}{g''})^2$, and it involves this very small factor 7×10^{-15} and then g'/g''. Fitting this ratio to the known bound of $\leq 2 \times 10^{-9}$ would imply that $g'>>g''$, which is probably unlikely in any theory at this time. It is more likely that the ratio of coupling constants be of the order of 1, in which case the branching ratio will be around 7×10^{-15} and the boson mass will be $m_X \sim 3 \times 10^4 m_W$.

Now there are other approaches that could be taken. For example, Kane and Thun[4] use the branching ratio of K_L and the mass difference to set limits on the two coupling constants g' and g'', and then they use these limits to make estimates about branching ratios for other processes such as the rare $\Sigma^+ \to p\mu^+e^-$ decay. Another thought is that Δm_X could actually exceed the experimental value as long as there are other contributions (coming from sources such as heavy Higgs or other bosons) to cancel out the excess. Alternatively, if there exist more than one of the heavy gauge bosons X, then they could in fact cancel out in the calculation of the K_L-K_S mass difference. In fact, an early example of that kind of cancellation was given twenty years ago by Lee and Yang[5] in the Schizon theory; they found that the two neutral bosons in that theory cancelled each other out in the mass difference for $K_L - K_S$. More recently, the cancellation has been discussed in the context of gauge theories by Kahn et al[6]. Thus, it is possible that, as the situation gets more complicated, this way of setting a bound on the branching ratio would go by the board.

We can also go through the same kind of analysis using Higgs bosons, and we can make similar estimates[7]. Use of the K_L-K_S mass difference is subject to the same caveat as before. Another possibility is technicolor and that whole range of ideas. It seems as though technicolor theories suggest that the branching ratio might be as large as 10^{-10}, which is just one order of magnitude lower than the present limit[8].

In conclusion, let me show a summary of the predictions in Table 1. If we see these modes, they would be a signal for new particles and new physics. If the branching ratios are in the ballpark of 10^{-14}, it would be hard to tell heavy leptons from gauge or Higgs bosons as the source of the new physics. If it is in the ballpark of 10^{-10}, that indicates technicolor and dynamical symmetry-breaking as the source. If it falls in between these extremes, then perhaps that might be a signal for new generations.

References

1. T. P. Cheng and L. F. Li, Phys. Rev. <u>D16</u>, 1565 (1977); B. W. Lee and R. Shrock, ibid 1444 (1977); J. Bjorken, K. Lane and S. Weinberg, ibid 1474 (1977); and references therein.
2. Lee and Shrock, reference (1).
3. P. Herczeg, Talk at Vancouver Workshop, June 1979.
4. G. Kane and R. Thun, Phys. Lett. (to be published).
5. T. D. Lee and C. N. Yang, Phys. Rev. <u>109</u>, 1410 (1960).
6. R. Kahn and H. Harari, to be published.
7. B. McWilliams and L. F. Li (to be published).
8. E. Eichten, Madison Workshop, October 1979.

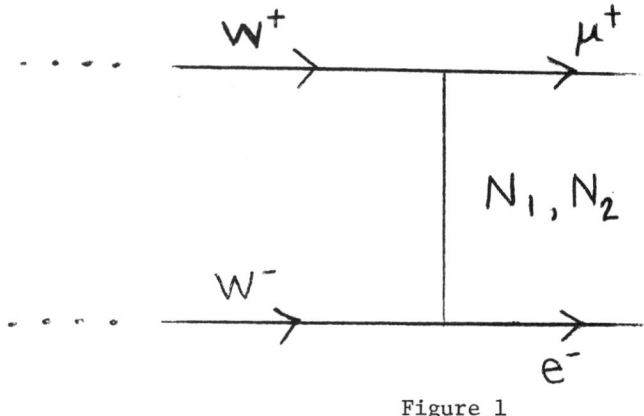

Figure 1

Table 1

Mechanism	New Leptons	Horiz. Gauge	Higgs	T-color
Mass range (GeV)	1-10	$\sim 2 \times 10^6$	$> 4 \cdot 10^3$	$3 \cdot 10^4$
B.R. ($K_L \to \mu e$)	$10^{-13} - 10^{-15}$	7×10^{-15}	$\sim 10^{-14}$	$\sim 10^{-10}$
B.R. ($K \to \pi \mu e$)	$10^{-11} - 10^{-13}$	$\sim 10^{-15}$	$10^{-15} - 10^{-17}$	$\sim 10^{-10}$
Exptl poss.	Sensitivities $\simeq 10^{-10} - 10^{-11}$		achievable-Kane & Thun	

DISCUSSION

KANE: Just two short comments. One is that I think you are too generous to the technicolor theories and I suspect that models around give numbers even bigger than 10^{-10} and there is probably a conflict or nearly one at present.

ROSEN: You mean that they may actually exceed the present bound?

KANE: I think that is probably the situation though it is a little hard to estimate. You are being fairly generous. A second comment is that there is a real possibility that the K decays are suppressed by extra things like the helicity suppression or small F_K or needing a vector current for the $K \to \pi$ vertex. An excellent decay to study is $\Sigma \to p\mu e$ because there is no mechanism that can suppress it and, in addition, experimentally one can achieve at least the sensitivity available for the K_L. So finally it may turn out to be the best test for this sort of thing.

CP VIOLATION

L. Wolfenstein
Carnegie-Mellon Univ., Pittsburgh, PA

Talks on CP violation have a tendency to seem the same every year. The subject is one on which the progress is somewhat limited. The first table has been the first table for a long time. It says that, from a phenomenological point of view, one can discuss three kinds of CP violation. I think this table first appeared in the paper T. D. Lee and I wrote in 1965. We can classify the different kinds of CP violation according to their different selection rules with respect to strangeness change, 0, 1 or 2. In particular we will emphasize the one and two, one being called milliweak of strength of order 10^{-3} weak, and two being superweak which is very weak, because it allows at tree level $\Delta S = 2$. The experimental consequences of superweak are well known; so far, they have turned out to be in agreement with what has been found. The best chances to find effects that are contrary to superweak in experiments now going on are: (1) the electric dipole moment of the neutron, for which the limit now is something like 10^{-24} e - cm, and for which an experiment now in progress hopes to get down to 10^{-26} and perhaps lower; (2) the famous η_{oo}/η_{+-} parameter (really magnitude), which is the ratio of the CP forbidden $\pi^{o}\pi^{o}$ mode to the forbidden $\pi^{+}\pi^{-}$ mode. The deviation of $|\eta_{oo}/\eta_{+-}|$ from unity is now known to be less than 6%. Two experiments, one by Cronin and collaborators and one by Adair and his collaborators, hope to see this at a level of 1%, if it exists.

Now, from the theoretical point of view, all the discussions are in terms of gauge theories, and gauge theories really don't provide any great insights. Rather the question is simply to find ideas about how to accomodate CP violation in gauge theories. There are two main ideas that have appeared in the last decade. One is the Kobayashi-Maskawa (KM) theorem, and the other is the idea of spontaneous CP violation. The KM theorem is a statement that the minimal SU(2) x U(1) model with only one Higgs doublet, and only two generations of quarks, just can't have CP violation. Actually, the theorem doesn't explain CP violation, but rather explains the apparent CP conservation, by saying that CP conservation is not really a law of nature, but just a good approximation in the region where a system of two generations is a good approximation. That suggests a variety of ways to construct models of CP violation. You either go to three generations of quarks, and that was the original idea of KM, or you go to two or three Higgs doublets, or you go to a larger group, where you have extra intermediate vector bosons which can mediate the CP violation.

Now there are a variety of reasons why any one of those may be desirable. Three generations of quarks may be desirable because there already appears to be a bottom quark, and everyone seems to believe there is a top quark. Extra Higgs doublets have been suggested in grand unified theories for other reasons, such as getting good ratios of lepton to quark masses, and also in some of the discussions of the baryon assymetry. Of course, one larger group is the grand unified theory group. But in the minimal grand unified theory, with the desert, the only new intermediate bosons are too heavy to help us with CP violation. There are other models with intermediate mass scales; for example, those using right-handed intermediate bosons, of Mohapatra and others, and then those that use horizontal gauge symmetries, like the model of Maehara and Yanagida.[1] Now in fact it is not unreasonable to imagine that more than one of these **are** operative and the CP violation in nature is not solely due to one of these theories. It is unfortunately not an either-or question.

The other thing that is a nice idea is the idea of the spontaneous CP violation, originally given in the gauge theory of T. D. Lee. That's the idea that one has CP invariance in the Lagrangian, but it's spontaneously broken. And what Lee showed was that you need at least two doublets of Higgs, so that in their vacuum expectation values, there would be a meaningful relative phase. There are two realizations of this I want to talk about. The first is the scheme that Steven Weinberg discussed, in which you adjoin to the Higgs model the concept of natural flavor conservation. That is, you demand that the neutral Higgs do not violate strangeness at the tree level. Then you need to invoke a discrete symmetry and you need three Higgs doublets to violate CP. In principle, this can be done with either hard or soft violation, but we do it with spontaneous CP violation, which is also called soft CP violation. If we do it that way and adjoin natural flavor conservation, then as a result, following the work of Branco,[2] we find there isn't any W-mediated KM CP violation in that case. So the only CP violation in this case is due to charged Higgs exchange.

There are some general features of this model. The first is that you need quite light charged Higgs bosons, because you now have natural flavor conservation, so if you want to get CP violation in the mass matrix, you have to again go to box diagrams with one Higgs and a W or two Higgs. Such a diagram is suppressed by a factor of M_s^2/M_H^2 as compared to the usual 2W exchange, and therefore the CP violation effect is of the order of M_s^2/M_H^2 where M_s is the strange quark mass. When you put it together, you need a Higgs on the order of 5 to 15 GeV of mass. That makes it interesting. It also gives a rather large electric dipole moment to the neutron, of the order of 10^{-24}, a somewhat rough number. And finally, of course, an interesting feature is that if you go to heavier quark systems, such **as** the B^0 system, you are dealing with Higgs couplings which are proportional to masses, you would expect to see larger amounts of CP violation.

Now the other model with spontaneous CP violation is the model that has only two doublets, like the original T. D. Lee model, which has been discussed by Sikivie and Lahanas and Vayonakis.[3] There, you have $\Delta S = 2$ at tree level; that is, you have an effective coupling that gives you $\Delta S = 2$ due to the exchange of the neutral Higgs, so this model is essentially a superweak theory. You need a very heavy Higgs, of the order of a thousand GeV, so you might not trust the calculations that much. The general features of this model are firstly, that you also have KM CP violations. You can't avoid it now. So you have both CP violations due to Higgs exchange and the KM CP violation due to W exchange, and you don't know their relative roles. In principle there is only one CP-violating parameter in the model, with only one complex phase from the vacuum expectation values, although there are a lot of other little mixing angles in the model. The electric dipole moment of the neutron in this model, I believe is of the order of magnitude of 10^{-29}, just as in the superweak model.

However, there are a number of very large uncertainties in the calculations. One uncertainty is that in the loop diagram that gives the dipole moment, you have intermediate states of heavy quarks and, of course, since this is a Higgs model, heavy quarks couple much more than light quarks. So if their mixing angles are not too small you can get an extra factor of one hundred from that. Charged Higgs might be lighter than the neutral Higgs, and might give you another factor of a hundred. So it is a very soft number, the dipole moment of the neutron, in this model.

There is also an interesting point about what is called the superweak model. I said this is superweak because of the effective $\Delta S = 2$ interaction at the tree level. However, because one of the reasons it is so weak is the Higgs coupling, which is proportional to quark masses, it changes very drastically as we go to heavier quarks. So what you mean by superweak, which is a phenomenological description, changes as you come to realizations of it. In this realization the superweakness begins to get less and less super as you go to a heavier quark.

Now finally, I just want to make a few remarks about the KM model, since that is still in some ways the most popular model. The first remark is that it does give a very small electric dipole moment. But there is the following point to make, I think: Suppose the electric dipole moment is discovered to be 10^{-25} e - cm and the experimentalist will say we finally rule out KM. They will immediately be told by theorists that there is the QCD θ which they had never heard about before. The QCD θ problem was discussed yesterday by Tye. From the phenomenological point of view the only thing it does is to give an electric dipole moment to the neutron. So that if the dipole moment is found, it can just always be blamed on the θ, until someone understands the whole issue better.

Now there has been a lot of discussion in the KM model of η_{+-}/η_{00}, written as $(1 + 3 \epsilon'/\epsilon)$. To get a non-zero value of ϵ'/ϵ you have to have some CP violation in a decay amplitude, not just in the mass matrix. Furthermore, in the KM model CP violation occurs only because you have coupling involving heavier quarks, because in the four-quark version you don't have any. Therefore, you have to have a decay amplitude involving the heavy quark, and that can be done by the so-called penguin graphs, in which a strange quark has a loop involving top and charmed quarks, and then emits a gluon or many gluons. Now that is a pure $\Delta I = 1/2$ amplitude. Now as we know that, the normal decays of conserved CP are not pure $\Delta I = 1/2$, and that means that the CP violating decays do not have exactly the same charge ratios as the CP conserving ones. That yields a formula for ϵ'/ϵ:

$$\left|\frac{\varepsilon'}{\varepsilon}\right| = .05 \frac{I_m A_o / R_e A_o}{(m'/\Delta m) + (I_m A_o / R_e A_o)}$$

where A_o is the amplitude for $s\bar{d} \to d\bar{d}$, including gluon corrections. $\Delta I = 1/2$ rule violation in the normal decays. Then ε'/ε depends on the dynamical contribution to CP violation being significant compared to $m'/\Delta m$, which is a CP violation in the mass matrix itself. Now the calculation, unfortunately, has a lot of uncertainties. It depends on the KM angles; it depends upon the calculations of Δm and how large the dispersive contributions are; it depends upon the parameter called B, which is the evaluation of the four-quark matrix element in calculating box diagrams. And finally it depends on the evaluation of the penguin. I followed the work of Gilman and Wise in evaluating the penguin and varied all other quantities and came then to the conclusion[5] that ε'/ε should be greater than .004. That means a 1% effect in η_{oo}/η_{+-}, or, since experiments measure the square of this, a 2% effect. However, other calculations of the penguin, particularly the work of Peccei and Gubererina give something almost a factor of three lower. Then it might be very hard to see the effect of the KM model from the proposed experiment.

I think the conclusion is that there is a fair chance that if the KM model is correct, one will see an effect from the experiment. But is is not known for sure. The second thing which has been noted by a number of people, and most recently in a preprint of Hagelin and Wise,[6] is that CP violation in the $B^o - \bar{B}^o$ mass matrix is small. The sort of simple picture is that when we are dealing with the b quark, the charm quark and the up quark are almost the same. And, of course, that makes it look like a four quark model and a four quark model has no CP violation in it.

The last remark I want to make about the KM model is the following: Why is CP violation small? That is one of the questions we always ask ourselves. One possibility is that CP violation is small in the KM model because one can almost get along with **two** generations, that is, there is very little coupling of the third generation with the other two. The question is: Is the coupling of the third generation to the other two really so weak that it provide the explanation of the smallness of CP violation. In the usual notation that has to do with s_3 and s_2, you get an effect proportional to $s_3 s_2 s$ (where $s_3 s_2$ is $\sin\theta_3 \sin\theta_2$) for those of you who know the formalism. In particular one could imagine, and I don't think that is ruled out by anything, that s_3 is just very, very small corresponding to the idea that the bottom quark almost doesn't couple at all to the up quark. T is a possibility for which $\sin\delta$ could be large. Most of the preferred solutions, preferred for whatever reason, give larger values of $s_3 s_2$ a have to require $\sin\delta$ to be very small, so that part of the explanation of the smaller CP violation in the KM model does come from the small amount of mixing, but you also have to assume at the same time that there is some rather small phase in the model as well.

TABLE 1
Classification of CP Violations
in Terms of ΔS

	$|\Delta S|$	Strength
Millistrong	0	10^{-3} x strong
Milliweak	1	10^{-3} x weak
Superweak	2	10^{-9} x weak

REFERENCES

1. T. Maehara and T. Yanagida, Prog. Theo. Phys. <u>61</u>, 1434(1979).

2. G. C. Branco, Phys. Rev. Lett. <u>44</u>, 504(1980); Phys. Rev. D. <u>22</u>, 2901(1980).

3. P. Sikivie, Phys. Lett <u>65B</u>, 141(1976); A. B. Lahania and C. E. Vayonakis, Phys. Rev. D <u>19</u>, 2158(1979).

4. F. Gilman and M. Wise, Phys. Rev. <u>D20</u>, 2392(1979); B. Guberine and R. D. Peccei, Nucl. Phys. <u>B163</u>, 289(1980).

5. L. Wolfenstein, Nucl. Phys. <u>B160</u>, 501(1979).

6. J. S. Hagelin and M. B. Wise, Harvard preprint HVTP-80/A070.

DISCUSSION

TSAO: I would like to comment. The Kobayashi-Maskowa model has a nontrivial θ. Any strong interaction that is CP violating but flavor-conserving will always depend on θ. Actually there are more than one experimental measurement results on CP violation than the neutron dipole moment. Other processes, like $\eta \to 2\pi$ or pion-nucleon interactions can also have a θ-dependent parity violation.

WOLFENSTEIN: The QCD θ is a P violating and T violating interaction. The electric dipole moment is so sensitive to simultaneous P and T violation that anything else is out of sight.

TSOU: You mean η and 2π would not have anything to do with that.

WOLFENSTEIN: Oh perhaps to the level of 10^{10} times better than a prospective experiment. $\eta \to 2\pi$ would measure simultaneous P and T violation but I don't think anyone could measure it in a relevant ballpark. The dipole moment is a good experiment.

PHENOMENOLOGY OF CP VIOLATION FROM THE KOBAYASHI-MASKAWA MODEL

Ling-Lie Chau Wang
Brookhaven National Laboratory, Upton, New York 11973

I. INTRODUCTION

In this talk I shall focus mostly on discussing the CP violation consequences of the K-M model, which Kobayashi, Maskawa[1] introduced in '77 for the purpose of incorporating CP violation via the complexity in the mixing matrix of the quarks. Much of the talk[2] is reviewing current work on the subject. Some new results of mine on the CP violation effects in exclusive and inclusive decays of bottom, charm and strange particles are also given.

II. THE MIXING MATRIX

In the K-M model, assuming the existence of the yet to be discovered top quark t, there are three doublets, $(u,d')_L$, $(c,s')_L$ and $(t,b')_L$, where $(d',s',b') = (d,s,b)V$. V is a 3 × 3 unitary matrix $V^+ V = 1$. In general for n doublets, the number of physically significant parameters in V is equal to the number of parameters for an n × n unitary matrix minus the relative phases of the doublets, i.e., $n^2 - (2n - 1)$. An orthogonal matrix can be characterized by $\tfrac{1}{2}n(n - 1)$ angles, thus the rest of the parameters $[n^2 - (2n - 1)] - \tfrac{1}{2}n(n - 1) = \tfrac{1}{2}(n - 1)(n - 2)$ has to be characterized by phases. For n = 2, V can be characterized by an angle θ_c and no phase. For n = 3, V is characterized by three angles and one phase

The V matrix is parametrized by Kobayashi and Maskawa[1] as

$$V = \begin{pmatrix} V_{ud} & V_{cd} & V_{td} \\ V_{us} & V_{cs} & V_{ts} \\ V_{ub} & V_{cb} & V_{tb} \end{pmatrix} = \begin{pmatrix} c_1 & -s_1 c_2 & -s_1 c_2 \\ s_1 c_3 & c_1 c_2 c_3 - s_2 s_3 e^{i\delta} & c_1 s_2 c_3 + c_2 s_3 e^{i\delta} \\ s_1 s_3 & c_1 c_2 s_3 + s_2 c_3 e^{i\delta} & c_1 s_2 s_3 - c_2 c_3 e^{i\delta} \end{pmatrix} \quad (2.1)$$

It is this complexity in V that provides the CP violation. Thus, the salient feature of the K-M model is that <u>the CP violation effect is tied with the nonvanishing of some of the matrix elements in the third row or third column, which means that the b and the t flavored particles must have pure hadronic decays.</u> Models with CP violation coming from the Higgs couplings, by having more Higgs doublets than the standard $SU(2)_L \times U(1)$ model, have no such correlation. Actually in many of these models, the b-flavored particles have only semileptonic decays though this is not imposed on by any first principles.[3]

Since the model is designed to provide CP violation, some of the parameters must be determined from the CP violation of the K_L, K_S system which, so far, is still the only experimentally established system having CP violation. The four parameters of the V matrix have been so far determined from four sets of experimental informations. The $0^+ \to 0^+$ nuclear β decay rates comparing to that of μ decay (assuming no effects from the mixing of the leptons) determines $|V_{ud}|$, and the hyperon semileptonic decays determines $|V_{us}|$. The results of Shrock and Wang's analysis[4] in '78 are $|V_{ud}| = .9737 \pm .0025$, $|V_{us}| = .219 \pm .003$, and $|V_{ud}|^2 + |V_{us}|^2 = .996 \pm .004$. The important point of the result is that the central value of $|V_{ud}|^2 + |V_{us}|^2$ is less than one, indicating that the old Cabibbo theory was not exactly true and there is "leakage" from the first two doublets. It allows the third doublet to decay, i.e., the b can decay into u.

The constraint the other two parameters V_{cs}, V_{cd} we use the two sets of experimental informations, i.e., the K_L, K_S mass difference and the CP violation parameter $|\varepsilon|$. To remind you[5] about the parameter ε, consider the mass matrix of $|K^0\rangle$ and $|\bar{K}^0\rangle$ states:

$$M = \begin{pmatrix} M_{11} - i\Gamma_{11}/2 & M_{12} - i\Gamma_{12}/2 \\ M_{21} - i\Gamma_{21}/2 & M_{22} - i\Gamma_{22}/2 \end{pmatrix}. \qquad (2.2)$$

where M_{ij}, Γ_{ij} are transition matrix elements from virtual and physical intermediate states respectively and can be complex numbers. CPT implies $M_{11} = M_{22}$, $\Gamma_{11} = \Gamma_{22}$, Hermiticity $M_{ij} = M_{ji}*$, $\Gamma_{ij} = \Gamma_{ji}*$, and CP invariance $M_{ij} = M_{ji}$, $\Gamma_{ij} = \Gamma_{ji}$. Thus CP invariance with CPT and hermiticity implies that all M_{ij}, Γ_{ij} are real. Therefore, imaginary parts M_{ij}^I and Γ_{ij}^I gives CP violation. After diagonalizing the mass matrix M, one obtains the eigenstates $|K_S\rangle = (1+\varepsilon)|K^0\rangle - (1-\varepsilon)|\bar{K}^0\rangle$, and $|K_L\rangle = (1+\varepsilon)|K^0\rangle + (1-\varepsilon)|\bar{K}^0\rangle$, where

$$\varepsilon = i(M_{12}^I - \tfrac{i}{2}\Gamma_{12}^I)/(M_{12}^R - \tfrac{i}{2}\Gamma_{12}^R), \qquad (2.3)$$

where the superscripts I, and R stand for imaginary and real parts respectively. The parameter ε can be measured by measuring

$$\eta_{+-} \equiv \langle \pi^+\pi^-|H_w|K_L\rangle / \langle \pi^+\pi^-|H_w|K_S\rangle = \varepsilon + \varepsilon',$$

and (2.4)

$$\eta_{00} = \langle \pi^0\pi^0|H_w|K_L\rangle / \langle \pi^-\pi^-|H_w|K_S\rangle = \varepsilon - 2\varepsilon',$$

where $\varepsilon' = \sqrt{2}\, e^{i(\delta_2 - \delta_0 + \pi/2)} \mathrm{Im}(A_2/A_0)$

The δ_2 and δ_0 are respectively the $I = 2$, $I = 0$ phase shifts of the $\pi\pi$ scattering amplitudes. The real part of the off diagonal matrix element is related to the eigenvalues M_S, M_L, Γ_S, Γ_L of the mass matrix M by $M_{12}^R = \tfrac{1}{2}(M_L - M_S)$, $\Gamma_{12}^R = \tfrac{1}{2}(\Gamma_S - \Gamma_L)$, where M_S, Γ_S, M_L, Γ_L are the mass and width of K_S, K_L respectively. The strategy here is to take $\Gamma_{12} = \tfrac{1}{2}\, 7.4 \times 10^{-15}$ GeV and $\Gamma_{12}^I \approx 0$ from experiment and calculate M_{12}^R, M_{12}^I from Fig. (2.1), which involves the mixing matrix.

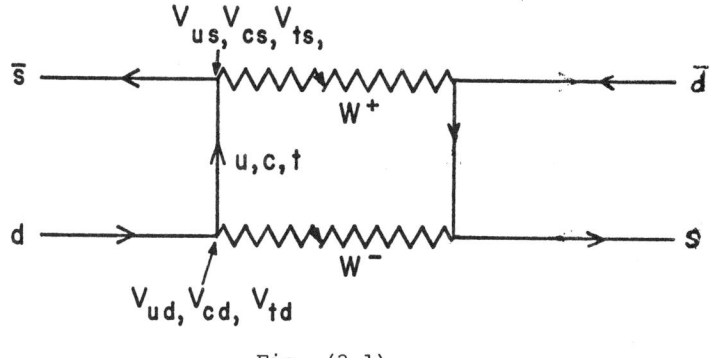

Fig. (2.1)

Fig. (2.1). The box graph for calculating the $K^0 - \bar{K}^0$ transition matrix

The imaginary part $M_{12}^I \propto s_1 s_2 s_3 s_\delta$ is directly from the complexity in the V_{ij}'s. Comparing the calculated M_{12}^R, M_{12}^I with experimental numbers $M_{12}^R = -\frac{1}{2} \times 3.52 \times 10^{-15}$ and

$$|\varepsilon| = |M_{12}^I|/\sqrt{(M_{12}^R)^2 + (\tfrac{1}{2}\Gamma_{12}^R)^2} = 2 \times 10^{-3},$$

we thus obtained V_{cs} and V_{cd}. There is one warning in calculating M_{12}: after abstracting all the known weak interaction information from Fig. (2.1), one still needs to estimate a strong interaction matrix element $\mathcal{M}_{12} = \langle\bar{K}^0|[\bar{s}\gamma_\mu(1-\gamma_5)d][\bar{s}\gamma^\mu(1-\gamma_5)d]|K^0\rangle$. Here the uncertainty can be as big as a factor of two from two different methods of calculations.[7,8] Another uncertainty is that we cannot fix the quadrants in which the angles θ_2, θ_3 and δ of Eq. (2.1) fall in; only $\zeta \equiv \text{sign}(\tan\theta_2 \cdot \tan\theta_3 \cdot \cos\delta)$ matters. The results are rather insensitive to the t quark mass. As an example we give one of the central[9] values of the V matrix determined in Ref. 8.

$$V = \begin{pmatrix} .97 & -.22 & -.046 \\ .22 & .85 - .66 \times 10^{-3}i & .48 + 3.2 \times 10^{-3}i \\ .068 & -.48 + 2.1 \times 10^{-3}i & -.88 - 1.0 \times 10^{-3}i \end{pmatrix} \begin{matrix} d \\ s \\ b \end{matrix} \quad (2.5)$$

with columns labeled u, c, t.

It is interesting to observe that the magnitude of the matrix element is the largest on the diagonal and decreases as the element moves away from the diagonal, i.e., there are flavor mixings but they like to keep the original identity. In physical terms, quarks decay in a cascade fashion. The b particles will prominently decay into charm particles, then charm to strange. This is now supported by experiment from CESR.[10] The t particles will decay mainly into b particles.

Though the central value of the V matrix, Eq.(2.1), has not been challenged by various considerations,[11] it is important to have independent determinations of V_{cs}, V_{cd} in a more model-independent[2] way similar to the determination of V_{ud}, V_{us}. Here I list a few of such possibilities:

(1) Obtain V_{cs} from $D \to \ell \bar{\nu}_\ell X$ (with K), and V_{cd} from $D \to \ell \bar{\nu}_\ell X$ (without K). It is desirable to study decay rates in $e^+e^- \to \psi(3770) \to D\bar{D}$ with one D or $\bar{D}$ explicitly selected from its exclusive decays.

(2) From the results of Ref. (12) $\Gamma(D^+ \to \pi^+\pi^0)/\Gamma(D^+ \to \bar{K}^0\pi^+)$ = $\frac{1}{2}|V_{cd}/V_{cs}|^2$, which, in addition, has the nice feature that both final states $\pi^+\pi^0$, $\bar{K}^0\pi^+$ are exotic, thus free from possible complications of final state interactions.

(3) Comparing the decays $b \to cW^+_{\to c\bar{s}}$ and $b \to cW^+_{\to \mu\bar{\nu}_\mu}$ ought to give information about V_{cs}.

It is interesting to note that if $V_{ud}V_{us} \neq -V_{cs}V_{cd}$, i.e., if the strangeness neutral current is not cancelled in the first two doublets then the t quark that so far eludes observation is needed. If $|V_{cs}|^2 + |V_{cd}|^2 < 1$, the b flavored particle must decay into charm.

III. CP VIOLATION FROM THE COMPLEXITY IN THE MASS MATRIX

As we have elaborated in the last section, the complexity in the mixing matrix gives rise to the CP violation effect in the K^0 system. The parameter ε_K specifies the deviation of K_S, K_L from CP eigenstates. It is Nature's magic that K has a mass so near the 3π threshold so that K_S (mainly goes to 2π) and K_L (mainly goes to 3π) can have such large time differences in life. Such wonder probably will not happen again in D^0, $\bar{D}^0$ system again. It probably will be hard to measure ε_c, ε_B using the same method as for ε_K. As pointed out a few years ago in Refs. (13) and (14), the transition of $D^0 \rightleftarrows \bar{D}^0$ (or $B^0 \rightleftarrows \bar{B}^0$) can give rise to the asymmetry δ of same sign double-lepton final state in $e^+e^- \to D^0\bar{D}^0X^0$ (or $\to B^0\bar{B}^0X^0$) $\to \ell^+\ell^+X^{--}$, $\ell^-\ell^-X^{++}$ is $\delta \equiv (N_{++} - N_{--})/(N_{++} + N_{--}) = 4\mathrm{Re}\varepsilon$, where ε is the CP violation parameter for D^0, or B^0 system. It was estimated[15] to be small, ($\delta \sim 10^{-3}$) for the K-M model, but bigger ($\delta \sim 10^{-2}$) for the Higgs CP violation. Thus a large double charge asymmetry in e^+e^- experiment can rule out the K-M model. However, such a double lepton charge asymmetry has sever contamination form the chain semileptonic decays of quarks.

IV. CP VIOLATION IN PARTIAL DECAY RATES

Besides contributing CP violation effects in the mass matrix, the complexity in the mixing matrix can also rise CP violation in the decay amplitudes. There have been many earlier studies[13,16,17,18] on the subject from various points of view. For convenience of discussion, I shall first use the quark-diagram scheme of Ref. (19) to give an overall view and also some new results. I shall comment on the known results where they fit.

The decays of a heavy-quark meson (the bottom, the charm, and the strange) can be described by six independent amplitudes, a, b, c, d, e, f, as shown in Fig. (4.1).

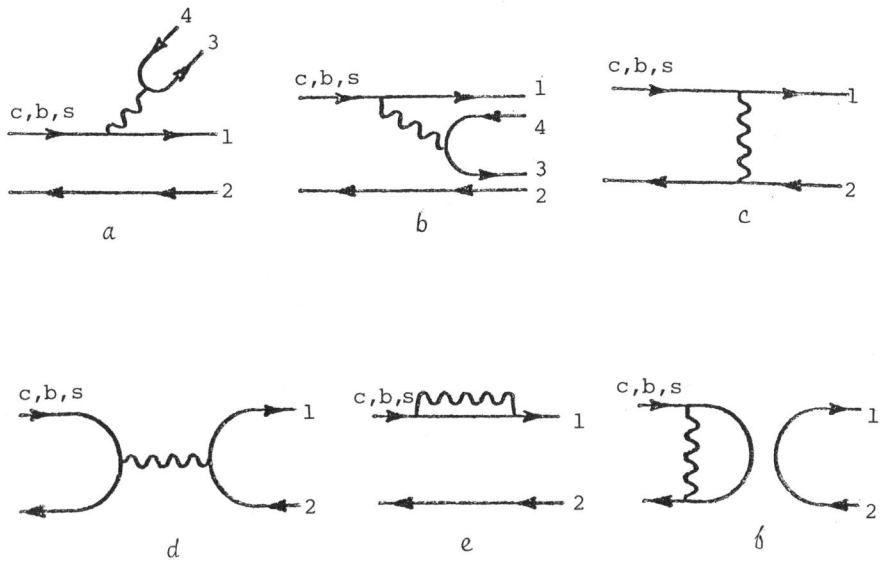

Fig. (4.1)

For a given final state of particles, we need only to add the appropriate $q\bar{q}$ lines (the hairpin quark lines) to each diagram and then project out the given final particles. In Ref. (19) the amplitudes of charm mesons D^0, D^+, F^+ decaying into two pseudo scalar mesons are given. These diagrams are meant to include all strong interaction effects (the gluon lines), which are, in general, not yet calculable. Thus we do not know the magnitude of each diagram. However, we can classify experimental results using the diagrams. Eventually, we can obtain the sizes and phases of these diagrams from decay rates and CP violation effects, which we shall elaborate.

It was discussed quite some time ago by the authors of Refs. (13) and (14) that, though CPT predicts equal total decay rate for particle and anti-particle, the partial decay rates of particle and anti-particle into CP conjugated final particles can be different if CP is not invariant. The quark-diagram scheme provides an easy way to sort out the decay channels where particle and anti-particle decay rates can be different.

a) CP violation in Charm decay.

In the following we list all the semi-mixing-angle-suppressed decays of D^0, D^+, F^+ into two pseudo mesons, taking from Ref. (19):

$$A(D^0 \to K^-K^+) = V_{us}V_{cs}(a + c + e + 2f) - V_{ud}V_{cd}(e + 2f), \quad (4.1a)$$

$$A(D^0 \to \pi^-\pi^+) = V_{us}V_{cs}(e + 2f) - V_{ud}V_{cd}(a + c + e + 2f), \quad (4.1b)$$

$$A(D^0 \to \bar{K}^0 K^0) = \tfrac{1}{2}(V_{us}V_{cs} - V_{ud}V_{cs})(2c + 4f),$$

$$A(D^0 \to \pi^0\pi^0) = \frac{1}{\sqrt{2}}[V_{us}V_{cs}(e + 2f) + V_{ud}V_{cd}(b - c - e - 2f)], \quad (4.1c)$$

$$A(D^0 \to \eta^0\eta^0) = \frac{1}{\sqrt{2}}[V_{us}V_{cs}(\tfrac{2}{3}c - \tfrac{1}{3}b + \tfrac{1}{6}e + f)$$
$$- V_{ud}V_{cd}(\tfrac{1}{6}b + \tfrac{1}{6}c + \tfrac{1}{6}e + f)], \quad (4.1d)$$

$$A(D^0 \to \pi^0\eta^0) = \frac{1}{\sqrt{3}}[V_{us}V_{cs}(-b + e) + V_{ud}V_{cd}(c - e)], \quad (4.1e)$$

and

$$A(D^+ \to \bar{K}^0 K^+) = V_{us}V_{cs}(a + e) - V_{ud}V_{cd}(d + e), \quad (4.2a)$$

$$A(D^+ \to \pi^0\pi^+) = \frac{1}{\sqrt{2}} V_{ud}V_{cs}(a + b), \quad (4.2b)$$

$$A(D^+ \to \eta^0\pi^+) = \frac{1}{\sqrt{2}} V_{us}V_{cs}(-2b + 2e) - V_{ud}V_{cd}(a + b + 2d + 2e), \quad (4.2c)$$

and

$$A(F^+ \to K^0\pi^+) = V_{us}V_{cs}(d + e) - V_{ud}V_{cd}(a + e), \quad (4.3a)$$

$$A(F^+ \to K^+\pi^0) = \frac{1}{\sqrt{2}}[V_{us}V_{cs}(d + e) + V_{ud}V_{cd}(b - e)], \quad (4.3b)$$

$$A(F^+ \to K^0\eta^0) = \frac{1}{\sqrt{2}}[V_{us}V_{cs}(2a + 2b + d + e) + V_{ud}V_{cd}(b - e)]. \quad (4.3c)$$

For $\bar{D}^0$, D^-, F^- decays, we replace V_{ij} in Eqs. (4.1), (4.2) and (4.3) by V_{ij}^*. That the amplitudes a, b, c, d, e, f do not change in particle and anti-particle decays is a consequence of CP invariance in strong interactions. I have not listed the mixing-angle nonsuppressed and doubly suppressed channels since they have the same decay probability for particle and anti-particles, see Ref. 19.

Typically, the decay amplitudes for particle, anti-particle are of the following form, e.g.,

$$A(D^+ \to \bar{K}^0 K^+) = V_{us}V_{cs} A_1 + V_{ud}V_{cd} A_2, \quad (4.4a)$$

$$\bar{A}(D^- \to K^0 K^-) = V_{us}^* V_{cs}^* A_1 + V_{ud}^* V_{cd}^* A_2, \quad (4.4b)$$

where $A_1 = a + e$, $A_2 = d + e$. For different decays, A_1, A_2 represents the corresponding combination of amplitudes a, b, c, d, e as given in Eqs. (4.1), (4.2) and (4.3). That the partial decay rates of particle and anti-particle can be different in the K-M model is due to the complexity in V_{ij},

$$\Delta_c \equiv \frac{\Gamma - \bar{\Gamma}}{\Gamma + \bar{\Gamma}} \equiv \frac{|A|^2 - |\bar{A}|^2}{|A|^2 + |\bar{A}|^2}, \text{ where } \Gamma \equiv |A|^2, \bar{\Gamma} = |\bar{A}|^2,$$

$$= \frac{4\text{Im}(V_{us}V_{cs}V_{ud}^*V_{cd}^*) \text{Im}(A_1 A_2^*)}{|A|^2 + |\bar{A}|^2}$$

$$= \frac{4s_2 s_3 s_\delta \, c_1 c_2 c_3 \, \text{Im}(A_1 A_2^*)}{(|A|^2 + |\bar{A}|^2) s_1^{-2}} \quad (4.5)$$

We divide the denominator by s_1^2 because both $|A|^2$ and $|\bar{A}|^2$ have a factor of s_1^2. Δ now is again proportional to $s_2 s_3 s_\delta$. The same combination contributes to the CP violation parameter ϵ in K_L decay. In addition to mixing angles and phases, Δ depends crucially on the phases and magnitude of A and $\bar{A}$. Δ is zero if A and $\bar{A}$ have the same phase. Unfortunately we do not have reliable ways to calculate A and $\bar{A}$. Therefore, it is extremely difficult to give an accurate prediction of Δ. The present scheme provides the information about what are the possible channels where particle-anti-particle decay rates can be different.

Using Fig. (4.1), we can work out decay amplitudes for higher multiplicty final states and for semi-inclusive decays. Here we list the channels for which particle and anti-particle can have different decay rates:

$$D^\pm \to \bar{K}^0 K^\pm, \eta^0 \pi^\pm, K^\pm K^\mp X(s = 0 \text{ states}), \eta^0 \pi^\pm X(s = 0 \text{ states}),$$

$$K^\pm X(s = \mp \text{ states}), K^\pm X(s = \mp \text{ states}), \eta^0 X(s = 0 \text{ states}), \text{ etc.} \quad (4.6a)$$

$$F^\pm \to K^0 \pi^\pm, K^\pm \pi^0, K^\pm \eta^0, K^0 \pi^\pm X(s = 0 \text{ states}), K^\pm \pi^0 X(s = 0 \text{ states}),$$

$$K^\pm \eta^0 (s = 0 \text{ states}), \pi^\pm X(s = \pm 1 \text{ state}), K^\pm X(s = 0 \text{ states}), \text{ etc.} \quad (4.6b)$$

$$\left.\begin{array}{c} D^0 \\ \bar{D}^0 \end{array}\right\} \to K^- K^+, \pi^+ \pi^-, \pi^0 \pi^0, \eta^0 \eta^0, \pi^0 \pi^0, \pi^0 \eta^0, \text{ and their inclusive states.} \quad (4.6c)$$

Here s denotes strangeness. The inclusive state X for decays of particle and anti-particle are CP conjugated. It is interesting to note from Eq. (4.2b), the mixing-angle-semisuppressed decay $D^\pm \to \pi^0 \pi^\pm$ has same decay rate, so do D, $\bar{D}^0 \to \bar{K}^0 K^0$.

Here we see a rich variety of channels where one can search for CP violation effects. Needless to say high experimental sensitivity, in the range of ε, is needed in such searches.

b) CP violation in B decays.

The B_{bu}^-, B_{bd}^0, B_{bs}^0 → ordinary (no charm) particle final states:

We first list the decay amplitudes of the B_{bu}^-, B_{bd}^0, B_{bs}^0 to two ordinary pseudo meson (no charm particle in the final states).

$$A(B_{bu}^- \to \pi^-\pi^0) = \frac{1}{\sqrt{2}} [V_{ub}V_{ud}(a+b+e+d) + V_{cb}V_{cd}\, e], \qquad (4.7a)$$

$$A(B_{bd}^0 \to \pi^+\pi^-) = V_{ub}V_{ud}(a+c+e+f) + V_{cb}V_{cd}(e+f), \qquad (4.7b)$$

$$A(B_{bs}^0 \to \pi^-K^+) = V_{ub}V_{ud}\, a + V_{cb}V_{cd}\, e. \qquad (4.7c)$$

We see that the interference can only come from the loop diagrams e and f, the so called "Penguin" diagrams. The partial decay rates can be different for particle and anti-particle for the following channels:

$$\left.\begin{array}{c} B_{bu}^- \\ B_{bu}^+ \end{array}\right\} \to \pi^\mp \pi^0,\ \pi^\mp X^0 (s=0), \qquad (4.8a)$$

$$\left.\begin{array}{c} B_{bd}^0 \\ \bar{B}_{bd}^0 \end{array}\right\} \to \pi^+\pi^-,\ \pi^+\pi^- X^0(s=0),\ \pi^+X^-(s=0),\ \pi^-X^+(s=0), \qquad (4.8b)$$

$$\left.\begin{array}{c} B_{bs}^0 \\ \bar{B}_{bs}^0 \end{array}\right\} \to \pi^\mp K^\pm,\ \pi^\mp K^\pm X^0(s=0),\ K^\pm X^\mp(s=0),\ \pi^\mp X^\pm(s=\mp 1). \qquad (4.8c)$$

The difference of partial decay rate in the CP conjugated decays are of the form

$$\Delta \equiv \frac{\Gamma - \bar{\Gamma}}{\Gamma + \bar{\Gamma}} = \frac{4\,\text{Im}(V_{ub}V_{ud}V_{cb}^*V_{cd}^*)\,\text{Im}(A_1 A_2^*)}{|A|^2 + |\bar{A}|^2}$$

$$= -2(s_2/s_3)s_\delta c_1 c_2 c_3\,\text{Im}(A_1 A_2^*)/[|A|^2+|\bar{A}|^2]\,(s_1)^{-2}(s_3)^{-2},$$

where $\tfrac{1}{2}[|A|^2+|\bar{A}|^2](s_1)^{-2}(s_3)^{-2} = |c_1 A_1|^2 + c_2[c_1 c_2 + (s_2/s_3)c_3 e^{i\delta}]A_2|^2$

$$- 2c_1 c_2 [c_1 c_2 + (s_2/s_3)c_3 c_\delta]\cdot\text{Re}(A_1 A_2^*). \qquad (4.9)$$

The important thing here is that Δ now is proportional to a factor of $(s_2/s_3)s_\delta$ different from that in charm decays, $s_2 s_3 s_\delta$ which is constraint to be small $\sim 10^{-4}$ by the observed CP violation in K_L decay. From the angle analysis of Ref. (8), we can, in principle, make s_3 very small and s_δ close to unity. For example, we can choose $s_2 = .3$, $s_\delta = 1$ and $s_3 \stackrel{\sim}{=} 0.005$, while still being consistent with all existing data, including the recent results of CESR.[10] Therefore, if the phases of A_1, A_2 are favorable, Δ can be large. We see that the study of CP violation in B decays will provide crucial information about the angles, phases, and strength of the amplitudes.

Earlier analysis of Bander, Silverman and Soni[16] estimated different partial decay rates for B and $\bar{B}$ from a time-like single gluon emission diagram.

The $B_{b\bar{u}}^-$, $B_{b\bar{d}}^0$, $B_{b\bar{s}}^0$ → double charm particle final states:

The mixing matrix and amplitude dependences of $B_{b\bar{u}}^- \to D^0 D^+$, $B_{b\bar{d}}^0 \to D^+ D^-$, $B_{b\bar{s}}^0 \to F^+ D^-$ are listed as follows:

$$A(B_{b\bar{u}}^- \to D^0 D^-) = V_{cb} V_{cd}(a + b + e) + V_{ub} V_{ud}(d + e), \quad (4.10a)$$

$$A(B_{b\bar{d}}^0 \to D^+ D^-) = V_{cb} V_{cd}(a + b + e) + V_{ub} V_{ud}\, e, \quad (4.10b)$$

$$A(B_{b\bar{s}}^0 \to F^+ D^-) = V_{cb} V_{cd}(a + b + e) + V_{ub} V_{ud}\, e. \quad (4.10c)$$

Again we see that there can be particle-antiparticle partial decay rate differences in

$$\left.\begin{array}{c} B_{b\bar{u}}^- \\ B_{\bar{b}u}^+ \end{array}\right\} \to D^0 D^\pm, \; D^\pm D^0 X^0 (s = 0), \; D^\pm X (c = \mp 1), \quad (4.11a)$$

$$\left.\begin{array}{c} B_{b\bar{d}}^0 \\ B_{\bar{b}d}^0 \end{array}\right\} \to D^+ D^-, \; D^+ D^- X^0 (s = 0), \; D^+ X^- (c = -1), \; D^- X^+ (c = \mp 1), \quad (4.11b)$$

$$\left.\begin{array}{c} B_{b\bar{s}}^0 \\ B_{\bar{b}s}^0 \end{array}\right\} \to F^\pm D^\mp, \; F^\pm D^\mp X^0 (s = 0), \; F^\pm X^\mp (c = \mp 1), \; D^\mp X^\pm (s = \mp 1, c = \pm 1). \quad (4.11c)$$

The partial decay rate is given by the same formula as in Eq. (4.9). Bernabeu and Jarlskog[18] discussed this situation. But only partial rate difference of $B_{b\bar{u}}^\pm \to D^0 D^\pm$ is predicted since the diagram e was ignored.

The dominant decay channels of $B_{b\bar{u}}$, $B_{b\bar{d}}$, $B_{b\bar{s}}$ are final states with $c = 1$. They, in this model, will in general have the same decay rates between particle and anti-particles, except the case considered in Ref. (17) where the final states can come from both D^0 and $\bar{D}^0$ state of the same B decay. The interference between D^0 and $\bar{D}^0$ provide CP violation effects. They considered the difference of the two decays

$$B^- \to {}^{D_0 K_s X^-}_{\bar{D}_0 K_s X^-} \to K_s K_s X^- \qquad (4.12a)$$

$$B^+ \to {}^{D_0 K_s X^+}_{\bar{D}_0 K_s X^+} \to K_s K_s X^+ \qquad (4.12b)$$

The rate difference again is of similar form to that of Eq. (4.9).

c) CP violation in the strange particle decay

Besides the CP violation effects in the K_s and K_L decays, we can also ask about partial rate differences: It is well known that $K^\pm \to \pi^\pm \pi^0$ must have the same decay rates from CPT. Our quark diagram scheme checks with that. We list the decay amplitudes of K into two mesons.

$$A(K^+ \to \pi^+ \pi^0) = \frac{1}{\sqrt{2}} V_{us} V_{ud} (a + b), \qquad (4.13a)$$

$$A(K^0 \to \pi^+ \pi^-) = V_{us} V_{ud} (a + c + e + 2\delta) + V_{cs} V_{cd} (e + 2\delta), \qquad (4.13b)$$

$$A(K^0 \to \pi^0 \pi^0) = \tfrac{1}{2} V_{us} V_{ud} (b + c + e) + V_{cs} V_{cd} e . \qquad (4.13c)$$

For $\bar{K}$ decays, same equations apply except V_{ij} replaced by V^*_{ij}. Here we see that the rate of $K^0 \to \pi^+\pi^-(\pi^0)$ can be different from $\bar{K}^0 \to \pi^+\pi^-(\pi^0)$ and $K^\pm \to \pi^\pm \pi^\pm \pi$ can differ in decay rates. Note that the differences here like in the B → ordinary particle case, come from the interference of the Penguin diagrams. The decay rate difference is again of the form of Eq. (4.5). They are always proportional to $s_2 s_3 s_\delta$, therefore of the same order of value as ε, depending on the phase and magnitude of A_1, A_2.

Based on the same quark diagram argument, it is easy to see that $\Lambda(\bar{\Lambda}) \to \pi^- p (\pi^+ \bar{p}^-)$ $\Sigma^+(\Sigma^-) \to p\pi^0 (\bar{p}\pi^0)$, $p\pi^+(\bar{p}\pi^-)$ can have different particle-anti-particle decay rates. The magnitudes of the differences are again proportional to $s_2 s_3 s_\delta$.

We see that the K-M model in our quark diagram formulation gives a systematic way of study the CP violation in partial decay rates. It is of interest to do experiments to check these partial decay rates systematically.

V. THE NEUTRON ELECTRIC-DIPOLE MOMENT

There are three form factors for the neutron, $<n|J_\mu^{e.m.}(0)|n>$ $\sim \bar{u}(p')[F_1(q^2)\gamma_\mu - F_2(q^2)\sigma_{\mu\nu}\gamma_\nu + F_3(q^2) i\gamma_5 \sigma_{\mu\nu} q^2]u(p)$, where $F_1(0) = 0$ the charge form factor, $F_2(0) = \mu_n$ the magnetic moment and $F_3(0) = d_n$ the electric dipole moment. Again the complexity in V_{ij} can give d_n of the neutron via the diagrams of Fig. (5a) with a photon attached in all possible ways. It was first estimated by Ellis,

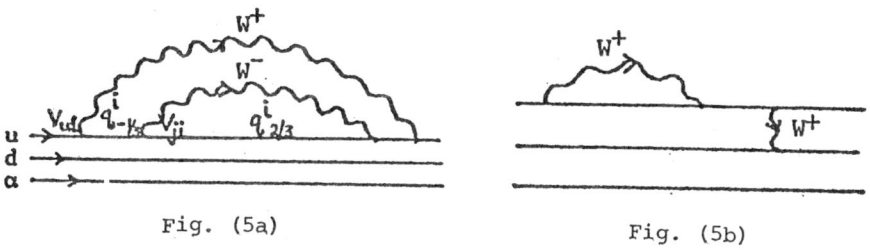

Fig. (5a)   Fig. (5b)

Diagrams considered for the neutron electro-dipole moment, where $q^i_{-1/3}$, $q^j_{2/3}$ are the quarks of charge of -1/3 and 2/3 respectively.

Gaillard, Nanopoulos[20] in '76, $d_n \sim 10^{-30}$ cm. Then Shabalin[21] showed that actually the sum of graphs in Fig. (5a) gives $d_n = 0$. Calculations have also been done including strong interations[22] and interquark exchange forces[23] Fig. (5b). The results are quite model dependent but they all give very small d_n in contrast to the result from Higgs CP violation, which is very close to the experimental[24] limit $d_n \leq 1.6 \times 10^{-24}$ cm.

VI. CONCLUDING REMARKS

To end the lecture, I would put these challenges to the experimentalists:

(1) "Direct" measurements of V_{cs}, V_{cd}: Inclusive and semileptonic decays of charm and B decays, $\Gamma(D^+ \to \pi^+\pi^0)/\Gamma(D^+ \bar{K}^0\pi^+)$.

(2) To narrow down alternatives to the K-M model it is crucial to know the B decay properties: Does B decay only semileptonically? Which decay of B is favored $b \to c \to s$ or $b \to u$? For these CESR already have an answer, yes and $b \to c \to s$ respectively. Is there b-changing neutral current, $b \to q\, \ell\bar{\ell}$, $B \to \ell\bar{\ell}$? Some limits are already given by the CESR Experiment.[10]

(3) CP properties of the charm and the B system: ε', $\frac{N^{++} - N^{--}}{N^{++} + N^{--}}$, differences of various partial decay rates of CP related channels.

(4) Better neutron electric dipole moment measurements.

The real challenge that confronts us is the "family" problem. How many generations of quarks are there? How does the mixing come about? What is the origin of CP violation? It is likely that the current distinction between the K-M origin and complex-Higgs origin may turn out to be a superfluous one.

References

1. M. Kobayashi and T. Maskawa, Prog. Theor. Phys. 49, 652 (1973).
2. For some earlier review and discussions on the subject, see L.L. Chau Wang, "Flavor Mixing and Quark Decay," Proceedings of the VIth Int. Conf. on Meson Spectroscopy, BNL, April 25-26, 1980; and "Quark Flavor Mixing and Its Physical Implications,"

Proceedings of the XXth Int. Conf. on High Energy Physics, Madison, July 17-23, 1980.

3. See L. Wolfenstein's talk at this workshop.
4. R. Shrock and L.-L. Wang, Phys. Rev. Lett., $\underline{41}$, 1692 (1978). The previous result was $|V_{ud}|^2 + |V_{us}|^2 = 1.004 \pm .004$, see M. Roos, Nucl. Phys. $\underline{B77}$, 420 (1974). The earliest attempt to evaluate this leakage was given by J. Ellis, M.K. Gaillard, and D.V. Nanopoulos, Nucl. Phys. $\underline{B109}$, 213 (1976).
5. R.E. Marshak, Riazuddin, C.P. Ryan, "Theory of Weak Interactions In Particle Physics," published by Wiley-Interscience, 1969.
6. T.D. Lee, R. Oeheme and C.N. Yang, Phys. Rev. $\underline{105}$, 1671 (1957). T.T. Wu and C.N. Yang, Phys. Rev, Lett. $\underline{13}$, 380 (1964).
7. R. Shrock and S.B Treiman, Phys. Rev. $\underline{D19}$, 2148 (1979); M.I. Vysotsky, "$K^0 - \bar{K}^0$ Transition in the Standard $SU(3) \times SU(2) \times U(1)$ Model." ITEP, Moscow preprint (1979).
8. R.E. Shrock, S.B. Treiman and L.-L. Chau Wang, Phys. Rev. Lett. $\underline{42}$, 1589 (1979). See also V. Barger, W.F. Long, and S. Pakvasa, ibid, $\underline{42}$, 1585 (1979).
9. For recent work on the mixing matrix analysis see J.S. Hagelin, "The Phase of ϵ'_K and the Sign of $\sin\delta$," Harvard University preprint HUTP-80/A018 (1980); and B.D. Gaisser, T. Tsao and M.B. Wise "Parameters of the Six-Quark Model," SLAC preprint. For $s_3 \gtrsim .5$, a small new allowed region was found for $\sin\delta < 0$, $\cos\delta < 0$. Part of this new region gives $b \to u > b \to c$. It is probably already ruled out by the CESR experiments. See Ref. 10.
10. C. Bebek, et al., Phys. Rev. Lett. $\underline{46}$, 84 (1981); K. Chadwick, et al., ibid, 88 (1981); B. Niczyporuk, et al., ibid, 92 (1981).
11. See Refs. 7, 8 and 9.
12. L.-L. Chau Wang and F. Wilczek, Phys. Rev. Lett. $\underline{43}$, 816 (1979).
13. A. Pais and S.B. Treiman, Phys. Rev. $\underline{12}$, 2744 (1975).
14. L.B. Okun, V.I. Zakharov, and B.M. Pontecorvo, Nuovo Cimento Lett. $\underline{13}$, 218 (1975).
15. A. Ali, Z.Z. Aydin, Nucl. Phys. $\underline{B148}$, 1651 (1979); J.S. Hagelin, Phys. Rev. $\underline{D20}$, 3893 (1979).
16. M. Bander, D. Silverman and A. Soni, Phys. Rev. Lett. $\underline{43}$, 242 (1979).
17. A.B. Carter and A.I. Sanda, Rockefeller University preprint, DOE/EY/2232B-203 (1980), to be published in the Phys. Rev. Lett.
18. J. Bernabeu and C. Jarlskog, "CP Violation in Decay Rates of Charged Bottom Mesons," CERN preprint Ref. TH.2929-CERN (1980).
19. T. Rizzo and L.-L. Chau Wang, "The Quark-Diagram Classification of Charm Decays," Brookhaven preprint BNL 27950 (1980).
20. J. Ellis, M.K. Gaillard and D.V. Nanopoulos, Nucl. Phys. $\underline{B109}$, 213 (1976).
21. E.P. Shabalin, Sov. J. Nucl. Phys. $\underline{28}$, 75 (1978).
22. J. Ellis, M.K. Gaillard, Nucl. Phys. $\underline{B150}$, 141 (1979), E.P. Shabalin, preprint ITEP-131-1979.
23. B.F. Morel, Harvard preprint HUTP-79/A009 (1979), D.V. Nanopoulos, A. Yildiz, and P.H. Cox, Harvard and University of New Hampshire preprint, HUTP-79/A024, E.P. Shabalin, Inst. of Theo. and Exp. Phys. preprint, ITEP-131 (1979).
24. N.F. Ramsey, Phys. Repts. $\underline{43}$, No. 10 (1978).

DISCUSSION

LANGACKER: In that last model that you just mentioned, surely there must be a strangeness-changing neutral current at some level.

WANG: No. It is designed not to have it.

LANGACKER: But is that just basically unnatural or is there some reason for that to happen?

WANG: Well, if you demand their absence, then you can avoid them.

B MESON PHYSICS AT CESR

W. A. Loomis
Harvard University
Cambridge, MA   02138

ABSTRACT

Experiments at the Cornell Electron Storage Ring (CESR) have given the first look at the weak decays of elementary particles carrying the b flavored quark. The experiments and their results are described.

The Cornell Electron Storage Ring (CESR) has been operating for the past year in the center of mass energy region of the upsilon (T) resonance system.[1] These resonances appear to indicate the threshold for the production of a new quark flavor in e^+e^- annihilation. The three lowest energy upsilon resonances, T(1S), T(2S), T(3S) are narrow resonances whose observed energy width is consistent with that of CESR. The fourth upsilon resonance T(4S) has a measurable width typical of a strong decay, suggesting that T(4S) decays directly into a pair of strongly interacting mesons each carrying a single b quark. We are thus provided with a source of B mesons whose weak decays we can observe.

There are two interaction regions at Cornell, each occupied by a detector. In the south interaction area there is the CLEO group, a collaboration of Cornell, Harvard, Rochester, Rutgers, Syracuse, and Vanderbilt Universities. The CLEO detector, Fig. 1, is a comprehensive detector with the capability of identifying and measuring charged and neutral particles. In the north interaction region is the Columbia-Stony Brook detector (CUSB), Fig. 2, which is optimized to measure the energies of photons and electrons while still being sensitive to pions and other particles.

Figure 3 shows the e^+e^- annihilation cross section in the region of the T(4S) obtained by CLEO. There are approximately 750 events in the T(4S) bump and the observed (uncorrected) average cross section of T(4S) is 0.56 nb over the mass range 10.528-10.568 GeV. CLEO finds a total width for the T(4S) of 19 ±3 MeV while CUSB finds 12 ±6 MeV. Since these widths are clearly in the strong interaction domain, it is likely that the T(4S) is decaying to pairs of B mesons

ISSN:0094-243X/81.720432-13$1.50 Copyright 1981 American Institute of Physics

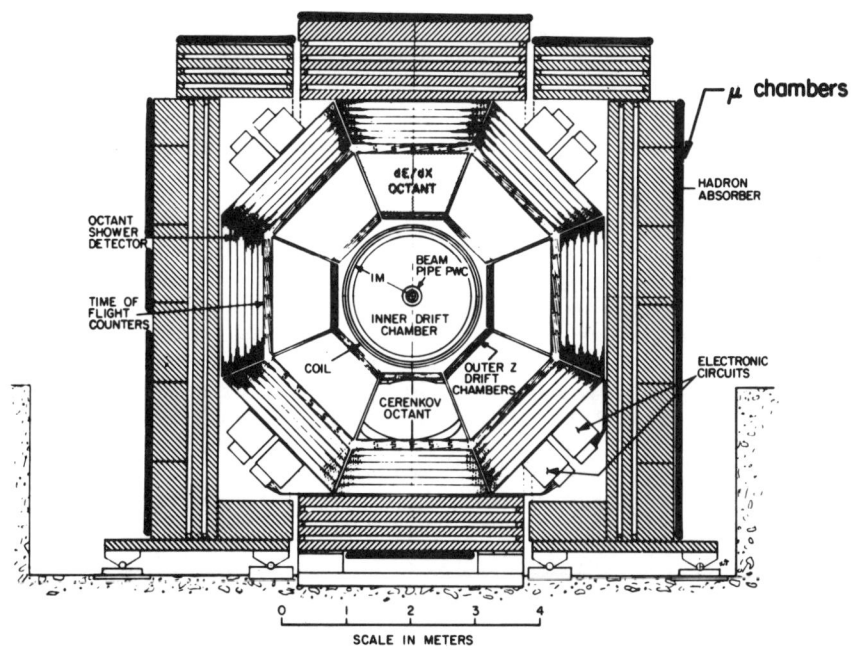

Figure 1. The CLEO detector

Figure 2. The CUSB detector

$$\Upsilon(4S) \xrightarrow{\text{strong}} B + \bar{B}$$

⟩ Weak Decays

$B, \bar{B}$ are mesons $B = b\bar{q}, \bar{B} = \bar{b}q$

Since the B mesons are carrying flavor we expect this flavor to disappear via the weak interaction. As there is no evidence to the contrary we assume we are producing unexcited B mesons. A quick look at the events in the bump region also reveals nothing in the multiplicities, lepton production rate, or missing energy that would imply that the produced B mesons are not decaying as per standard weak interaction theory. Of course this does not rule out branching ratios to exotic decay modes. To begin to understand B decays, we assume the events in the $\Upsilon(4S)$ bump are pairs of B mesons and turn to the standard theory to analyze these decays.

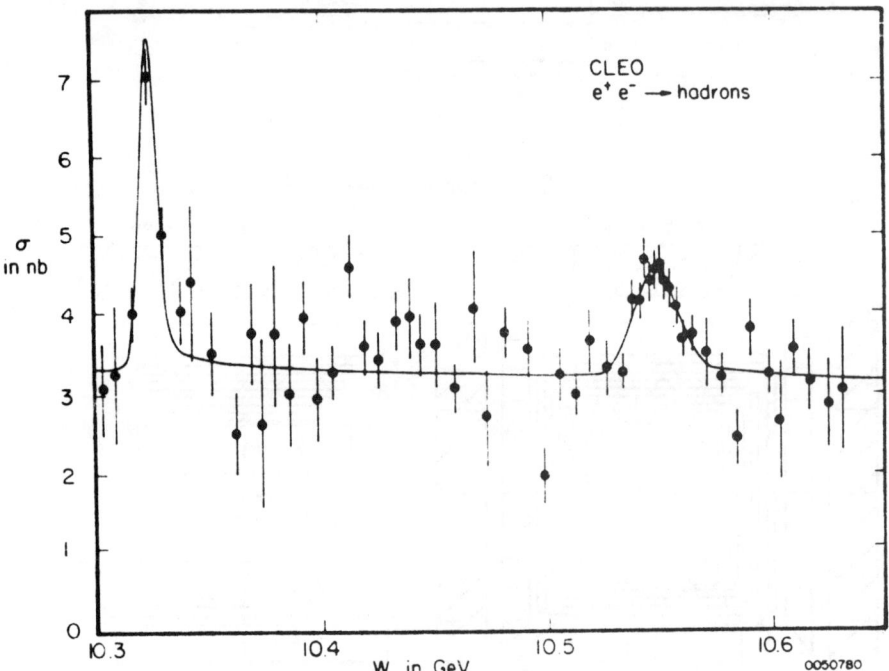

Figure 3. The e^+e^- annihilation cross section in the $\Upsilon(3S) - \Upsilon(4S)$ region from the CLEO collaboration.

The quark spectator model (QSM) implies that the weak interaction diagram in Fig. 4 will dominate B decays.

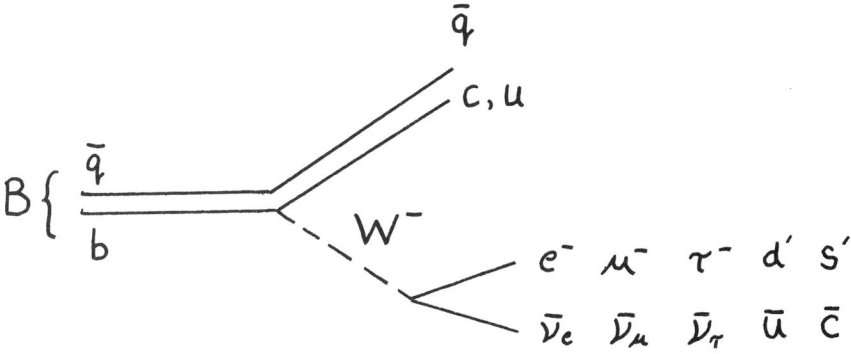

Figure 4. The dominant Feynman diagram for B decay in the Quark Spectator Model.

In this model the b quark decays as if it were a free lepton (like a muon) and the $\bar{q}$ quark making up the B meson is a spectator. The final state interactions which produce the observed decay products have negligible influence on the decay. This model ignores some other first order weak diagrams and with our awareness of the anomalies of charmed meson decay is probably naive. The QSM however makes very simple predictions that we can compare to.

The matrix element for the decay is proportional to the strength of the weak current between the initial and final state of the decaying quark

$$M \propto (\bar{u}, \bar{c}, \bar{t})(1 - \gamma_5) \gamma_\mu \{M\} \begin{pmatrix} d \\ s \\ b \end{pmatrix}$$

$$M = \begin{Bmatrix} c_1 & s_1 c_3 & s_1 s_3 \\ -s_1 c_2 & c_1 c_2 c_3 + s_2 s_3 e^{i\delta} & c_1 c_2 s_3 - s_2 c_3 e^{i\delta} \\ -s_1 s_2 & c_1 s_2 c_3 - c_2 s_3 e^{i\delta} & c_1 s_2 s_3 + c_2 c_3 e^{i\delta} \end{Bmatrix}$$

M is the Kobayashi-Maskawa[2] matrix and describes the mixing of the strong interaction conserved flavor currents into the weak interaction conserved currents which allow the decay of flavor.

The study of B decays should be an important source of information on the matrix elements of {M}. Some of the predictions of this model are that the B lifetime should be

$$\tau_B \sim 4 \times 10^{-15}/(s_2^2 + s_3^2 + 2 s_2 s_3 \cos \delta) \quad (1)$$

$$\text{set } c_i = 1$$

The decay rate ratio of the decays where the b quark goes to the charm quark to states where the b quark goes to the u quark is:

$$\frac{\Gamma(b \to c)}{\Gamma(b \to u)} \sim \frac{|c_1 c_2 s_3 + s_2 c_3 e^{i\delta}|^2}{|s_1 s_2|^2} \quad (2)$$

The expectation is that the $b \to c$ decay modes should predominate over the $b \to u$ decay modes because the former mode is less suppressed in the "Cabibbo" manner.

The QSM also makes predictions for the semi leptonic branching ratios of the B mesons and the number of kaons per B decay. For the semi-leptonic decays the branching ratio is just the fraction of the time the virtual W boson can decay leptonically. This fraction comes from the relative number of quark and lepton pairs available to a virtual W whose energy is that lost in the quark transition $b \to c$ or $b \to u$. Since the u quark is lighter than the c quark the virtual W boson is more energetic in $b \to u$ decays and this increases the W's chance to decay to quark pairs as well as making the final state leptons more energetic. A Monte Carlo calculation based on Sjostrand's Monte Carlo[3] gives

B.R.$(B \to X \ell \nu) = 15\%$ if $b \to u$ predominates

$= 17\%$ if $b \to c$ predominates

The average number of kaons in the final state should equal the average number of strange quarks which occur directly or from the decay of charm quarks at the rate of one strange quark per charm quark. For example, if a b quark decays exclusively to a c quark we get one strange quark from this c quark and about 0.60 strange quarks from the fraction of W^- boson decays that go to $\bar{c}s$. Using this reasoning the pure $b \to c$ decay mode gives 1.60 kaons per B decay while pure $b \to u$ gives 0.90 kaons per B decay.

The experimental procedure has been to measure the prompt lepton and kaon production rates of the events in the 4S region and surrounding continuum and by subtraction estimate the rates for the 4S bump itself. The 4S bump events have a slightly higher multiplicity than the continuum so CLEO has used this fact in defining one cut for the events we analyze. CLEO demands that events have 5 or more charged tracks and visible charged energy, $E_{charged} \geq 3.0$ GeV. These cuts suppress the continuum hadronic events somewhat and remove $\tau^+\tau^-$ pair production.

CLEO has the capability of seeing prompt electrons and muons in hadronic events and can also detect inclusive neutral and charged kaons. CUSB is able to detect prompt electrons in hadronic events.

In Table 1 the sensitivity of the CLEO detector for the various weak decay modes is outlined. Table 2 describes the CLEO analysis criteria for the weak decay signals. Despite the fact that we might expect 2-3 kaons and about 0.6 leptons per $B\bar{B}$ $\Upsilon(4S)$ event these analyses are difficult because the total acceptances for

TABLE 1

Sensitivity of CLEO Detector to

Weak Decay Channels

Signal	p range (model)	p range Observable	Geometrical(X) Reconstruction Efficiency	Model Dependent Momentum Fraction (b → c)
$e^{\pm}$	0.5-2.5GeV/c	$\geq$1.0GeV/c	0.24±.04	0.75±.05
$\mu^{\pm}$	0.5-2.5GeV/c	$\geq$1.3GeV/c	0.59±.06	0.50±.09
$K^{\pm}$	0.1-1.3GeV/c	0.5-1.0GeV/c	0.1±.02	0.37±.04
K^{o}	0.1-1.3GeV/c	>0.3GeV/c	.023 combined	

the various inclusive decays are less than 20 percent. Figure 5 shows a prompt muon event in the CLEO detector with the muon clearly registering beyond the steel shield outside the detector.

TABLE 2

CLEO ANALYSIS CRITERIA

Signal	Criteria
$e^\pm$	Fires Cerenkov counter. Shower energy matches drift chamber momentum to ~60%. Shower distribution OK for $e^\pm$
$\mu^\pm$	Drift chamber track extrapolates to hit in chambers outside of steel shell. Punch through checked on $\Upsilon(1S)$.
$K^\pm$	TOF time matches K hypothesis to 2σ, differs from π or p by 3σ. DE/DX ionization matches K hypothesis σ_{TOF} ~400 psec.
K^o	Vertex K^o > 0.7 cm from event vertex pair mass; δm^2 < .025 GeV^2. Drift chamber resolution ~240 μm.

Figure 6 is the $\pi^+\pi^-$ mass distribution of neutral charged particle pairs which pass all the other K^o criteria including separated vertex and planarity. Figure 7 shows the visible signal seen by CLEO for four weak decay inclusive channels. The visible cross section for prompt electrons seen by CUSB is shown in Figure 8.

The momentum spectrum of the prompt electrons in the 4S region is shown in Figure 9. This is compared to the expected spectrum for the decay $B \to De\nu$ with the D subsequently decaying to leptons. A similar comparison for the CLEO muon spectrum is shown in Figure 10. Both spectra seem to support the expectation that B mesons decay predominantly to charm final states although the muon spectrum gives pause for thought.

The various decay rates are similarly calculated:

$$\text{Rate} = \frac{\Delta\sigma_{vis}/(\epsilon \cdot f(b \to c))}{\sigma\Delta_{4S}}$$

σ_{vis}, σ_{4S} are the respective visible and total cross sections for the signal and the $\Upsilon(4S)$ bump. ϵ is the geometrical acceptance for the visible cross section and $f(b \to c)$ is the fraction of the momentum spectrum of the decay mode visible to the apparatus. f is somewhat mode dependent and has been calculated under the assumption that B's decay exclusively to charm.

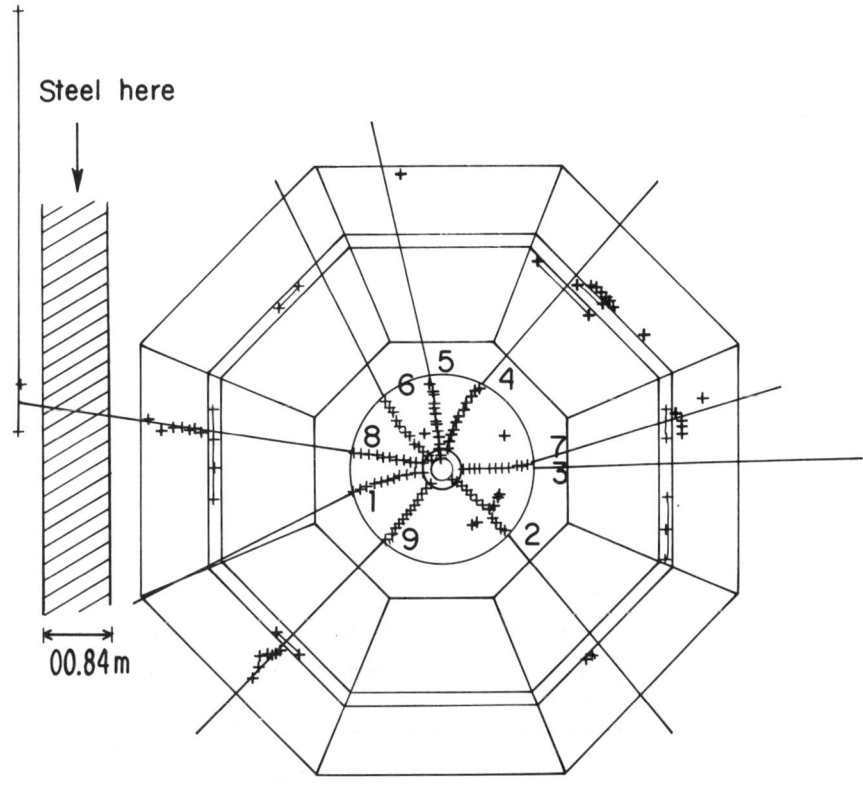

Figure 5. A prompt muon event in the CLEO detector. The steel outer shield of the detector is indicated.

The branching ratios for $B \to Xe^{\pm}\nu$ and $B \to X\mu^{\pm}\nu$ are shown in Table 3. These data have been corrected for a small amount of charm decay to leptons in the final state. The numbers are not inconsistent with the naive Q S M predictions but do tend to be lower. The CLEO number of charged and neutral kaons per B decay in the 4S region and per event in the continuum region are given in Table 4. These numbers support the predominance of charm final states in B decay, as per expectations. Because of the variety of D decays one cannot easily predict the relative rates of neutral and charged kaon production. CLEO finds 1.7 ± 0.5 statistical ± 0.4 systematic kaons per B decay.

CLEO has in the T(4S) events:

    2 e^+e^-
    1 μ e
    0 μ μ

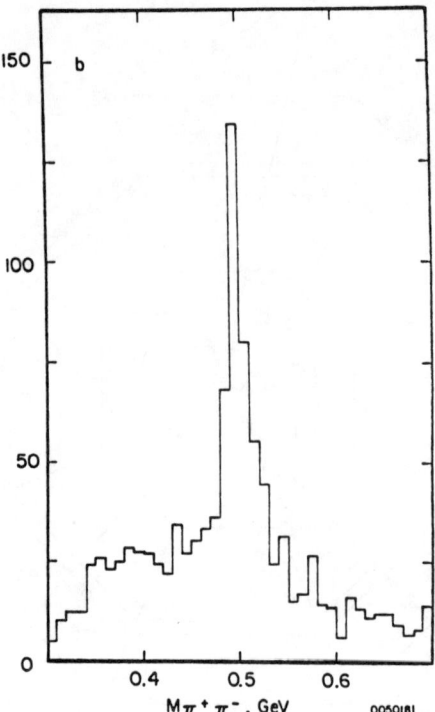

Figure 6. Mass distribution of pion pairs that verticize away from the primary interaction point and satisfy planarity and other cuts for CLEO data.

One of the e^+e^- events is consistent with a ψ decay. In the other event the net momentum of the e^+e^- precludes their both coming from one B meson. We find that the branching ratio for $B \rightarrow X\,e\,e$ is less than 1.3% (at 90% confidence level). Theorists[4] have pointed out that it is possible to have a consistent theory of weak interactions in which the b quark is not part of a quark doublet with the top quark. Rather b is mixed in with the existing currents of weak decay. If this is the case the branching ratio for B decays to neutral currents has a lower bound and

$$\frac{\text{B.R.} \ (B \rightarrow X\ell^+\ell^-)}{\text{B.R.} \ (B \rightarrow X\ell\nu)} > 0.125$$

CLEO's present 90% confidence limit on this ratio is 0.25, about two times higher than the lower bound. This is not quite the ratio of our branching ratios above because the semi leptonic branching ratio must be corrected for the neutral current decays and we also are giving confidence levels rather than best estimates. We are

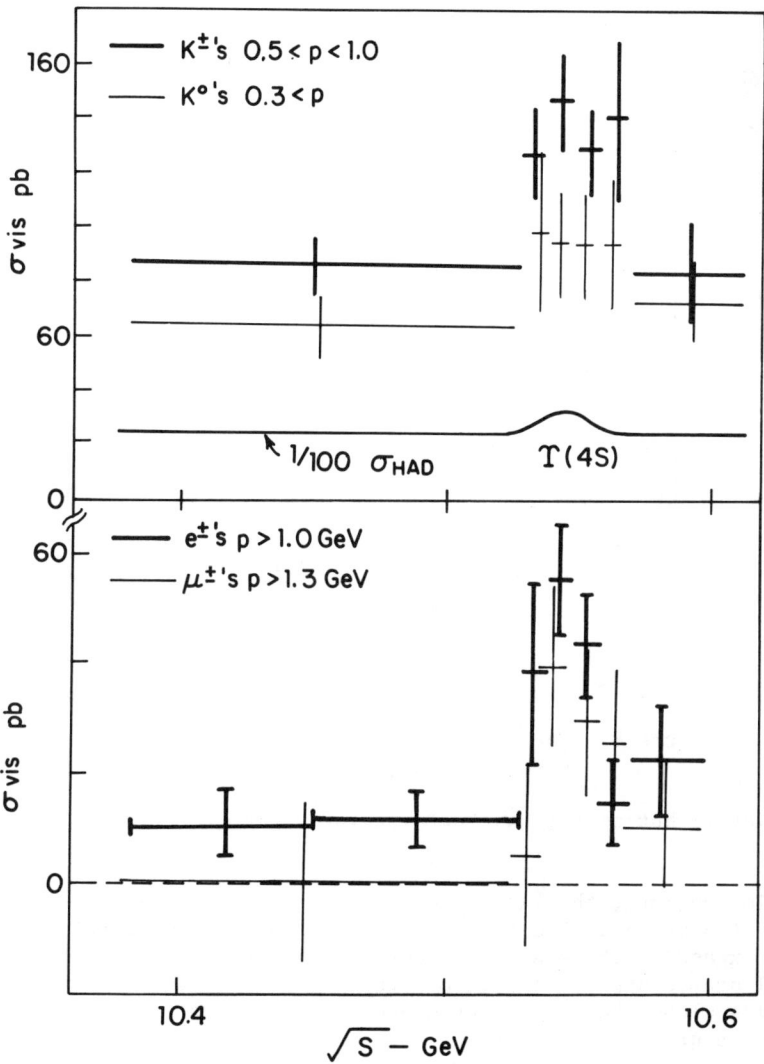

Figure 7. The visible cross sections for the weak decay channels of the CLEO experiment in the T(4S) region.

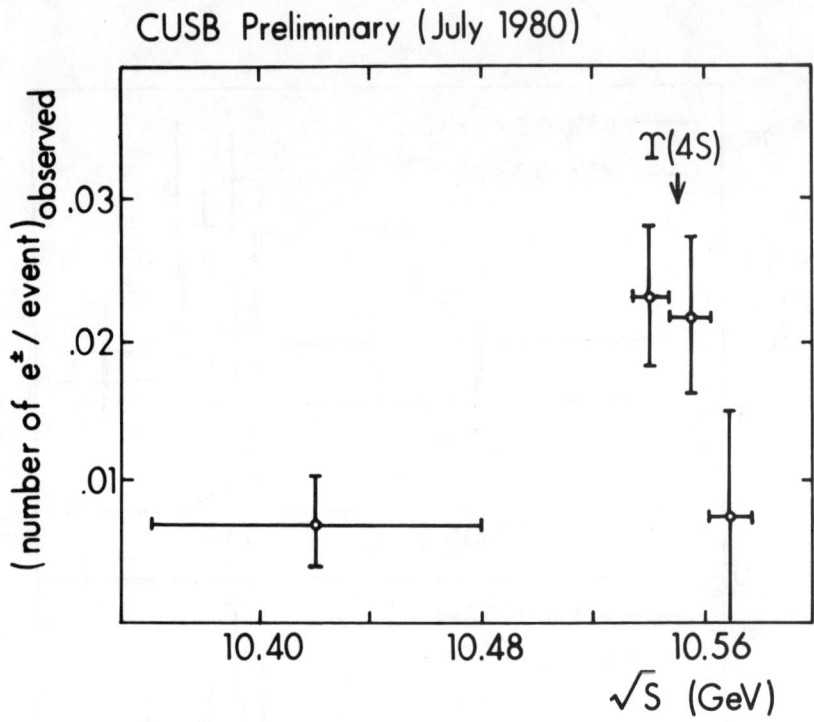

Figure 8. The visible cross section for prompt electrons seen by CUSB.

closing in but as yet have nothing to say about the existence of the t-quark.

To conclude, the experiments at CESR have established the existence of B mesons. The B mesons seem to decay as we might expect from extending the standard theory of the weak decays. Charmed final states appear to predominate in the decays. While we are getting close there is no evidence for or against the existence of the t-quark.

Looking to the future we are beginning to analyze our data to test some of the more exotic weak decay models and to look for constrained $B\bar{B}$ events. We will also increase our $\Upsilon(4S)$ data samples and upgrade the detectors. B decays at CESR should continue to be exciting physics for some time.

443

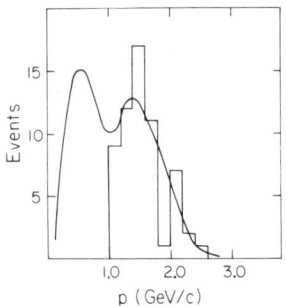

Figure 9. The visible momentum spectrum of the prompt electrons from the $\Upsilon(4S)$ peak for CLEO. The curve is the electron spectrum expected from $B \to De\nu$ and $B \to D^*(2000)e\nu$ and $D \to Xe\nu$ decays. The data are cut at 1.0 GeV.

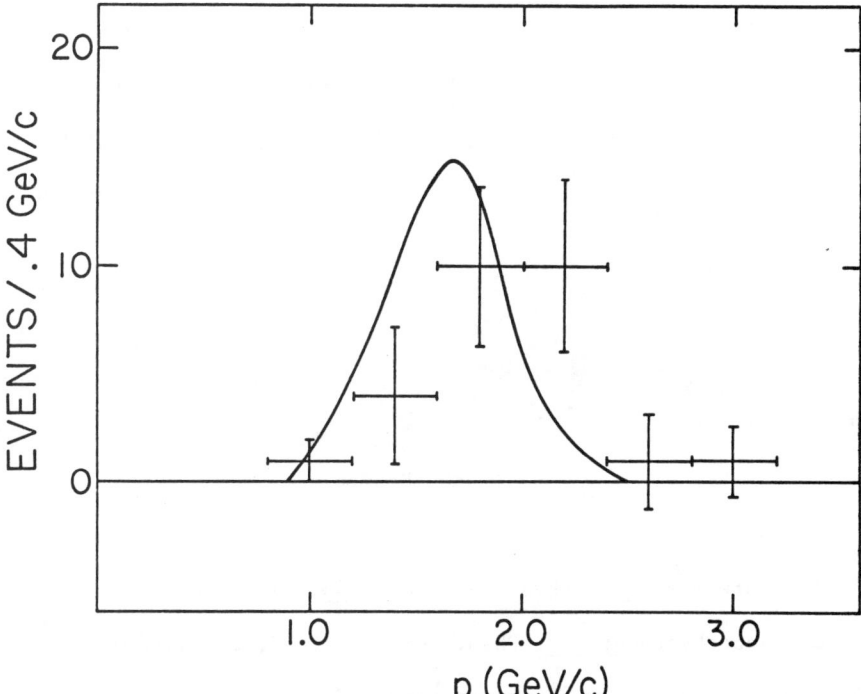

Figure 10. Visible muon-momentum distribution for events from the $\Upsilon(4S)$. The curve is that expected for the decay $B \to XD\mu\nu$ followed by $D \to X'\mu\nu$. The lower momentum cut off of this spectrum is determined by the requirement that muons penetrate the steel shield of the apparatus (CLEO data).

TABLE 3

WEAK DECAY RATES FOR B MESONS

Signal	B.R. ($B \to X\ell\nu$)	
$e^{\pm}$	$13 \pm 3_{stat.}\ 3_{syst.}$ %	CLEO
	$15 \pm 5_{stat.\ +\ syst.}$ %	CUSB
$\mu^{\pm}$	$9.4 \pm 3.6_{syst.\ +\ stat.}$ %	CLEO

TABLE 4

KAON RATES

	Total Acceptance		Kaons/Event	
	Continuum	(4S)	Continuum	(4S)
K^o	.03	.023	$0.9\pm.1\pm.18$	$1.6\pm0.8\pm0.3$
K^+	.04	.037	$0.92\pm.1\pm.18$	$1.8\pm0.6\pm0.4$

$$N_{K/B} = 1.7 \pm 0.5 \pm 0.4$$

REFERENCES

(1) D. Andrews et al., Phys. Rev. Lett. **45**, 219 (1980).

G. Finocchiaro et al., Phys. Rev. Lett. **45**, 223 (1980).

(2) M. Kobayashi and T. Maskawa, Prog. of Theo. Phys. **49**, 652 (1973).

(3) T. J. Sjostrand, LUND preprint LU TP80-3.

(4) G. L. Kane, Michigan Report No. UMHE80-18. This report has a fairly complete list of references.

HEAVY QUARK DECAYS

N. Cabibbo

New York University, New York, NY

We will discuss the decay of heavy hadrons, those with charm and bottom quarks. Starting from the most fundamental, we would like to measure the mixing angles, investigate new tests of QCD, and study the structure of hadrons. In the last couple of years, charm decay has presented us with two small mysteries. The first is in the inclusive kaon decay of the D. The ratio of this inclusive mode for D_0 to the total D^0 decay rate is ~95%. The main decay mode of charm is into strange. $D^0 \to K$ has a strange particle in the final state, so it is O.K., but the same mode for D^+ has a ratio something like 50-80%, substantially lower than we expect. We shall see that this mystery is related to other aspects of the problem. The second mystery, although a little minor, has been written about a lot. It is the comparison between the KK, Kπ and $\pi\pi$ modes of the D^0. On the basis of weak interaction theory and because they are forbidden by U spin (we are using SU(3) here), you would expect the $D^0 \to \pi\pi$ mode and $D^0 \to KK$ mode both to be ~5%. Experimentally you find the KK mode is higher and the $\pi\pi$ lower than this. You can think of a number of explanations: it could be coming from mixing angles; you could try to understand it through penguin diagrams. (If you don't know about them, you can forget about them. For those who do know about them, it is very hard to explain this anomaly with penguins.); or this discrepancy could be an ordinary SU(3) symmetry breaking effect. We have an example of this for instance with $F_K/F_\pi = 1.25$ The D $\to$ K discrepancy could be due to something like the symmetry breaking that causes this anomaly. The SU(3) prediction for the $K_{\mu\nu}$ to $\pi_{\mu\nu}$ ratio is ~60% of the prediction.

We feel some things are very firm; let me talk about them first. One thing we feel we understand very well is semi-leptonic decay. The $\Delta S = -1$ are the predominant decay modes; the D^0 and D^+ $\Delta S = -1$ decay rates are related by isospin. The equality of these two rates means practically an equality between the total rates, apart from $\sin^2\theta$ factors. You also have two other relations predicted by U spin, relating these to the forbidden decays, the $\Delta S = 0$ decays, which have a $1/\tan^2\theta$ factor in them. The other relationship, oddly enough, is between $\Delta S = -1$ F^+ decay and $\Delta S = 0$ D^+ decay. The origin of this equality follows simply from the decay mechanism. Decay can only occur for these two cases through the gluon emission from the quark annihilation diagram. The other processes occur through straight decay. This accounts for the lack of relationship between the last two decay rates and the previous four.

Now we can consider the semi-leptonic decay rate by supposing that the main diagram involves the direct decay of the charm into the strange quark. This has been computed taking into account virtual and real gluon corrections. This yields an estimate for the semi-leptonic decay which is fairly reliable, save the mass of the charmed quark, which is unknown in the calculation. The result

$$\Gamma_{SL} = \frac{G^2 M_c^5}{192\pi^2}\left[\begin{array}{c}\text{Phase Space}\\ \text{Corrections}\end{array}\right]\left\{1 - \frac{2\alpha_s}{3\pi}\left(\pi^2 - \frac{25}{4}\right)f\left(\frac{M_s}{M_c}\right)\right\}\cos^2\alpha, \quad (1)$$

is somewhat like muon decay with an m^5 factor. We have a handle on phase space through the spectrum of the decay electron. The observed spectrum depends on the charmed quark mass. The expression in the brackets of (1) above modifies the phase space factor. This has been plotted on fig. 1. From these curves and the semi-leptonic branching ratios, we can actually calculate the lifetime. The cos α factor, coming from extra mixing angles, is very close to 1, within a few per cent. This comes from the rather well known B lifetime.

So we know the semi-leptonic rates. What can we say about non-leptonic rates? The problem here is divided into two parts. We are very confident about one, computing an effective Lagrangian for the non-leptonic interactions. This, however, has a very large gluon correction. The leading corrections can be easily estimated by renormalization group arguments, and involve the quanitites

$$\left[\alpha_s \ln\frac{M_W^2}{M_c^2}\right]^n \sim O(1) \qquad \text{1 loop (leading log)} \quad (2)$$

$$\alpha_s\left[\alpha_s \ln\frac{M_W^2}{M_c^2}\right]^n \sim O(\alpha_s) \quad \left\{\begin{array}{l}\text{1 loop exact}\\ \text{Gluon-strahlung}\\ \text{2 loop (leading log)}\end{array}\right. \quad (3)$$

All corrections are potentially large and could in effect become the leading corrections. The next order of order α_s is indicated in eq. (3). This is not at all an easy computation, but it has been recently attempted by Altarelli, Maiani, Martinelli and Petrarca. It involves computing corrections up to two loops. As a result we have acquired a clear idea of what the effective Lagrangian for charmed decay looks like.

$$\mathcal{L} = \frac{G}{\sqrt{2}}(f_+ O_+ + f_- O_-) \quad (4)$$

$$O_{\pm} = \tfrac{1}{2}[(\bar{s}c)(\bar{u}d) \pm (\bar{s}d)(\bar{u}c)]$$

O_- dominance $\equiv$ '20' dominance (5)

$\Delta I = \tfrac{1}{2}$ in strange decays;

$$f_+ = \left(\frac{\alpha(M_c^2)}{\alpha(M_w^2)}\right)^{-0.29} \left[1 - 0.41\frac{\alpha(M_c^2) - \alpha(M_w^2)}{\pi}\right] \quad (6)$$

~ 0.6 ;

$$f_- = \left(\frac{\alpha(M_c^2)}{\alpha(M_w^2)}\right)^{0.57} \left[1 + 1.59\frac{\alpha(M_c^2) - \alpha(M_w^2)}{\pi}\right]. \quad (7)$$

~ 2.3, $(\Lambda = 0.25, \alpha(m_c^2) = 0.4,$ [ ] $=$ 2 loop corrections).

The structure involves the product of two operators which are essentially products of Fermi operators with several γ_μ, γ_5 factors. The two loop corrections tend to confirm the full result – they decrease f_+ and increase f_-.

We think this structure for the Lagrangian is firmly established, but what do these operators do to the quarks inside the hadrons? We need to know the structure of the hadrons before we can use this effective Lagrangian. In the case of semi-leptonic decays, the charm decays into other quarks, but this doesn't work here. For example, the rate for the non-leptonic decays should be the same for all the mesons, whereas experimentally it is known that the non-leptonic decay rate of the D^0 is 5 times that of the D^+. Another prediction, this time on an exclusive channel, is $\Gamma(D^0 \to \bar{K}\pi)/\Gamma(D^0 \to K^-\pi^+) = .04$. Experiments yield a ratio of 4. Even if this simple mechanism will not work, maybe another will. The first mechanism of which one thinks of is something which involves the spectator quark, something like fig. 2. This kind of a diagram would be nice because it can occur for D^0 but not for D^+. This could help the situation for the D^0 decays. The main trouble is the pseudoscalar combination decaying into light quarks is forbidden by helicities so if we compute the contribution of this diagram we find it is off by a factor of 100. However, after the experiment indicated this discrepancy, many people, some of them here, proposed that one can really eliminate the helicity problem by having in the final states not only these two light quarks, but also one or more gluons. Then we have the possibility of making up a pseudoscalar state. With gluons in the final state, of course, the two quarks could also form a vector state. Color also helps the situation somewhat. You can try to compute this very naively by having a vector decaying into quarks, in which case the decay rate is

$$\Gamma_{D^0} = \left[\frac{(GM_D^3)^2}{8\pi^2 \alpha^2}\right] \Gamma(\rho \to \mu^+\mu^-) \frac{(f_+ + f_-)^2}{12} F.$$

The vector decay is similar to $\rho \to \mu^+\mu^-$, so the estimate includes that particular width. We have to have an extra factor F here to convert the electromagnetic decay of the ρ to this weak decay. So we put in all the angular factors, and we get the factor F, our fudge factor, to be ~2 in order to explain $\Gamma(D^0)/\Gamma(D^+) \simeq 5$.

There is another mechanism which has been proposed. The mechanism we have just discussed enhances the D^0 decay but does not effect the D^+. This second mechanism tends to decrease the D^+ decay. In the naive mechanism of direct charm decay, you end up with two identical fermions in the final state, namely two $\bar{d}$ states, and thus a Pauli suppression factor. This was pointed out by Guberina, Nussinov, Peccei, Rücke, Jaganathen and Mathur. In this way we can decrease the D^+ decay by a factor of 2. Essentially here we have two operators. One puts 2 fermions in the same state and the other puts them in different states. Table I shows the different results of these two cases due to interference.

<u>TABLE I</u>

	No Interference	Total Interference
$\Gamma_{D^0, NL}$	$(2f_+^2 + f_-^2)\Gamma_{SL}$ $\sim (5.5)\Gamma_{SL}$	$4f_+^2 \Gamma_{SL}$ $\sim (1.9)\Gamma_{SL}$
$(B.R)_{SL}$	$\frac{1}{\frac{1}{2} + 2f_+^2 + f_-^2}$ $\sim 13\%$	$\frac{1}{\frac{1}{2} + 4f_+^2} \sim 26\%$

One easily sees you lose a factor of about 2. So for example, the semi-leptonic branching ratio of the D^+ with no interference you get about 13%. If you have interference then this branching ratio is increased to about 26%. Some of the results in the papers I have quoted above are slightly different from those listed here, because of the differences in the f_+ and f_-, but the computations should be fairly accurate.

Putting all of this together, you obtain a fairly good fit to essentially all experimental data, but this is not surprising. In the beginning we computed a beautiful effective Lagrangian with a simple mechanism, thinking that the measurements would give a test to the QCD corrections. Since we have three different mechanisms to play around with, it is clear that we can fit everything in a natural way, but we are not testing much. There is one factor we didn't mention. Although the allowed decay of the D^+ cannot go by annihilation, the forbidden one can, as shown in fig. 3. The D^+ ratio of $\Delta S = 0$ to $\Delta S = 1$ can be anomalously large. This would be indicated by finding a small number of K mesons experimentally. Now this is a real prediction because very little is known about the D^+ decay experimentally. The prediction is also that F^+ decay by annihilation shown in fig. 4. There are no strange particles in the final state. This is very different from the simple mechanism, which has two strange quarks in the final state. So if the annihilation diagram dominates over the simple diagram you will have no kaons in the final state.

Will things get any better when we move to higher flavors? Presumably, yes. We expect the reduced mass to approach a constant as one of the quark masses becomes very heavy, so the probability of annihilation should also become a constant. The relative probability for decay should increase. If it is 1/5 in the case of charm (17.5 GeV) it would be 2 in the case of the bottom (5.0 GeV) and 20 in the case of the top (17.5 GeV). And so finally for a very heavy quark, the interference should disappear, leaving only quark decays.

In conclusion let me try to give a sort of overview. With charm we do not know what is going on; we have several different mechanisms. The main test would be in F decays. Then we have the b decays of course, the principal problem being measurement of the mixing angles, but the b decay situation should be simpler theoretically. Now that we have information of b decay into charm, we are waiting for the results on b decays into non-charm states, e.g. the $b \to u$ amplitude.

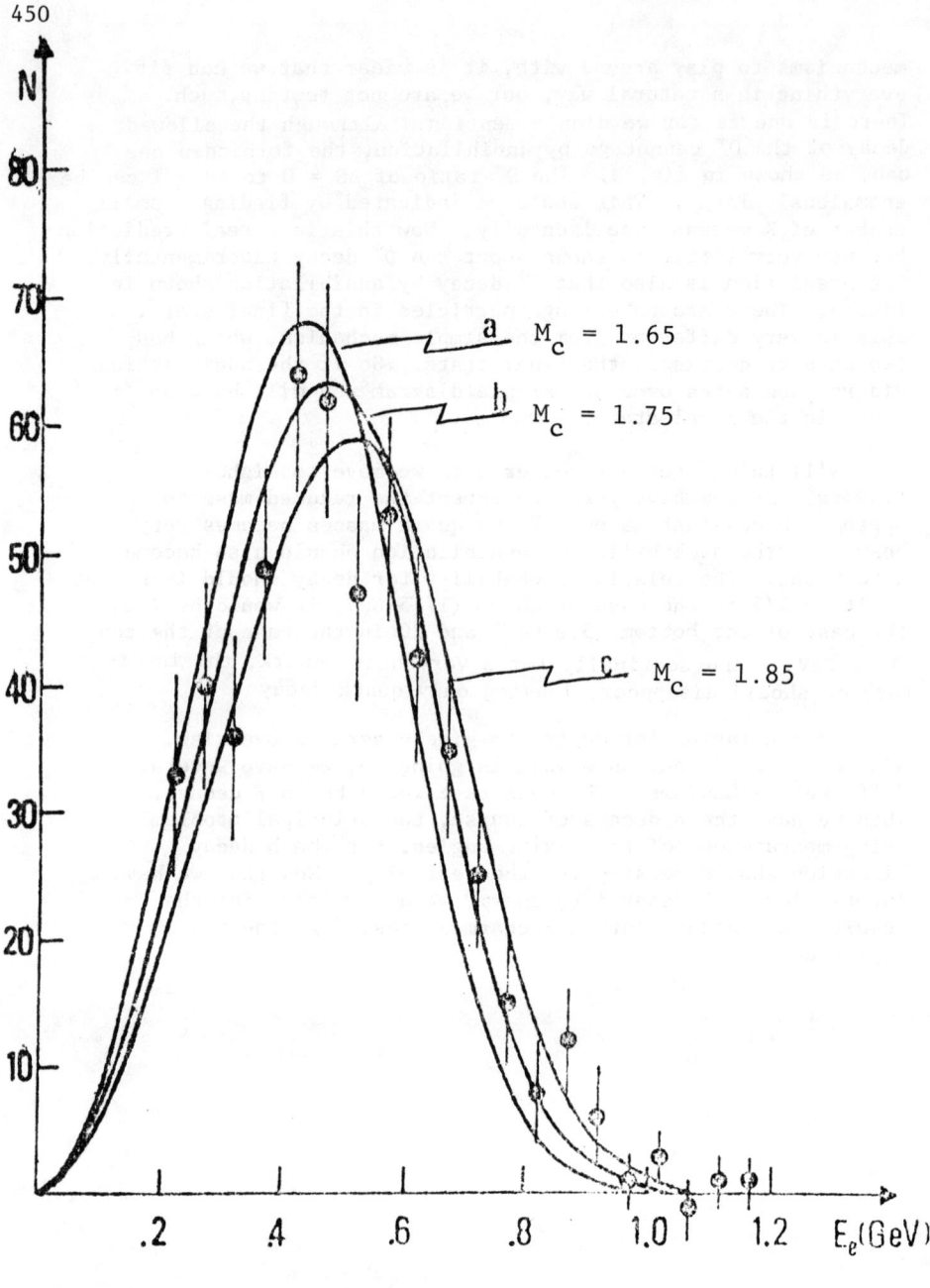

For $M_c = 1.75$, $M_s = .5$
$$\tau_c = \frac{1}{\cos \alpha} \cdot \left[\frac{B_{SL}}{0.10}\right]^{-1} 3 \cdot 10^{-13}$$

Fig. 1 Plot of dependence of semi-leptonic rates on charmed quark mass.

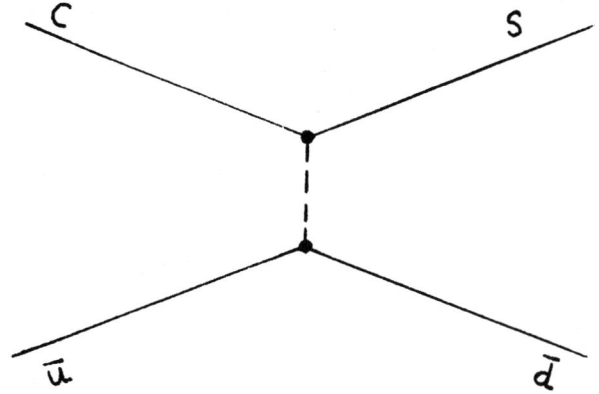

Fig. 2 Diagram for D^o Decay

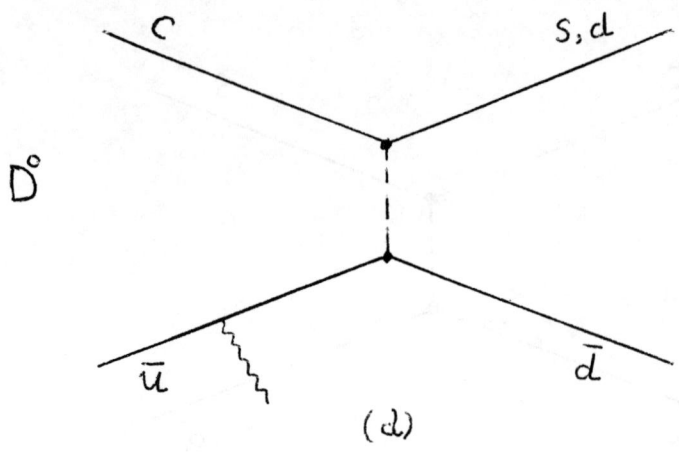

(a)

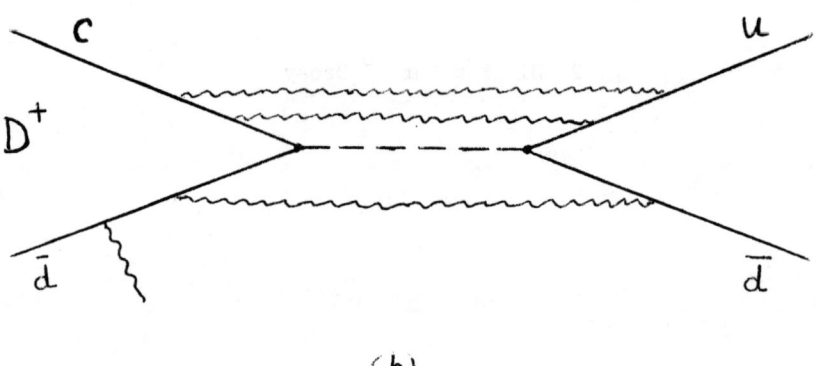

(b)

Fig. 3 a. Gluon corrections for D^o-decay; $(f_+ + f_-)^2 \sim 2.16 - 0.68$ for allowed and forbidden modes.

b. Gluon corrections for D^+-decay; $(f_+ - f_-)^2 \sim 2.16 - 0.68$ for forbidden modes.

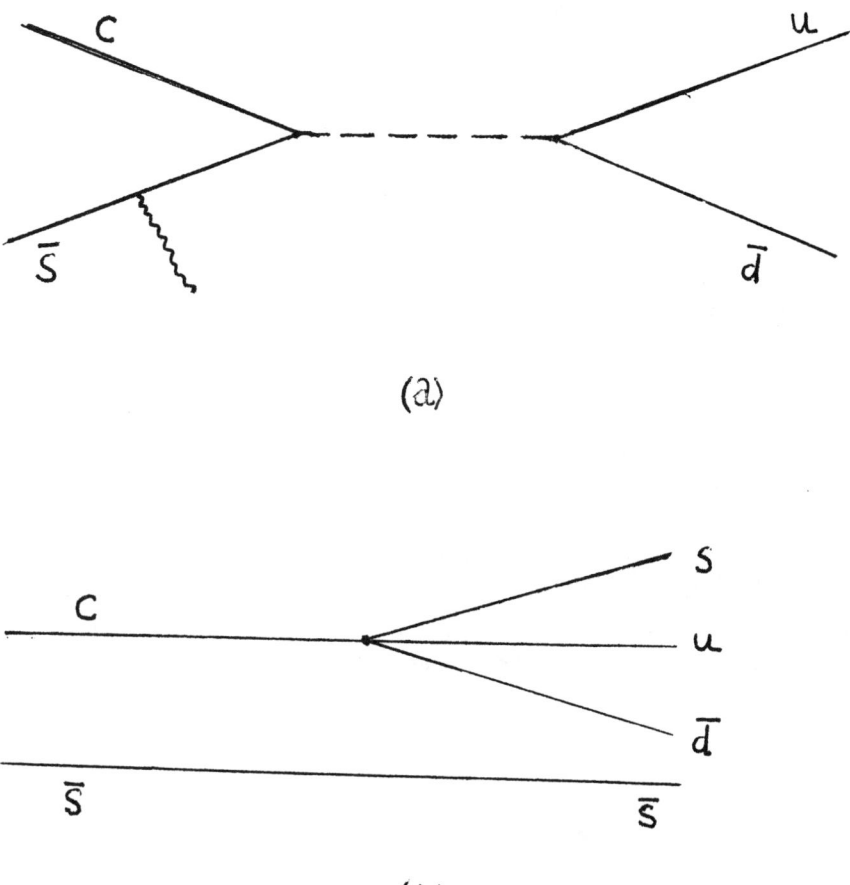

Fig. 4 Diagram for F^+ Decays

DISCUSSION

ROSEN: I wonder if it is a serious problem in the case of the D^+ and the F^+ diagrams that you wrote up before, where the initial quarks annihilate into a W, that you should really have two gluons coming off in order to get a color singlet in the final state.

CABIBBO: One gluon is enough.

ROSEN: No, because the W carries no color.

CABIBBO: The corrections are built into the effective Lagrangian. Your remark is completely right, however, for example the $(f_+ - f_-)$ terms without gluon corrections would be zero. However, if f_+ and f_- are very different, the correction is enormous. You are right, these are mainly carried by diagrams which have a gluon correction.

WANG: Then wouldn't that be another factor of a α down? You have another gluon there.

CABIBBO: Oh, could they make two gluons you mean? Yes.

WANG: So, therefore it is not clear whether this process would be dominated through this mechanism. This is certainly different from the exchange graph; for an exchange graph you need only one gluon in the D^0 decay.

CABIBBO: The exchange is really a sum of many, many terms with any number of gluons exchanged here in any way.

Multi-Generational Models of Quarks and Leptons
Horizontal Symmetries

Kameshwar C. Wali

Physics Department, Syracuse University,
Syracuse, N.Y., 13210

1

One of the deep mysteries of nature - we used to be told - is the e-μ puzzle. If we ignore the fact that the muon is approximately 200 times greater in mass than the electron, the two are identical in all other respects. In the energy regions we are familiar with, both participate only in the weak and electromagnetic interactions. Their weak interactions have identical structures and are governed by a universal strength. Their electromagnetic properties obey the predictions of quantum electrodynamics to a fantastic degree of accuracy. No wonder then, that one thought there must be some deep reason underlying this replication. For years theorists sought an explanation in terms of dynamics - maybe the QED propagator has a 'hidden' second pole. Experimenters thought in terms of more accurate experiments which would reveal differences between them. It has not turned out that way. The puzzle has remained, and as often happens in Particle Physics, one way of solving a puzzle (and thereby diffuse its fundamental significance) is to make it part of a bigger puzzle. The e-μ puzzle recently has become a puzzle involving the 'replication' of families or generations of fundamental fermions (quarks and leptons). Horizontal symmetries deal with this problem, by studying classification schemes, the nature of intra-generational interactions, and their experimental consequences. During the past few years, a considerable amount of work has been done in this area. In what follows, I shall discuss some of the main points which I hope will serve as starting points for more detailed discussions.

2

With the recent discovery of the B-meson and observation of its weak decays[1], one might as well assume the existence of the t-quark. Its actual discovery, in my opinion, is only a matter of time and energy. With this assumption, we have at least three generations or families:

1st Family	2nd Family	3rd Family	
$\begin{bmatrix} u^i \\ d^i \end{bmatrix}_L$	$\begin{bmatrix} c^i \\ s^i \end{bmatrix}_L$	$\begin{bmatrix} t^i \\ b^i \end{bmatrix}_L$	Quarks
d_R^i, u_R^i	s_R^i, c_R^i	b_R^i, t_R^i	

i=1,2,3 color index

$\begin{bmatrix} \nu_e \\ e^- \end{bmatrix}_L$	$\begin{bmatrix} \nu_\mu \\ \mu^- \end{bmatrix}_L$	$\begin{bmatrix} \nu_\tau \\ \tau^- \end{bmatrix}_L$	, Leptons
e_R^-	μ_R^-	τ_R^-	

where the quarks (anti-quarks) belong to a triplet (anti-triplet) representation and the leptons to the singlet representation of the color group $SU(3)_c$. There are fifteen chiral states in each family and the three families have identical [SU(2)xU(1)] quantum numbers. Hence, separately they have the same weak and electromagnetic interactions. Presumably, they all are spin 1/2 fermions; they are, therefore, very unlike the hadronic excited states which exhibit higher spin and higher internal symmetry representations. Consequently, if quarks and leptons are the fundamental set of fermions, all the three families have to be treated as equally fundamental. If they are composites of still elementary objects, the composite model has to explain the existence of such identical families and their exact number. In any composite model with conventional dynamical schemes, there are bound to exist, in addition, higher spin quarks and leptons. Until they are discovered, it is safe to assume that the quarks and leptons we presently think exist, are fundamental.

3

This 'superfluous' replication, so named by Glashow[2], finds no explanation in the framework of the standard theory. Associated with this problem are also the problems of, for example, a) quark masses and mixing angles, b) CP-violation, c) number of flavors and Astrophysics, d) quark-lepton correspondence.

The currently believed quark masses display a bewildering hierarchy. The generalized Cabibbo-type angles and phases which come into play when one goes from the gauge eigenstates to mass eigenstates govern the charm, top, botton decays, CP-violation and give rise to quantitative estimates of the rare decay modes. The number of flavors, especially the number of light neutrinos, has great implications in astrophysics. Thus generally speaking, a whole range of phenomena depend on our understanding of the horizontal or intra-generational interactions. A complete theory should incorporate the generation structure in a non-trivial way, and predict the form of the mass matrices in different charge sectors so that some of the above questions can be answered. Without such a theory, there will be too many arbitrary parameters.

4

A more complete theory may be, as a first step, approached from the point of view of purely electro/weak interactions. In such an approach,

$$[SU(2) \times U(1)] \longrightarrow G_W \supset SU(2) \times U(1) \times G_H,$$

where G_H describes 'horizontal' or intra-generational interactions. G_H can be chosen to be discrete or continuous. Discrete symmetries[3] reduce the number of parameters, mainly in the Yukawa sector and restrict the form of the mass matrices. However, choice of discrete symmetries is itself quite arbitrary and contrary to the spirit of the gauge era. Consequently, various gauge groups have been proposed[4] in the literature. They all have varying degrees of success to their credit.

However, the main drawback of this approach based on the considerations of purely electroweak interactions is that <u>the number of generations</u> is not predictable. Also there is considerable arbitrariness in the choice of the Higgs system - one or several multiplets? What is the nature of the Higgs potential which gives the desired hierarchy in the vacuum expectation values at its absolute minimum, etc.? Consequently one has at ones disposal quite a number of free parameters. It is not difficult to accommodate (or even fit) all the relevant experimental data.

5

This brings us to Grand Unified Theories, which alone have the potential of predicting the number of generations and assigning quantum numbers which distinguish the different generations. In this respect one should note that the currently popular Grand Unified Theories based on [5]SU(5), [6]SO(10) or the exceptional group [7]E_6 are all single generation theories. They are, therefore, exact analogues of the standard theory for electro/weak interactions, but include strong interactions. The number of generations is still a mystery within the framework of such theories.

The minimal theory is the theory based on SU(5). The 15 chiral states of each family referred to earlier fit nicely into the anomaly free combination of representation 5*+10. The theory is asymptotically free, explains charge quantization, gives $\sin^2\theta_w \sim .2$ and yields an acceptable value for the b-quark mass.[8] However, in spite of these nice features, the theory is incomplete, since it is a single generation scheme. The mass hierarchies between different generations, the Cabibbo-type mixing angles between particles of different generations, CP-violating effects and so on are unexplained. The theory contains 23-30 parameters to accommodate three generations. Clearly this cannot be a fundamental theory. A Grand Unified Theory must account for the generation structure and must include intra-generations interactions.

Multi-generational Grand Unification schemes can be classified into two main categories:

1) those in which the conventional Higgs mechanism is used to break the symmetry

2) those in which dynamical symmetry breaking is invoked, necessitating the introduction of Hypercolor or Technicolor.

In both the classes of theories, one makes certain assumptions which have become, so to say, rules of the game. They are

i) the Grand unification group G is simple, or semi-simple. If semi-simple, a discrete symmetry is assumed to ensure a single coupling constant.

ii) No bizarre fermions

With respect to $SU(3)_{color}$, the fermionic representations include only color singlets (leptons), color-triplets (quarks), and color-anti-triplets (anti-quarks).

iii) renormalizability, which means anomaly free. If

$f_L \equiv$ set of all left-handed fermions

$= \Sigma n_r \underline{r}$,

where n_r are integers and $\underline{r}$ denote representations of G,

$\Sigma n_r A_r = 0$,

where A_r is the anomaly associated with the representation $\underline{r}$.

(iv) the representation is complex

(v) no replication

$n_r = 0, 1$.

(vi) the number of generations $N_g \geq 3$.

(vii) the theory be asymptotically free.

Is there a solution that satisfies all these criteria? The answer is

Simple Groups

Unitary Groups	SU(N)	NO
Orthogonal Groups	SO(4N+1)	NO
[9]Exceptional Group	E_8	?

If we relax one or more assumptions, solutions are possible. For example [10]SU(11) provides a solution with exactly three generations, if we take the following anomaly free combination of spinor representations

$$[4]+[8]+[9]+[10]$$

of SU(11) to represent the fundamental set of fermions. There are altogether 561 fermions in the above representation, of which only 45 survive, giving the three families in the low energy region. The others can be made to acquire superheavy masses. The model, however, does not satisfy the criterion of asymptotic freedom. Likewise if we relax the requirement (v) and allow a replication of representations in the choice of anomaly free combination, one can obtain a model based on [11]SU(9) with three generations, whose only virtue is that it contains a fewer number of fermions to start with than the SU(11) model. However, if we take this approach, the number of generations is not really predicted; solutions with any number of generations can be found. Consequently the model is not very attractive.

Next, if we consider the technicolor alternative, we are required to assume that

$$G \supset [SU(3)_c \times SU(2) \times U(1)] \times G_{TC},$$

where G_{TC}, the group that describes technicolor must be bigger than $SU(3)_c$ in order to provide superstrong forces which form condensates in the TeV region. If $G_{TC} \equiv SU(n)$, then $n \geq 4$. With this proviso, if we examine simple groups, once again we face difficulties. There is no solution for [12] $SU(N)$ groups for any N. In the case of the orthogonal of the class [13] $O(4n+2)$, $SO(18)$ provides a solution with three generations, but the technicolor group is Sp_4 which is only as big as $SU(3)$. Technicolor forces then become strong only in the GeV region. The smallest representation of the next possible candidate, $SO(22)$, contains 1024 fermions! With this huge number of fermions, the asymptotic freedom is lost.

Thus simple groups do not appear to provide a satisfactory answer for a multi-generational, Grand Unified Theory with or without technicolor. The next simplest possibility is to consider semi-simple groups

$$G \equiv G_I \times G_{II},$$

with a discrete symmetry $I \leftrightarrow II$, to ensure a single gauge coupling constant instead of two, as would be the situation in the general case. In the class of models based on $SU(N) \times SU(N)$, $SU(5) \times SU(5)$ is the minimal multi-generational model[14] and this has five generations. With the assumptions listed before, one can restrict such models[15] to $5 \leq N < 10$. Such models do have some attractive features, but very few details have been explored as yet to come to any definitive conclusion. Also the incorporation of the technicolor alternative is not possible in such models.

6

Finally, recently a very promising and interesting model[16] based on the $SO(10)_V \times SO(10)_H$ group structure has emerged. The $SO(10)_V$ describes a single family and hence has all the nice features associated with SO(10). The other SO(10) includes technicolor, extended technicolor and horizontal interactions and gives rise to a multi-generational structure for the fermions which are chosen to belong to an irreducible representation

$$(\underline{16},\underline{10}) + (\underline{10},\underline{16})$$

of SO(10)xSO(10). At present there appear to be two possibilities: a) the second SO(10) is broken down to SO(7)xSO(3) and the groups SO(7) and SO(3) are identified to be, respectively, the technicolor and horizontal symmetry groups; then the number of generations (i.e. SO(7) technicolor singlets) is exactly 3.
b) the second SO(10) is broken down to $SO(6) \times SO(4) \equiv SU(4) \times SU(2) \times xSU(2)$. SU(4) can then be identified as the technicolor group, with SU(2)xSU(2) providing the horizontal symmetry group. This breakdown predicts 4 generations of the $\underline{16}$-plet of SO(10) and has a nice symmetry between color (Pati-Salam)-technicolor at medium energies. In incorporating technicolor, it automatically brings in the so-called 'extended technicolor' interactions or 'sideways interactions' which provide a mechanism for generating masses for the gauge bosons and the ordinary fermions. It appears to be a very tight scheme, and although details have yet to be worked out, it has all the earmarks of a realistic, multi-generational flavor-color-technicolor unification scheme.

Summarizing, the repeated occurrence of generations or

families is a mystery which has to be reckoned with in order to have more complete unified theories. Intragenerational interactions govern a wide range of phenomena and their specification and understanding is absolutely necessary for a quantitative understanding of the 'New Physics'. They also contain the clues for such long standing problems as CP-violation, precise magnitudes of the rare decay modes of some old and new particles.

Acknowledgment

This work was supported in part by the U.S. Department of Energy, under contract no. DE-AS02-65ER03533.

References and Footnotes

1. See the talk by Loomis in these proceedings.
2. S.L. Glashow, The Future of Elementary Particle Physics, HUTP-79/A059 (Papers based on a lecture series given at the 1979 Cargèse Summer Institute).
3. A fairly complete list of references to be found in reference of A. Davidson and K.C. Wali, Phys. Rev. $\underline{D21}$, 787 (1980). See also H. Harari, SLAC Report No. SLAC-PUB-2363, 1979.
4. S. Barr and A. Zee, Phys. Rev. $\underline{D17}$, 1854 (1978); F. Wilczek and A. Zee, Phys. Rev. Lett. $\underline{42}$, 421 (1979); C.L. Ong, Phys. Rev. $\underline{D19}$, 2738 (1979); H. Harari, in Proceedings of the XIX International Conference on High Energy Physics, Tokyo, 1978, edited by S. Homma, M. Kawaguchi, and H. Miyazawa (Phys. Soc. of Japan, Topkyo, 1979); T. Maehara and T. Yanagida, Prog. Theor

Phys. 60, 822 (1978); 61, 1434 (1979); A. Davidson, M. Koca and K.C. Wali, Phys. Rev. Lett. 43, 92 (1979); Phys. Rev. D20, 1195 (1979); A. Davidson, M. Koca and K.C. Wali, Phys. Lett. 86B, 47 (1979); A. Davidson and K.C. Wali, Phys. Rev. D21, 787 (1980). In addition there are left-right symmetric models, see for instance G. Senjanovic and R. Mohapatra, Phys. Rev. D12, 1502 (1975). For a reivew of the CP-violation problem in gauge theories see R. Mohapatra, Proceedings of the 19th International Conference on HEP, Tokyo, 1978, p.604. See also H. Fritzsch, ibid. p.593.

5. H. Georgi and S.L. Glashow, Phys. Rev. Lett. 32, 438 (1974).
6. H. Georgi, in Particles and Fields (American Institute of Physics, New York, 1975), p.575; H. Fritzsch and P. Minkowski, Ann. Phys. (N.Y.) 93, 193 (1975).
7. See F. Gürsey, Second Johns Hopkins Workshop on current problems in High Energy Physics, p.3.
8. Such deductions also depend upon the use of Renormalization Group equations.
9. See. I. Bars and M. Gunaydin, Phys. Rev. Lett. 45, 859 (1980).
10. H. Georgi, Nucl. Phys. B156, (1979) 126.
11. P. Frampton and S. Nandi, Phys. Rev. Lett. 43, 1460 (1979).
12. P. Frampton, Phys. Rev. Lett. 43, 1912 (1979).
13. M. Gell-Mann, P. Ramond and R. Slansky, in Supergravity, edited by P. Van Nieuwenhuizen and D.Z. Freedman (North-Holland, Amsterdam, 1979).
14. A. Davidson, Phys. Lett. 90B, 87 (1980).

15. A. Davidson and K.C. Wali, Multi-generation SU(N)xSU(N) flavor unifying schemes, Syracuse University Report SU-4217-170, Phys. Rev. (to be published).
16. A. Davidson, K.C. Wali and P.D. Mannheim, Phys. Rev. Lett. <u>45</u>, 1135 (1980).

Calculable Masses in GUTS: An E_6 Example

P. RAMOND
Physics Department, University of Florida,
Gainesville, Florida 32611

ABSTRACT

The importance of a cubic Higgs coupling for obtaining calculable masses in Grand Unified Theories is emphasized. An illustrative E_6 example is presented where some of the masses are radiatively generated.

INTRODUCTION

The formulation of Grand Unified Theories[1,2] has now reached a high level of sophistication. We have minimal enlargements of the low energy group $SU_2 \times U_1 \times SU_3^c$ such as SU_5,[2] SO_{10},[3] and E_6.[4] The simplest such model based on SU_5 has yielded quantitative successes[5] [$\sin^2\theta_w$, $\frac{m_b}{m_\tau}$] as well as, remarkably enough, testable predictions[6] [proton decay rate]. Its generalizations to SO_{10} and E_6 have led to a revival of theoretical interest in massive neutrinos and neutrino oscillations[7]. Yet these models do not explain the bureaucracy of nature, apparent in the triplication of families (asuming the existence of the t-quark). By enlarging these gauge groups to say SU_8,[8] $SO_{18,22}$,[9] E_8[10] respectively, one can incorporate the number of families and introduce a family group. However an $SU(3)$ family group brings anomalies which must be dealt with either by a judicious choice of representations or by straightforward doubling. One exceedingly exciting possibility has been envisaged by Gaillard, Ellis and Zumino (GEZ) who present an SU_8 model[11] extracted from $N = 8$ supergravity with the anomaly problem expurgated by dogma[12]. With this notable exception it is fair to say that we have no fundamental understanding of the fermion multiplicity. In the following we ask a humbler question: how can we understand the actual values of the fermion masses and mixing angles? While we do not solve this problem it is appropriate in this workshop to describe some attempt in this direction.

The known fermion spectrum is very rich -- yet no pattern of masses has yet proven successful as evidenced by our inability to predict the mass of the t-quark. We have the leptons e(.5), μ(105), τ(1780), the quarks u(4.5), d(7.5), s(150), c(1200), b(4400), t(>17000) and the neutrinos ν_e(<40ev), ν_μ(<500Kev), ν_τ(<250Mev), where their masses appear in Mev's, unless otherwise indicated. In each charged sector we are struck by the appearance of very small interfamily mass ratios $\frac{e}{\mu}$, $\frac{d}{b}$, $\frac{u}{t}$, etc... and in general slightly larger intrafamily ones. Thus it would seem reasonable to expect that some mass ratios appear as radiative correction in a Grand Unified Theory of all fermions. For this reason it would be desirable to devise some restrictions on model building based on such considerations. For instance an ideal scenario would be to have "primary" (i.e. uncalculable) masses for the τ-family only, calculable one loop level μ-family mass, and calculable two-loop level e-family masses. No such model has yet been realized,[13] but it is a worthy aim.

In a renormalizable field theory, a calculable[14] mass will be generated in the following way: call f a fermion and h a Higgs; then the renormalizable Yukawa coupling will be of the form ffh and will yield uncalculable masses when evaluated in a non-trivial vacuum. Calculable masses will be generated by couplings in the effective Lagrangian of the form ffh^2, ffh^3, etc... if the channels explored by h^2, h^3, etc... are not already present in h. In Grand Unified Theories where we have in general $\Delta I_w = 0$, $\Delta I_w = 1/2$, $\Delta I_w = 1$ masses these calculable radiative corrections can lead to unexpected surprises. For instance, consider a theory with two Higgs h and H, both with Yukawa couplings. Assume that H acquires a $\Delta I_w = 0$ value and h a $\Delta I_w = 1/2$ value. Then the radiative corrections will generate - $\Delta I_w = 0$ masses coming from Hh, H^2h, etc... (and also hh if there are several h), $-\Delta I_w = 1/2$ masses coming from Hh, H^2h, etc... and $\Delta I_w = 1$ masses coming from hh,.... The relevant question is the strength of such contribution relative to the primary renormalizable couplings. However in view of the smallness of many of the fermion masses as compared with the mass of the W boson we expect such contributions to play a significant role. The general contributions coming from one loop diagrams are of the form

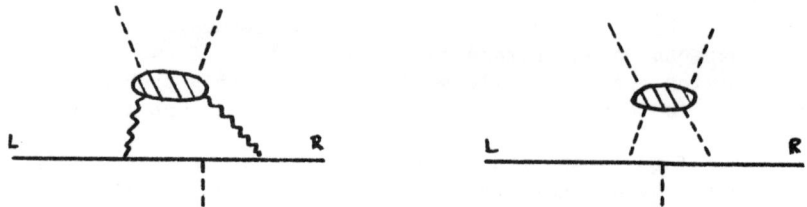

where the upper blobs are tree diagrams - a dashed (wavy) line represents a Higgs (gauge) particle; a solid line is a fermion; we also have diagrams with any odd number of Higgs lines stemming from the fermion line. The contributions to ffh^2 come at this level from taking the upper blobs to be

The first one does not exist but the second one does if there is a fundamental Higgs cubic coupling. The contributions to ffh^3 come when the blobs are

 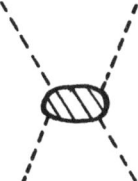

and both correspond to fundamental couplings in the Lagrangian. Now at the two-loop level we can have

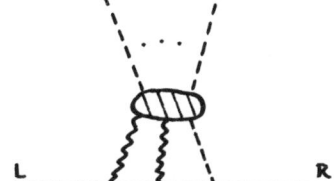 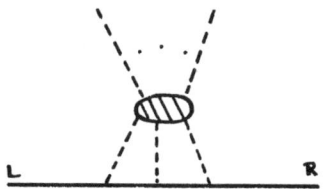

and such diagrams will contribute to ffh^2 if there exists in the theory a cubic Higgs coupling which enables the expansion

 = 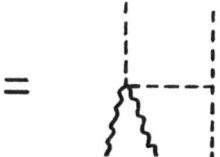 +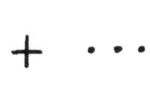

This corresponds to the Witten diagram[15]. This short survey of radiative corrections emphasizes the importance of Higgs cubic couplings. Now the question is to find models where such couplings are naturally found.

As an illustration of the above remarks, consider a model based on the group E_6.[16] The left-handed fermions are assigned to the fundamental 27_f, and the Higgs are assigned to the 27, and to the 78. For simplicity consider a model with one family of fermions, three 27_H of Higgs and one 78 Higgs. The primary couplings are of the form $27_f 27_f 27_H$. Such a coupling is invariant under a global phase symmetry X with value 1 for 27_f and -2 for 27_H. When the 27_H acquire vev's this phase symmetry will be broken, thereby creating a Goldstone-Nambu boson unless the Higgs potential V breaks it explicitely. The most general terms in V are $27 \cdot \overline{27}$, $78 \cdot 78$, $27 \cdot 27 \cdot 27$, $(27 \cdot 27 \cdot 27)_S$, $(27 \cdot 27 \cdot 27)_A$, $(27)^3 \cdot 78$, $(78)^4$, $(78)^4 \cdot 3$. Of these terms only the cubic term 27^3 and the quartic term $(27)^3 \cdot 78$ can give mass to the X Goldstone-Nambu boson. The quartic term can be made naturally absent by the discrete symmetry $78 \to -78$. Thus we have a model with a natural cubic coupling which is needed in order to avoid a massless particle. This cubic term breaks X down to a discrete symmetry mod. 6. This symmetry in turn allows the calculable couplings $27_f 27_f \overline{27}_H \overline{27}_H$ and $27_f 27_f \overline{27}_H \overline{27}_H 27_H$. Do these necessarily correspond to calculable channels not explored in the primary couplings? The group theory tells us that

$$27 \times 27 = (\overline{27} + \overline{351}')_S + (\overline{351})_A$$

$27 \times 27 \times \overline{27} = \overline{27} + 2(27) + \overline{351}' + 2(\overline{351}) + 2(\overline{1728}) + \overline{7371} + \overline{7722}$.
Thus we note that starting from only 27_H's we excite the 27, 351' and 351 (if several families are present) channels by calculable amounts. This feature of E_6 is a natural consequence of the cubic coupling. The content of each 27 is

$$27 = 16 + 10 + 1 \quad \text{under } SO_{10}$$
$$= (\overline{5} + 10 + 1) + (5 + \overline{5}) + 1 \quad \text{under } SU_5.$$

The 27_H can have the following vev's:

$$\Delta I_w = 0 \;:\; \begin{cases} 1 \text{ of } SO_{10} \\ 16 \text{ of } SO_{10}, \text{ but } 1 \text{ of } SU_5 \end{cases}$$

$$\Delta I_w = 1/2 : \begin{cases} 16 \text{ of } SO_{10}, \; \overline{5} \text{ of } SU_5. \\ 10 \text{ of } SO_{10}, \; \overline{5} \text{ of } SU_5. \\ 10 \text{ of } SO_{10}, \; 5 \text{ of } SU_5. \end{cases}$$

Now suppose that the Higgs acquire the following vev's:

$\Delta I_w = 0$: $<27_H> \sim 1$ of SO_{10}

$<27'_H> \sim 1$ of SU_5, but 16 of SO_{10}.

$\Delta I_w = 1/2$: $\langle 27_H'' \rangle \sim 5$ of SU_5.

Each E_6 fermion family consists of one charge 2/3 quark, two charge -1/3 quarks, two charge -1 leptons and five neutral leptons. When 27_H acquires its vev along the SO_{10} singlet the fermions that belong to the 10 of SO_{10} all get a $\Delta I_w = 0$ mass from the primary coupling $27_f 27_f 27_H$. This leaves 17 massless degrees of freedom. But the non-renormalizable induced couplings give mass to one additional neutral lepton (the 351′ contribution) which is the SO_{10} singlet in 27. Such a term is induced at the one loop level by

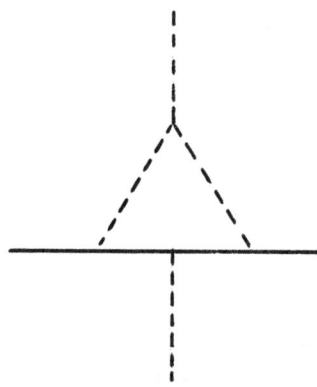

and at the two loop level by the Witten[15] diagram.

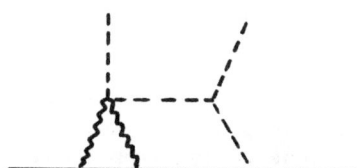

This leaves us with 16 massless degrees of freedom, corresponding to the 16 of SO_{10}. At the same time, the 27_H vev breaks E_6 down to SO_{10}.

In order to break SO_{10} down to SU_5 we need to have an extra 27 with $\Delta I_w = 0$ vev. So we now consider the effect of the $27_H'$ with vev along the SU_5 singlet direction. Its effect at the tree level is to mix the hitherto massless charged fermions with the massive ones without altering the number of massless particles. However the radiative contribution $16_f 16_f 16_H 16_H$ now gives mass to the extra neutral fermion which was in the 16_f. The mechanism is the same as that just described. The remarkable fact is that we have, at the cost of two 27_H been able to reduce the E_6 model to an effective SU_5 model with $\bar{5} + 10$ content. This can even be done for three families of fermion 27's and two 27_H by having primary couplings only along $27_e \cdot 27_\mu$ and $27_\tau \cdot 27_\tau$ so that the family mass matrix be non-singular. This takes care of all the $\Delta I_w = 0$ masses. The 78 of Higgs is assumed to break SU_5 down to $SU_2^w \times U_1 \times SU_3^c$.

Now suppose that the only $\Delta I_w = 1/2$ vev comes from another $27''_H$ with vev along the 5 of SU_5. The effect of the primary coupling $27_f 27_f 27''_H$ is to give a mass to the charge 2/3 quark, leaving one charge -1/3 and one lepton massless. However the radiative corrections can open up the 5 channel necessary to give these a mass, by means of an interplay between the

$\Delta I_w = 0$ and $\Delta I_w = 1/2$ channels coming from $27_f 27_f \overline{27''_H} 27_H 27'_H$, $27_f 27_f \overline{27''_H} 27_H$, $27_f 27_f \overline{27''_H} 27'_H$, $27_f 27_f \overline{27''_H} 27_H 27_H$, $27_f 27_f \overline{27''_H} 27'_H 27'_H$.
For example the diagram

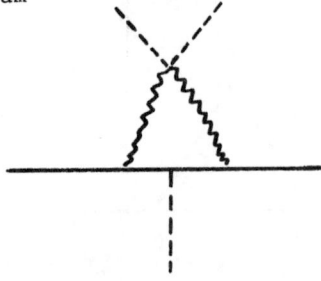

gives a one loop contribution. For instance if we apply this scenario to the τ-family, we will obtain a calculable b-quark mass of the form

$$\frac{m_b}{m_W} \sim \frac{\alpha}{\pi} \times \text{dimensionless number.}$$

The estimate of such diagrams in unified theories is made difficult by the gauge hierarchy problem. However we feel that the presence of such corrections coming from the interplay of the $\Delta I_w = 0$ and $\Delta I_w = 1/2$ sectors is exciting and comes free of charge in this model[13]. Perhaps this is the way in which particles in the lighter families acquire masses.

To summarize this embryonic model, we have shown that the E_6 model with a modest expenditure in Higgs particles can be made to look like the SU_5 model. Nevertheless, it does not solve the family problem. It would be desirable to introduce a family group in such a way that the Higgs cubic couplings are still present. At this stage we just remark that if the three fermion families formed a triplet of SU_3^f (f=family), then in E_6 we would need Higgs that transform as (27,6). It is remarkable that the $\underline{6}$ of SU_3 obeys our constraint, namely the symmetric cubic coupling $\underline{6}.\underline{6}.\underline{6}$ contains one SU_3^f singlet! Of course the problem with a family SU_3 is that it brings anomalies, and we do not know how to surmount this difficulty except by introducing new fermions.

ACKNOWLEDGEMENTS

I would like to thank my collaborator M. Bowick. Also at this very informative workshop, I learned from Prof. F. Gursey that he and Prof. M. Serdaroglu have also noticed the natural presence of the Witten diagram in the E_6 model. I also thank Prof. Curtright for useful discussions.

REFERENCES AND FOOTNOTES

1. J. C. Pati and A. Salam, Phys. Rev. $\underline{D8}$ (1973) 1240; $\underline{D10}$ (1974) 275. See the recent review of P. Langacker, SLAC-PUB 2544, to be published in Physics Reports. <u>All</u> references not mentioned in the following are contained in this review.
2. H. Georgi and S. L. Glashow, Phys. Rev. Lett. 32 (1974) 438.
3. H. Fritzsch, P. Minkowski, Ann. of Phys. 93 (1975) 193; H. Georgi, Particles and Fields, 1974 (APS/DPF Williamsburg) ed. C. E. Carlson (AIP NEW YORK, 1975), p. 575.
4. F. Gursey, P. Ramond, P. Sikivie, Phys. Lett. $\underline{B60}$ (1975) 177.
5. A. J. Buras, J. Ellis, M. K. Gaillard and D. V. Nanopoulos, Nucl. Phys. $\underline{B135}$ (1978) 66.
6. T. J. Goldman, D. A. Ross, Phys. Letters $\underline{84B}$ (1979) 208; W. J. Marciano, Phys. Rev. $\underline{D20}$ (1979) 274.
7. S. M. Bilenky and B. Pontecorvo, Physics Reports 41C (1978) 225.
8. See for instance Th. L. Curtright and P. G. O. Freund in "Supergravity", ed. by P. Van Nieuwenhuizen and D. Z. Freedman (North-Holland, 1979, Amsterdam).
9. M. Gell-Mann, P. Ramond, R. Slansky in "Supergravity", op. cit.
10. F. Gursey, private communication.
11. J. Ellis, M. K. Gaillard, B. Zumino, Phys. Lett. $\underline{94B}$ (1980) 343.
12. Veltman's theorem, unpublished.
13. R. Barbieri and D. V. Nanopoulos, CERN Preprint TH 2870 (1980).
14. See J. C. Taylor, "Gauge Theories of Weak Interactions", Cambridge University Press, Cambridge 1976, chapter 15 and references contained therein.
15. E. Witten, Phys. Letters $\underline{91B}$ (1980) 81.
16. This work has been done with M.Bowick and will be presented in a separate publication in more detail.
17. F. Gursey, Invited Talk at the First Workshop on Grand Unification, Durham, April 1980. F. Gursey and M. Serdaroglu, Yale preprint 1980.

DISCUSSION

VELTMAN: You seem to rely on a 2 loop calculation of Witten et al. who use that to make an estimate of certian masses. Now this two-loop thing is generally wrong if your vacuum is in the context of an unnatural theory, that is, a theory which has two length scales and one made to hold them together.

RAMOND: I am aware of that, but I do not know what to do about it except the näive evaluation.

VELTMAN: Use the Veltman theorem.

RAMOND: Well, I suppose that is always a way out. I would rather think that if you can feed back a $\Delta I_w = \frac{1}{2}$ mass to the lower sector, the gauge hierarchy might just tell you that you cannot feed too much of it back, otherwise the vector bosons would know about it and become fat on it as well. So I would think the gauge hierarchy would just forbid you from having too large a $\Delta I_w = \frac{1}{2}$ feedback but I don't see how it will force you to be smaller than the näive expectation.

EXTENDED UNIFICATION MODELS

A. Zee
Univ. of Washington, Seattle, WA

I am afraid that I have to disappoint the audience, I've just decided not to talk about SO(big) and I also will not talk about SO(big) x SO(big). Now recently I decided that we already have reached a stage when we should think about general structural properties of grand unified theories rather than specific models. So even though once in a while I still think about SO(big), in this talk I'll tell you about some of my recent thoughts about the structure of grand unified theories. I'd like to ask myself some structural questions, questions that are inspired by the structures of gauge theories. As you will see, these questions are essentially unanswerable but by talking about them, perhaps one may get some insight on such questions as how Nature decides on a certain gauge group and a particular number of fermion fields. So this discussion has to be very tentative and exploratory.

My first remark is this: Nature is presumably described by a gauge theory, but we notice a very striking feature, namely that the gauge groups are very small. Let me explain what I mean by small. Suppose we have n fermion fields, and by fermion fields I mean two-component Weyl fields (so that in this way of counting one family contains 15 fermion fields). Now, if I write down the Lagrangian for n fermion fields, it has a very large global symmetry of $SU(n)$ x $U(1)$. A very interesting question, I think, is why does Nature gauge only a tiny subgroup? For example, with all the particles that we know now, including the yet-to-be observed top quark, we have 45 fields, so the global symmetry is $SU(45)$ x $U(1)$. So why does Nature pick out something like $SU(5)$ or $SO(10)$ which is, compared with $SU(45)$, very small? Many physicists beginning with Einstein have felt that local symmetries are more attractive than global symmetries. So why doesn't Nature turn this very large global group into a gauge group of like size? The second question that one could perhaps try to ask is: what principles determine which subgroup should be gauged?

Now, as I said, these are really unanswerable questions at this stage of development of physics, but I would like to discuss them nevertheless. The first question is actually easy to answer; it is easy to answer after 1969. Before 1969 it was a mystery. The answer is that if you gauge $SU(n)$ x $U(1)$, putting the n fermions into the fundamental representation, then, of course, you would have anomalies and the theory would not be renormalizable. In order to explore the second question, I'm going to pose for myself the following group-theoretic question, a purely group theory problem at this state.

Given SU(n) x U(1), what is the largest simple subgroup G such that
the theory obtained by gauging the subgroup G is free from anomaly
and free from bare masses. I won't go into any details but I will
try to explain briefly the second assumption. The conjecture is that
if bare masses are allowed by the gauge symmetry, then in general
there will be some mass scale for these masses and presumably this would
establish the mass scale of the gauge theory. The second condition
would rule out, for example, the type of theory which Pati and Salam
favor and which Jogesh Pati talked about yesterday. He would say,
in the kind of theory he talked about yesterday, that if you have 45
fields then you should gauge SU(45). Of course, you would have an
anomaly. To avoid this, what they suggest is simply to introduce
mirror fermions, namely fermions belonging to $\overline{45}$. And, of course with
45 and $\overline{45}$, the group theory allows you to form a bare mass. (Of course,
you don't have to. You could just make the $\overline{45}$ heavy, as they do.)
We also rule out the type of theories which may be referred to as
iceberg theories or theories with particles under the rug. An example
is the SU(11)-type theory of Howard Georgi, which a previous speaker
mentioned. In these theories you would start out with 579 particles,
say, and then remove all of them leaving you with 45.

Anyway, we have formulated a purely mathematical problem which
one could easily work out. Group theoretically, you could write down
a bare mass if the fermions are in a real representation. So given
n, you have to find a simple subgroup G of SU(n) x U(1) and a set of
representations that would add up to n, in such a way that the resulting
thoery is anomaly-free (so that it is renormalizable) and furthermore
(by the condition of no bare masses) that no subsets of these
representations are real. This is a perfectly well-defined mathematical
problem. Even with a very minimum knowledge of group theory, one can
easily work out the answer for any given n. So I worked out a table (Table
ble 1) for all integers up to 80 or something like that and it actually
goes very fast, because it turns out, for many integers, there are no
solutions at all. Games like this may actually shed some light on
some of the questions that eventually, perhaps sometime in the far
future, we as physicists would like to answer. Namely, why does Nature
pick certain groups, why does Nature pick a certain number of fields,
why are there three families and things like that. Because, if Nature
had picked some integers there will be no solution according to the
criteria that I mentioned. We only have to look at groups with complex
representations. Now, mathematicians have classified all such groups,
and it turns out that there are only a limited number of them;
they are SU(n), SO(4n + 2) and E(6). One could examine all the
subgroups of SU(n) x U(1) that belong to this set. And as I said
one could work out such a table. And the table would look something
like Table 1; for a given number, there would be a certain group. So
for instance, suppose Nature decided not to include the top quark. The
nature would have to construct a world with 39 fields. Well, according
to the criteria that I mentioned, that is not possible.

Another very striking property of the physical world that we all believe in, and this is one experimental fact which I think is certainly firmly established, is that electric charge is vectorlike. What I mean by that is that for every fermion field with charge +q, there is one with charge -q, if q is not equal to zero. Again, by fermion field, I mean a two-component Weyl field. Now the real world, without question, is vectorlike because if it is not vectorlike then there will be severe infrared difficulties, obviously. Those fermions with charge +q would not have a partner, would not have another field with charge -q to combine with to form a mass; so those fermions would be massless. And there would be a massless photon coupled to these fermions and there would be such infrared difficulties that the S matrix as usually defined would not exist. Now quantum field theory, of course, exists even if the S matrix does not exist. Our knowledge of quantum field theory tells us that the only condition that is really necessary is that the trace of the charge operator must be zero in order not to have an anomaly. Remember, all my fields are left-handed. So, for instance, I could have ordinary quantum electrodynamics, but instead of having a positron and an electron field, I could have a field with charge three and twenty-seven fields of charge -1. This theory would certainly satisfy the anomaly-free condition. Unfortunately, at this point, we cannot answer the question of what happens in such a bizarre field theory. We just don't know enough about quantum field theory. Perhaps it is reasonable to conjecture that the U(1) of electromagnetism is dynamically broken by the theory. So the question is why does Nature arrange things so that the charge is vectorlike?

This is probably an unanswerable question and so I have no answer for it, but let me point out that in the context of these grand unified theories it is very mysterious. In grand unified theories, we start with some big group and we break it down to some subgroup, SU(3) x SU(2) x U(1) or something. In any case, we break it down to something containing a U(1) generator which, when evaluated over fermions, must be vectorlike. This poses an argument against elementary Higgs theory--because, if you have elementary Higgses, you put the Higgs fields in some representation and, in general, there is no connection between two representations. Which U(1) is electromagnetism depends on the Higgs representation, and so, when that charge is evaluated over the fermions there is absolutely no reason for it to be vectorlike. In a dynamical theory, which U(1) is picked out depends on the fermion condensate so there is at least a chance that the U(1) is vectorlike.

Let me restate the problem as I see it. It is that experimentalists have observed that the trace of any odd power of the charge operator is zero. Now, after many years of physics, we understand the first two conditions in this hierarchy of conditions. The first one, that trace Q is zero, is necessary if electromagnetism is part of a unified theory. The second condition is necessary if electromagnetism is to be renormalizable. But why does the 5th power of Q have a trace of zero? That is a deep question which I don't have any answer for. If you take the

Georgi-Glashow SU(5) for example, and you choose the fermions to be in some anomaly-free representation and put the Higgses in 24 and 5 (as usual), then charge in general would not be vectorlike. So let me end again with a group theory problem. One can't answer the previous question, but one can answer a much simpler question. Given a simple group G and a bunch of fermion representations R so that it is anomaly-free, could there exist a vectorlike charge? Now I have studied this question recently with Steve Barr and we find that one could prove a variety of theorems with various specific assumptions, but in general there are many examples, many groups and many representations for which there is no vectorlike charge operators.

Table 1

N_f	G
15	SU(15)
16	SO(10)
27	E(6)
30	SU(5)
32	SO(10)
34	SU(4)
37	SU(3)
40	SU(3)
42	SU(7)
43	SU(3)
45	SU(5)
48	SO(10)
49	SU(7)
54	SU(3)
57	SU(3)
60	SU(8)
63	SU(7)
64	SO(14)
67	SU(3)
68	SU(4)
70	SU(3)
72	SU(4)
75	SU(5)
76	SU(4)
77	SU(3)
81	SU(9)

DISCUSSION

MARSHAK: In the spirit of your talk, let me ask a question. When you consider more than one family, you found as the maximal gauge group for 3 families: SU(5) in the absence of a right-handed neutrino (per family) and SO(10) in the presence of ν_R. For 4 families, you found SU(8) and SO(14) respectively. It seems to me that if one is searching for an extended GUT that will explain the number of families, SU(5) and SO(10) require replicated representations for 3 families, SU(8) is rather clumsy for 4 families while SO(14) breaks down into subgroups that can accomodate well to 4 family representations. Do you have any comment?

ZEE: No, I don't. In the kind of thing I am doing, within the scope of the discussion, there is no way to tell what the group breaks into eventually. For example, given 15 fermions, according to this, if God would have agreed to our criteria, He would have picked SU(5). But then I see no reason why SU(5) should be broken to SU(3) x SU(2) x U(1). That, too, is a mystery.

WEINBERG: I didn't get the bare mass condition. Of course, if there are bare masses, the reasonable thing, as we said, is that these particles would be very heavy. But if you are talking about grand unification, it is reasonable to believe that there is a family of very heavy particles like, for example, the other gauge bosons that we don't see at low energy. So what is wrong with that?

ZEE: The only honest answer to your question is that that is a condition which I have to impose to get somewhere. It is based on a notion which various people have emphasized. It has been pointed out that in the standard SU(3) x SU(2) x U(1) theory there is no bare mass. In the SU(5) theory there are also no bare masses. In the standard SU(5) theory there are $\bar{5}$ and 10, but not a 5. The explanation is that if you put in a 5, then you have to put in another $\bar{5}$ also, and the reason you don't see a 5 is that the 5 and a $\bar{5}$ combine to form five massive Dirac fields. Subscribing to some sort of totalitarian (or libertarian) principle in which any allowed bare mass is actually present, one would write the theory without the 5 and the $\bar{5}$ if one only looks at low energy physics. In fact, there again, you asked the right question, but that is contained in one of the slides that unfortunately I had to skip. Obviously, the possibility you alluded to may exist and is illustrated by the recent discussions on the right-handed neutrinos. That is the question; we don't know whether we live in a world with 45 or 48 fermions.

YANG: In the spirit of your discussion and at the risk of repeating myself, I would like to point out that the symmetries that we know in nature derive from first space-time symmetry and then there are the internal symmetries. And, of course, in general relativity there are coordinate transformation symmetries. But I would like to emphasize that there is another symmetry which is so far quite unique and that is charge conjugation symmetry which derives from an algebraic origin. It exists because we formulate the theory with complex numbers, and that is the real reason for charge conjugation. Also, let us notice it is in quantum mechanics where complex numbers come to the fore. Before quantum mechanics you could solve matrix equations using complex numbers, but it was not necessary. It was just a tool. In quantum mechanics nobody can break quantum mechanics into real and imaginary parts and write down sensible equations. So charge conjugation really exists because that operation is the isomorphism of the complex algebra. The next complicated algebra is quarternion algebra and the isomorphism of this algebra is $SU(2)$. So I would think that there is a very high possibility that another direction to look for the structure of symmetries is in the algebra. It is a formidable task, but I think it is worthwhile. I may be wrong.

ZEE: That is certainly worthwhile, of course; that is something Feza Gursey has been working on for 20 years and octonions, Feza tells me, are related to various exceptional groups. So that is why he and his collaborators are motivated to look at $E(6)$.

YANG: Yes, then you go to octonions and the invariance group of octonions is $G(2)$. That is, I think, going a bit too far, because the octonion algebra is not only not commutative it's not even associative; it's just a horrible mess.

ZEE: I hope that tomorrow we will have an opportunity to argue about this with Feza in public and I am certainly intrigued, of course. Let me make a comment that is not in answer to your question but that is along the line I discussed. There is always some global symmetry left and that is very disturbing. For example, the original symmetry $U(1)$ is always present at the end. Also, if fermions are put into several representations, there are residual global symmetries, abelian or even non-abelian.

BEYOND GRAND UNIFICATION

E.C.G. Sudarshan[†]
Lyman Laboratory of Physics
Harvard University
Cambridge, Massachusetts

and

J. C. Ward[*]
Center for Particle Theory
University of Texas
Austin, Texas

Ever since the nature of the weak four-fermion reaction was identified by Sudarshan and Marshak to be chiral (left-handed) V-A, there have been attempts to incorporate strange particle weak interactions and to combine them with Maxwell-Dirac electrodynamics of fermion fields. The latter was already known to be the minimal gauge invariant coupling. It was then natural to seek and to find a unification of electrodynamics and weak interactions in terms of a single theory and to find that this had to be a gauge theory also.[2] But the electroweak theory did contain a new ingredient in contrast to earlier gauge theories like electromagnetism and gravitation, namely the concept of spontaneous symmetry breakdown and the associated nonvanishing vacuum expectation values of suitable (scalar) fields.

When one studies the electroweak interactions of leptons of the first family, one is struck by the chiral nature of both weak and electromagnetic interactions with only the electron mass term breaking chirality invariance.[1] It would be desirable to obtain this breakdown also to be generated by nonvanishing vacuum expectation values.

In the "standard model", the electron mass as well as the vector meson masses are brought about by the vacuum expectation values of certain scalar (Higgs) fields which are introduced for this purpose only. They were not an essential part of the gauge structure of the theory.

We have investigated the possibility that the electric, weak, strong and gravitational interactions are all aspects of a unified gauge theory[3] which, under suitable dynamical breakdown, would generate fermion and vector boson masses; and that at the same time gravitation can be treated on a par with the other gauge interactions.

The Riemannian Bundle and the Fibre

We start with a Riemannian manifold admitting an internal symmetry. The fully covariant derivative is then given by an operator of the form

$$D_\mu = \partial_\mu + C_\mu \tag{1}$$

where C_μ is a suitable connection which takes account of the Riemannian connections as well as those due to the "internal" symmetries, but which itself involves no derivatives. The nonabelian nature of the local groups is reflected in the presence of noncommuting operators making up the connection. Since we are dealing with a general Riemannian space, the spinor and tensor quantities transform under a group distinct from the coordinates. Since we are eventually interested in obtaining the electron by a suitable limit of the equations of motion, we would like to consider a form of the equations of motion linear in the derivatives.

On the other hand, the simplest invariant that we can construct from the gauge invariant derivative operator (1) is the quadratic Laplace-Beltrami operator: it does not reduce to an expression linear in the derivatives. We are therefore led to higher invariants of even order. The next simplest are the quartic invariants:

$$g^{\mu\lambda} g^{\nu\rho} D_\mu D_\lambda D_\nu D_\rho \tag{2a}$$

$$g^{\mu\lambda} g^{\nu\rho} D_\mu D_\nu D_\rho D_\lambda \tag{2b}$$

$$g^{\mu\lambda} g^{\nu\rho} D_\mu D_\nu D_\lambda D_\rho \tag{2c}$$

Since we need to eliminate the fourth order derivative terms, we may consider any of a two-parameter family of linear combinations of the expressions:

$$g^{\mu\lambda} g^{\nu\rho} [D_\mu, [D_\lambda, D_\nu]] D_\rho = D_1 \tag{3a}$$

and

$$g^{\mu\lambda} g^{\nu\rho} [D_\mu, D_\nu][D_\lambda, D_\rho] = D_2 \tag{3b}$$

It is clear that (3a) is linear in spatial derivatives while (3b) contains no derivative.

Accordingly, we start with a fundamental action principle with the fully gauge invariant Lagrangian density:

$$L = \bar{\psi} (a_1 D_1 + a_2 D_2) \psi . \tag{4}$$

This is the universal Lagrangian with which we start. As we consider different theories we will choose ψ differently and, of course, the group of local transformations. But in all cases we choose the same action:

$$A = \int d^4x \sqrt{-\det g(x)}\, L(x) . \qquad (5)$$

While this action (4),(5) appears familiar, we stress the fact that no separate kinetic terms are introduced for the gauge fields, no mass terms are introduced for the field ψ and no scalar Higgs fields are introduced. All these are to emerge from this theory under suitable limiting conditions. There are no gauge fields without interaction and no equations of motion. <u>No kinematics without dynamics</u>! The entire action vanishes if there is no interaction. Note that the ψ field has no dimension.

At this juncture it is advantageous to introduce coordinates into the discussion[4] and write:

$$C_\mu = \Gamma^{\rho}_{\mu\sigma} X^{\sigma}_{\rho} + i A^{\alpha}_{\mu} \gamma_\alpha + \frac{1}{2} B^{\alpha\beta}_{\mu} \Sigma_{\alpha\beta} \qquad (6)$$

where $i X^{\sigma}_{\rho}$ are GL(4,R) generators, γ_α are the Dirac vector matrices and $\Sigma_{\alpha\beta}$ the corresponding antisymmetric tensor matrices. In particular the X^{σ}_{ρ} satisfy the commutation relations:

$$[X^{\sigma}_{\rho}, X^{\sigma'}_{\rho'}] = \delta^{\sigma}_{\rho'} X^{\sigma'}_{\rho} - \delta^{\sigma'}_{\rho} X^{\sigma}_{\rho'} . \qquad (7)$$

By direct calculation we obtain:

$$[D_\mu, D_\nu] = R_{\mu\nu\sigma}{}^{\rho} X^{\sigma}_{\rho} + i V_{\mu\nu}{}^{\alpha} \gamma_\alpha \qquad (8)$$
$$+ \frac{1}{2} T_{\mu\nu}{}^{\alpha\beta} \Sigma_{\alpha\beta}$$

with

$$R_{\mu\nu\sigma}{}^\rho = \Gamma^\rho_{\nu\sigma,\mu} - \Gamma^\rho_{\mu\sigma,\nu} + \Gamma^\tau_{\mu\sigma}\Gamma^\rho_{\nu\tau} - \Gamma^\rho_{\mu\tau}\Gamma^\tau_{\nu\sigma}$$

$$V^\alpha_{\mu\nu} = A^\alpha_{\nu,\mu} - A^\alpha_{\mu,\nu} + 2A^\beta_\mu B^{\beta\alpha}_\nu - 2A^\beta_\nu B^{\beta\alpha}_\mu$$

$$T^{\alpha\beta}_{\mu\nu} = B^{\alpha\beta}_{\nu,\mu} - B^{\alpha\beta}_{\mu,\nu} + 2A^\alpha_\mu A^\beta_\nu - 2A^\alpha_\nu A^\beta_\mu \qquad (9)$$

Consistent with this choice, ψ carries a spinor index also. These spinor labels are the only "internal" symmetries at this level. For simplicity of presentation, we wish to outline the strategy of our approach to a cosmic theory at this level.

Einstein, Dirac and Higgs Limits

Consider the limit

$$<\bar\psi\psi> = 0 \quad , \quad <D_\mu\psi, D_\nu\psi> \neq 0 \qquad (10)$$

which we call the Einstein limit. By reason of local Lorentz invariance we must then have:

$$<D_\mu\psi, D_\nu\psi> = \frac{(a_1+a_2)\Lambda}{4} g_{\mu\nu} \qquad (11)$$

where Λ is a suitable quantity with the dimensions of an inverse area. Consequently, the action now becomes:

$$-\int d^4x \sqrt{-\det g(x)} \text{ to } <D_\mu\psi, D_\nu\psi> g^{\mu\lambda}g^{\nu\rho}[D_\lambda, D_\rho]$$

$$= -\int d^4x \sqrt{-\det g(x)} \,\Lambda\, g^{\nu\rho} R_{\lambda\nu}{}^\lambda{}_\rho \qquad (12)$$

Comparing this with the standard Einstein Lagrangian we identify the

Newtonian constant:

$$\kappa = (\Lambda(a_1+a_2))^{-1} \quad . \tag{13}$$

The second limiting form is one where we neglect the Riemannian curvature: we may therefore make the restriction:

$$\begin{aligned} A_\mu^\alpha(x) &= g_\mu^\alpha \phi(x) \\ B_\mu^{\alpha\beta}(x) &= 0 \end{aligned} \tag{14}$$

We consider the Dirac limit in which $\psi(x)$ is independent of x. Then:

$$[D_\mu, D_\nu] = 2i \Sigma_{\mu\nu} \psi^2$$

and hence the Lagrangian density becomes:

$$L = 12\psi^3 \bar\psi (ia_1\phi + 4(a_1-a_2)\phi)\psi \tag{15}$$

By rescaling ψ according to:

$$2\sqrt{3\phi^3 a_1} \; \psi \to \psi$$

and writing:

$$m = 4(a_1-a_2)\phi \tag{16}$$

the Lagrangian density assumes the standard Dirac form.

When $\phi(x)$ is considered space-dependent, we have a proto-Higgs field, a scalar field which is selfcoupled by virtue of the non-abelian gauge couplings of the connection fields. Taking the limit:

$$<\psi,\psi> \neq 0 \tag{17}$$

we can get a kinetic term of the form;

$$-3 g^{\mu\nu} (2a_2 - a_1) <\psi,\psi> \partial_\mu \phi \, \partial_\nu \phi \qquad (18)$$

together with the interaction terms:

$$18 <\psi,\phi\psi> \phi^2(x) + 48(a_1 - a_2) <\psi,\psi> \phi^4(x) \qquad (19)$$

Internal Symmetries

We now proceed to incorporate internal symmetries like isospin. There are two ways to do this: the first is to choose an internal symmetry group which does not mix spin and "isospin". The local group in this case would be a direct product non-simple group. This would, however, not be a truly unified theory. Rather than pursue this method, we consider the possibility that the "isospin" and "spin" do mix: only one internal symmetry group which includes the spinor transformations and the "isospin".

To make this combination, it is easiest to deal with the fundamental field ψ realizing a spinor representation of the "isospin" group. This would prefer a theory in which the "isospin" group is an <u>orthogonal</u> group. Present indications on grand unification[5] lend some support to this choice. With this choice in mind, we may write the connection in the form ($1 \leq j, k \leq 2n$):

$$\begin{aligned} C_\mu &= \Gamma^\rho_{\mu\sigma} X^\sigma_\rho + i A^\alpha_\mu \gamma_\alpha K + \frac{1}{2} B^{\alpha\beta}_\mu \Sigma_{\alpha\beta} \\ &+ i F^j_\mu \Lambda_j + \frac{1}{2} G^{jk}_\mu \Lambda_{jk} + i E^{\alpha j}_\mu L_{\alpha j} \end{aligned} \qquad (20)$$

Here K is the matrix which satisfies:

$$K\Lambda_j = -\Lambda_j K, \quad K^2 = +1 \tag{21}$$

and the Λ_j and Λ_{jk} among themselves obey the commutation relations of the Lie algebra of $O(2n+1)$. Then the $\gamma_\alpha K$, $\Sigma_{\alpha\beta}$, Λ_j, Λ_{jk} together obey the commutation relations of $O(2n+2,3)$. In particular:

$$\Lambda_j \Lambda_k = \delta_{jk} 1 + i \Lambda_{jk}$$
$$L_{\alpha j} = -i\gamma_\alpha K \Lambda_j \tag{22}$$

The calculations are more complicated in this case but the Einstein limit is essentially the same. In the interest of brevity we omit the derivation.

New features emerge with regard to the limit in which:

$$\Gamma \to 0 \quad B_\mu^{\alpha\beta} \to 0,$$
$$F_\mu^j \to 0, \quad G_\mu^{jk} \to 0$$
$$A_\mu^\alpha \to g_\mu^\alpha \phi(x)$$
$$E_\mu^\alpha \to g_\mu^\alpha \chi^j(x) \tag{23}$$

If we introduce the matrix field:

$$\Omega = K\phi - iK\Lambda_j \chi^j \tag{24}$$

then, by direct calculation, in this Higgs limit:

$$[D_\mu, D_\nu] = -i(\Sigma_{\mu\nu}\Omega^2 - \gamma_\mu \Omega,_\nu + \gamma_\nu \Omega,_\mu)$$
$$L = \bar\psi \, (48(a_1-a_2)\Omega^4 + ia_2 \Sigma_{\mu\nu}[\Omega,_\nu, \Omega,_\mu]$$
$$-3(2a_2-a_1) \, \Omega,_\nu \, \Omega,_\nu)\psi \tag{25}$$

At this stage of the theory, in addition to the fundamental spinor "isospinor" field and the gravitational field, we have 2n+1 scalar fields which transform as a scalar and a 2n-vector under the O(2n) group; this limit identifying, as it does, the α and μ indices, breaks the O(2n+2,3) symmetry and even the O(2n+1) symmetry. Along with this, we have a number of vector gauge fields F_μ^j, G_μ^{jk}, $B^{\mu\beta}$ and tensor fields $A_\mu^\alpha - \frac{1}{4}g_\mu^\alpha \phi^\mu$, $E_\mu^{\alpha j} - \frac{1}{4}g_\mu^\alpha \chi^j$ and $B^{\alpha\beta\mu} - \frac{1}{8}g^\alpha{}_\mu B^{\sigma\beta} - \frac{1}{8}g^\beta{}_\mu B^{\alpha\sigma}$.

Chirality and Internal Symmetry

We have gone into some detail on this model to demonstrate the possibility that the scalar (Higgs) fields could arise, along with their dynamics, from the gauge fields themselves; and thus be an intrinsic part of the gauge theory. But we cannot yet take these models seriously since they are not chiral. Since we know the weak interactions are definitely chiral, we must therefore consider the interplay of chirality and internal symmetry.

As a first step, we recall the well known fact that the fundamental spinor representation of orthogonal groups in odd dimensions split into two inequivalent representations when restricted to the orthogonal group in one dimension lower. In the present case, this reduction is obtained by diagonalizing the product of all the Clifford elements $\gamma_\alpha K$, Λ_j which happens to be $\gamma_5 K$. The representations are therefore pseudochiral rather than chiral. This symmetry realized by:

$$\left(\frac{1 \pm \gamma_5 K}{2}\right) \Sigma_{\alpha\beta} \, , \, \left(\frac{1 \pm \gamma_5 K}{2}\right) \Lambda_j \text{ and } \left(\frac{1 \pm \gamma_5 K}{2}\right) L_{\alpha j}$$

would certainly not be realized by the particles since the kinematics of the Dirac particles already involve the breaking of this symmetry by identifying spinor and space rotations. However, we expect the fields to be classifiable under this extended "internal" symmetry scheme.

Instead, we look for a lower level of symmetry which is obtained by taking the non-spinorial internal symmetries alone. We note that the group generated by $\Sigma_{\alpha\beta}$ and Λ_{jk} is contained in the group of all internal symmetries, and that $\gamma_\alpha K$, Λ_j, $L_{\alpha j}$ all intertwine between the irreducible representations of the smaller group. Hence a convenient way of breaking the symmetry is to have these intertwining operators have associated fields which develop vacuum expectation values. These are the scalar fields ϕ and X_j we introduced above. The X_j, transforming as a vector with respect to $O(2n)$ and $O(4)$, couples to the direct product of a chiral (left-handed) field to the chiral (left-handed) antiparticle field. By suitable choosing such a scalar field vacuum expectation value we can give appropriate masses for the fermion fields.

The choice of left-handed or right-handed components of the fermion fields to furnish the representations of the internal symmetry group should be supplemented by fields of the opposite chirality by virtue of the combined internal and spinor symmetry assumed above. It has been standard practice to assume that the masses may be quite different for these chiral conjugute fields. This doubling of the representations has important bearing on the question of anomalies.

Remarks

In this presentation we have not made any quantitative calculations or worked out the Higgs phenomenology. We may, however, point out that apart from the scalar mesons as suitable restrictions of the connection fields, we may also have scalar fields as the fundamental fields on which the fully covariant derivatives act. But such scalar fields cannot enter alone; they must have vector and other tensor fields along with them. Such scalar fields will couple to the connection fields and thus indirectly to the scalar fields already considered. Consequently, we will have mixing of these two kinds of scalar fields; and the "observed" Higgs fields could be suitable linear combinations of these.

We hope to elaborate on the theory outlined elsewhere. We conclude by pointing out that the problem of families of quarks and leptons would be handled in this approach by a generalization of the same combined spinor-isospinor representation; and hence family recurrence is to be understood within an extended orthogonal group framework.[6]

We wish to thank Sheldon Glashow and Narasimha Mukunda for counsel and criticism.

[†]Permanent Address: Center for Particle Theory and Department of Physics, University of Texas, Austin, Texas 78712.

[*]Permanent Address: Macquarie University, West Ryde, New South Wales, Australia.

1. E.C.G. Sudarshan and R. E. Marshak, Proceedings of the Padua-Venice Conference on Mesons and Newly Discovered Particles, (Societa Italiana di Fisica 1957); Phys. Rev. 109, 1860 (1958).
2. S. Weinberg, Rev. Mod. Phys. 52, 515 (1980); A. Salam, Rev. Mod. Phys. 52, 525 (1980); S. L. Glashow, Rev. Mod. Phys. 52, 539 (1980) for an authoritative account.
3. J. C. Ward, Proc. Nat. Acad. Sciences (US) 75, 2568 (1978)
   C. N. Yang, Phys. Rev. Letters 33, 445 (1974).
4. The calculations in this section have been done in collaboration with N. Mukunda, J. Pasupathy and R. Kaul.
5. H. Georgi and S. L. Glashow, Phys. Rev. Letters 33, 438 (1974)
   J. Pati and A. Salam, Phys. Rev. D10, 275 (1974)
   H. Fritzsch and P. Minkowski, Ann. Phys. (N.Y.) 93, 193 (1975).
6. F. Wilczek, Proc. Lepton-Photon Conference (Fermi Lab 1979).

DISCUSSION

FRITZSCH: Suppose one generates the Higgs fields in your way, i.e. dynamically. What kind of constraints does one get?

SUDARSHAN: You notice I have been very cagey about specifying what are the internal symmetry fields, what are the gauge fields. It seems natural to have all fields which would saturate the spinor representation. This would mean that you would have all the bilinears which can be obtained by multiplying the fundamental spinors with themselves, so you would get the $\underline{45}$, for example; all kinds of things which can be obtained from multiplying the $\underline{16}$ by the $\underline{16}$.

FRITZSCH: Yes, but what about the scalars? The Higgs scalars?

SUDARSHAN: Higgs scalars come from the connections. Among the connections is a connection field which corresponds to internal symmetry transformations, and at the same time contains the space time spinor labels. It's a vector field that contains two indices α and μ. At the level at which the Dirac equation and the conventional dynamics are assumed, by obtaining the breakdown of the thing, the two indices become the same. At that point, this thing, which looks like a vector connection field, becomes a scalar field. And therefore every connection term that appears here has a Higgs partner along with it. So Higgs is carried along with the gauge field. So, depending upon how many gauge fields you have of this particular kind, there will be corresponding Higgs fields.

Gauge Hierarchy and Decoupling

York-Peng Yao

Department of Physics
University of Michigan
Ann Arbor, Michigan 48109

I. Introduction

I want to talk about two problems in grand unification and indicate their solutions. These are the decoupling problem and the gauge hierarchy problem. They are interrelated. This work has been performed with Kazama and Unger[1].

As is well-known, in grand unification[2], we assume the existence of a symmetry G, such that some scalar acquires a big vacuum expectation value V and then subsequently another scalar develops a vacuum expectation value v to bring the symmetry to $SU(3)_{color} \times (SU(2) \times U(1))_{E.M.\,weak}$

$$G \xrightarrow{V} G' \xrightarrow{v} SU(3)_{color} \times (SU(2) \times U(1))_{E.M.\,weak}$$

Now, the problems are:

(1) Gauge hierarchy issue:[3] There are two aspects:

(a) why is V >> v. We have no deep answer for this. It is in our opinion the same kind of question "why is m_μ >> m_e?". They are just input parameters.

(b) To us, a more serious problem is "can we define a light sector"? In other words, can we maintain light particles to have small mass to all orders in perturbation without fine tuning. The answer is yes, to all orders in coupling expansion.

(2) Decoupling theorem:[4] Having said that we know how to separate particles naturally into light and heavy sectors, we may ask "does there exist an effective local Lagrangian theory, such that we can use it to reproduce all the light particle physical matrix elements at energy and $|p_i|$ << M? The answer is again yes, to all orders of perturbation in couplings, i.e.

$\mathcal{L}(g, M, m, \mu) \to \Gamma^n$ (1 light particle irreducible)

$\mathcal{L}^*(g^*, m^*, \mu) \to \Gamma^{*n}$ (1 particle irreducible) we can show that

(1) $\Gamma^n = Z^{n/2} \Gamma^{*n} + O(\frac{1}{M^2})$

(2) $g^* = g^*(g, \ln M/\mu)$

$m^* = m\, f(g, \ln M/\mu)$

Besides, we know how to calculate, to all orders in g, g^* and m^* via improved perturbation, i.e. renormalization group equation. The operator structure of $\mathcal{L}^*$ is obtained from $\mathcal{L}$ by deleting all terms involving heavy fields.

Note that we do our physics in the low energy region throughout. We consider $\ln M/\mu$ effects to be radiative correction due to heavy particles. This is in contradistinction to some other people's attitudes[5], in which they boost up the energy of the external particles and devise methods so that their effective theories, which may not be local, will join smoothly with the full theory.

II. Model

We need a model to make my statements more concrete. Let me now be more specific. Let us consider an O(3) gauge model with two scalar triplets

$$\mathcal{L} = -\frac{1}{4}(\partial_\mu \vec{A}_\nu - \partial_\nu \vec{A}_\mu - e\vec{A}_\mu \times \vec{A}_\nu)^2$$

$$-\frac{1}{2}(\partial_\mu \vec{\phi}_1 - e\vec{A}_\mu \times \vec{\phi}_1)^2 - \frac{1}{2}(\partial_\mu \vec{\phi}_2 - e\vec{A}_\mu \times \vec{\phi}_2)^2$$

$$-\frac{1}{2}m_1^2 \vec{\phi}_1^2 - \frac{1}{2}m_2^2 \vec{\phi}_2^2 - \frac{1}{4}\lambda_1(\vec{\phi}_1^2)^2$$

$$-\frac{1}{4}\lambda_2(\vec{\phi}_1^2)^2 - \frac{1}{2}\lambda_3 \vec{\phi}_1^2 \vec{\phi}_2^2 - \frac{1}{2}\lambda_4(\vec{\phi}_1 \cdot \vec{\phi}_2)^2$$

We assume that $\lambda_1, \lambda_2 > 0$ so that there is a lower bound for the potential. Now, we shall assume that the vacuum is unstable, such that

$$\vec{\phi}_1 = \begin{pmatrix} v_1 + \sigma \\ \pi_2 \\ \pi_3 \end{pmatrix} \qquad \vec{\phi}_2 = \begin{pmatrix} \psi_1 \\ v_2 + \phi \\ \psi_3 \end{pmatrix}$$

with $v_1 \gg v_2$ (renormalized v_1 and v_2 are input parameters). The other parameters are adjusted to make the potential reach its absolute minimum. At the tree level, we have $\lambda_4 > 0$ and

$$\max\left(-\sqrt{\lambda_1 \lambda_2},\ \frac{-\lambda_1 v_1^2}{v_2^2},\ \frac{-\lambda_2 v_2^2}{v_1^2}\right) < \lambda_3 < \sqrt{\lambda_1 \lambda_2}$$

The minimum vacuum conditons are

$$-m_1^2 = \lambda_1 v_1^2 + \lambda_3 v_2^2$$

$$-m_2^2 = \lambda_2 v_2^2 + \lambda_3 v_1^2$$

In this way, m_1^2 and m_2^2 are determined parameters. In fact, we can show that a pertrubation series can be organized such that m_1^2 and m_2^2 never appear in any calculation, to all orders.

Let us now talk about the spectrum which will bring out another problem which must be treated in discussing decoupling. We find that σ and ϕ mix. The mass eigenstates are

$H = \sigma \cos\theta - \phi \sin\theta$

$h = \sigma \sin\theta + \phi \cos\theta$

with masses

$$m_H^2 \cong 2(\lambda_1 v_1^2 + (\lambda_3^2/\lambda_1) v_2^2) \qquad \text{Heavy}$$

$$m_h^2 \cong 2(\lambda_2 - \lambda_3^2/\lambda_1) v_2^2 \qquad \text{Light}$$

$$\sin\theta \cong -\lambda_3 v_2/(\lambda_1 v_1)$$

Now, we quantize the theory in the 't Hooft-Feynman gauge

$$\mathcal{L}_{gauge} = -\frac{1}{2\alpha}(\partial_\mu A_1^\mu - \alpha e v_2 \psi_3)^2$$

$$-\frac{1}{2\alpha}(\partial_\mu A_2^\mu - \alpha e v_1 \pi_3)^2$$

$$-\frac{1}{2\alpha}(\partial_\mu A_3^\mu - \alpha e(v_1 \pi_2 - v_2 \psi_1))^2$$

$$+ \mathcal{L}_{ghost}, \qquad\qquad \alpha_{tree} = 1$$

Then we have the following spectrum

Physical Particles	(mass)^2	would be Goldstone partner
A_1	$e^2 v_2^2$	ψ_3
A_2	$e^2 v_1^2$	π_3
A_3	$e^2(v_1^2 + v_2^2)$	$\xi = (v_1\pi_2 - v_2\phi_1)/(v_1^2+v_2^2)^{1/2}$
H	$\cong 2(\lambda_1 v_1^2 + (\lambda_3/\lambda_1)v_2^2)$	
h	$\cong 2(\lambda_2 - \lambda_3^2/\lambda_1) v_2^2$	
$\eta = (v_2\pi_2 + v_1\psi_1)/(v_1^2+v_2^2)^{1/2}$	$\lambda_4(v_1^2+v_2^2)$	

The light particles are A_1, h, and ψ_3 and the ghost of A_1. Let us generically call these tree (masses)^2 m^2. Our solution to the gauge hierarchy problem is

$$m_{true}^2 = m^2 (1 + \sum_{i=loop} [a_i (\ell n\, v_1^2/\mu^2)^{n_i} + b_i (\ell n\, v_1^2/\mu^2)^{n_i - 1} + \ldots])$$

to all orders.

Now h and H will mix further when we carry out the loop expansion. We shall devise a set of Green's functions, such that mixing is automatically taken into account. To illustrate the natural choice, let us consider h-h

scattering. In the full theory we have

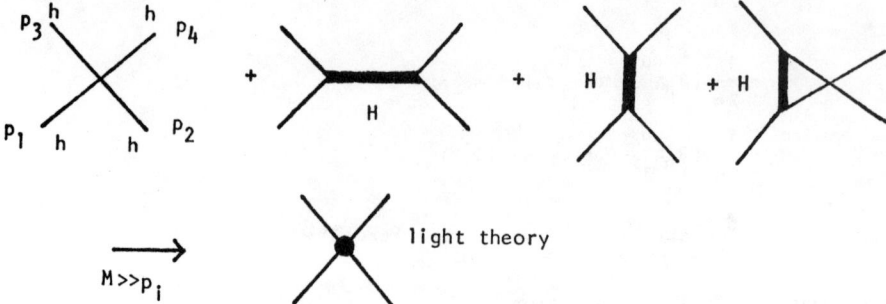

So, the Green's functions in the full theory that we want are the one light particle irreducible Green's functions. The decoupling theorem that we can prove to all orders in coupling expansion is the following

$$\Gamma^n_{1LPI}(p_i) = Z^{n/2} \Gamma^{*n}_{1PI}(p_i) + O(1/v_1^2)$$

$$p_i \ll M$$

in which Γ^n is calculated via the full Lagrangian $\mathcal{L}(g,\lambda,v_1,v_2,\mu,\alpha)$, while Γ^* is calculated from a light Lagrangian, the operator structure of which is that obtained when all heavy fields in $\mathcal{L}$ are all deleted. The coupling, mass and gauge parameters are all functions of $g, \lambda, v_1, v_2, \mu$ and α.

$$\mathcal{L}^*_{light} = \mathcal{L}^*(g^*, \lambda^*, v^*, \mu, \alpha^*)$$

$$g^* = g^*(g, \lambda, \alpha, \ln v_1^2/\mu^2)$$

$$v^* = v_2\, f(g, \lambda, \alpha, \ln v_1^2/\mu^2)$$

$$\lambda^* = \lambda^*(g, \lambda, \alpha, \ln v_1^2/\mu^2)$$

$$\alpha^* = \alpha^*(g, \lambda, \alpha, \ln v_1^2/\mu^2)$$

We do minimal subtraction in both theories.

III. Methodology

How do we prove such results?

(1) We first show that given any diagram with heavy internal lines, the corresponding integral can be rearranged so that the O(1) terms all have the heavy lines shrunken into vertices with no more than 4 light lines entering and/or leaving. The rest are negligible. This establishes the local renormalizable nature of the effective vertices. In fact, we can identify what these effective vertices are in relation to the light Lagrangian. However, a more economical way to show what the resulting local theory ensues is via BRS identities.

(2) We can show that the relevant BRS identities in the full theory $(v_1 \gg |p_i|, v_2) \to O(2)$ spontaneously broken BRS identities.

Thus, the limit of this O(3) theory is just the Abelian Higgs model in its asymmetrical phase. Besides, only v_2 appears explicitly in the resulting identities, which is a confirmation of the stability of gauge hierarchy.

To illustrate the first part, it is best to give an example. Consider a three point function

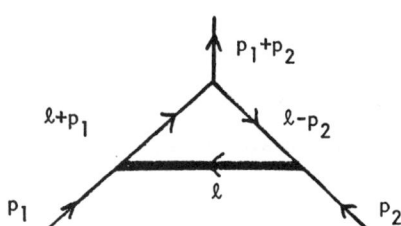

$$I = M^2 \int d^4\ell \; \frac{1}{(\ell+p_1)^2+m^2} \; \frac{1}{(\ell-p_2)^2+m^2} \; \frac{1}{\ell^2+M^2}$$

we define

shrinking a (sub) graph to a point

≡ setting all external momenta which go into/out of this graph to zero
≡ localization

There are two (sub)graphs which contains a heavy line

1. (shrink operator τ_1) $\dfrac{1}{\ell^2+M^2} = \dfrac{1}{M^2}$

2. (shrink operator τ_2) $\dfrac{1}{\ell^2+M^2}\dfrac{1}{(\ell+p_1)^2+m^2}\dfrac{1}{(\ell-p_2)^2+m^2}$

$= \dfrac{1}{\ell^2+M^2}\dfrac{1}{\ell^2+m^2}\dfrac{1}{\ell^2+m^2}$

Then, we have an identity

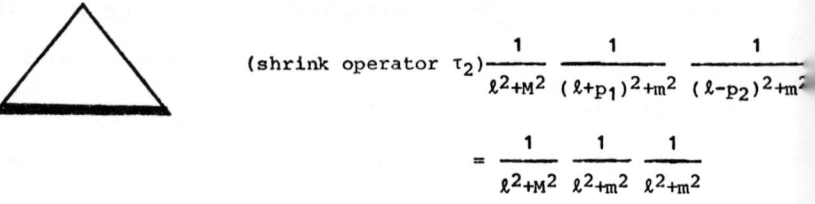

Note that the resulting integrals are automatically renormalized. Only the second and third terms, which have been localized, contribute to order 1. The first term is negligible.

The proof of reduciton of BRS identities depends heavily on power counting.

IV. Renormalization Group Equations

I want to devote the remainder of the discussion to the renormalization group equations which will be used to perform leading $\ln$ sum of the dependence on $\ln v_1^2 / \mu^2$ of the effective parameters. As I said earlier, we stay in low energy region to do our physics. Now when we work with the full theory, we have the renormalization group equation

$$(\mu \frac{\partial}{\partial \mu} + \beta_g \frac{\partial}{\partial g} + \gamma_{v_1} v_1 \frac{\partial}{\partial v_1} + \gamma_{v_2} v_2 \frac{\partial}{\partial v_2} + \gamma_\alpha \frac{\partial}{\partial \ln \alpha} - n\gamma) \Gamma^n = 0$$

with

$$\beta_g = \mu \frac{d}{d\mu} g, \quad \gamma_{v_1} = \mu \frac{d}{d\mu} \ln v_1, \text{ etc.},$$

where Γ^n has only light external lines.

When we work with the effective theory, we have

$$(\mu \frac{\partial}{\partial \mu} + \beta^*_{g^*} \frac{\partial}{\partial g^*} + v^* \gamma^*_{v^*} \frac{\partial}{\partial v^*} + \gamma^*_{\alpha^*} \frac{\partial}{\partial \alpha^*} - n\gamma^*) \Gamma^{*n} = 0$$

Now, we put the decoupling equation, which is

$$\Gamma^n = z^{n/2} \Gamma^{*n} + 0(\frac{1}{M^2}),$$

into the renormalization group equation of the full theory and demand that the resulting equation agree with that from the effective theory. We obtain many relations, which are basically chain rules for differentiation. For example

$$\beta^*(g^*) = (\mu \frac{\partial}{\partial \mu} + \beta_g \frac{\partial}{\partial g} + \gamma_{v_1} \frac{\partial}{\partial \ln v_1} + \gamma_\alpha \frac{\partial}{\partial \ln \alpha}) g^*$$

This rather humble equation in fact allows us to sum up the leading contributions. For example, returning to 0(3) model, we have for the gauge coupling,

$$e^* = e \, f(e^2 \ln v_1^2/\mu^2) + e^3 \, f'(e^2 \ln v_1^2/\mu^2) + \cdots$$

e is considered to be small and $e^2 \ln v_1^2/\mu^2$ is 0(1). In leading order

$$e^* \cong e \, f(e^2 \ln v_1^2/\mu^2)$$

$$\beta^*(e^*) \cong 1/3 \frac{e^{*3}}{16\pi^2}$$

$$\beta(e) \cong -20/3 \frac{e^3}{16\pi^2}$$

we have an equation

$$(\mu \frac{\partial}{\partial \mu} + \beta \frac{\partial}{\partial e}) e^* = \beta^*(e^*)$$

or

$$(\mu \frac{\partial}{\partial \mu} + \beta \frac{\partial}{\partial e}) (-24\pi^2/e^{*2}) = 1$$

To solve this equation, we define a new running coupling constant with respect to the full theory where $\kappa = \ln v_1/\mu$.

$$\frac{d}{d\kappa} \bar{e} = \beta(\bar{e})$$

The solution is

$$\frac{24\pi^2}{e^{*2}} = \frac{24\pi^2}{\bar{e}^2} + \ln v_1/\mu$$

or

$$e^{*2} = \frac{e^2}{1 + \frac{7}{16\pi^2} e^2 \ln v_1^2/\mu^2}$$

This result, when generalized to an appropriate group, agrees with solutions of other people.

In conclusion, we have shown that

(a) Gauge hierarchy is a non-issue, by which we mean that there is no need for fine tuning to separate out the light sector. Once we divide heavy and light sectors at the tree level, then at every loop order only $g^2 \ln v_1^2/\mu^2$ corrections contribute to the light effective parameters.

(b) We have a decoupling theorem. i.e. there exists a light Lagrangian, $\mathcal{L}^*(g^*, \lambda^*, v^*, \alpha^*, \mu)$, which can be used to deal with low energy physics. Besides, the dependence of g^*, v^*, λ^*, α^* on the parameters in the full theory $(g, \lambda, v_1, v_2, \alpha)$ can be determined by staying in low energy regions. There exist natural renormalization group equations to sum up the $\ln v_1^2/\mu^2$ powers.

This work is supported partially by the U.S. Department of Energy.

REFERENCES

1. Yoichi Kasama, David G. Unger and York-Peng Yao, UM HE 80-36; Yoichi Kazama and York-Peng Yao, in preparation.
2. J.C. Pati and A. Salam, Phys. Rev. D$\underline{8}$, 1240 (1973); H. Georgi and S.L. Glashow, Phys. Rev. Letts. $\underline{32}$, 438 (1974); H. Georgi, H.R. Quinn and S. Weinberg, Phys. Rev. Letts. $\underline{33}$, 451 (1974).
3. The original gauge hierarchy problem was pointed out be E. Gildener, Phys. Rev. D$\underline{14}$, 1667 (1976). For further discussion and references see I. Bars, invited talk at Orbis Scientiae 1979, Coral Gables, Florida; E. Gildener, Phys. Letts. $\underline{92B}$, 111 (1980); K. T. Mahanthappa and D. G. Unger, preprint COLO-HEP-14 and UM HE 79-31, (revised 1980).
4. K. Symanzik, Comm. Math. Phys. $\underline{34}$, 7 (1973); T. Appelquist and J. Carrazone, Phys. Rev. D$\underline{11}$, 2856 (1975). There has been a flurry of activity in demonstrating decoupling at the one loop or two loop level. A partial list is B. Ovrut and H. Schnitzer, Phys. Rev. D$\underline{21}$, 3369 (1980), D$\underline{22}$, 2518 (1980) and Brandeis preprints; T. Hagiwara and N. Nakazawa, preprint HUTP-80/A012; F.T. Hadjioannou, A.B. Lahanas and C.E. Vayonakis, Phys. Lett. $\underline{84B}$, 427 (1979); G. Senjanovic and A Sokorac, Nucl. Phys. B$\underline{164}$, 305 (1980); M. Yoshimura, Prog. Theor. Phys. $\underline{64}$, 353 and 594 (1980). For different approaches, see S. Weinberg, Phys. Rev $\underline{22}$, 1414 (1980); P. Binetruy and T. Schucker, CERN preprints TH-2802-CERN, TH-2857-CERN; L. Hall, preprint HUTP-80/A024.
5. A partial list is: D.A. Ross, Nucl. Phys. B$\underline{140}$, 1 (1975); A.J. Buras, J. Ellis, M.K. Gaillard, and D.V. Nanopoulos, Nucl. Phys. B$\underline{135}$, 66 (1978); T. Goldman and D.A. Ross, Phys. Letts. $\underline{84B}$, 208 (1979); C.H. Llewellyn Smith, G.G. Ross and J.F. Wheater, Oxford Preprint 1980; I. Antoniadis, C. Bouchiat, J. Iliopoulos, LPTENS 80/21 (1980). For a review and further references, see J. Ellis, M.K. Gaillard, D.V. Nanopoulos and S. Rudaz, Nucl. Phys. B$\underline{176}$, 61 (1980).

Report No. DOE/EY/2232B-220

PHYSICAL IMPLICATIONS OF
DYNAMICAL SYMMETRY BREAKING

M. A. B. Bég

The Rockefeller University
New York, NY 10021

ABSTRACT

Some model-independent physical implications of a class of hypercolor-based theories of dynamical symmetry-breaking are described and discussed. The role which e^+e^- colliders can play, in distinguishing between such theories and the canonical methodology, is underlined.

ISSN:0094-243X/81/720505-10$1.50 Copyright 1981 American Institute of Phyics

1. I shall speak about some work done in collaboration with Dr. Ahmad Ali of DESY.[1] This work is a logical continuation of work published earlier[2,3,4] and very closely related work has been done by Ellis, Gaillard, Nanopoulos and Sikivie at CERN.[5] What I have to say will compliment the report of Kane to this meeting.

Let me begin by reminding you that spontaneous dynamical breaking of chiral symmetry was first studied in a seminal[6] paper by Nambu and Jona-Lasinio. That dynamical symmetry-breaking (or DSB, for short) could be used to provide an alternative to the canonical methodology in weak interaction theory was suggested in an article[7] which Sirlin and I wrote in 1973-74. Weinberg[8], in 1976, proposed to make use of the expected breakdown of chiral symmetries in a QCD-like theory (hereinafter called QC´D) to implement the dynamical Higgs mechanism; this proposal, rediscovered independently by Susskind[9] in 1979, has been the object of much attention during the preceding year. Following ref. 2 we shall use the term hypercolor to describe the C´ degree of freedom.

2. Hypercolor-based DSB has been afflicted from the very beginning with a rather serious problem, the problem of generating current or "ultra-violet" fermion masses. The first attempt at resolution[10], due to Dimopoulos and Susskind, introduced and utilized the notion of extended hypercolor (EHC); theories with EHC, however, run into a variety of difficulties, most important of these appears to be a conflict with the sharp experimental limits on flavor-changing neutral currents.[11] The word "monophagy", which permeates the recent literature, embodies a concept evolved to handle this last problem, but no solution is at the moment in sight. To be able to discuss hypercolor-based DSB one must, of course, assume that a solution to the mass problem exists; I shall make this assumption, and try to focus on those

implications of DSB which do not depend on any deep commitment to specific models of mass-generation.

3. The DSB theories under discussion are easily distinguishable[3] from the normal Higgs field scenario at energies greater than one to two TeV. If nature makes use of hypercolor, a new hadronic spectroscopy--rich in content and carrying the signature of this brand of DSB--will become manifest. The challenge is to pinpoint the nature of the Higgs mechanism via experiments in the 10GeV to 100GeV range, i.e., at energies which are either accessible or likely to be in the near future.[2-4] The only signals of DSB in this energy regime are pseudo-Goldstone bosons[12] (PGB's); these are artifacts of too much symmetry in the uhr-Lagrangian, they are found in almost all models and they can mock elementary Higgses. One must therefore tackle the problem of distinguishing between π'^0, a light neutral PGB, and ϕ^0, the Higgs scalar of the Weinberg-Salam (W-S) theory. Two handles are available:

(a) In processes which involve only <u>flavor-diagonal couplings of spin-0 particles to fermions</u> there are the parity tests of ref. 2. Here it is worth pointing out that the considerations of ref. 2 are model-independent if π'^0 is a flavor-neutral PGB. Such PGB's transform as $^1S_0$ bound states of hyperquarks q'_i and antihyperquarks $\bar{q}'_i$ and are therefore odd under CP. Thus CP invariance, which may be deemed to be good for our purposes, is sufficient to guarantee: $\phi^0 \to D\bar{D}$ (yes), $\pi'^0 \to D\bar{D}$ (no), etc.

(b) <u>There are no PGB-gauge-field couplings at the no-QFD-loop level.</u>[2] This is to be contrasted with the situation in the W-S theory which predicts a large ϕZZ coupling at the tree level.

To take advantage of (b) and (a) above, in the most efficient way possible, it is necessary to consider experiments with e^+e^- colliding beams.

4. The particles of the DSB scenario which e^+e^- colliders will first render visible would be charged hyperpions[2-4]: $\pi'^{\pm}$. Methods for identifying $\pi'^{\pm}$ have been discussed elsewhere[2-4,13] and I shall not go into it here. I would like to comment, however, on an interesting little side effect which one can study if one has energies sufficient to excite toponium. For $m_{\pi'^+} < m_t - m_b$,

$$\Gamma(J_T \to \pi'^+ b\bar{t} \ [\to \pi'^+\pi'^- \ b\bar{b}]) >> \Gamma(J_T \to \ell\bar{\ell} \text{ or } q\bar{q} \text{ or } 3g \ldots) \quad (1)$$

In other words, toponium will go almost exclusively into hyperpion channels [see Fig. 1].

If charged hyperpions are seen at a mass ~10GeV, it would be a strong indication that we are not on the wrong track. However, one cannot really tell a π'^+ from ϕ^+, one of the charged Higgs particles which will exist in any extension of the W-S model. To obtain more insight into the nature of the Higgs mechanism, from experimental data, it is necessary to look at neutrals.

5. Both π'^o and ϕ^o are expected to be produced in the decay of toponium[1]:

$$\Gamma(J_T \to \pi'^o \gamma)/\Gamma(J_T \to \phi^o \gamma) = (n_{F'})^{-1} \quad (2)$$

where $2n_{F'}$ is the number of hyperflavors. To tell a π'^o from a ϕ^o it may be necessary, therefore, to invoke the parity tests discussed earlier.

Other means of ϕ^o production which have been discussed in the literature are the reactions: $Z \to \phi^o \mu^+\mu^-$, $Z \to \phi^o \gamma$ and $e^+e^- \to \phi^o Z$. To calculate the rates of the corresponding reactions involving hyperpions, $Z \to \pi'^o \mu^+\mu^-$, etc., we shall make the simplifying assumption that π'^o is an isotriplet in hyperflavor space. Then the relevant amplitudes are expressible in the form

$$A[V^i(k_1) \to V^j(k_2) + \pi'^o(q)] = f\left(\frac{k_1^2}{M^2}, \frac{k_2^2}{M^2}, \frac{q^2}{M^2}\right) \varepsilon_{\mu\nu\rho\sigma} k_1^\mu k_2^\nu \varepsilon_i^\rho \varepsilon_j^\sigma \quad (3)$$

where $V^i(k) \equiv \gamma$ or Z with momentum k, $q \equiv k_1 - k_2$ and ε_i and ε_j are the polarization vectors of V^i and V^j respectively. M is the <u>constituent</u> hyperquark mass ~ 1TeV. For k_1^2, k_2^2, and q^2 all $<< M^2$, one may use the Bell-Jackiw-Adler anomaly to calculate the various f's. Thus

$$f_{ZZ\pi'^o} = - \frac{2\alpha}{3\pi} \frac{g_A'}{f_{\pi'}} \cdot \frac{\sin^2\xi(1-2\sin^2\xi)}{(\sin 2\xi)^2} \cdot N_{C'} \cdot n_{F'} \quad (4)$$

with similar expressions for $f_{Z\gamma\pi'^o}$ and $f_{\gamma\gamma\pi'^o}$. Here α is the fine structure constant, $f_{\pi'}$ is the hyperpion Goldberger-Treiman constant, g_A' is the hyperquark analogue of G_A/G_V and ξ is the Glashow-Weinberg-Salam angle. $N_{C'}$ and $n_{F'}$ are the number of hypercolors and hyperflavor-doublets respectively. Note that the f's calculated in the limit $k_1^2 = k_2^2 = q^2 = 0$ have the added and all important virtue of being exact to all orders[14] in QC′D (as well as QCD, a relevant point if hyperquarks carry the attribute of ordinary color as well). Eq. (4) is to be compared with the relationship in the W-S model

$$g_{ZZ\phi^o} = 2(G_F \sqrt{2})^{1/2} m_Z^2 . \quad (5)$$

The ratio $\Gamma(Z \to \pi'^o \mu^+ \mu^-)/\Gamma(Z \to \phi^o \mu^+ \mu^-)$ is plotted in Fig. 2 for equal values of π'^o and ϕ^o mass. It is evident that unless one is willing to consider absurdly large values of $N_{C'}$ and $n_{F'}$, a positive signal in $Z \to \mu^+ \mu^- +$ (spin 0) would be indicative of ϕ^o. Even if one is willing to consider values of $N_{C'}$ and $n_{F'}$ such that the signals for π'^o and ϕ^o production are of comparable magnitude, one can tell which particle is being produced; the dimuon invariant mass distribution in the two cases is very different [Fig. 3].

6. Let me conclude by stating the morals which can be drawn from our discussion.

Firstly, PEP and PETRA, which have the energy to produce relatively light charged hyperpions, can tell us if--in taking hypercolor-based DSB seriously--we are on the right track.

Secondly, LEP, with the energy and the luminosity required to produce Z-bosons at a rate of about 10^5 per day, could provide information about reactions such as $Z \to \mu^+\mu^- +$ (spin-0) and thereby unravel the nature of the Higgs mechanism.

Thirdly, nature could choose to be perverse and make use of spin-0 particles whose masses are such as to give us no relevant positive signal either at PEP, or at PETRA or at LEP. Experiment would then yield no information about the mechanism of symmetry-breaking until TeV machines are built.

Acknowledgements

This work was supported in part by the U. S. Department of Energy under Contract Grant Number DE-AC02-76ER02232B. It is a pleasure to thank Professor A. I. Sanda for enjoyable discussions.

References

1. A. Ali and M. A. B. Bég, Rockefeller University Report No. DOE/EY/2232B-212 (1980).
2. M. A. B. Bég, H. D. Politzer and P. Ramond, Phys. Rev. Lett. 43, 1701 (1979).
3. M. A. B. Bég in "Recent Developments in High Energy Physics", Edited by B. Kursunoglu, A. Perlmutter and L. F. Scott (Plenum Publishing Corporation, New York, 1980) p. 23.
4. M. A. B. Bég, Proceedings of the XXth Int. Conf. on High Energy Physics, Madison, Wisconsin (1980); G. L. Kane, ibid.
5. J. Ellis, M. Gaillard, D. Nanopoulos, and P. Sikivie, CERN Preprint TH.2938 (1980).
6. Y. Nambu and G. Jona-Lasinio, Phys. Rev. 122, 345 (1961).
7. M. A. B. Bég and A. Sirlin, Annu. Rev. Nucl. Sci. 24, 379 (1974).

8. S. Weinberg, Phys. Rev. $\underline{D13}$, 974 (1976).

9. L. Susskind, Phys. Rev. $\underline{D20}$, 2619 (1979).

10. S. Dimopoulos and L. Susskind, Nucl. Phys. $\underline{B155}$, 237 (1979). See also E. Eichten and K. D. Lane, Phys. Lett. $\underline{90B}$, 125 (1980).

11. S. Dimopoulos and J. Ellis, CERN Preprint TH.2949 (1980).

12. S. Weinberg, Phys. Rev. Lett. $\underline{29}$, 1698 (1972).

13. A. Ali, H. B. Newman and R. Y. Zhu, DESY Preprint (1980).

14. S. Y. Pi and S.-S. Shei, Phys. Rev. $\underline{D11}$, 2946 (1975).

Figure Captions

Fig. 1: The rate for the semiweak decay of toponium, J_T, as a function of the charged hyperpion mass. n_F is the number of hyperflavor doublets.

Fig. 2: The ratio $\Gamma(Z \to \pi'^0 + \mu^+\mu^-)/\Gamma(Z \to \phi^0 + \mu^+\mu^-)$ for equal values of m_{ϕ^0} and $m_{\pi'^0}$. n_F is the number of hyperflavor doublets and N_C, the number of hypercolors. We have used $\sin^2\xi = 0.20$ corresponding to $m_Z = 94$ GeV.

Fig. 3: The dimuon invariant-(mass)^2 distribution from the decays $Z \to \pi'^0 + \mu^+\mu^-$ and $Z \to \phi^0 + \mu^+\mu^-$. We have normalized both the distributions to the same area $[=\Gamma(Z \to \phi^0 + \mu^+\mu^-)]$ for equal m_{ϕ^0} and $m_{\pi'^0}$. The relative scales can be read off Fig. 2.

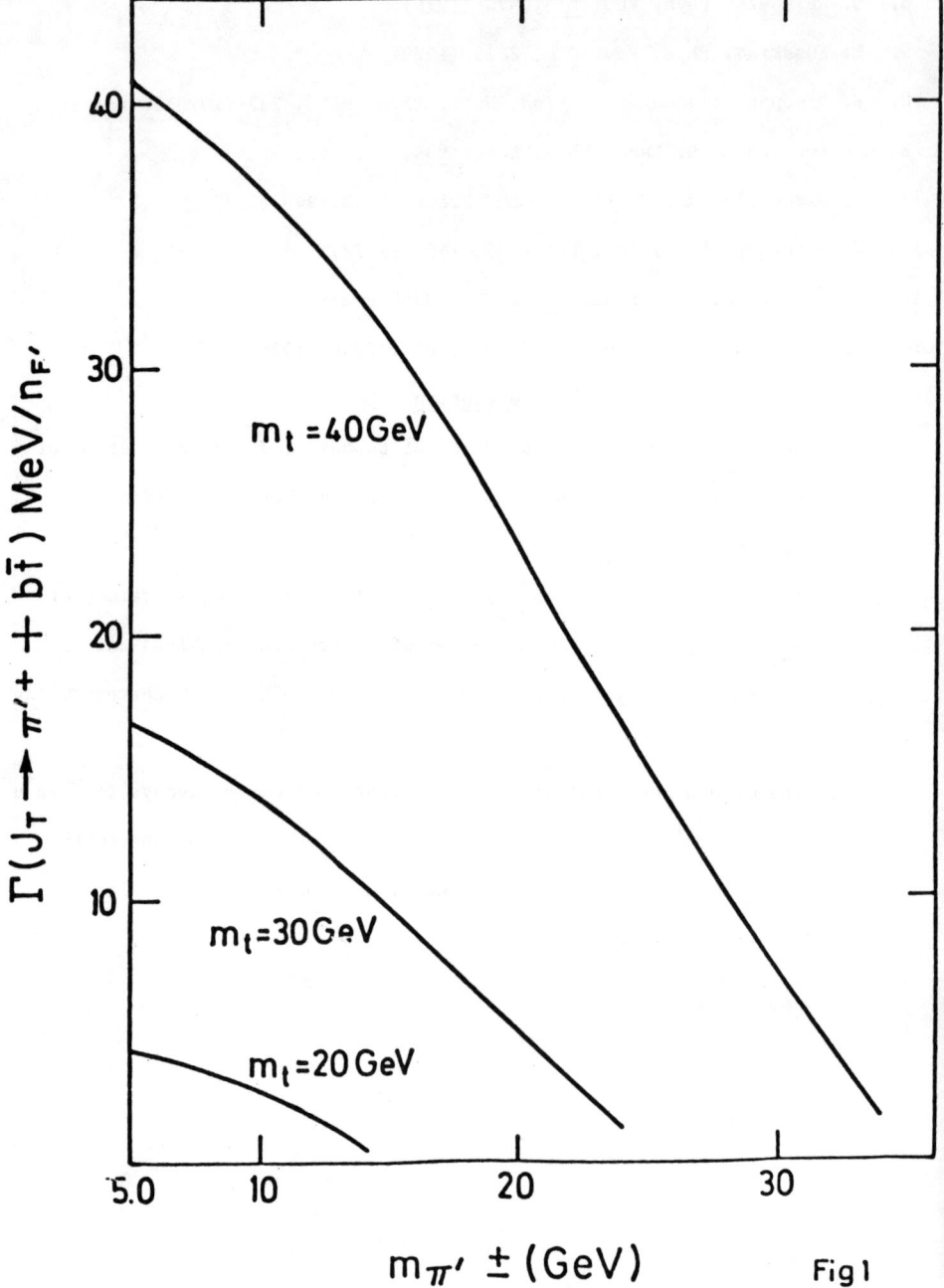

Fig 1

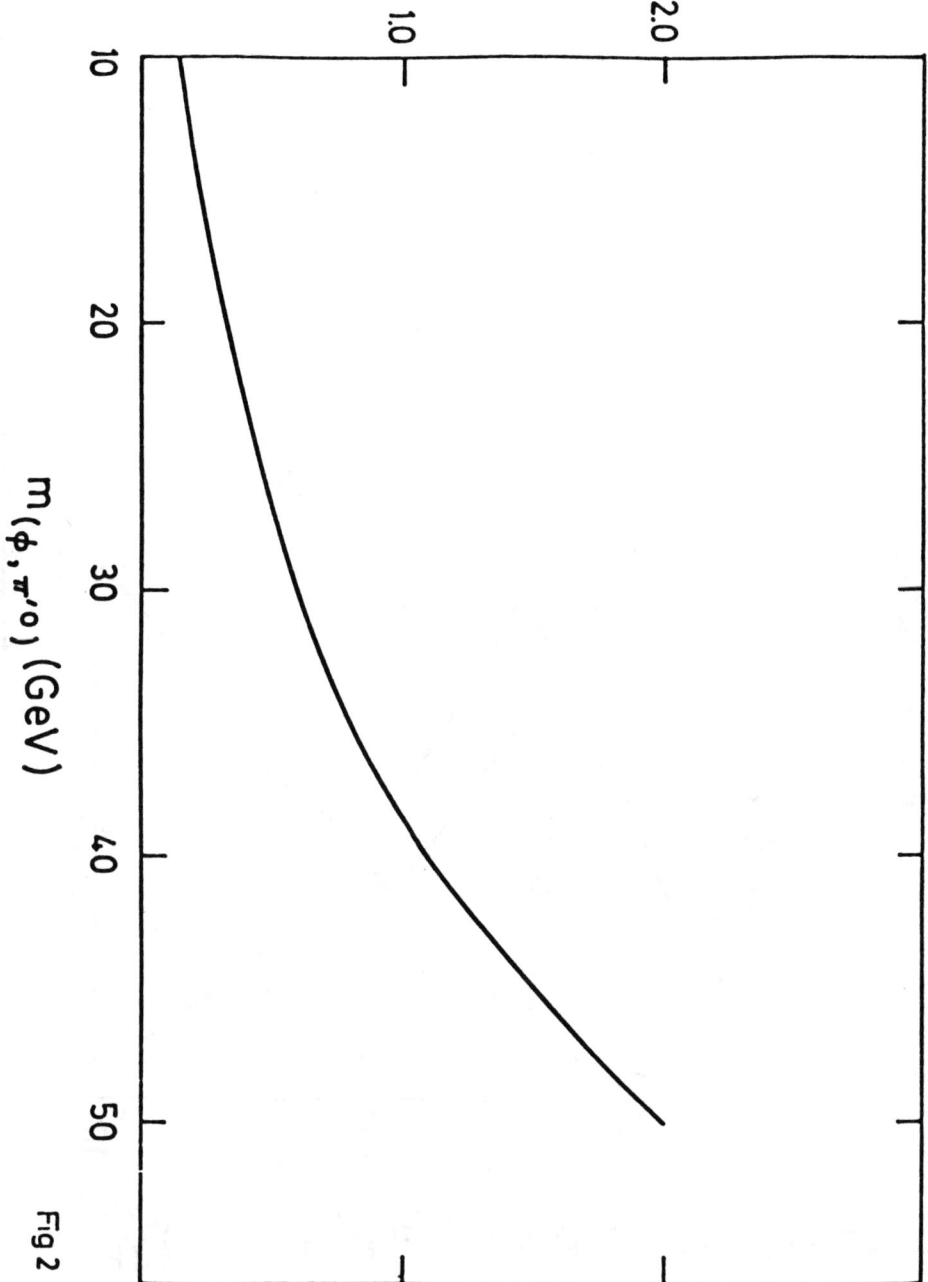

Fig 2

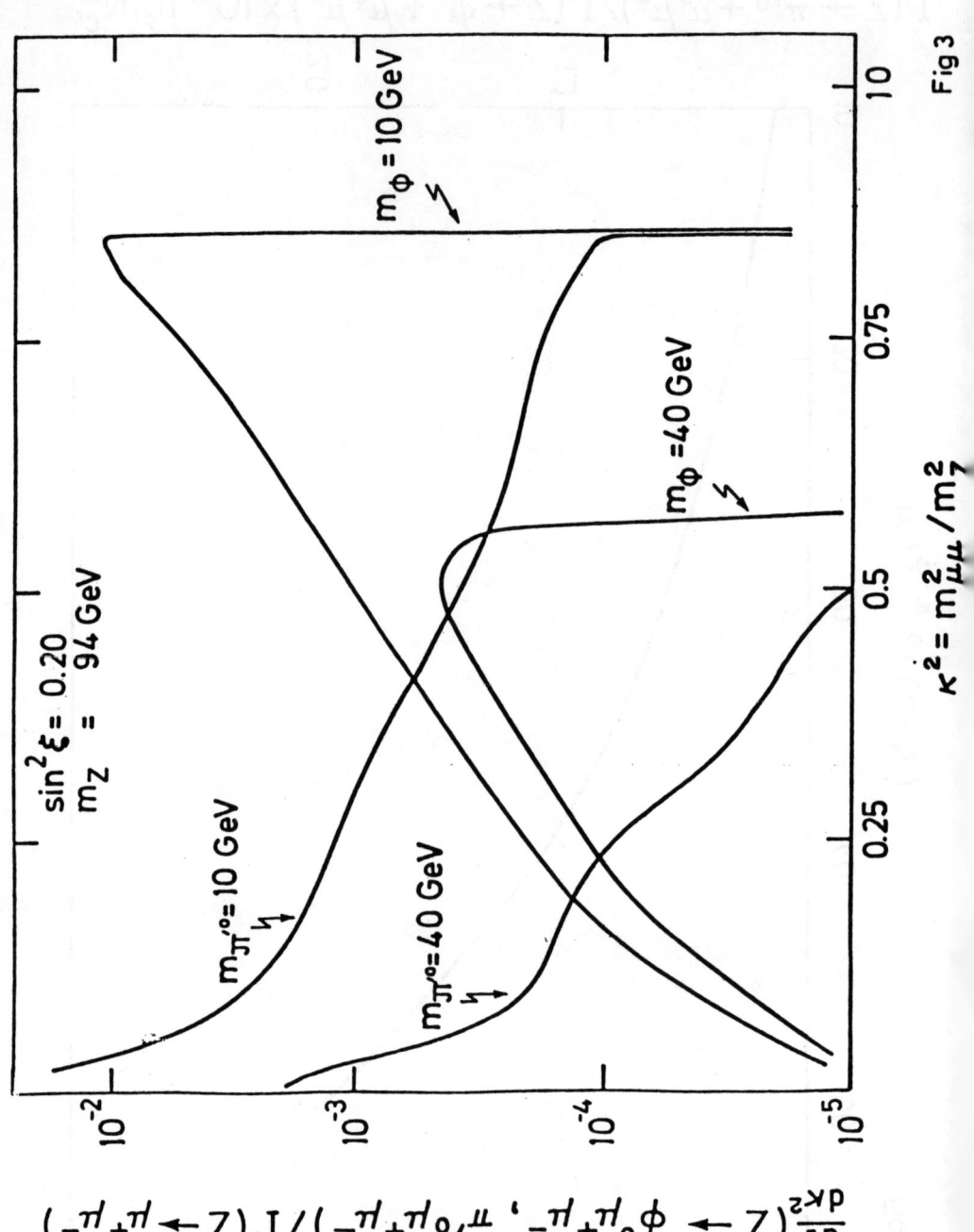

Fig 3

SOME IDEAS ON COMPOSITE QUARKS AND LEPTONS

N. Srednicki
Princeton University, Princeton, NJ

I want to explore in very general terms the possibility that quarks and leptons are actually composite particles. The reason for asking this question is that if they are indeed composite particles, one can introduce new kinds of mass generation mechanisms for the quarks and leptons. This question is a subset of technicolor questions; in technicolor theories, we want to replace the standard Higgs sector of weak interactions by gauge fields and fermions. That means we can't give masses to the ordinary quarks and leptons by the usual Higgs mechanism and so we have to find other mechanisms. I will talk about one possible mechanism that involves compositeness of quarks and leptons. Therefore, we need a theory that will give us composite fermions.

We know of one such theory: quantum chromodynamics (QCD) has the proton and all the other baryons as bound states. Baryons are fermions, so there is no problem in principle with getting composite fermions. One property of the proton, however, is that its mass is more or less the same as its inverse radius. If you want a theory where the electron is composite, you can't have that property; the mass of the electron is much less than its inverse radius. (For a recent discussion, see Ref. 1). So we need a theory which produces a composite particle whose mass is extremely light on the scale of compositeness. The only way, in general, we can have such a theory is if there is a symmetry which prevents us from having a mass term for the electron which is the same order of magnitude as its inverse radius. In QCD, as we understand it, there is no unbroken symmetry which keeps the proton from having a mass the same order of magnitude as its inverse radius, and so it does.

Now I would like to present an example of a theory in which we can have massless composite fermions. The theory I'll use is a very familiar one, but I'll make some non-standard assumptions about it. The model theory is quantum chromodynamics with three colors and two flavors of massless quarks. The analysis that I am going to give was first done by the Stanford group, S. Raby, S. Dimopoulos, and L. Sussking.[2] A different kind of analysis (which I will also discuss) was given by G. 't Hooft.[3] This specific example is due to Susskind and myself.[4] The full symmetry group of the theory is $SU^c(3) \times SU(2)_L \times SU(2)_R \times U(1)_B$. In the theory we have a set of fermions.

$$\psi = (\underline{3},\ \underline{2},\ \underline{1},\ +1/3)_L$$
$$\bar{\psi} = (\underline{\bar{3}},\ \underline{1},\ \underline{2},\ -1/3)_L$$
$$\psi^* = (\underline{\bar{3}},\ \underline{2},\ \underline{1},\ -1/3)_R \qquad (1)$$
$$\bar{\psi}^* = (\underline{3},\ \underline{1},\ \underline{2},\ +1/3)_R$$

These ψ's are all two-component Weyl fields. The two left-handed fields are a color triplet, $SU(2)_L$ doublet carrying baryon number +1/3 and a color anti-triplet, $SU(2)_R$ doublet with baryon number -1/3. The right-handed fields are the charge conjugates of the left-handed fields. (Charge conjugation changes left to right and complex conjugates change the group representations.)

Now, in QCD, we believe that the composite operators $\bar{\psi}\psi$ and $\bar{\psi}^*\psi^*$ develop nonzero vacuum expectation values. One consequence of this is that $SU(2)_L \times SU(2)_R$ symmetry breaks down to an $SU(2)$ vector symmetry. This is because we have an object with a non-zero vacuum expectation value that carries quantum numbers of the $SU(2)_L \times SU(2)_R$ groups. Another consequence is the presence of massless pions in the theory. These are the Goldstone bosons of the broken generations. (In real QCD, pions are not massless because small bare quark masses spoil the original $SU(2)_L \times SU(2)_R$ symmetry, turning the pions into pseudo-Goldstone bosons.) This symmetry-breaking is also responsible for the proton mass. Most nonperturbative phenomenology in QCD can be traced back to this effect. Theorists have never been able to demonstrate from first principles that this nonzero vacuum expectation value actually occurs. There are indications that it is at least reasonably likely. For example, you can show that one-gluon exchange forces between a ψ spinor and a $\bar{\psi}$ spinor are attractive, and furthermore that for those two to bind into a color singlet is the most attractive scalar channel for two fermions.

There is one other channel that is attractive: two ψ spinors in a color triplet bound state. Now I want to explore the consequences of assuming that in QCD the detailed dynamics works out so that $\langle\bar{\psi}\psi\rangle = 0$, but $\langle\psi\psi\rangle \neq 0$. Again, we don't believe this really happens in the theory because we see a reasonable version of it in the real world, and it doesn't happen. But just to illustrate the composite fermion mechanism that I want to discuss, I am going to assume that $\langle\bar{\psi}\psi\rangle = 0$, but $\langle\psi\psi\rangle \neq 0$. So let me write out what I mean by $\langle\psi\psi\rangle$ in more detail:
$$\langle \psi_i^\alpha \psi_j^\beta \varepsilon^{ij} \varepsilon_{\alpha\beta\gamma}\rangle = \langle \bar{\psi}^{*\alpha}_a \bar{\psi}^{*\beta}_b \varepsilon^{ab} \varepsilon_{\alpha\beta\gamma}\rangle = V_\gamma \neq 0.$$
ψ carries a color index given by the Greek letters and an $SU(2)_L$ singlet (ε^{ij} is the antisymmetric tensor symbol). We also antisymmetrize on the color indeces, leaving one color index left over. Similarly, the $\langle\bar{\psi}^*\bar{\psi}^*\rangle$ condensate also forms with contracted $SU(2)_R$ indices. Thus, we have a vacuum expectation value which carries a color index. The condensates also carry baryon number. This signals

the spontaneous breakdown of both color and baryon number. Color
in this version of QCD would break down from SU(3) to SU(2), and
five gluons would gain mass. Baryon number, a global symmetry, is
broken, but there is a new baryon number which is generated by the
old baryon number, minus a broken generator of the color group.

$$B_{new} = B - \lambda_8 \qquad (2)$$

where λ_8 is the usual Gell-Mann matrix with a different normalization:

$$\lambda_8 = \begin{pmatrix} 1/3 & 0 & 0 \\ 0 & 1/3 & 0 \\ 0 & 0 & -2/3 \end{pmatrix} \qquad (3)$$

Now we want to see how the fermions transform under the new
symmetries $SU(2)^C \times SU(2)_L \times SU(2)_R \times U(1)_{new\ B}$

$$\psi = (\underline{2}, \underline{2}, \underline{1}, 0) + (\underline{1}, \underline{2}, \underline{1}, +1)$$

$$\bar{\psi} = (\underline{2}, \underline{1}, \underline{2}, 0) + (\underline{1}, \underline{1}, \underline{2}, -1) \qquad (4)$$

The former color triplets split up into color doublets and singlets.
The color doublet (which consists of the first two components in
a vector that λ_8 would act on) has new baryon number zero. This
ψ had old baryon number +1/3, minus 1/3 from λ_8 gives new baryon
number zero for the color doublet. The color singlet (which is
the third component of the vector) has new baryon number 1/2-(-2/3),
or +1. And similarly for the antiparticles. Now the color doublets
should still be confined by $SU(2)^C$ forces. They have a dynamical
effective mass which comes from their couplings to the nonzero
vacuum expectation values of the condensates. There is an
effective term in the theory which gives a mass to these particles,
just like in the usual scenario for QCD where there is an effective
mass term for quarks coming from the $\bar{\psi}\psi$ vacuum expectation value.
The color SU(2) singlets, however, don't couple to the condensate
and remain massless. Their masslessness is enforced by the un-
broken $SU(2)_L \times SU(2)_R$ symmetry. In this scenario, $SU(2)_L \times SU(2)_R$
remains completely unbroken, as opposed to the usual $<\bar{\psi}\psi> \neq 0$
scenario where it breaks down into a vector SU(2) symmetry. So
in this hypothetical version of QCD, we have found observable
(by which I mean unconfined) massless fermions which happen to
carry the quantum numbers of the proton and neutron. They are
SU(2) isospin doublets and they carry baryon number $\pm$ 1.

What, you may ask, does this have to do with composite
fermions? There is a principle called Higgs/confinement com-
plementarity. It is not at the level of a rigorous theorem,
although a version of it has been rigorously proved in lattice
gauge theory by Fradkin and Shenker[5]. 't Hooft discussed it
in his Cargese summer lectures.[3] The Stanford group has discussed

it in detail[2], and various other versions of it have been mentioned
in the literature. It says that in certain theories you cannot tell
whether or not a gauge symmetry undergoes a spontaneous breakdown.
For example, in an SU(2) gauge theory with a Higgs doublet, you
could assume that the massive vector particles in the theory are
gauge bosons that get mass via the Higgs mechanism, or you could
assume that they are bound states of confined Higgs particles:

$$W_\mu^+ = \phi_i^+ \overleftrightarrow{D}_\mu \phi_j^+ \varepsilon^{ij}$$
$$W_\mu^o = \phi_i^+ \vec{D}_\mu \phi_i \qquad (5)$$

where ϕ is the Higgs doublet and D_μ is the covariant derivative.
If you take these expressions and replace ϕ by its vacuum expectation value you recover just the gauge fields from the covariant
derivative (apart from the overall scale). Thus, we are led to
a general principle that says that the spectrum is the same whether
you assume the gauge symmetry breaks down and gauge bosons gain
mass or whether you assume that the vector particles in the theory
are composite particles made out of confined fields. A technical
point is that this is not always true; you can't always make this
reinterpretation. The Higgs fields (or the composite operators
that have vacuum expectation values) must be in faithful
representations of the gauge group. I won't try to explain what
that means. If you don't know, just remember that it's not
always true that the reinterpretation is possible. In our
previous theory - QCD with the unusual condensate assumption -
the reinterpretation is possible and those fermions that look
like massless protons and neutrons can indeed be reinterpreted as
if color symmetry and the original baryon number were unbroken,
but that the theory formed massless composite fermions which are
Wigner mode realizations of the $SU(2)_L \times SU(2)_R$ symmetry.

This really isn't good enough to get composite quarks and
leptons. I have told you how to get massless fermions, but
quarks and leptons aren't massless, at least not all of them.
We need a mechanism that can generate masses for them. Let me
give a generic discussion of how that could happen in these
kinds of models. Let's suppose we start with a gauge group G
which at some energy scale M_1 undergoes a dynamical breakdown
to a new gauge group G_1. Further, suppose that at some lower
energy M_2, the group breaks again to a new group G_2. Suppose
at the end of all this symmetry-breaking, we have a fermion ψ,
which did not couple directly to any of the condensates and so
stayed apparently massless. However, because there are two
stages of symmetry breakdown, you can draw this Feynman diagram:

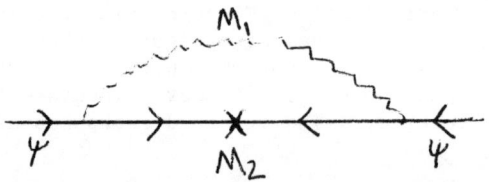

In this Feynman diagram, ψ emits a gluon with mass $\sim M_1$, and turns into another fermion, with dynamical mass $\sim M_2$. (In other words, it couples to the M_2 condensate). The diagram gives a small mass to ψ. The small mass is of the order of magnitude M_2^3/M_1^2. The M_1^2 comes from the boson propagator and the M_2^3 from the vacuum expectation value. If I try to reinterpret these things as composite states, I have to draw all kinds of horrendous diagrams, usually involving instantons, to get the mass.

If I have a theory with several stages of symmetry breaking, I can get a broad spectrum of light fermion masses because the various M_i^3/M_j^2 will, in general, span several orders of magnitude. This could explain the wide variation in quark and lepton masses.

I'm out of time but I can't quit without briefly mentioning the very beautiful analysis of 't Hooft which is complementary to this. 't Hooft wanted to know under what general circumstances a gauge theory could produce massless fermions. So he imagined starting with some fermions strongly coupled to a non-Abelian gauge group G_S. In general, the fermions will have some global symmetries, which gives you a flavor group. Now 't Hooft imagines weakly gauging the flavor group - so weakly that you don't disturb the spectrum of bound states of the strong group. However, this new theory may not be renormalizable because of anomalies. 't Hooft then puts in "spectator" fermions which are G_S singlets, but carry just the right flavor quantum numbers to cancel the anomalies. Then he requires that the massless sector of the theory form a consistent, renormalizable quantum field theory all by itself. The massless particles are the G_S gauge bosons, the spectator fermions, and any massless composites. In general, he finds that it is impossible to find a set of massless composites which cancel the anomalies of the spectators, and therefore it is necessary to have a spontaneous breakdown of at least some of the flavor symmetry.

Let me sum up these ideas. Composite quarks and leptons require the underlying theory to have a chiral symmetry, unbroken at the binding scale, which enforces their lightness. This kind of scheme could explain quark and lepton mass hierarchies. Unfortunately, no realistic model exists as yet.

REFERENCES

1. S. Drell and S. Brodsky, Phys. Rev. D22, 2236 (1980).

2. S. Raby, S. Dimopoulos, and L. Susskind, Nucl. Phys. B169, 373 (1980). S. Dimopoulos, S. Raby, and L. Susskind, Nucl. Phys. B173, 208 (1980).

3. G. 't Hooft, lecture given at the Cargèse Summer Institute (September, 1979).

4. M. Srednicki and L. Susskind, Princeton University preprint (February, 1980).

5. E. Fradkin and S. Shenker, Phys. Rev. $\underline{D19}$, 3682 (1979).

DISCUSSION

WEINBERG: The comment has to do with the last point you mentioned about the 't Hooft conditions. I don't disagree with what you say but I think something more may be said about it. It really is not true that the 't Hooft analysis shows that in a wide class of theories you must have spontaneous symmetry breaking and in fact it is the exception rather than the rule. For example, in quantum chromodynamics, if you have any number of flavors that is not a multiple of three, SU(N) x SU(N) does not have to be spontaneously broken; it is only in the case where you have three or six etc. of flavors where it does. The confusion may arise from the fact that in addition to this condition about anomalies, and by the way, I recommend dropping the spectator fermions which is just the way 't Hooft happened to think about it, and just say that the bound states have to reproduce the anomalies of the original theory. In addition to that condition which is completely reasonable and probably can be proved rigorously, 't Hooft used another condition which he called the decoupling condition and many people have gotten the impression that it is a straightforward application of the Appelquist-Carazone theorem. The decoupling condition, as 't Hooft uses it, says that if you imagine having N quarks so that there is a flavor symmetry SU(N) x SU(N), then when you give one of them a large mass and remove it from the theory, the remaining symmetry SU(N-1) x SU(N-1) must not be able to prevent the composites which contain that quark from also getting a large mass; this follows because if you are going to remove it from the theory you have got to remove all the composites that contain it. That is a completely reasonable hypothesis.

The trouble is that in order to use it the way 't Hooft uses it, and a large number of other people following 't Hooft have used it, you have to make an additional assumption, which is that as any of the quark masses go to infinity the pattern of spontaneous symmetry breaking does not change, an assumption I find totally unreasonable although in special cases it may happen to be true. And therefore, you cannot derive these conclusions. There is an alternative assumption that you might make which is not subject to that criticism. You might suppose that if you have a theory with N massless quarks so that the flavor symmetry is SU(N) x SU(N) and then you give one of them an infinitesimal mass, in which case you know that the remaining flavor symmetry is SU(N-1) x SU(N-1), that any composite containing that massive quark must get a little mass. The representations to which the particles belong must not be such as to prevent composites which contain that massive quark from becoming massive. According to this general principle, the things that contain massive constituents ought to be massive. That is a stronger assumption; it can't be derived from the Appelquist-Carazone theorem but it immediately leads to all the algebraic results derived from 't Hooft and others without having to make assumptions about the continuity of symmetry-breaking as the mass goes to infinity.

The trouble is that the assumption is so strong that it would, for example, forbid the spontaneous breaking of any non-chiral symmetries whatever. For example, if you have SU(3) and if you have 3 quarks in quantum chromodynamics which are massless, then you have an SU(3) x SU(3) symmetry; if you give those three quarks a small equal mass then you still have an SU(3) symmetry. If that is spontaneously broken, there must be massless scalar Goldstone bosons like the old kappa of 15 years ago, but they would have to be composed of massive constituents. So you would have a wonderful rule that tells you the eight-fold way cannot be spontaneously broken, which would be great because in fact it is not spontaneously broken. Maybe that would explain why the eight-fold way is not broken spontaneously.

The trouble is that if you look at models and, in particular, go back to the Nambu-Jona-Lasinio-type model, you find that the assumption just is not true and in the Nambu-Jona-Lasinio-type model, it is just as easy for non-chiral symmetries to be spontaneously broken as chiral symmetries and, in general, it is perfectly possible in these theories to have massless particles made out of massive constituents. So I don't believe the 't Hooft decoupling condition in either of its forms, his or mine, and therefore, I think in fact you can not draw such strong conclusions as spontaneous symmetry breaking almost always occurs. (My work on these matters has been done in collaboration with John Preskill.)

SREDNICKI: My only comment is that I agree with you completely about the decoupling theorems. I was not aware that throwing them away changed the results you can get from 't Hooft's equations so drastically, but I certainly agree that they are now nearly as believable as 't Hooft's anomaly requirements.

$M_Z\cos\theta = M_W$ AND THE QUANTUM NUMBER OF DYNAMICAL HIGGS BOSONS[*]

Hung-Sheng Tsao
Rockefeller University
New York, N.Y. 10021

Weinberg[1] and Susskind[2] have observed that in the theory of dynamically broken electro-weak interactions due to extra strong interaction[3,4,5] (technicolor, hypercolor, metacolor etc.), the relation $M_Z\cos\theta = M_W$ follows naturally if the $U(1) \times SU(2)$ quantum number of the technifermions are the same as that of the ordinary fermions, because the dynamical Higgs bosons composed of the technifermions are weak doublets.

I would like to point out in this talk that in the dynamically broken symmetry theory $M_Z\cos\theta = M_W$ is a very general phenomena. Actually it is a consequence of the maximum parity violating nature of the technifermion representation.

Since the ordinary color interaction plays no direct role in the following discussion, I will simply omit it. For definiteness I take $SU(N)$ as the technicolor gauge group. The model gauge group for the dynamically broken electroweak interaction is $U(1) \times SU(2) \times SU(N)$.

The left-handed fermions are in (Y,I,N) representation and the right-handed fermions are in $(Q,0,N)$ representation, with the arbitrary isospin I and the weak hypercharge Y.[6] Let's denote e the generic coupling constant of the $U(1) \times SU(2)$ gauge group and g the coupling constant of the $SU(N)$ gauge group. The standard dynamical symmetry breaking scenario says[3], when e is turned off and g is small, the fermionic system had a global $U(1) \times SU_L(2I+1) \times SU_R(2I+1)$ symmetry. As g becomes of order one in the TeV

---

[*] Work supported in part by the Department of Energy under Contract Grant Number DE-AC02-76ER02232B.000

energy scale ($\Lambda_s \sim$TeV), the strong SU(N) interaction causes the $SU_L(2I+1) \times SU_R(2I+1)$ symmetry to be spontaneously broken down to the $SU_V(2I+1)$ symmetry, with the appearance of $4I(I+1)$ Goldstone bosons. As we turn on e, the $U(1) \times SU(2)$ gauge symmetry will be dynamically broken down to the $U_Q(1)$ gauge symmetry with the appearance of the massive charged gauge boson W and the neutral gauge boson Z and some charged pseudo-Goldstone bosons.

In terms of the effective dynamical Higgs bosons ϕ^i, composed of technifermions ψ_L and ψ_R:

$$\phi^i = \bar{\psi}_L \psi^i_R , \tag{1}$$

which have the $U(1) \times SU(2)$ quantum number

$$(Y_i, I), \quad Y_i = -I, -I+1, \ldots I-1, I , \tag{2}$$

the spontaneous breakdown of the $SU_L(2I+1) \times SU_R(2I+1)$ symmetry to the $SU_V(2I+1)$ symmetry means:

$$<\phi^i> = <\bar{\psi}_L \psi^i_R> = V \qquad \text{for all i} \tag{3}$$

Using the well known Lee-formula[7]

$$\rho \equiv \left(\frac{M_W}{M_Z \cos\theta}\right)^2 = \frac{\sum_{I,Y} 2Y^2 |V_{I,Y}|^2}{\sum_{I,Y}(I^2+I-Y^2)|V_{I,Y}|^2} , \tag{4}$$

we have in our situation

$$\rho = \frac{\sum_{-I}^{I}(2Y_i^2)}{\sum_{-I}^{I}(I^2+I-Y_i^2)} . \tag{5}$$

Now using the relations

$$Q = I_3 + Y \tag{6}$$

and

$$\sum_{-I}^{I} Y_i^2 = \sum_{-I}^{I} I_3^2 = \frac{1}{3}\sum_{-I}^{I}(\vec{I})^2 = \frac{1}{3}\sum_{-I}^{I}(I^2+I)$$

we get

$$\rho = 1 \ ! \tag{7}$$

We can also derive this interesting result by using Weinberg's formalism[3]. So we have demonstrated that as long as the technifermion belongs to a maximally parity-violating representation, independent of its weak isospin, $\rho=1$ follows. The observation of Weinberg and Susskind for the I=1/2 case is just a special case of our result.

Next let's briefly discuss some properties of the Goldstone bosons. The 4I(I+1) Goldstone bosons consist of 2I+1-Q charged Q and 2I neutral bosons. Among them, as we turn on e, one charged and one neutral Goldstone boson are the unphysical Higgs boson, the rest of the charged bosons are pseudo-Goldstone bosons[1,5,8] with the masses[9]

$$M_Q^2 = Q^2 \frac{3\alpha}{4\pi} M_Z^2 \ln\left(\frac{\Lambda_S}{M_Z}\right)^2 \tag{8}$$
$$= (8.6Q \text{ GeV})^2 \text{ for } \Lambda_S = 1\text{TeV}$$

These multi-charged Higgs bosons will give a spectacular rise in R (the e^+e^- to hadron to muon ratio):

$$R = \frac{1}{4} Q^2 \ . \tag{9}$$

As for the 2I-1 neutral Goldstone bosons, they will remain massless in this model theory. We could take either one of the following three different points of view:

(1) One could use their existence to rule out the consideration of the I≠1/2 representation. In this aspect our observation demonstrates that the condition $\rho=1$ and the requirement of the absence of the massless Goldstone

bosons in the model gauge theory U(1)xSU(2)xSU(N) limits the weak isospin of the technifermions to 1/2.

(2) Since there also exists the possibility that the fermion pair could create some dynamical gauge boson bound states,[10] these massless Goldstone bosons could become the unphysical Higgs bosons of these gauge bosons. So the presence of the massless Goldstone bosons is just a signal of the existence of some massive gauge bosons with masses of order Λ_s.

(3) It is possible that in a truely grand unified theory one will not encounter the massless Goldstone bosons, they all become unphysical Higgs bosons of some elementary gauge bosons.

Details and other related points will be discussed elsehwere.[11]

ACKNOWLEDGEMENT

I wish to thank my colleagues at the Rockefeller University, especially W. Marciano and A. Sanda for discussions on the ideas presented in this talk.

REFERENCES

1 S. Weinberg, Phys. Rev. D1, 1277 (1979).

2 L. Susskind, Phys. Rev. D20, 2619 (1979).

3 S. Weinberg, Phys. Rev. D13, 974 (1976), and the references within.

4 S. Dimopoulous and L. Susskind, Nucl. Phys. B155, 237 (1979).

5 E. Eichten and K. Lane, Phys. Lett. B90, 125 (1980).

6 We will take $K < \frac{11}{2} N$ for maintaining the asymptotic freedom of the SU(N) gauge group.

7 B.W. Lee, Proceedings of the XVI International Conference on High Energy Physics (1972) Batavia, Illinois.

8   M.A.B. Beg, H.D. Politzer and P. Ramond, Phys. Rev. Lett. <u>43</u>, 1701 (1979).
9   M.E. Peskin, preprint DPh-T/80/46.   J. Preskill, preprint HUTP-80/A033.
    S. Dimopoulos, S. Raby and G.L. Kane, preprint UMHE 80-22.
10 F.A. Bais and J.-M. Frere, preprint CERN-TH 2911, 1980.
11 H.-S. Tsao, to be published. H.-S. Tsao in the Proceedings of the 1980 Guangzhou Conference on Theoretical Particle Physics.

GENERATING MASS WITHOUT HIGGS FIELDS

Kerson Huang

Center for Theoretical Physics
Laboratory for Nuclear Science and Department of Physics
Massachusetts Institute of Technology
Cambridge, Massachusetts USA 02139

In the Weinberg-Salam model, as in the various grand unification models, Higgs fields were introduced to endow particles with mass in a gauge-invariant way. These Higgs fields seem to be undesirable features of the theory, for they bring with them a large number of arbitrary parameters. Most likely, they are phenomenological order parameters, not unlike the Ginsberg-Landau order parameter in superconductivity. I shall describe some work done, in collaboration with my graduate student Roberto Mendel[1], that attempts to generate masses spontaneously, without Higgs fields. The principles are not new, and have been suggested sometime ago by Jackiw and Johnson[2], and Cornwall and Norton.[3] What we did was to apply them specifically to the Glashow-Weinberg-Salam model[4] without Higgs. It is to be emphasized that the mechanism generates masses in a way similar to the generalization of the energy gap in the BCS theory of superconducting. It does not produce a "dynamical Higgs field".

The renormalized full fermion propagation is a diagonal matrix in flavor space:

$$S(p) = \frac{1}{\not{p} - \phi(p^2)}. \qquad (1)$$

A theorem due to Goldstone[5] states that $\{\gamma_5, \phi(p^2)\} = 2\gamma_5 \phi(p^2)$ satisfies a Bethe-Salpeter (BS) equation for total four-momentum zero, and relative form-momentum p^μ. A non-trivial solution to the BS equation would mean that the fermions acquire mass through spontaneous breaking of chiral symmetry. Physically, it would mean that fermion-antifermion pseudoscalar bound states of <u>zero total four-momentum</u> exist, which are degenerate with the bare vacuum. They would form a condensate in the physical vacuum, similar to the condensate of Cooper pairs in superconductivity.

The neutrinos are all massless by construction, for no right-handed neutrinos are put into the theory. Thus, $\phi_\nu \equiv 0$. The quark masses generated here will be the "current quark" masses.

Although one expects, by continuity, that the same BS equation admits a zero-mass (but not zero total four-momentum) solution,

---

*This work is supported in part through funds provided by the U.S. Department of Energy (DOE) under contract DE-AC02-76ER03069

ISSN:0094-243X/81.720528-06$1.50 Copyright 1981 American Institute of Physic corresponding to a massless pseudoscalar particle, this state actually decouples from the theory.[2,3] Without solving the equation in detail, one cannot say whether or not there exist massive bound states; but they have little to do with the Higgs field in the Weinberg-Salam sense, for they are not directly tied to the spontaneous generation of mass.

The renormalized BS equation for the $\phi_n(p^2)$, where n labels the flavor, is of the form

$$\phi_n(p^2) = \int \frac{d^4q}{(2\pi)^4} K_{nm}[p,q;\phi] \phi_m(q^2), \qquad (2)$$

where the BS kernel is considered to be a functional of all the proper self-energies ϕ_1, ϕ_2, ... It is represented graphically in Fig. 1, where p is positive along the direction of the arrow. Solid lines represent full propagators.

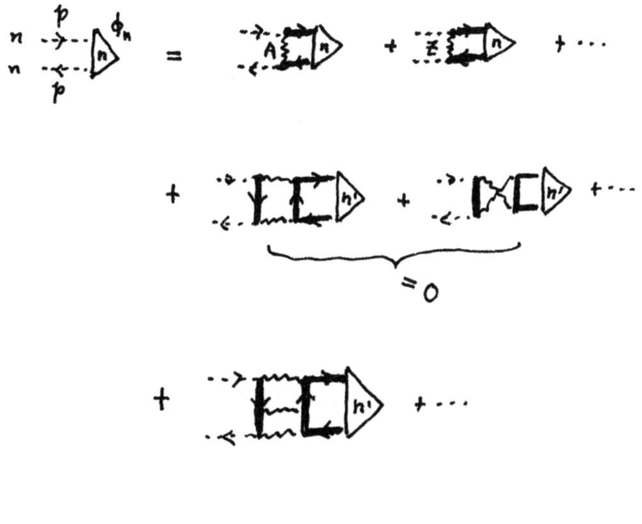

Fig. 1. Expansion for the BS equation, treating the kernel as a functional of the fermion self-energy points.

It has the following features:
(a) To lowest order only photons and Z exchange contribute. The contributions of W exchange vanish to this order because of the pure left-handedness of the coupling. Thus, to this order, there is no coupling among the different flavors.
(b) The lowest order graphs that can couple all flavors

involve two vector exchanges; but they turn out to vanish. Thus, the lowest non-vanishing order in which all flavors are coupled is $O(e^6)$, a representative graph of which is the last one in Fig. 1.

(c) The asymptotic behavior of $\phi(p^2)$, as $p^2 \to \infty$, is determined by the lowest order graphs, because the higher order ones all contain enough propagators, so as not to rely on $\phi(p^2)$ to supply the necessary convergent factor. Thus, a necessary condition for dynamical mass generation is that the lowest order BS equation has a non-trivial solution.

(d) In the lowest order approximation, which amounts to a modified ladder approximation, in which full fermion propagators replace base order, the asymptotic behavior of $\phi_n(p^2)$ is the same as that of the Goldstein solution[6]:

$$\phi_n(p^2) \xrightarrow[p^2 \to \infty]{} m_n (-p^2/m_n^2)^{-\epsilon_n},$$
$$\epsilon_n = \tfrac{1}{2}[1 - (1-4\lambda_n)^{1/2}] \approx \lambda_n, \qquad (3)$$
$$\lambda_n = \frac{3e^2}{16\pi^2}(Q_n^2 + Q'_{nL} Q'_{nR}),$$

where $e^2/4\pi \cong 1/137$, Q_n is the electric charge, and Q'_{nL}, Q'_{nR} are respectively the neutral charge of the left- and right-handed particle, all in units of e. The parameters m_n are arbitrary mass scales, which are uncoupled to this order, but which are coupled through graphs of order e^6, so that eventually only one overall work scale remains arbitrary. The condition that a non-trivial solution exist is that

$$0 < \lambda_n \leq \tfrac{1}{4}. \qquad (4)$$

The vector masses are obtained by calculating the vacuum polarization tensors $\pi_V^{\mu\nu}(k)$ near $k=0$. They receive additional contributions from all flavors, as shown in Fig. 2.

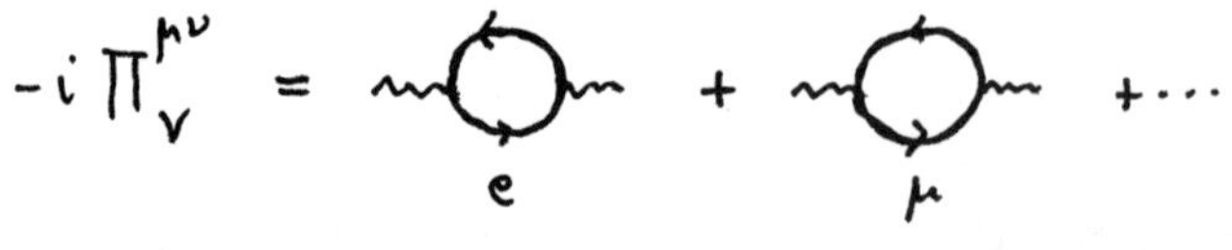

Fig. 2. Vector-boson proper self-energies.

The photon remains massless because it has only vector interactions, and is therefore not directly coupled to the condensate of pseudoscalar bound states. For W and Z, we obtain the sum-rules[7]

$$M_W^2 = \frac{g^2}{64\pi^2} \sum_n \frac{m_n^2}{\epsilon_n}[1 + O(g^2)], \qquad \frac{M_W}{M_Z} = \cos\theta, \qquad (5)$$

where θ is the Weinberg angle, and $g = e/\sin\theta$, with $\sin^2\theta \cong 1/4$.

To see whether the condition (4) is fulfilled, we look at the following table:

	Q^2	$Q'_L Q'_R$ (for $\sin^2\theta=1/4$)	mass?
e,μ,τ	1	$-1/3$	yes
u,c,t	4/9	$-8/27$	yes
d,s,b	1/9	$-5/27$	no

It is seen that (4) is satisfied for all leptons and all the charge 2/3 quarks, but fails for all the charge -1/3 quarks. The latter therefore remain massless in this model.

The above conclusion is in contradiction with semi-empirical evidence, particularly for the b quark. Therefore we propose to amend the model. To do this in a simple-minded way, we note that the form of λ_n in (3) reflects the fact that the bound-states in the condensate are bound by the net Coulomb force arising from short-distance photon and Z exchange:

$$ F \xrightarrow[r \to 0]{} - \frac{Q^2 + Q'_R Q'_L}{r^2} . \tag{6}$$

The contribution of Z exchange is always repulsive, because $Q'_R Q'_L < 0$. The reason that d,s,b fail to develop mass is that the Z repulsion overcomes the electric attraction between the particle-antiparticle pair. To remedy this, we propose to simply cut off the Z contribution at small distances. From a more detailed study of the BS equation, this cutoff has the comfortable upper bound

$$ \Lambda^2 \ll 10^{25} m_e^2, \tag{7}$$

which corresponds to a distance

$$ a \gg 10^{-27} \text{ cm}. \tag{8}$$

In practice, the effect of doing this is simply to omit the term $Q'_{nL} Q'_{nR}$ in (3), and take

$$ \lambda_n = \frac{3e^2}{16\pi^2} Q_n^2 . \tag{9}$$

With this modification of the model, we obtain

$$M_W^2 = \frac{1}{12 \sin^2\theta} \left[m_e^2 + m_\mu^2 + m_\tau^2 + \frac{27}{4}(m_u^2 + m_c^2 + m_t^2) + 27(m_d^2 + m_s^2 + m_b^2) \right]. \tag{10}$$

Using $M_W = (37/\sin\theta) \text{GeV}/c^2$, we obtain

$$m_t = 48 \text{ GeV}/c^2. \tag{11}$$

If, for some reason, this model can be so modified that we can generate masses for d,s,b which keeping λ_n intact in the form given in (4), then we obtain $m_t = 25 \text{GeV}/c^2$.

As remarked before, mass ratios can be calculated only through inclusion of graphs of $O(e^6)$ in the BS equation. Such a calculation is being attempted.

The moral we draw from this work is twofold:
(a) The Higgs field may be a phenomenological order parameter. In particular, the excitation spectrum may be quite different from that of the Weinberg-Salam model with Higgs.
(b) The gauge principle may only be approximate, in that Z may be less "elementary" than the photon. What happens at extremely short distances may be quite different from what is imagined in the so-called grand unified theories.

REFERENCES

1. K. Huang and R. Mendel, MIT preprint CTP #876 (1980).
2. R. Jackiw and K. Johnson, Phys. Rev. D8, 2386 (1973).
3. J. M. Cornwall and R. E. Norton, Phys. Rev. D8, 2338 (1973).
4. S. L. Glashow, Nucl. Phys. 26, 579 (1961); S. Weinberg, Phys. Rev. Lett. 19, 1264 (1964); A. Salam, in Elementary Paricles Theory ed. N. Svarthholm (Almquist and Wiksell, Stockholm, (1968)); S. L. Glashow, J. Iliopoulis, and L. Maiani, Phys. Rev. D2, 185 (1970).
5. J. Goldstone (unpublished), quoted in M. Baker, K. Johnson, and B. W. Lee, Phys. Rev. 133, B209 (1964).
6. J. Goldstone, Phys. Rev. 91, 1516 (1953).
7. Carter and H. Pagels, have suggested a sum-rule of a similar form.

DISCUSSION

BEG: There is a published paper two colleagues of mine, Ashton Carter and Heinz Pagels. They have a sum-rule similar to the one you showed.

HUANG: Yes, Carter and Pagels [Phys. Rev. Lett. 43, 1845 (1979)] suggested a sum-rule of the same qualitative form. In our work we calculated the coefficients appearing in the sum-rule explicitly, in the GWS model. There is also a difference in point of view, and in the conclusions we draw.

COMPOSITE MODELS

S. Adler
Institute for Advanced Study
Princeton, NJ

I want to talk about composite models. In forming composite models there are two basic stratagies one can follow. One is that you can try to include the color group $SU(3)_c$ kinematically. For example in a model which was constructed by Greenberg and Sucher, $SU(3)_c$ is still present at the fundamental field level. (See Greenberg's report to this workshop. Many other people have similar models.)

The other idea, which is what I want to talk about, is to try and generate the $SU(3)_c$ group dynamically, without putting it into the Lagrangian to start with. This is suggested by a scheme that is really just a set of rules without dynamics which Harari and Shupe proposed. I will discuss their rule, then suggest what I call U(2) algebraic chromodynamics, or a quaternionic dynamics, as a possible dynanics for the Harari-Shupe rule. I will briefly discuss the ideas behind that and then Biedenharn will discuss some work he, Horwitz, and Sepunaru have done in connection with the U(2) theory and a quantum mechanics based on quaternions.

What Harari and Shupe suggested was a heuristic composite model to account for the charge-multiplicity regularities of the known fermions. You have three generations with 3Q going from -3 to 0, then back up to +3 in each generation, as shown in Table 1.

Table 1

Charge Multiplicities of Known Fermions

GENERATION	3Q							
	-3	-2	-1	0	0	+1	+2	+3
3rd	T^-	$\bar{t}$	b	$\bar{\nu}_T$	ν_T	$\bar{b}$	t	T^+
2nd	μ^-	$\bar{c}$	s	$\bar{\nu}_\mu$	ν_μ	$\bar{s}$	c	μ^+
1st	e^-	$\bar{u}$	d	$\bar{\nu}_e$	ν_e	$\bar{d}$	u	e^+

The idea of the Harari-Shupe scheme was to try and obtain the whole set of states within a generation as three-particle bound states of a fundamental two-component constituent, that Harari calls the rishon. The rishon has two components: a T component with a charge of 1/3, and a V component with a charge of zero. The higher generations will be regarded as internal dynamical excitations of a type not yet understood. In other words, one won't know precisely the excitation mechanism until one knows the dynamics. The idea is to try to explain what is going on in a horizontal generation.

The Harari-Shupe scheme is motivated by the following picture, which I first saw in a paper by Glashow, and that is if you plot the charge multiplicity of the known fermions within a given generation, ignoring for the moment the neutrino state-counting, you get a cube turned on end (fig. 1). Furthermore the triangles of the u and $\bar{d}$ states are oriented just like weight diagrams for $SU(3)$ 3 and $\bar{3}$.

There is something suggestive about this diagram. In the past when such diagrams have appeared, as in the eight-fold way, they have usually indicated a composite structure, and the Harari-Shupe scheme is an attempt to realize this structure as a 2^3.

That is the basic idea. Let me say that when you put the neutrino helicity-counting in, I believe a more correct idea is something like fig. 2, with the neutrino and the antineutrino identified, and then of course you need a helicity-selection rule to make the weak interactions come out right, and presumably the second state is driven up to a very high mass. So you start out with 16 states, one is split off, and the rest form a pattern like fig. 2.

Then the Harari-Shupe assignments of the states for generation one are as follows:

Table 2

Harari-Shupe Assignment for Generation 1

e^+					T_1	T_2	T_3			
u_1, u_2, u_3		V_1 T_2 T_3 ,			T_1	V_2	T_3 ,	T_1	T_2	V_3
$\bar{d}_1, \bar{d}_2, \bar{d}_3$		T_1 V_2 V_3 ,			V_1	T_2	V_3 ,	V_1	V_2	T_3
ν_e					V_1	V_2	V_3			

I will show in a moment, in terms of what are essentially spin-1/2 column vectors, what the order means. It is essentially just like composing three angular momenta, but in order to get color out of it you have to do something non-trivial, and that is what I am going to emphasize. And then, as I said, the higher generations are regarded as internal dynamical excitations.

For this scheme to work the dynamics have to give in the end an $SU(3)_c \times [SU(2) \times U(1)]_{\text{Weak-EM}}$ for the secondary interactions of the composites and that is a very non-trivial task for any composite model. But what I want to focus on here is whether you can get $SU(3)_c$ out from a scheme starting with two constituents.

So we have this kind of picture. We have at the level of the composites the secondary interactions described by the known phenomenological Lagrangian, based on $SU(3) \times SU(2) \times U(1)$. Now we know that this is a gauge theory, and for color singlets a local field theory. The question is what is down below this level, what is a reasonable dynamics for the Harari-Shupe rule?

First of all, I think it has to be a gauge theory. In general, one finds that if you form composites from gauge particles you see less gauge symmetry at the composite level than you do at the original level. A good example is nuclear physics. You start with QCD; you form composites, protons and mesons, and the effective theory for that is essentially the chiral σ-model which is not a gauge theory of the conventional type at all. In atomic physics, you start with QED, and end up with intermolecular forces described by intermolecular potentials that are not gauge potentials at all. So it is hard to see how you can get all this gauge structure at the composite level without starting from a gauge theory.

The fact that there are two constituents suggests that you have to start with either SU(2) or U(2). Conventional SU(2) QCD cannot generate $SU(3)_c$ because SU(2) is self-conjugate. However, it turns out that a U(2) theory of a peculiar variety, the variety of theory I'm calling "algebraic chromodynamics" can generate $SU(3)_c$ at the composite level. I'll explain in a moment what this theory is. It is not a local field theory; I'm going outside the usual rules of local field-theory, which brings an extra handedness into the formulation which allows you to build up non self-conjugate group structures.

The basic strategy, then, will be the following. I will discuss theories where you drop the microscopic causality postulate. I'll argue that you have to have operators for which [A(x), B(y)] is non-zero for space-like separated arguments. But then I'll replace this by a strong invariance requirement to get physical sense, and that invariance requirement is that the theory should be invariant under local operator-valued gauge transformations. The idea here is very much like the transition from special to general relativity. In special relativity you have a flat background metric, and in general relativity you drop that requirement but you replace it by the statement that the laws of physics should be covariant with respect to general coordinate transformations.

That was the philosophy, now let me show you what happens with the Harari-Shupe model that requires non-locality to get color out.

We will take the rishon $\binom{r_1}{r_2}$ or $\binom{T}{V}$ to be a two-component spinor. The U(2) charges acting on these spinors are just the unit matrix and the three Pauli-matrices; these are just another notation of course for Hamilton's quaternions. The reason why it is better to write them in the Pauli matrix form is because if you use the abstract notation, you don't see that the spinors play a role in the theory as essentially "square roots" of the quaternions, whereas if you use the Pauli matrix notation the spinors appear in a natural way.

The three rishon states are formed as follows from the "up" or "down" states of the constituents: e^+ is $\binom{1}{0}_1$, $\binom{1}{0}_2$, $\binom{1}{0}_3$. The u_1 would look like $\binom{0}{1}_1$, $\binom{1}{0}_2$, $\binom{1}{0}_3$. $\binom{0}{1}_1$ for the first rishon and $\binom{1}{0})$ for the second and third; u_2 as $\binom{1}{0}_1$, $\binom{0}{1}_2$, $\binom{1}{0}_3$ with $\binom{0}{1}$ as the second rishon and so forth (see table 3). Now let's look

Table 3

$$r = \begin{pmatrix} r_1 \\ r_2 \end{pmatrix} = \binom{T}{V}$$

U(2) charges acting on these are

$$T^0 = \begin{pmatrix} 1 & 0 \\ 0 & 1 \end{pmatrix}, \quad T^1 = \begin{pmatrix} 0 & 1 \\ 1 & 0 \end{pmatrix}, \quad T^2 = \begin{pmatrix} 0 & -2 \\ 2 & 0 \end{pmatrix}, \quad T^3 = \begin{pmatrix} 1 & 0 \\ 0 & -1 \end{pmatrix}$$

(Just the quaternions)

3 r-states are

$$e^+ = \binom{1}{0}_1 \binom{1}{0}_2 \binom{1}{0}_3$$

$$u_1 = \binom{0}{1}_1 \binom{1}{0}_2 \binom{1}{0}_3, \quad u_2 = \binom{1}{0}_1 \binom{0}{1}_2 \binom{1}{0}_3, \quad u_3 = \binom{1}{0}_1 \binom{1}{0}_2 \binom{0}{1}_3$$

$$\bar{d}_1 = \binom{1}{0}_1 \binom{0}{1}_2 \binom{0}{1}_3, \quad \bar{d}_2 = \binom{0}{1}_1 \binom{1}{0}_2 \binom{0}{1}_3, \quad \bar{d}_3 = \binom{0}{1}_1 \binom{0}{1}_2 \binom{1}{0}_3$$

$$\nu_e = \binom{0}{1}_1 \binom{0}{1}_2 \binom{0}{1}_3$$

at what a typical color transition would look like. It is a horizontal transition between u_1 and u_2. I want to argue now that non-locality is needed. The typical color transition $u_1 \to u_2$ can be written as an exchange operator τ^+ acting on the spinor subspace 1, τ^- acting on subspace 2, times a unit operator acting on subspace 3 of u_1.

$$u_1 = \begin{pmatrix} 0 \\ 1 \end{pmatrix}_1 \begin{pmatrix} 1 \\ 0 \end{pmatrix}_2 \begin{pmatrix} 1 \\ 0 \end{pmatrix}_3 \qquad u_2 = \begin{pmatrix} 1 \\ 0 \end{pmatrix}_1 \begin{pmatrix} 0 \\ 1 \end{pmatrix}_2 \begin{pmatrix} 1 \\ 0 \end{pmatrix}_3$$

$$= \underbrace{(\tau_1^+ \, \tau_2^- \, \tau_3^0)}_{\text{Exchange operator}} u_1$$

This exchange operator is bilinear in the charge matrices of the individual rishons.

Well, now if you look at a conventional canonical field theory this is essentially a matrix-valued structure. What you find is that you can get matrix structure like this in the constrained components (those components are related by Gauss' law to the charges), but not in the dynamical, or transverse components. Whereas if you want to get $SU(3)_c$ coming out, you have to have such exchange matrix structures appearing in all of the components, transverse as well as longitudinal gauge fields, and that is where you have to go outside the formalism of local quantum field theory.

Now in order to consistently put matrix structure into all the gauge theory components, you have to enlarge your invariance group. The idea here is to extend the gauge principle to require an operator-valued gauge invariance with no factor reordering. This is not present in conventional Yang-Mills theory.

Let me illustrate that with a simple SU(2) example. Let us consider SU(2) QCD and consider two Yang-Mills vectors $\vec{A}$ and $\vec{B}$. The gauge transformations on $\vec{A}$ and $\vec{B}$ are:

$$\delta_g \vec{A} = \vec{U} \times \vec{A} - \vec{A} \times \vec{U} \qquad (1)$$

$$\delta_g \vec{B} = \vec{U} \times \vec{B} - \vec{B} \times \vec{U} \qquad (2)$$

We can work out the gauge transformation on $(\vec{A} \times \vec{B})$; it is

$$\delta_g(\vec{A} \times \vec{B}) = \vec{U} \times (\vec{A} \times \vec{B}) - (\vec{A} \times \vec{B}) \times \vec{U} + \vec{\Delta} \qquad (3)$$

where the remaining factor $\vec{\Delta}$ may be calculated using vector identities. It is

$$\vec{\Delta} = (\vec{U} \cdot \vec{A} - \vec{A} \cdot \vec{U}) \cdot \vec{B} + \vec{A}(\vec{U} \cdot \vec{B} - \vec{B} \cdot \vec{U}) + (\vec{A} \cdot \vec{B}) \vec{U} - \vec{U}(\vec{A} \cdot \vec{B}) \qquad (4)$$

and it vanishes only if you assume that all the commutators of things that are space-like separated vanish. It is not identically zero. So in general in standard gauge theories you have to do operator reorderings to verify the gauge invariance.

Now if you assume microscopic causality, there is no problem because fields at a single point can be thought of as a limit of fields with spacelike separated arguments, which commute or anticommute. But if you want to start putting non-trivial matrix structures in the fields these reorderings of course have to be eliminated.

The basic construction works only for U(N) gauge theories, but what you do is start from a Hermitian basis for U(N) including the unit matrix λ^0. These are trace-normalized to obey

$$tr(\lambda^a \lambda^b) = 2 \delta^{ab}, \quad a,b = 0,1,2,\ldots,(N^2-1) \qquad (5)$$

Now the key property for U(N) is a completeness relation; when you sum over all N^2 matrices of the basis, the result is a factored product of Kronecker deltas on the spinor indices:

$$\lambda^a_{AB} \lambda^a_{CD} = 2 \delta_{AD} \delta_{BC} \qquad (6)$$

where A, B, C and D are spinor indices and a sum over a = 0 to $(N^2 - 1)$ is understood. A brief proof of (6) goes as follows.

Any N by N matrix may be written as $M_{AB} = C_a \lambda^a_{AB}$. But by the trace-orthonormality property (5),

$$C_a = \tfrac{1}{2} \text{tr}(\lambda^a M) = \tfrac{1}{2} \lambda^a_{CD} M_{DC}$$

so

$$M_{AB} = (\tfrac{1}{2} \lambda^a_{AB} \lambda^a_{CD}) M_{DC}$$

which requires that

$$\lambda^a_{AB} \lambda^a_{CD} = 2 \delta_{AD} \delta_{BC}$$

An alternative form of this completeness relation is

$$\lambda^a_{AB} \lambda^{*a}_{CD} = 2 \delta_{AC} \delta_{BD} \tag{7}$$

What I will now do is go very fast because the equations look almost like the Yang-Mills equations and I will just focus on the differences from a conventional Yang-Mills theory. The field equations in the gluon sector look like

$$f^a_{\mu\nu} = \partial_\nu b^a_\mu - \partial_\mu b^a_\nu - ig P^a_q(b_\mu, b_\nu) \tag{8a}$$

$$D_\nu f^{a\mu\nu} = \partial_\nu f^{a\mu\nu} + ig P^a_q(b_\nu, f^{\mu\nu}) = g J^{a\mu} \tag{8b}$$

$$J^{a\mu} = \tfrac{1}{2} \bar{r} \gamma^\mu \lambda^{*a} r \tag{8c}$$

and

$$P^a_q(u,v) = q^{abc}(u^b v^c - v^b u^c) \tag{9}$$

where

$$q^{abc} = \tfrac{1}{2}(d^{abc} + i f^{abc}),$$
$$\lambda^a \lambda^b = 2 q^{abc} \lambda^c$$

in terms of a component form for the fields, where I have eliminated the U(N) basis matrices, which serve as carrier matrices. Eqs. (8) look a lot like Yang-Mills equations, except that in the outer product (9), which would usually have f^{abc} times an anticommutator of the fields, uses instead q^{abc}, which has got the d-symbol involved as well as the f-symbol. The d-symbol multiples a commutator rather than the anticommutator. Putting that d-symbol in has the nice effect of deleting all those terms which I had to commute to verify gauge invariance in the Yang-Mills case. So this is where the component form differs from QCD.

If I define matrix fields, as in the original Yang-Mills paper, then in fact until I get to the spinor source terms, the equations look just like the equations of the Yang-Mills theory written in abstract form:

$$F_{\mu\nu} = \tfrac{1}{2} \lambda^a f^a_{\mu\nu} \tag{10a}$$

$$B_\mu = \tfrac{1}{2} \lambda^a b^a_\mu \tag{10b}$$

$$J_\mu = \tfrac{1}{2} \lambda^a J^a_\mu \tag{10c}$$

$$F_{\mu\nu} = \partial_\nu B_\mu - \partial_\mu B_\nu - ig[B_\mu, B_\nu] \tag{11}$$

$$D_\nu F^{\mu\nu} = g J^\mu \tag{12}$$

where for any $W = \tfrac{1}{2} \lambda^a w^a$,

$$D_\mu W = \partial_\mu W + ig[B_\mu, W]. \tag{13}$$

Note that

$$\delta F_{\mu\nu} = D_\nu(\delta B_\mu) - D_\mu(\delta B_\nu)$$

The analog of the second Maxwell equation comes from the Jacobi identity for the commutators:

$$D_\lambda F_{\mu\nu} + D_\nu F_{\lambda\mu} + D_\mu F_{\nu\lambda} = 0. \tag{14}$$

Now for what is different. Let's now look for the field equations of the spinor sources. It must guarantee the covariant current conservation as an identity, because

$$g D_\mu J^\mu = D_\mu D_\nu F^{\mu\nu} = \tfrac{1}{2}[D_\mu, D_\nu] F^{\mu\nu}$$

But

$$[D_\mu, D_\nu] W = -ig[F_{\mu\nu}, W]$$

so

$$D_\mu J^\mu = -\tfrac{1}{2} i [F_{\mu\nu}, F^{\mu\nu}] = 0. \tag{15}$$

So let's look for an expression for J^μ. What I will do is take the conventional form from QCD for the components of $J^{\mu a}$; this gives

$$J^\mu_{AB} = \tfrac{1}{2} \lambda^a_{AB} (\tfrac{1}{2} \bar{r}_C \gamma^\mu \lambda^{*a}_{CD} r_D) \tag{16}$$

$$= \tfrac{1}{2} \bar{r}_A \gamma^\mu r_B = \left[\begin{pmatrix}\text{internal}\\ \text{column}\\ \text{vector } \bar{r}\end{pmatrix} \otimes \begin{pmatrix}\text{internal}\\ \text{row}\\ \text{vector } r\end{pmatrix}\right]_{AB}$$

(It turns out to be convenient to use λ* instead of λ). Now when I form the matrix-valued current J^μ, I sum over the λ-matrix outside. But here something special to U(N) happens. I can use my completeness relation (7) and this sum just becomes $\delta_{AC} \delta_{BD}$, with indices in the correct matrix order because I used λ* instead of λ in the definition of the current. (You get a mirror-image theory if you used λ, it is just less convenient to write it down.) What happens is that the matrix-value current factors into a product of the spinor $\bar{r}_A$ times the Dirac matrices times the spinor r_B, so the internal index structure of this matrix term completely factors onto the two spinors. If the messy projection operator is present in the completeness relation, that does not happen, and you end up with δ_{AB}, say, times spinors contracted with each other.

Now it is the factorization property that allows you to write down the analog Dirac equation and get a full operator gauge symmetry without ever doing any factor reordering. We take as a fermion field equation,

$$\bar{r}_A (\overleftarrow{\partial\!\!\!/} + i m_0) = -i g B_{\mu AB} \bar{r}_B \gamma^\mu \quad (17a)$$

$$(\partial\!\!\!/ - i m_0) r_B = i g \gamma^\mu r_A B_{\mu AB} \quad (17b)$$

$$(\partial\!\!\!/ = \gamma^\mu \partial_\mu)$$

(m_0 is a bare mass; it can be zero). Note that I have got a definite ordering here with the gauge field to the right of r, and the indices in correct matrix-product order. If you think of B_μ as an $\bar{r}r$-term, the ordering rule is that you never order two r's or $\bar{r}$'s together. You can easily check that the divergence of the current is just its commutator with the B-field and therefore its covariant divergence is zero:

$$\partial_\mu J^\mu = \partial_\mu (\tfrac{1}{2} \bar{r} \gamma^\mu r) \quad (18)$$
$$= \tfrac{1}{2} (\bar{r} \overleftrightarrow{\partial\!\!\!/} r + \bar{r} \partial\!\!\!/ r)$$
$$= -ig [B_\mu, \tfrac{1}{2}(\bar{r} \gamma^\mu r)].$$

Hence

$$\partial_\mu J^\mu = -ig [B_\mu, J^\mu], \quad (19)$$

$$D_\mu J^\mu = \partial_\mu J^\mu + ig[B_\mu, J^\mu] = 0.$$

Now you can show, and I'll just sketch it for you, that the above set of equations are covariant under operator-valued gauge transformations. I mean that if I write down gauge variations by analogy, just as you do in the Yang-Mills case defining δ_g by

$$\delta_g r = -i\, r\, U$$
$$\delta_g \bar{r} = i\, U\, \bar{r}$$
$$\delta_g B_\mu = -g^{-1} D_\mu U$$

then

$$\delta_g F_{\mu\nu} = D_\nu(\delta_g B_\mu) - D_\mu(\delta_g B_\nu)$$
$$= g^{-1}[D_\mu, D_\nu] U$$
$$= i[U, F_{\mu\nu}]$$
$$\delta_g J^\mu = i[U, J^\mu]$$

so for the gluon sector,

$$\delta_g (D_\nu F^{\mu\nu}) = (\delta_g D_\nu) F^{\mu\nu} + D_\nu (\delta_g F^{\mu\nu})$$
$$= ig[(\delta_g B_\nu), F^{\mu\nu}] + i D_\nu [U, F^{\mu\nu}]$$
$$= -i[D_\nu U, F^{\mu\nu}] + i[D_\nu U, F^{\mu\nu}] + i[U, D_\nu F^{\mu\nu}]$$
$$= ig[U, J^\mu] = g\, \delta_g J^\mu \qquad (20)$$

while for the fermion sector,

$$\delta_g [(\slashed{\partial} - i m_0) r] = [\slashed{\partial} - i m_0](-i\, r\, U)$$
$$= -i[(\slashed{\partial} - i m_0) r] U - i \gamma^\mu r\, \partial_\mu U$$
$$= g \gamma^\mu r\, B_\mu U - i \gamma^\mu r\, \partial_\mu U$$
$$= -i \gamma^\mu r\, (\partial_\mu + i g B_\mu) U.$$

But if we evaluate $\delta_g[ig\gamma^\mu r B_\mu]$ we get

$$ig\gamma^\mu(\delta_g r)B_\mu + ig\gamma^\mu r(\delta_g B_\mu)$$
$$= g\gamma^\mu r U B_\mu + ig\gamma^\mu r(g^{-1}D_\mu U)$$
$$= g\gamma^\mu r U B_\mu - i\gamma^\mu r\{\partial_\mu U + ig[B_\mu, U]\}$$
$$= g\gamma^\mu r U B_\mu - g\gamma^\mu r U B_\mu$$
$$\quad + g\gamma^\mu r B_\mu U - i\gamma^\mu r \partial_\mu U$$
$$= -i\gamma^\mu r(\partial_\mu + ig B_\mu)U$$

so

$$\delta_g[(\not{\partial} - im_0)r] = \delta_g[ig\gamma^\mu r B_\mu]. \tag{21}$$

You can verify just by going through the equations that there is covariance of the equations of motion without ever interchanging any operator factors. Therefore by the simple expedient of going from SU(N) to U(N), I have, in effect, gained an infinite enlargement of the invariance group.

The calculation in this form basically involves repeated application of the Jacobi identity and the definitions. I'll show you an alternative way of deriving the same statement by starting from an action principle. If you write an action principle involving a total trace of the following form,

$$\delta S = \int d^4x\, \mathcal{L} = 0$$
$$\mathcal{L} = tr\left\{-\tfrac{1}{4}F_{\mu\nu}F^{\mu\nu} - \tfrac{1}{2}i[\bar{r}(\not{\partial}-im_0)r] - \tfrac{1}{2}g B_\mu[\bar{r}\gamma^\mu r]\right\}. \tag{22}$$

(This is an implicit trace over all operator structures. The square brackets are understood to be internal index matrices, i.e.

$[\bar{r}\ldots r]_{AB} = \bar{r}_A\ldots r_B)$, and then do arbitrary operator-valued variations and use the cyclic invariance property of the trace, what you find is that $\delta S = 0$ for arbitrary operator variations implies just the operator equations of motion I had before.

$$\delta S = \int d^4x \, tr \left\{ -(D_\nu \delta B_\mu) F^{\mu\nu} - \right. \tag{23}$$

$$- \tfrac{1}{2} g \delta B_\mu [\bar{r} \gamma^\mu r] - \tfrac{1}{2} g B_\mu [\delta \bar{r} \gamma^\mu r]$$
$$- \tfrac{1}{2} g B_\mu [\bar{r} \gamma^\mu \delta r] - \tfrac{1}{2} i [\delta \bar{r} (\slashed{\partial} - i m_0) r]$$
$$\left. - \tfrac{1}{2} g B_\mu [\bar{r} (\slashed{\partial} - i m_0) \delta r] \right\}$$

$$= \int d^4x \, tr \left\{ \delta B_\mu \left(D_\nu F^{\mu\nu} - \tfrac{1}{2} g [\bar{r} \gamma^\mu r] \right) - \right.$$

$$- \tfrac{1}{2} i [\delta \bar{r} ((\slashed{\partial} - i m_0) r - i g \gamma^\mu r B_\mu)] +$$

$$\left. + \tfrac{1}{2} i [(\bar{r} (\slashed{\partial} + i m_0) + i g B_\mu \bar{r} \gamma^\mu) \delta r] \right\}$$

But then it is very easy to verify (again using the cyclic permutation invariance of the trace) that $\mathcal{L}$ is invariant under the gauge variation I gave.

$$\delta_g \mathcal{L} = tr \left\{ -\tfrac{1}{2} i F^{\mu\nu} [U, F_{\mu\nu}] - \right. \tag{24}$$

$$- g \left(-g^{-1} \partial_\mu U + i [U, B_\mu] \right) J^\mu +$$
$$+ \tfrac{1}{2} U [\bar{r} (\slashed{\partial} - i m_0) r] - i g B_\mu [U, J^\mu] -$$
$$\left. - \tfrac{1}{2} [\bar{r} (\slashed{\partial} - i m_0)(r U)] \right\} = 0$$

However, if you derive operator equations from an invariant action

principle, the equations have to be gauge covariant. The only
operator reordering that I have done here, I have done using the
cyclic invariance of the trace. That means that there is no
operator reordering necessary to verify the gauge covariance.

Let me now conclude. The operator gauge invariance is
valid for the U(N) theory with any N. If you add an additional
requirement, that there be a natural conjugation operation (that's
necessary to get a reasonable quantum mechanics), then you find
you are left with just two theories. One is the U(1) theory
which is just a rewriting of the usual quantum electrodynamics,
the other is the U(2) theory which is essentially a quaternionic
dynamics. Well, we know that the U(1) theory is the basis for all
atomic physics. I think it is an interesting question to ask
whether the U(2) theory, which seems to tie in naturally with the
Harari-Shupe scheme, is perhaps the basis for all of subnuclear
physics.

What one has at the moment, as I have shown in the Lagrangian
and the equations of motion, is a formal gauge invariance. As
Dr. Biedenharn will explain, there is a construction for an
n-particle space, but it is different from what you do in ordinary
complex quantum mechanics. But what is needed to really try
and test these ideas in detail is a full quantum theory in the
quaternionic case and an analog of path-integrals and Feynman
rules, and these are open questions for the future.

e^+ 3Q = 3, N = 1

u   3Q = 2, N = 3

$\bar{d}$ 3Q = 1, N = 3

ν_e 3Q = 0, N = 1

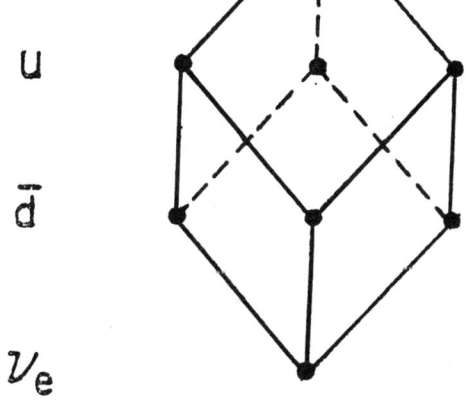

Figure 1

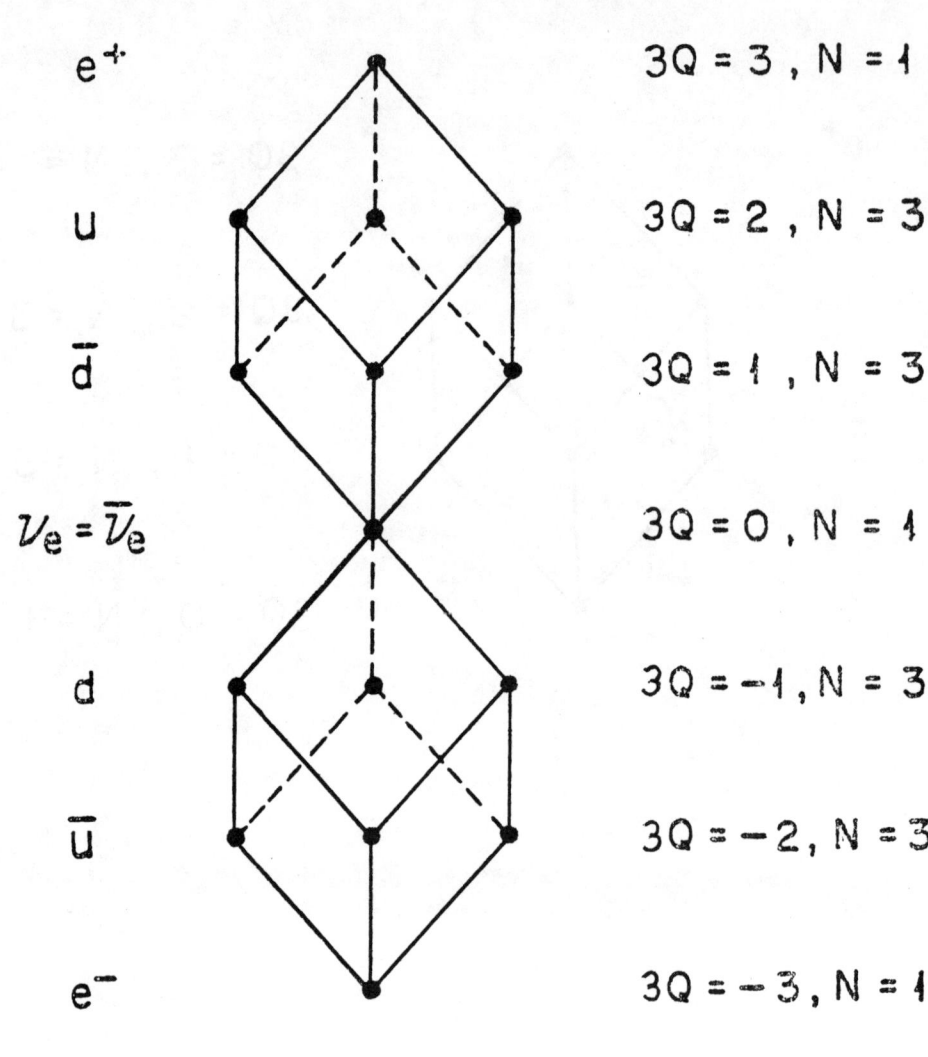

Figure 2

DISCUSSION

YANG: I am kind of confused, Steve. Is your Lagrangian a field theory or not a field theory?

ADLER: Essentially at this point the Lagrangian is simply a formal Lagrangian.

YANG: So it is not a field theory.

ADLER: No, but it simply is a c-number Lagrangian which I vary under the rule that I can make arbitrary operator value variations and use cyclic permutation under the trace.

YANG: So you have not second quantized it?

ADLER: I have not. That, as I say, is the key problem.

YANG: Okay, let's take the c-number theory.

ADLER: So far the operator structures are implicit; it is not c-number in the sense I have never had to assume the various factors commute. So far, any possible quantization is implicit.

YANG: Well, we'd better talk later.

VELTMAN: I am used to computing radiative corrections on, let's say, the one loop level to keep things simple, so I have here the $SU(2) \times U(1)$ theory. Let's keep this in mind for a moment. I would, in theory, compute radiative corrections, with loops and all that. In some form or other I assume that the theory that you are discussing will boil down to that in some approximation. In the theory that you have in mind are the vector bosons composite?

ADLER: The vector bosons would have to come from a three gluon exchange. It would, in effect, be composite.

VELTMAN: I am speaking about the vector bosons of weak interactions.

ADLER: Yes, they would have to be composite.

VELTMAN: Then the next question: if I compute a closed-loop, at some point, at some high momentum, would the calculations deviate from the usual calculations? Let's say as your momentum gets in the neighborhood of the mass of the rishons. Would calculations become different from the usual calculations, in the sense that you get forms factors, and the things dissolve?

ADLER: I really don't know the answer. I don't think I can make an honest statement about computing loops until I understand what the Feynman rules of the thing are.

VELTMAN: I assume that certainly for low energies you have to reduce to the effective theory, don't you somehow?

ADLER: That is right.

VELTMAN: So at higher momenta there would probably be a change. I would assume you have deviations at high energy, and yet in the theory, there is no provision that at higher momentum there would still be the full $SU(2) \times U(1)$ symmetry; that is not guaranteed since the original theory does not have it. If he would have said yes, I would have said his theory is wrong and I would have to go as far as that.

ADLER: The answer is that I do not know though, because the original theory is based on an SU(2) gauge group. It is just that color and flavor are mixed up in a funny way. What we see is color and flavor as being distinct low energies but combined at very high energies.

VELTMAN: You have to understand that theories with composite vector boson are, if the symmetry does not persist at infinitely high energies, infinitely unnatural.

ADLER: I don't see why.

VELTMAN: I will explain it to you when we have more time.

COMMENTATORS

AN UNDERLYING THEORY BASED ON QUATERNIONS FOR ALDER'S ALGEBRAIC CHROMODYNAMICS

L.P. Horwitz*
Syracuse University, Syracuse, New York 13210

and

L.C. Biedenharn**
Duke University, Durham, North Carolina 27706

Abstract

It is shown that the complex-linear tensor product for quantum quaternionic Hilbert (module) spaces provides an algebraic structure for the non-local gauge field in Adler's algebraic chromodynamics (for U(2) underlying algebra).

---

* On sabbatical leave from Tel Aviv University, Ramat Aviv, Israel. Research supported in part by the Binational Science Foundation (BSF), Jerusalem, Israel, and Syracuse University Physics Research Subsidy.
** Work Supported in part by the National Science Foundation.

In his report to this conference, Adler has discussed the motivations and techniques for his algebraic approach, in a semi-classical framework, to chromodynamics. Adler's procedure[1-4] imbeds the algebraic structure of the gauge field in a tensor product space constructed from the algebra generated by the source charges. In particular he has shown[5] that this procedure can be applied to the construction, from pre-quarks, of the eight composite systems of a quark-lepton generation, providing a constructing theoretical basis for the mo of Harari[6] and Shupe[7].

The correspondence between Adler's construction and the usual approach as taken in quantum field theory has not yet been clarified. We shall show that the form of the tensor product for a quantum quaternionic Hilbert (module) space le to the algebraic structure of the gauge field in Adler's chromodynamics. The mo unusual feature of Adler's construction -- namely that the dimensionality of the gauge field space increases with the number of pre-quarks -- is reproduced in th present theory.

Let us consider a space of elements $f,g,\ldots \in H_Q$ which we shall call vector right linear over quaternion $q \in Q$, where Q is the real algebra generated by the elements $1, e_1, e_2$ of the Clifford algebra $C_2(e_1^2 = e_2^2 = -1, e_1 e_2 + e_2 e_1 = 0, e_3 \equiv e_1 e_2)$; i.e., if $f, g \in H_Q$ $fq_1 + gq_2 \in H_Q$. A (linear) scalar product

$$(f,g) = (g,f)^* \in Q, \quad (f,gq) = (f,g)q \qquad (1)$$

is defined, where * is the quaternion involutary automorphism for which

$$\|f\|^2 = (f,f) \geq 0 \qquad (2)$$

and the equality is valid only if $f = 0$. We shall call a space of the type H_Q a module space. The sets of projection operators defined on the closed line manifolds of this space form a weakly modular orthocompleted lattice, and there satisfy the axioms of quantum theory.[8-10] Defining left multiplication by quate (which may be a _different_ realization than is used for right multiplication), it can be shown[10] that every element $f \in H_Q$ has the decomposition $f = f_0 + f_1 e_1 + f_2 e_1 + f_3 e_2$, where the f_j, $j = 0,1,2,3$, are _formally_ real, i.e., $e_i f_j = f_j e_i$. An alternative representation is

$$f = \psi_0 + \psi_1 e_2, \tag{3}$$

where ψ_0, ψ_1 are formally complex valued. In addition to the scalar product (1), it will be useful to define a real scalar product

$$(f,g)_R = tr(f,g) \tag{4}$$

and a complex scalar product (with values in $\mathbb{C}(1, e_1)$ and linear over this subfield)

$$(f,g)_c = tr(f,g) - e_1 tr[(f,g)e_1], \tag{5}$$

where the trace is over the 4×4 real representation of C_2 normalized to tr 1 = 1. In terms of the representation (3),

$$(f,g)_c = (\psi_0, \chi_0) + (\psi_1, \chi_1)^*, \tag{6}$$

where χ_0, χ_1 are the complex components of g. Along with this hierarchy of scalar products (with the _same_ norm), one may define quaternion, complex and real linear manifolds in H_Q, and the corresponding projection operators. Self-adjoint operators can be shown to have a spectral resolution[11] (with appropriate linearity), as in complex Hilbert space. Although the standard proof fails (both the commutator and anti-commutator may have zero expectation value!) nonetheless we have shown[10] that the uncertainty relation $\Delta x \Delta p \geq \frac{\hbar}{2}$ still obtains.

We now turn to the study of many body systems. The significant fact here is that: _quaternion linear tensor products cannot be constructed._[12] (The non-commutativity of the quaternion scalars means that a multi-linear product is impossible.) We shall therefore consider a _complex linear tensor product_, based on the decomposition of Equation (3) and the complex scalar product of Equation (5.) (Recall that this preserves the topology of the original quaternion structure.) Note that $fz=\psi_0 z+\psi_1 z^* e_2$, $z\epsilon\mathcal{C}(1,e_1)$, and hence multiplication by a complex number is an <u>antilinear operati</u> on the second (quaternionic) component of f. The linear functional

$$\Psi_0(f) = \psi_0, \qquad \Psi_1(f) = \psi_1^*, \tag{7}$$

can be used as a (two-component) basis for H_Q. In fact,

$$\sum_{\alpha=0,1} (\Psi_\alpha(f), \Psi_\alpha(g)) = (f,g)_c \tag{8}$$

and we remark that the action of left multiplication by a quaternion is given by

$$\Psi_0(e_1 f) = e_1 \psi_0 \qquad \Psi_1(e_1 f) = -e_1 \psi_1^*$$
$$\Psi_0(e_2 f) = -\psi_1^* \qquad \Psi_1(e_2 f) = \psi_0$$
$$\Psi_0(e_3 f) = -e_1 \psi_1^* \qquad \Psi_1(e_3 f) = -e_1 \psi_0 \tag{9}$$

and hence

$$\Psi(e_3 f e_1) = \tau_1 \Psi(f) \equiv \Psi(\tau_1 f)$$
$$\Psi(e_2 f e_1) = \tau_2 \Psi(f) \equiv \Psi(\tau_2 f)$$
$$\Psi(-e_1 f e_1) = \tau_3 \Psi(f) \equiv \Psi(\tau_3 f), \tag{10}$$

where τ_1, τ_2, τ_3 are the usual Pauli spin matrices.

The N body BE and FD generalization of (7),

$$\Psi_{\alpha_1 \alpha_1 \ldots \alpha_n}(f_1 f_2 \ldots f_n) = \frac{1}{\sqrt{N!}} \sum_P (\pm)^P P(\psi_{\alpha_1}^n \psi_{\alpha_2}^{r_2} \ldots \psi_{\alpha_n}^{r_n}), \tag{11}$$

where to maintain linearity for complex right multiplication, the $\psi_{\alpha_i}^{r_i}$ compone of f_{r_i} appear conjugated if $\alpha_i = 1$, has the same properties with respect to norms and scalar products as the corresponding quantities in the usual complex theory but with all one-body scalar products of the complex linear form (6). Annihilation-creation operators can be defined[10] in the usual way, and satisfy $[a(f),a^\dagger(g)]_\pm = (f,g)_c$. Note that, in the sum (15), the indices $r_1, r_2 \ldots r_n$ are permuted, but not the component labels $\alpha_1, \alpha_2 \ldots \alpha_n$. Working in the x-representation, we see that the complex scalar product $(f,g)_c$ is invariant under the local (gauge) U(2) transformation $f(x) \to q(x) f(x) z(x)$, $g(x) \to q(x) g(x) z(x)$; since all scalar products of the function (15) can be expressed in terms of complex one-particle scalar products, we may induce in this way gauge transformations for the N body states. The local gauge transformation on Ψ induced by transformations on the constituent particle states is:

$$\bar\Psi(qf_1, qf_2, \ldots qf_n)(x_1, \ldots x_n) = E_1(q(x_1)) E_2(q(x_2)) \cdots E_n(q(x_n))$$
$$\cdot \Psi(f_1, \ldots f_n)(x_1 \ldots x_n), \qquad (12)$$

where we use the notation $qf(x) = \sum_n q_n(x) f(x) z_n(x)$, and $E_{\alpha\beta}(q) \Psi_\beta(f)(x)$
$= \sum_n z_n(x) E_{\alpha\beta}(q_n(x)) \Psi_b(f)(x) = \Psi_\alpha(qf)(x)$ defines the complex linear representations of the U(2) gauge algebra on the two component constituent indices, as in (10) $(E(q)^\dagger E(q) = 1)$.

Infinitesimal translation of the state (12) results in a sum of derivatives acting on each term of the product, and it appears that a gauge compensation field restricted in form to the direct sum of the algebras acting on each index would be adequate. Although the gauge compensation field, which we shall call $B^\mu(x)$, acts locally at each point, its action may also alter the structure of the tensor product. It is this mechanism which opens the possibility of maintaining essential non-linearity in the Yang-Mills equations on a semi-classical level. The tensor products $\Psi(f, \tau ag) = E_2(\tau a)\Psi(f,g)$ have the same linearity properties as $\Psi(f,g)$; they differ by relative phases and permutations on the components of g.

We therefore define the general action of a gauge field as:

$$\sum_j B^\mu(x_j) \Psi(f_1, f_2 \ldots f_n)(x_1 \ldots x_n) \equiv \sum_{j,a_1 \ldots \hat{a}_1 \ldots a_n} B^{\mu a_1 \ldots \hat{a}_j \ldots a_n}(x_j)$$

$$\cdot \tau_{a_1} \ldots \tau_{a_n} \Psi(f_1 \ldots f_n)(x_1 \ldots x_n) . \qquad (13)$$

where we have, for convenience, defined the local field $B^{\mu a_1 \ldots \hat{a}_j \ldots a_n}$ (with index a_j deleted) with values in the jth constituent algebra. The operator $B^\mu(x_j)$ is in the tensor product space of the algebras associated with each particle, in agreement with the general structure postulated by Adler.

We define the covariant derivative as (we call $i \equiv e_1$, the representation of the universal right multiplication by e_1)

$$D^\mu \Psi(f_1 \ldots f_n)(x_1 \ldots x_n) = \sum_j (\partial_j^\mu - i g B^\mu(x_j)) \Psi(f_1 \ldots f_n)(x_1 \ldots x_n)$$

$$(14)$$

Under the local transformation $f_i(x) \to q(x_i) f_i(x)$, one finds that it is sufficient that $B^\mu(x_j)$ transform as

$$B'^\mu(x_j) = E_1(q(x_1)) \ldots E_n(q(x_n)) B^\mu(x_j) E_1(q(x_1))^{-1} \ldots E_n(q(x_n))^{-1}$$

$$- ig^{-1} \partial_j^\mu E_j(q(x_j)) E_j(q(x_j))^{-1} \qquad (15)$$

The inhomogeneous term involves only the jth constituent algebra, and hence only affects components of $B^{\mu a_1 \ldots a_n}(x_j)$ for which $a_i = 0$, $i \neq j$. In terms of the constit algebra valued fields $B^{\mu A_j}(x_j)$, $A_j \equiv \{a_1, a_2 \ldots \hat{a}_j, \ldots a_n\}$, (15) can be written (for arbitrary x)

$$B'^{\mu A_j}(x) = E_j(q(x)) B^{\mu A_j}(x) E_j(q(x))^{-1}$$

$$-ig^{-1} \partial^\mu E_j(q(x)) E_j(q(x))^{-1} \delta_{A_j, 0} , \qquad (16)$$

defining the transformation properties of a set of general fields.

It is essential to remark that the notation in Equations (13-16) is overly brief. The field $B^{\mu A_j}(x_j)$ for $A_j=0$, is indeed a local field, and Equation (16) is inhomogeneous. However, the fields with $A_j \neq 0$ are non-local in the sense that the orientation of the operators $E(q)$ depends on all the other points x_i, $i \neq j$. This is the new feature of the construction and is responsible for the non-locality found by Adler.

As a special case, suppose $E_j(q(x)) = z(x) \, E_j(q(x))$; the inhomogeneous term in (16) then has the form

$$\partial^\mu E_j(q(x)) \, E_j(q(x))^{-1} + \partial^\mu z(x)(z(x))^{-1} \qquad (17)$$

i.e., it contains compensation for the action of the universal local phase transformations on the right (as for a local electromagnetic potential).

It is straightforward to verify that the field strength tensor

$$F^{\mu\nu A_j}(x) = \partial^\nu B^{\mu A_j}(x) - \partial^\mu B^{\nu A_j}(x) - ig\{B^{\mu B_j}(x) \, B^{\nu C_j}(x)$$

$$- B^{\nu B_j}(x) \, B^{\mu C_j}(x)\} \, q^{A_j B_j C_j} \, , \qquad (18)$$

where $q^{A_j B_j C_j} = q^{a_1 b_1 c_1} \ldots \widehat{q^{a_j b_j c_j}} \ldots q^{a_n b_n c_n}$ is the product of the multiplication rule coefficients defined by $\tau^b \tau^c = q^{abc} \tau^a$, transforms as

$$F'^{\mu\nu A_j}(x) = E_j(q(x)) \, F^{\mu\nu A_j}(x) \, E_j(q(x))^{-1} \qquad (19)$$

Although one can construct a field strength in the full tensor product algebra

$$F^{\mu\nu}(x) = \partial^\nu B^\mu(x) - \partial^\mu B^\nu(x) - ig[B^\mu(x), B^\nu(x)] , \qquad (20)$$

by multiplying (18) by $\tau_{a_1} \ldots \tau_{a_j} \ldots \tau_{a_n}$ and summing over A_j, gauge transformations are determined, as in (15), by the particle configuration in the many body wave function. For any x_j one obtains

$$F'^{\mu\nu}(x_j) = E_1(q(x_1)) \ldots E_n(q(x_n)) \, F^{\mu\nu}(x_j) (E_1(q(x_1)) \ldots E_n(q(x_n)))^{-1}$$
$$\qquad (21)$$

The covariant derivative on the local fields associated to the constituents may be identified from the relation

$$\delta F^{\mu\nu A}j(x) = D^{\nu}\delta B^{\mu A}j(x) - D^{\mu}\delta B^{\nu A}j(x), \quad (22)$$

from which one obtains ($W^A j \epsilon$ j^{th} constituent algebra)

$$D^{\mu}W^A j(x) = \partial^{\mu}W^A j(x) - ig\{W^B j(x) B^{\mu C}j(x) - B^{\mu B}j(x) W^C j(x)\}.$$
$$\cdot q^{AjBjCj}. \quad (23)$$

Assuming the equations of motion

$$D_{\nu} F^{\mu\nu A}j(x) = J^{\mu A}j(x), \quad (24)$$

it follows that

$$D_{\mu} J^{\mu A}j(x) = 0. \quad (25)$$

It is also an identity (from the Jacobi identity) that

$$D_{\mu} F^{Aj}_{\nu\lambda} + D_{\nu} F^{Aj}_{\lambda\mu} + D_{\lambda} F^{Aj}_{\mu\nu} = 0. \quad (26)$$

For two particle states, the gauge field B^{μ} is in the space of the fundamental representation of U(4), (the complex extension of the representation of the Clifford algebra C_4) and for three particle states of U(8) (the complex extension of the Clifford algebra C_6). These representations (complex 4 and 8 dimensional) are too large for application to presently known phenomena. The procedure introduced by Adler[1-4] of separating out a set of universal generators in the fundamental representation has the effect of reducing the potentially large algebra of gauge fields. Adler's procedure in effect introduces a "carri algebra τ_0" (his phrase) to construct scalars in the product algebra from U(2) and τ_0. (This procedure is discussed in detail in reference 14).

We have shown previously[9] that the Clifford algebra C_6, when a minimal ideal and one generating element are held fixed, has the automorphism group G_2, and that products of generating elements obey the Cayley (octonion) algebra as equivalence relations in this ideal. Considering the three body state to be a new "particle", a lepton or a quark, the gauge fields could couple in a way somewhat different from that described above (i.e., to the (commuting) constituent charges). Let us construct the Clifford generators as follows

$$\varepsilon_1 = E_1(e_1)$$
$$\varepsilon_2 = iE_1(e_2)E_2(e_1)$$
$$\varepsilon_3 = iE_1(e_2)E_2(e_2)$$
$$\varepsilon_4 = E_1(e_2)E_2(e_3)E_3(e_1)$$
$$\varepsilon_5 = E_1(e_2)E_2(e_3)E_3(e_2)$$
$$\varepsilon_6 = E_1(e_2)E_2(e_3)E_3(e_3)$$
$$\varepsilon_7 = -\varepsilon_1 \varepsilon_2 \cdots \varepsilon_6 = -E_1(e_3) \quad (27)$$

The minimum ideal in which the Cayley algebra emerges is generated by

$$P_o = P_{123} P_{516} P_{624} = \frac{1}{8}(1-\varepsilon_1\varepsilon_2\varepsilon_3)(1-\varepsilon_5\varepsilon_1\varepsilon_6)(1-\varepsilon_6\varepsilon_2\varepsilon_4) \quad (28)$$

The three anticommuting second degree nilpotents[15] (fermions) contained in this algebra are (all other sets are equivalent to these)

$$\eta_1 = P_{123}\varepsilon_5 = \frac{1}{2}(\varepsilon_5 + \varepsilon_4\varepsilon_6\varepsilon_7)$$
$$\eta_2 = P_{516}\varepsilon_4 = \frac{1}{2}(\varepsilon_4 + \varepsilon_2\varepsilon_3\varepsilon_7)$$
$$\eta_3 = P_{624}\varepsilon_3 = \frac{1}{2}(\varepsilon_3 + \varepsilon_1\varepsilon_5\varepsilon_7) \quad (29)$$

Harari and Zeiberg[16] have found the subgroups generated by $[\eta_i, \eta_j], [\eta_i, \eta_j^\dagger] [\eta_i^\dagger, \eta_j^\dagger]$ form a scheme with many useful features.

Since the gauge fields carry charge, it is possible that the effective charge carried by each of the constituents is strongly modified by the presence of the others. In terms of the representation (27), for example,

$$\eta_1 = \frac{1}{2} E_3(e_3)(E_1(e_3) + E_1,(e_2)E_2(e_3)), \tag{30}$$

contains correlations between the constituent states. If the anticommuting elements η_i were to play the role of charges on the constituents (i.e, $\eta_i + \eta_i^\dagger$, $i(\eta_i - \eta_i^\dagger)$), the algebras mentioned above would be generated by means of the commutator term in the Yang-Mills Equations (22-26).

Acknowledgements

One of us (L. Horwitz) would like to thank H. Harari and N. Zeiberg for discussing their ideas with him, and S. Nussinov for a helpful discussion on the notion of charges in non-Abelian theories.

References

1. S.L. Adler, Phys. Rev. $\underline{D18}$, 411 (1978).

2. do., Phys. Lett. $\underline{86B}$, 203 (1979).

3. do., Phys. Rev. $\underline{D19}$, 1168 (1979).

4. do., Phys. Rev. $\underline{D20}$, 1386 (1979).

5. do., "Quaternionic Chromodynamics As A Theory of Composite Quarks and Leptons", preprint, December (1979).

6. H. Harari, Phys. Lett. $\underline{86B}$, 83 (1979).

7. M.A. Shupe, Phys. Lett. $\underline{86B}$, 87 (1979).

8. C. Piron, "Foundations of Quantum Physics", (W.A. Benjamin, Reading, Mass.) 1976.

9. L.P. Horwitz and L.C. Biedenharn, Helv. Phys. Acta, $\underline{38}$, 385 (1965).

10. do., "Quaternion Quantum Mechanics and Second Quantization", submitted to Comm. Math. Phys. (Dec. 1980).

11. H.H. Goldstine and L.P. Horwitz, Math. Ann. $\underline{164}$, 291 (1966).

12. We thank Professor Jacques Tits (I.H.E.S.) for a discussion of the mathmathics of quaternions.

13. See, for example, Landau and Lifshitz, "Quantum Mechanics", (Addison-Wesley, Reading, Mass.) 1958.

14. L.C. Biedenharn, D. Sepunaru and L.P. Horwitz, invited paper at the IX International Colloquium on Group Theoretical Methods in Physics, Cocoyoc, Mexico, June 1980. (To be published by Springer-Verlag.)

15. L.P. Horwitz and H.H. Goldstine, Math. Ann. $\underline{154}$, 1 (1964).

16. Private communication.

COMPOSITE MODEL OF QUARKS AND LEPTONS WITH STRONG AND WEAK INTERACTIONS ON A PARALLEL FOOTING

O. W. Greenberg* and Joseph Sucher*
Center for Theoretical Physics, Department of Physics and Astronomy
University of Maryland, College Park, Maryland 20742

ABSTRACT

We discuss a composite model of quarks, leptons and other "elementary" particles in which strong and weak interactions are on a parallel footing.

INTRODUCTION

Why composite models? Quarks and leptons occur in a parallel way, with the same $SU(2)_{flavor}$ structure in each generation, and with the same generation structure (assuming the top quark is found). Composite models can account for this by using the same flavor-carrying field to construct both quarks and leptons, and by interpreting generations as an internal excitation. At the same time, composite models allow a reduced number of fundamental fields. The program of grand unified theory leaves many parameters arbitrary. Composite models offer a way to reduce the number of such parameters.

Hadrons and leptons enter elementary particle physics in very different ways at present: there is a dense spectrum of hadrons, hadronic flavor is not gauged, the strong hadronic interactions are a residual effect of a confining gauge group (quantum chromodynamics or QCD) which commutes with flavor, while, by contrast, there is a sparse spectrum of leptons, and the weak interactions arise by gauging the flavor degree of freedom. Composite models,[1] such as the one described below, can allow both strong and weak interactions to arise as residual effects of confining local gauge symmetries, without gauging the flavor degree of freedom in either case, and, at higher masses, allow a more complex lepton spectrum than appears in the present mass range. The low mass of both quarks and leptons compared to the Planck mass and to the unifying mass in grand unified theories is not well understood in the standard scheme; neither is the peculiar mass spectrum of the quarks and leptons. Although we do not provide an understanding of these matters in this article, we believe that the idea of compositeness of quarks and leptons provides new avenues of approach to these problems. The shortcoming of our present motivations for compositeness is that, up to now, there is no "smoking gun," no compelling direct evidence for compositeness. We suggest three examples of

---

*Supported in part by the National Science Foundation.

direct evidence for compositeness: (i) the discovery of a system of excited leptons (or quarks) which decay electromagnetically to the ground-state leptons (or quarks) with a mass formula and widths which can be calculated in a composite model (and which agree with experiment), (ii) the derivation of a mass formula for the present quarks and leptons together (hopefully) with correct predictions for quarks and leptons as yet unseen, and (iii) the observation of leptonic (or quark) form factors which could not be understood as radiative effects.

THE MODEL[2]

To carry flavor, which in the standard model is associated with the left-handed group $(SU(2))_L$, we introduce a flavor-doublet of spin-1/2 fields, "flavons," $F = (F_u, F_d)$. We assume the flavons carry flavor for both quarks and leptons. To avoid quarks and leptons of spin higher than one-half due to spin rearrangement, we assume quarks and leptons are two-body composites of a flavon and a scalar: a color-antitriplet scalar "chromon" for quarks and a "leptoscalar" for leptons. If the particles associated with these fundamental fields are of very high mass, so that they would not yet have been seen, then there are severe dynamical problems associated with the formation of quarks and leptons as light-mass bound states. If the fundamental particles have light or zero mass, then they would have been seen, unless they are permanently confined. We assume the latter, and take as the confining gauge group either $SU(N)$ or $SO(N)$. We will say something about the value of N later; however, without experimental data, we cannot fix N. Table 1 lists the fundamental fields and their quantum numbers for

	Spin	Charge	$dimSU(3)_c$	$dimSU(N)_s$
F_u	$\frac{1}{2}$	q+1	1	N
F_d	$\frac{1}{2}$	q	1	N
C	0	$q+\frac{1}{3}$	3*	N
S	0	q+1	1	N
A_μ	1	0	1	1
G_μ^a	1	0	8	1
F_μ^α	1	0	1	N^2-1

Table 1. Fundamental fields

the case that the confining group is $SU(N)$. The dynamics of confinement by $SU(N)$ gauge bosons is what gives structure to the quarks and leptons; for this reason we call this $SU(N)$ confining interaction "quantum structuredynamics" or, for short, "QSD." We can either keep the photon and/or the SU(3) color gluons as elementary gauge bosons, using the idea that exact local symmetries are associated with elementary fields, or construct as composite the photon from $(q+1) \bar{F}_u \gamma_\mu F_u - q \bar{F}_d \gamma_\mu F_d + (q+1/3) i C^\dagger \overleftrightarrow{\partial_\mu} C + (q+1) i S^\dagger \overleftrightarrow{\partial_\mu} S$ and/or the color gluons from $i C^\dagger \overleftrightarrow{\partial_\mu} \lambda^a C$. In this latter case, the QSD annihilation graphs would have to give mass to the SU(3)-singlet gluon, leaving the color octet massless. With elementary color gluons, the Lagrangian for QSD is

$$L = -\frac{1}{4} G^a_{\mu\nu} G^{a\mu\nu} - \frac{1}{4} F^\alpha_{\mu\nu} F^{\alpha\mu\nu} + \bar{F}_i (i\slashed{\partial} + g_S \slashed{G}) F_i$$
$$+ |(\partial_\mu - ig_S G_\mu - ig_c F_\mu) C|^2 + |(\partial_\mu - ig_S G_\mu) S|^2 \quad (1)$$
$$- \frac{1}{2} m_c^2 C^\dagger_\alpha C^\alpha - \frac{1}{2} m_S^2 S^\dagger S - \lambda_c (C^\dagger_\alpha C^\alpha)^2 - \lambda_c C^\dagger_\alpha C^\alpha S^\dagger S - \lambda_S (S^\dagger S)^2 ,$$

where the $SU(N)_S$ and $SU(3)_C$ gauge fields and potentials are related by

$$G^a_{\mu\nu} = \partial_\mu G^a_\nu - \partial_\nu G^a_\mu + g_S f^{abc} G^b_\mu G^c_\nu ; \quad a,b,c = 1,2,\ldots,N; \quad (2)$$

$$G_\mu = \frac{1}{2} \lambda^a G^a_\mu, \text{ where } \lambda^a\text{'s are } SU(N) \lambda\text{'s}; \quad (3)$$

$$F^\alpha_{\mu\nu} = \partial_\mu F^\alpha_\nu - \partial_\nu F^\alpha_\mu - g_c f^{\alpha\beta\gamma} F^\beta_\mu F^\gamma_\nu ; \quad \alpha,\beta,\gamma = 1,2,3; \quad (4)$$

$$F_\mu = \frac{1}{2} \lambda^\alpha F^\alpha_\mu, \text{ where } \lambda^\alpha\text{'s are } SU(3) \lambda\text{'s}. \quad (5)$$

We assign the first generation of quarks u,d, and leptons ν_e,e to the lowest J=1/2 states of $(F_u \bar{C})$, $(F_d \bar{C})$ and $(F_u \bar{S})$, $(F_d \bar{S})$, respectively. Higher generations can be incorporated in two distinct ways: (i) as radial space excitations, or (ii) as internal excitations. (i) The first case is analogous to the assignments made in the Bohr model or the quark model. However it suffers from two problems. The first is that the spacing between radial excitations is expected to be of order r_0^{-1}, where r_0 is the characteristic size of the bound state. The present bound on r_0 from e^+e^- annihilation to electron or muon pairs is $r_0 \leq 10^{-16}$ cm, which corresponds to a spacing of $\gtrsim 200$ GeV, much too large to be associated with the higher generations. The second problem is that in most models orbital excitations occur at energies lower than or, at best, comparable to those of radial excitations. Orbital excitations are very likely to lead to states with J > 1/2 at energies comparable to those of states with J = 1/2, in contradiction with the fact that all quarks and leptons found so far have spin 1/2. Later we will make some comments about potentials which can elevate orbital excitations above radial ones, by way of background information; however, we emphasize that such potentials will not arise in renormalizable theories. (ii) To incorporate higher generations as internal excitations, we can make use of $SU(N)_S$ singlets which are many-body, rather than two-body.[3] Two possibilities to do this come to mind: to make use of the ε symbol in SO(N), and introduce states such as $F^a S^b S^c S^d \varepsilon_{abcd}$,

$F^a S^b C^{\bar{s}} \bar{s}^d \epsilon_{abcd}$, $F^a S^b \bar{s} C^{\bar{s}} \bar{s}^d \epsilon_{abcd}$, $F^a \bar{s}^b \bar{s} C^{\bar{s}} \bar{s}^d \epsilon_{abcd}$, and then also $F^a S^a$ for N=4; or to introduce extra $\bar{S}S$ or $\bar{C}C$ pairs to make states analogous to the "exotic" states in the quark model. In the former case, the charge parameter q must be chosen to be -1. The latter case is similar to the proposal of Derman[4] and Veltman[4] except that they use extra elementary Higgs bosons, instead of extra pairs. These schemes will stand or fall depending on whether or not they yield good mass formulas for the higher generations.

In addition to the local $H = SU(3)_c \times SU(N)_S$ symmetry, L has global $G = SU(2)_L \times SU(2)_R \times U(1)$ flavor symmetry which gives a chiral classification of the $\bar{F}F$ bound states. There will be two-body scalar composites $\bar{F}_L F_R = \psi$ which we identify as Higgs fields of the effective low-energy theory, as well as vector composites $\bar{F}_L \gamma_\mu \tau^i F_L = W_{L\mu}^i$ and $\bar{F}_R \gamma_\mu \tau^i F_R = W_{R\mu}^i$, i=1,2,3 and $\bar{F}\gamma_\mu F$, which we identify, before spontaneous symmetry breaking, as massless gauge bosons of the effective low-energy theory which have V∓A and pure V couplings, respectively.[5] There will also be N-body scalar composites which we identify as additional Higgs fields of the low-energy theory; for example for N=4,

$$\phi_L = [(F_{La}^T C^{-1} \tau^i \gamma_\mu F_{Lb})(F_{Lc}^T C^{-1} \tau^j \gamma^\mu F_{Ld})$$

$$- \frac{1}{3} \delta^{ij} (F_{La}^T C^{-1} \tau^k \gamma_\mu F_{Lb})(F_{Lc}^T C^{-1} \tau^k \gamma^\mu F_{Ld})] \epsilon^{abcd}, (C = i\gamma^2 \gamma^0) \quad (6)$$

with an analogous expression for ϕ_R; both ϕ's are flavor-spin-2 composite scalar fields.[6] We accept the conjecture[7] that, when the size r_0 of the composites is much less than their Compton wavelength $\hbar/mc$, the Lagrangian L (Eq. 1) of the fundamental fields will give rise to a renormalizable local effective low energy Lagrangian L_{eff} for the composite fields, which include the different generations of quarks and leptons, the $SU(2)_L \times SU(2)_R \times U(1)$ vector bosons, and the Higgs mesons ψ, ϕ_L and ϕ_R. Renormalizability implies that L_{eff} is a spontaneously broken local, G gauge-invariant Lagrangian with left-right symmetry, whose reduction to the standard $SU(2)_L \times U(1)$ theory has been discussed in Ref. 8.

R. N. Mohapatra, in a contribution to this workshop, has given an argument that $\sin^2\theta_w = 1/4$ in our model.[9]

All the two-body $SU(N)_S$ singlets in our model, except $\bar{C}C$, $\bar{S}S$, $\bar{S}C$, and $\bar{C}S$, which we assume lie high in mass, correspond to quarks, leptons, weak vector bosons, or Higgs mesons.

A major feature of our point of view is that the weak interactions of quarks and leptons should follow as residual effects of flavor-independent QSD, just as the strong interactions are considered to follow as residual effects of flavor-independent QCD. For an example, see Fig. 1. Thus the parallelism in which both the strong and weak interactions are associated with flavor-independent local gauge symmetries supplements the quark-lepton parallelism (in the counting of states) which follows from the introduction of the flavons.[10]

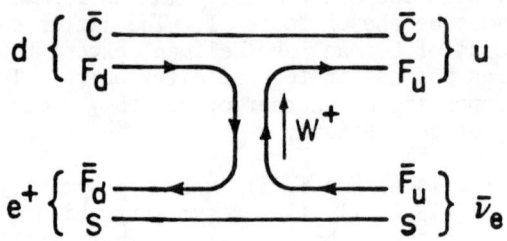

Fig. 1. The reaction $de^+ \to u\bar{\nu}_e$ (a crossed reaction of $d \to ue^-\bar{\nu}_e$) mediated by composite W^+ exchange.

Whether the scheme presented above is really capable of leading to the observed mass spectrum of the leptons and the "observed" mass spectrum of the quarks and weak bosons is an open question.[11] The virtue of a composite model such as ours based on a confining gauge theory is that, in principle, the spectrum and interactions of the composites are determined by the gauge theory and if the conjecture of Ref. 7 holds the effective theory of the composites at low energy will be a local renormalizable gauge theory. The drawback is that we are no better able to do non-perturbative calculations of the properties of the composites in our model than we can for hadrons in QCD. We remind the reader that the derivation of the pion-nucleon and nucleon-nucleon interactions from QCD is a central, open problem now under intensive study by our nuclear colleagues. We also do not know how the conjecture of Ref. 7 can be derived as a property of the vertex functions of the effective low-energy theory, and if α_s, α_{em}, and G_F will have their observed magnitudes. We discussed our composites in a qualitative way, gave conditions which, if met, would insure that our model agrees with known weak interaction phenomenology, and checked that no contradictions with experiment arise at a kinematic level in Ref. 2.

We now discuss the constraints on the size r_0. For models in which no selection rule based on internal symmetry prohibits it, the strongest constraint comes from the observed suppression of the decay $\mu \to e\gamma$: $r_0 \lesssim (10^2 \text{ TeV})^{-1}$ if we estimate the matrix element as $e(m_\mu r_0)^2$, and $r_0 \lesssim (10^8 \text{ TeV})^{-1}$ if we estimate it as $em_\mu r_0$.[12] The bound $r_0' \lesssim (5 \text{ TeV})^{-1}$ which follows from the experimental bound on g-2 for e and μ is weaker.

We now give aruments for Cabibbo structure and for the GIM mechanism in the context of our model. Similar arguments have been given independently by Visnjić-Triantafillou[13] and by Terazawa.[13] We introduce phenomenological $SU(2)_f$ flavor currents,

$jL_\mu^i = 1/2 \bar{F}_L \tau^i \gamma_\mu F_L$, and their associated charges which satisfy the SU(2)$_L$ algebra, $[Q_L^+, Q_L^-] = 2Q_L^3$. The charge-raising zero momentum-transfer Cabibbo matrix is $C_{ij} = \langle u_i | Q_L^+ | d_j \rangle$, where $i, j = 1, 2, 3, \ldots$ label the generations. Since the Cabibbo matrix acts only on the F's, we don't expect it to be diagonal. Provided single-quark intermediate states dominate, the SU(2)$_L$ algebra leads to the sum rule

$$\delta_{ij} = \langle u_i | [Q_L^+, Q_L^-] | u_j \rangle = \sum_k C_{ij} C_{jk}^*, \qquad (7)$$

which states that C is approximately unitary. If, in addition, the Q_L^+ operators change the generation i by at most one and the matrix elements $\langle u_i | Q_L^+ | d_j \rangle$ are real, we recover the GIM structure of weak interactions. The elements of the momentum-transfer dependent Cabibbo matrix, which are determined by the overlap of the quark vertex functions for the different generations, are analogous to quasi-elastic form factors. We expect the diagonal elements to decrease with increasing momentum transfer.

The size r_0 of the composites can be much smaller than the typical size $r_h = \hbar/m_\rho c$ of hadrons or the Compton wavelength $r_\ell = \hbar/m_\ell c$ of leptons provided that in SU(N)$_S$ x SU(3)$_c$ N is greater than three. We assume SU(3)$_c$ and SU(N)$_S$ have the same strength at a large mass scale.[14] Various extended color schemes have been suggested to achieve small ratios of size (or, equivalently, large mass scale ratios) in Ref. 15. Assuming that F_u and F_d are light, and S and C are very heavy, and only the gauge multiplet of the unifying group U contains both light and heavy particles, the renormalization group equations lead to $N \approx 3 + \ln(M_N/M_3)/\ln(M_u^2/M_N^2)$, where M_3 and M_N are the masses at which SU(3)$_c$ and SU(N)$_S$ become confining (i.e., the relevant $\alpha=1$) and M_u is the mass at which the coupling constants coalesce. For example, the minimum value N=4 requires $M_u = 10^{10}$ GeV if we choose $M_N = 10^5$ GeV and $M_3 = 1$ GeV. Larger values of N allow smaller M_u. We do not think it essential to introduce a scheme of unification at the present stage of development of our model; none-the-less, we comment briefly on the possible particle assignments for the choice of SU(N+3) as the unifying group U. We decompose representations of SU(N+3) as (dim SU(N)$_S$, dim SU(3)$_c$) under the subgroup SU(N)$_S$ x SU(3)$_c$. The antiflavons (rather than the flavons) $\bar{F}_u$ and $\bar{F}_d$ could each be an (N,1) part of an N+3 fundamental representation, accompanied in each case by a (1,3) color antiflavon. Similarly, the antileptoscalar $\bar{S}$ could be an (N,1) part of a fundamental representation, accompanied by a (1,3) color antileptoscalar. The antichromon could be an (N,3) part of the rank-two antisymmetric representation (N+3)(N+2)/2, accompanied by an (N(N-1)/2,1) and a (1,3*). The SU(N)$_S$ and SU(3)$_c$ gluons could appear as (N^2-1,1) and (1,8), respectively, in the adjoint representation of SU(N+3), together with an (N,3*), an (N*,3) and a (1,1). (Since both spin-0 and spin-1/2 fields are present in our model, an alternative approach to unification might be via supersymmetry.)

BACKGROUND COMMENTS ON RADIAL AND ORBITAL STATES IN NON-RELATIVISTIC POTENTIAL MODELS

The applicability of non-relativistic potential models to the structure of quarks and leptons is problematic, because the bound states are likely to be relativistic. In particular, we see no reason for our QSD model to lead to a non-relatistic picture. None-the-less, as background information we comment briefly on such a picture. For the explicit example of a non-relativistic particle bound in a potential $\sim r^\alpha$, in the semiclassical approximation[16] for the $E_{n\ell}$, where n is the principal quantum number and ℓ is the orbital angular momentum, the requirement $E_{11} > E_{30}$, i.e., the first orbital excitation lies above the first three quark or lepton generations, requires $\alpha < -3/2$.[17] To have n generations below the first orbital excitation requires $\alpha < -(2n-3)/(n-1)$. A simple way to see that the elevation of orbital excitations relative to $\ell = 0$ states imposes special constraints on the binding interaction is to note that the uncertainty principle applied to the radial coordinate, $rp_r \sim \hbar$, leads to an expression for the radial kinetic energy, $\hbar^2/2m_{red} r^2$, which has the same form as the centrifugal barrier, with an "ℓ" value ~ 0.6. Thus $\ell=0$ and $\ell=1$ states tend to be close in energy.[18] The masses of the quarks and of at least the charged leptons increase rapidly with generation. In the semiclassical approximation for the energy levels E_n of radial states of a non-relativistic particle in a potential $\sim r^\alpha$, the most rapid increase with n, $E_n \propto n^2$, occurs for $\alpha \to \infty$, which corresponds to an infinite square well, an opposite limit from the limit $\alpha \to -2$ which allows many radial excitations below the first orbital excitation.

CONCLUDING REMARKS

In summary, we have proposed a composite model of quarks, leptons, weak vector bosons, and Higgs mesons which has certain elements of simplicity: it has few fundamental fields and many composite states, quark-lepton parallelism holds, and the strong and weak interactions are on a parallel footing -- both are residual effects of flavor-independent local gauge interactions. The model accounts, in a schematic way, for the generations of quarks and leptons, and predicts generations of narrow states up to the W mass. Under assumptions stated in the text, the model has Cabibbo structure, GIM mechanism, and spontaneously-broken left-right symmetry for the weak interactions.

We thank our colleagues both at Maryland and elsewhere for stimulating discussions and useful suggestions. We are particularly indebted to Jogesh Pati for his trenchant comments and good advice, and to Roberto Casalbuoni and Rabi Mohapatra for helpful remarks.

REFERENCES

1. Composite models of leptons were considered in O. W. Greenberg and C. A. Nelson, University of Maryland Technical Report No. 74-006 (1973) (unpublished) and, in a revised version of this article, Phys. Rev. D 10, 2567 (1974). The constrast between hadronic and leptonic physics was emphasized and an attempt was made to place them in closer analogy in O. W. Greenberg and G. B. Yodh, Phys. Rev. Lett. 32, 1473 (1974). J. C. Pati and A. Salam, Phys. Rev. D 10, 275 (1974), footnote 7, noted the possibility of expressing their unified model in terms of subconstituent fields. They suggested that four flavors are carried by a spin-1/2 fermion and that color and lepton-number are carried by four spin-0 fields. Thus they use the same constituents to carry flavor for both quarks and leptons. O. W. Greenberg, Phys. Rev. Lett. 35, 1120 (1975) independently suggested a composite quark model in which flavor is carried by a spin-1/2 fermion and color is carried by a spin-0 boson. A meson is then a four-body system. He attempted to use the subquarks in a concrete way to explain the J/ψ and associated resonances as excitations of this four-body system. J. C. Pati, A. Salam, and J. Strathdee, Phys. Lett. B 58, 265, (1975), discussed variants of their preon scheme, including detailed gauge models for them. For further references, see Ref. 2.
2. O. W. Greenberg and J. Sucher, Phys. Lett. B 99, 339 (1981). A preliminary report of this work was given at the 20th International Conference on High Energy Physics, at Madison, Wisconsin, July, 1980 (to appear in Proceedings of the Conference).
3. We thank R. Casalbuoni and R. Gatto for pointing out this possibility.
4. E. Derman, Phys. Lett. B 95, 369 (1980), and Univ. of Colorado Report COLO-HEP-23 (1980). M. Veltman, in Proc. 1979 Int. Symp. on Lepton and Photon Interactions at High Energies, ed. T. B. W. Kirk and H. D. I. Abarbanel (Fermi National Accelerator Lab., Batavia, 1979), p. 529.
5. We thank J. C. Pati for suggesting that we generate an effective $SU(2)_L \times SU(2)_R \times U(1)$ gauge theory with left-right symmetry spontaneously broken by composite Higgs fields. Similar ideas appear in J. C. Pati, Proceedings of the 1980-ν Conference (Erice, June 1980), edited by E. Fiorini (Plenum Press, New York, 1981).

6. The form (6) is an $SU(N)_S$ singlet, a Lorentz scalar, and a flavor non-singlet. Simpler expressions constructed from the F_L's fail to have these three properties.

7. M. Veltman, quoted in J. Ellis, M. K. Gaillard, and B. Zumino, Phys. Lett. B 94, 343 (1980).

8. J. C. Pati and A. Salam, Phys. Rev. Lett. 31, 661 (1973); Phys. Rev. D 10, 275 (1974); G. Senjanović, and R. N. Mohapatra, Phys. Rev. D 12, 1502 (1975); G. Senjanović, Nucl. Phys. B 153, 334 (1979); and references cited in these articles.

9. R. N. Mohapatra, Max-Planck-Institute für Physik und Astrophysik Report MPI-PAE/Pth 1/81 (1981); to appear in these Proceedings.

10. Alternatively, the local gauge symmetry for the weak and electromagnetic interactions could be introduced at the level of L by gauging the flavor degree of freedom and introducing fundamental Higgs mesons. Then this parallelism would not hold, unless the fundamental gauge bosons were to acquire much larger masses (via spontaneous symmetry breaking) than the composite vector bosons.

11. The problem of the mass spectrum includes the following issues: why are quarks, leptons, weak vector bosons, etc., so light in mass compared to the large mass scales of the grand unified theories and of the inverse size r_0^{-1} of the composites; why are the mass spacings between quarks and leptons so small compared to r_0^{-1}; why are weak vector bosons much more massive than quarks and leptons; why do the quark and lepton mass spacings increase rapidly with generation; why do all quarks and leptons have spin 1/2 (assuming the spin of τ, which is not yet established, is 1/2); why do all neutrinos have very small or vanishing masses; why are not electromagnetic mass differences very large, of order e^2/r_0, if the photon is elementary? A problem specific to our model is: why are $\bar{S}C$, $\bar{C}S$, etc. very massive?

12. To see that the matrix element could be as large as $\sim em_\mu r_0$, consider a model in which the muon is an excited state of a bound spin-1/2-spin-0 two-body system with only the spin-1/2 constituent charged. Then the matrix element for radiative decay without a change in parity, is $M = e<\phi_f|\exp(-i\vec{k}\cdot\vec{r})\vec{\alpha}\cdot\hat{\epsilon}|\phi_i> \sim e<\phi_f|\vec{k}\cdot\vec{r}\,\vec{\alpha}\cdot\hat{\epsilon}|\phi_i> \sim ekrA$, $A = <\phi_f|\hat{k}\cdot\hat{r}\,\vec{\alpha}\cdot\hat{\epsilon}|\phi_i>$. For $\mu \to e\gamma$, $k \sim m_\mu$ and $r \sim r_0$, so that $M \sim em_\mu r_0 A$. Now $A \sim v/c$, and for a non-relativistic system with constituents not differing vastly in mass, $k \sim p^2/M_{red} \sim p\,v/c$ so that $kr_0 \sim v/c$ and $M \sim e(m_\mu r_0)^2$. (See R. Barbieri, L. Maiani, and R. Petronzio, Phys. Lett. B 96, 63 (1980) for a similar estimate.) However, for a relativistic system one can have $v/c \sim 1$ so that $M = em_\mu r_0$.

13. V. Visnjić-Triantafillou, Fermilab Pub-80/34 Thy (1980); H. Terazawa, Prog. Theor. Phys. 64, 1763 (1980).

14. H. Georgi, H. Quinn, and S. Weinberg, Phys. Rev. Lett. $\underline{33}$, 451 (1974).

15. S. Weinberg, Phys. Rev. D $\underline{13}$, 947 (1976); $\underline{19}$, 1277 (1979); L. Susskind, Phys. Rev. D $\underline{20}$, 2619 (1979).

16. C. Quigg and J. L. Rosner, Phys. Rep. $\underline{56}$, 167 (1979).

17. E. Nowak, J. Sucher, and C. H. Woo, Phys. Rev. D $\underline{16}$, 2874 (1977) pointed out that a singular potential can elevate orbital excitations.

18. A more radical and speculative way of elevating orbital excitations is to imagine that, phenomenologically, a continuum degree of freedom is present. For example, the QSD gauge flux which permanently confines the quark and lepton constituents may be equivalent to a "rigid" string or "rod." We can then identify the longitudinal excitations of the rod, which have angular momentum zero, with the higher generations, and compare the energy, E_n, associated with longitudinal excitations, with E_ℓ, the energy associated with rotational excitations of the rod which have non-zero angular momentum ℓ. We estimate $E_n = (2n+1) \times \pi \hbar v/L$, and $E_\ell = 5\ell(\ell+1)\hbar^2/2ML^2$, where v is the velocity of longitudinal waves in the rod, and L and M are the length and the mass of the rod, respectively. The condition for $E_n \ll E_\ell$ is $\pi MLv \ll 25\hbar$.

How to Combine the Idea of Subquarks with the Idea of Grand Unification — a spinor subkoma model —

Masaki YASUE[*]

Institute for Nuclear Study, University of Tokyo
Tanashi, Tokyo 188, Japan

Abstract

A *subkoma*: a fundamental constituent of quarks and leptons is introduced in order to unify two ideas: subquarks and grand unifications. The subquarks respect the $SU(3)_c$ symmetry arranged à la Harari and Shupe. A basic interaction of *subkomas* can be regarded as an emission of a quasi-spin wave carrying an internal quantum number, which produces SO(6) for subquarks (chroms) and SO(10) for quarks and leptons formed by chroms and wakems responsible for SO(4). All (gauge) bosons are composites of four *subkomas*; coherent emission of two quasi-spin waves. Applying the spin and statistics to the *subkoma* system with subcolors taken as an origin of binding forces, I find 2+2+1 generations of quarks and leptons depending on three different configurations of spins of wakems and chroms. An SO(10) grand unified model with "technicolor" mechanisms is also suggested.

---

[*] Fellow of the Japan Society for the Promotion of Science.

Composite models of quarks and leptons have received much attention in order to find a clue of solving the generation puzzle, the weak mixing and so on.[1] This report is organized on the basis of one published work[2] and one recent preprint.[3] I briefly describe a standard subquark model which contains two kinds subquarks, source of electroweak interactions and source of strong interactions. I proceed to discuss a possible connection between the standard subquark model and the SO(10) unified model.[4] A new product called *subkoma* is introduced into our formulation of composite quarks and leptons. The idea of *subkoma* unifies two ideas: subquarks and grand unification. The rishion model[5] can be regarded as a *subkoma* model based on SO(6)×U(1) instead of SO(10). To say something about the number of generations and the issue of the weak mixing, I will construct a dynamical *subkoma* model where confining forces acting on *subkomas* are produced by "subcolor" gluons of a subcolor gauge theory.[6] I follow the naming of subquarks made by Terazawa:[7] "wakem" source of electroweak interactions, "chrom" source of QCD interactions and "hakam" source of horizontal interactions (generations)[8]

§1 Standard Subquark Model

A standard subquark model I rely upon contains fermionic wakems (w) and bosonic chroms (B):

$$\nu = (w_1 B_0) \quad u_j = (w_1 B_j)$$
$$\ell = (w_2 B_0) \quad d_j = (w_2 B_j) \quad (j = R, B \text{ and } Y), \tag{1}$$

where B_0 & B_j carry the lepton & quark number. The prototype model of this kind was first proposed by Prof. Sakata and his collaborators,[9] stimulated by the Kiev symmetry which was found by Prof. Marshak and Prof. Okubo with Prof. Gamba,[10] in order to understand the universality of weak interactions. Nowaday, we know that quarks and leptons are well described by the gauge theories based on $SU(2)_L \times U(1)$ for electroweak interactions and $SU(3)_C$ for strong interactions. To reproduce these symmetries in the quark-lepton world, I arrange wakems and chroms in such a way that wakems are sources of $SU(2)_L \times U(1)$ and chroms are sources of $SU(3)_C$ and $U(1)$ of $SU(2)_L \times U(1)$ (Table 1).

$(w_1)_L$, $(w_2)_L$	$\cdots SU(2)_L$
$(w_1)_R$, $(w_2)_R$, B_0, B_j	$\cdots U(1)$
B_j	$\cdots SU(3)_C$

Table 1 Quantum numbers of chroms and wakems in $SU(2)_L \times U(1) \times SU(3)_C$

I myself am not satisfied with this arrangement because chrom B_j is both sources of $U(1)$ and $SU(3)_C$. I believe that a subquark must carry a definite quatnum number or a symmetry, in other word, a subquark is introduced as a carrier of a separable symmetry of quarks and leptons. It is the working hypothesis which Prof. Sakata invoked.[11] Accepting this working hypothesis, I can readily reach the gauge symmetry of $SU(2)_L \times U(1)_R \times SU(3)_C \times U(1)_{B-L}$,[12] various aspects of which have been extensively studied by Prof. Marshak with Dr. Mohapatra.[13]

It is now clear that wakems carry $SU(2)_L \times U(1)_R$ and chroms are sources of $SU(3)_c \times U(1)_{B-L}$ (Table 2). I would like to conclude

| $(w_1)_L, (w_2)_L, (w_1)_R, (w_2)_R$ | $\cdots SU(2)_L \times U(1)_R$ |
| B_0, B_j | $\cdots SU(3)_c \times U(1)_{B-L}$ |

Table 2 Quantum numbers of wakems and chroms in $SU(2)_L \times U(1)_R \times SU(3)_c \times U(1)_{B-L}$

that the idea of subquarks favors $SU(2)_L \times U(1)_R \times SU(3)_c \times U(1)_{B-L}$.

Let me touch on the generations. In the standard subquark model, the generations of quarks and leptons can be understood in three different ways:

(1) to introduce another subquark called "hakam", carrying probably the horizontal gauge symmetry;[8] $B_j = C_j h^{(N)}$,

(2) to exploit excited states of the first generation,[14]

and (3) to be ascribed to different configurations of wakems and chroms.[15]

My attempt discussed here will give the third possibility.

§2 "Unification" of Two Ideas: Subquarks and Grand Unification

Now, I ask myself, "What comes into our scope if we combine the idea of subquarks with the idea of grand unification?" I think that there are two ways to unify these two ideas. The first possibility is to put wakems and chroms together with hakams into one multiplet of, say, SU(8) for example[16]:

$$\underline{8}: (w_1, w_2, C_R, C_B, C_Y, h^{(1)}, h^{(2)}, h^{(3)}). \tag{2}$$

It is on the same line as in case of quarks and leptons. In this sense, the idea of subquarks is under the command of the idea of grand unification (GUT >> Subquarks). The second possibility, which I would like to describe in detail, makes the situation reversed. It is the idea of subquarks that prevails the idea of grand unification (Subquarks >> GUT). What it means is that wakems: sources of $SU(2)_L \times U(1)_R$ and chroms: sources of $SU(3)_C \times U(1)_{B-L}$ are nothing but the same ones, which obey some unknown symmetry corresponding to or yielding a grand unified symmetry in the quark-lepton world.

§3 Fundamental Constituent: "Subkoma"

Next job is to find such an entity. Let us look into the structure of a spinorial representation of $SO(10)$:[17]

$$u_{\alpha L} = (\xi_1 B_\alpha) \qquad d^c_{\alpha L} = (\xi'_1 \bar{D}_\alpha)$$
$$d_{\alpha L} = (\xi_2 B_\alpha) \qquad u^c_{\alpha L} = (\xi'_2 \bar{B}_\alpha) \qquad (3)$$

where the lepton number is regarded as the fourth color ($\alpha = 0$), ξ & ξ' are two real spinors of $SO(4)$ and B_α is a complex spinor of $SO(6)$. You will find that there is a very attractive coincidence between the subquark model (1) and the $SO(10)$ model (3). Furthermore, it has been demonstrated that the spinor of $SO(2N)$ can be described by N fermi oscillators each of which possesses two degree of freedoms denoted by ↑ and ↓,[18] which we refer to as a quasi-spin. I shall identify these oscillators with real objects inside quarks and leptons. This new product is named as "*subkoma*". What is expected if the *subkoma* is

[Koma] A top: a commonly cylindrical or conoidal device that

has a tapering usually steel-shod point on which it is made to spin and that is used especially as a toy (in Japanese).

regarded as a fundamental entity? For instance, quarks and leptons in SO(10) are given by composite states of five *subkomas* expressed as

$$< \overset{\uparrow}{①} \overset{\uparrow}{②} \overset{\downarrow}{③} \overset{\uparrow}{④} \overset{\downarrow}{⑤} > . \qquad (4)$$

I am now in position to explain properties of the *subkoma*. The properties of the *subkoma* can be examined in relation to the SO(10) symmetry of quarks and leptons, five *subkoma* systems, in the similar way to that I employed to find the necessary properties of wakems and chroms, combined into quarks and leptons in $SU(2)_L \times U(1) \times SU(3)_C$. Main results are summarized as follows:

(1) A basic interaction is assumed to be based on an emission of a quasi-spin wave carrying an internal quantum number. You can replace

$$\quad (\overset{a}{\uparrow}\overset{a}{\uparrow}) \quad \text{a quasi-spin wave} \qquad (5)$$

the quasi-spin wave by a composite of two *subkomas*.

(2) Wakems are composites of two *subkomas* and chroms are composites of three *subkomas*:

$$W_{1L} = (\text{①} \text{②}) \qquad W_{1R} = (\text{①} \text{②})$$
$$W_{2L} = (\text{①} \text{②}) \qquad W_{2R} = (\text{①} \text{②}) \qquad (6a)$$

$$C_0 = (\text{③} \text{④} \text{⑤}) \qquad C_R = (\text{③} \text{④} \text{⑤})$$
$$C_B = (\text{③} \text{④} \text{⑤}) \qquad C_Y = (\text{③} \text{④} \text{⑤}) \qquad (6b)$$

and quarks and leptons are expressed by (3) with $\xi \to W_L$, $\xi' \to W_R$ and $B \to C$. The emergence of $SU(3)_c$ respected by C_R, C_B and C_Y is just in the same way as in the rishion model.

(3) All interactions between quarks and leptons are described by the simultaneous emission of two quasi-spin waves. There shows up an "effective" $SO(10)$ symmetry of quarks

$$\begin{array}{c} \text{diagram} \end{array} \qquad \begin{array}{c} aa\ bb \\ (\uparrow\uparrow \cdot \downarrow\downarrow) \end{array} \qquad (7)$$

and leptons.

(4) If we trace a change of the quasi-spins before and after the interactions, we will find the following selection rules. Let $m_z^{(a)}$ be a variable behaving such as

$$m_z^{(a)} = +1 \quad \text{for } \uparrow \quad \text{and} \quad -1 \quad \text{for } \downarrow . \qquad (8)$$

(i) Exact quasi-spin conservation

$$\Delta(\sum_a m_z^{(a)}) = 0 \qquad (9)$$

leads to an "effective" $SU(5) \times U(1)$ symmetry.

(ii) No quasi-spin correlation between wakems and chroms leads to at most SO(4)×SO(6) and at least $SU(2)_L \times U(1)_R \times SU(3)_C \times U(1)_{B-L}$, the seocnd of which is subject to the condition (i).

(iii) The proton decays take place if there exists the quasi-spin correlation between wakems and chroms.

§4 Rishion Model in A Subkoma Model

The relationship between the rishion model and my *subkoma* model now becomes clear. Because in the rision model, quarks and leptons are composites of three rishions, an "effective" SO(6) is expected for the *subkoma* model. In fact, the rishion model can be formulated in a *subkoma* model based on SO(6)×U(1). Quarks and leptons are expressed as

$$\bar{e} = \uparrow_1 \uparrow_2 \uparrow_3$$
$$u = \uparrow_1 \uparrow_2 \downarrow_3, \; \uparrow_1 \downarrow_2 \uparrow_3, \; \downarrow_1 \uparrow_2 \uparrow_3$$
$$\bar{d} = \downarrow_1 \downarrow_2 \uparrow_3, \; \downarrow_1 \uparrow_2 \downarrow_3, \; \uparrow_1 \downarrow_2 \downarrow_3 \qquad (10)$$
$$\nu = \downarrow_1 \downarrow_2 \downarrow_3 .$$

The charge is defined by

$$Q^\gamma = T_z + \frac{Y}{2} \quad \text{in the unit of } \frac{1}{3}, \qquad (11)$$

where T_z is nothing but $m_z^{(a)}/2$ (a = 1,2,3) and Y is for U(1) (Table 3).

subkoma	rishion	T_z	Y	Q^Y
ⓐ↑	T	$\frac{1}{2}$	1	$\frac{1}{3}$
ⓐ↓	V	$-\frac{1}{2}$	1	0

Table 3   *subkomas* v.s. *rishions*

The U(1) of SO(6)×U(1) can be absorbed into SO(8): SO(8) ⊃ SO(6)×U(1). But quarks and leptons are composites of four *subkomas*; therefore, the extra *subkoma* must be boson. Regarding the extra *subkoma* as a *di-subkoma*, I would like to insist that SO(10) is the only candidate for the spinor-*subkoma* model. Because the U(1) factor is involved, all bosons; photon, W, Z and gluons, can be expressed by composites of four *subkomas* or coherent states of two quasi-spin waves as mentioned in §3. The original charge assignment is taken to be

$$Q^Y = (\tfrac{1}{2}, 0, -\tfrac{1}{6}, -\tfrac{1}{6}, -\tfrac{1}{6}) \quad \text{for} \quad (①↑, ②↑, ③↑, ④↑, ⑤↑) \quad (12)$$

Let me summarize the discussion presented so far. It is clear that the standard subquark model with wakems and chroms and the rishion model in which the emergence of $SU(3)_c$ is sigfnified are related to each other in the *subkoma* model based on the effective SO(10) symmetry, which has been derived by imposing the "unification" of two main ideas: subquarks and grand unification, in a non-trivial way.

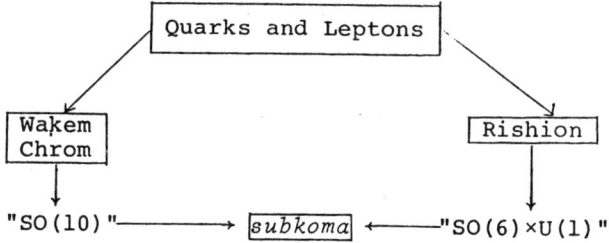

§5 Subkoma Model with Subcolors

Along the scenario described above, we will construct a dynamical *subkoma* model by introducing a subcolor as a binding force among the *subkomas*. The scenario is as follows: *subkomas* are combined into wakems and chroms at the early stage of the formation of composites and subsequently wakems and chroms form quarks and leptons. Therefore, at most three kinds of subcolors are required to form wakems (two *subkomas*), chroms (three *subkomas*) and quarks and leptons (five *subkomas*). I will adopt $SU(2)$, $SU(3)/O(3)$ and $SO(5)/O(5)$. But the condition for getting confining forces only allows me to employ $SU(2)$ and $SO(5)/O(5)$. Furthermore to guarantee to get the combinations of $(w_L C)$ and $(w_R \bar{C})$ but not of $(w_L \bar{C})$ and $(w_R C)$, I must choose $O(5): SO(5) \times Z_2$. The five *subkomas* transform as vector $\{5\}$ or (axial) vector $\{5'\}$ under $O(5)$:

$$\begin{array}{ll} \{5\} & \text{for } S_\alpha^{(a\uparrow)} \\ \{5'\} & \text{for } S_\alpha^{(a\downarrow)} \end{array} \quad (a = 1 \sim 5) \tag{13}$$

where $S^{(a\uparrow)}$ ($S^{(a\downarrow)}$) denotes $\overset{\uparrow}{\underset{\downarrow}{\text{ⓐ}}}$ ($\overset{\downarrow}{\underset{}{\text{ⓐ}}}$) and α runs from 1 to 5, specifying the O(5) vector. The extra subcolor SU(2) is also carried by the first two *subkomas* (a = 1 and 2), which will form wakems. In this sense, I have implicitly introduced the difference between the *subkomas* to be combined into wakems and those into chroms. I use i and j describing the SU(2) subcolor:

$$\begin{matrix} i & \text{for} & S_{\alpha j}^{(a\uparrow)} \\ i & \text{for} & S_{\alpha}^{(a\downarrow)j} \end{matrix} \quad (a = 1 \text{ and } 2) \quad (14)$$

My scenario with subcolors is shown as follows:

<u>Scenario</u>

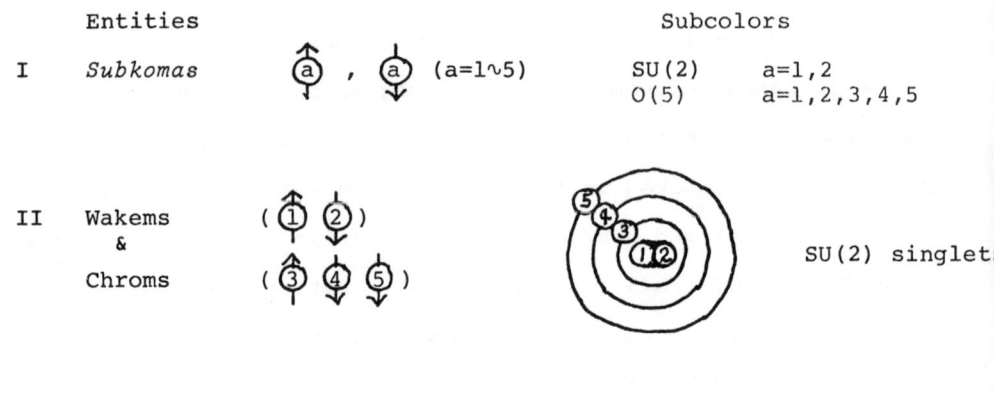

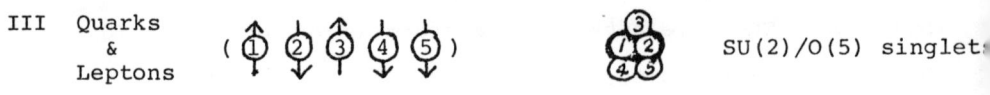

Although there is expected to show up various exotic states of the five *subkomas* other than ordinary quarks and leptons, I content myself at the present stage with showing that the composites include ordinary quarks and leptons. Assuming the *subkomas* obey

585

the Fermi statistics, I can examine spins and internal symmetry described by a (= 1∼5) and ↑ & ↓ of the composites of the *subkomas*.

Results are summarized below:

1. Composites of $S^{(1)}$ and $S^{(2)}$ with $\begin{array}{c}\alpha\\\beta\end{array}$ including wakems

wakems	1 2 (↑ ↑)	1 2 (↓ ↓)	1 2 (↑ ↓)	1 2 (↓ ↑)
scalar	a↑ b↑		a↑ b↓	
vector	a↑ b↑		a↑ b↓	

Table 4   Composite states containing wakems

2. Composites of $S^{(3)}$, $S^{(4)}$ and $S^{(5)}$ with $\begin{array}{c}\alpha\\\beta\\\gamma\end{array}$ including chroms

chroms	3 4 5 (↑ ↑ ↑)	3 4 5 (↓ ↓ ↑)	3 4 5 (↓ ↑ ↓)	3 4 5 (↑ ↓ ↓)
spin 3/2	a↑ b↑ c↑	a↓ b↓ c↑		
spin 1/2	a↑ b↑ c↑	a↓ b↓ c↑	a↓ b↓ c↑	a↓ c↑ b↓

Table 5   Composite states containing chroms

It is interesting to note that $(\overset{a}{\uparrow}\overset{a}{\uparrow})$ and $(\overset{a}{\downarrow}\overset{a}{\downarrow})$ (a = 1∼5) belong to {10} of the effective SO(10) symmetry and furthermore these states can be O(5) invariants. However, $(\overset{a}{\uparrow}\overset{a}{\downarrow})$ cannot be O(5) invariants because it is (pseudo) scalar. As you might notice, these states can be identified with the quasi-spin wave (5).

Quarks and leptons are formed out of wakems and chroms due to the O(5) confining force:

$$\epsilon_{\alpha\beta\gamma\delta\sigma}\underbrace{(S_\alpha^{(1)} S_\beta^{(2)})}_{\text{wakems}} \underbrace{(S_\gamma^{(3)} S_\delta^{(4)} S_\sigma^{(5)})}_{\text{chroms}}, \tag{15}$$

where the formation of the SU(2) singlet wakems are understood. Therefore, I can have three groups of quarks and leptons (A, B and C) corresponding to three different configurations of wakems and chroms (Table 6).

composite sub-quarks	Quarks and Leptons		
	A	B	C
wakem	scalar	vector	
chrom	spin 1/2		spin 3/2
multiplicity	2	2	1

Table 6 Three Groups of Quarks and Leptons

The multiplicity is due to the internal symmetry denoted by (a↑) and (a↓): 2 from $\begin{array}{|c|c|}\hline 3 & 4 \\\hline 5 & \\\hline\end{array}$ and 1 from $\boxed{3\,4\,5}$. Let the *subkomas* be massless L-handed states and then I find that spin-configuration are

$$\begin{array}{|c|c|c|}\hline 1 & 3 & 5 \\\hline 2 & 4 \\\hline\end{array} \quad \text{for A,}$$

$$\begin{array}{|c|c|c|}\hline 3 & 1 & 2 \\\hline 4 & 5 \\\hline\end{array} \quad \text{for B,} \tag{16}$$

$$\text{and} \quad \begin{array}{|c|c|c|}\hline 3 & 4 & 5 \\\hline 1 & 2 \\\hline\end{array} \quad \text{for C.}$$

Therefore, altogether I have 2+2+1, 5 generations of the {16} L-handed massless quarks and leptons.

The weak mixing is in general expected to occur among the states containing the same constituent: spin 1/2 chrom for A and B, and vector wakem for B and C:

$$A \longleftrightarrow \underset{\substack{\text{spin 1/2} \\ \text{chrom}}}{B} \longleftrightarrow \underset{\text{vector wakem}}{C} \tag{17}$$

If I neglect the prolification due to the internal symmetry, then I feel like identifying

$$(ud e\nu_e) \quad \text{with} \quad A,$$
$$(cs\mu\nu_\mu) \quad \text{with} \quad B, \tag{18}$$
and $\quad (tb\tau\nu_\tau) \quad \text{with} \quad C.$

The most faithful scenario to the SO(10) grand unified model is to activate the O(5) confining force at the early stage of the formation of composites. Thus, I have the composites of two *subkomas*: $(\overset{aa}{\uparrow\uparrow})$ & $(\overset{aa}{\downarrow\downarrow})$ and those of five *subkomas*, which include would-be quarks and leptons. At the next stage, the SU(2) confining force gets strong and the SU(2)/O(5) singlet composites are formed. These include ordinary quarks and leptons and lots of scalars such as

$$(\nu^c \nu^c), \tag{19}$$

belonging to {126} and $\underset{\sim}{1}$ of SU(5) and

$$(\overset{a}{\uparrow}\overset{a}{\uparrow}\overset{a}{\downarrow}\overset{a}{\downarrow}), \tag{20}$$

belonging to {45} and (6,1) for a = 1 and 2 and (1,15) for a =

3, 4 and 5 under SO(4)×SO(6). Therefore, I may conclude that SO(10) can break down to $SU(2)_L \times SU(3)_c \times U(1)$ if SO(10) is gauged. Thus, "technicolor" mechanisms are naturally expected.

§6 Summary

I have demonstrated an intuitive description of <u>a spinor-subkoma model</u> of composite quarks and leptons which is extracted from the "unification" of two ideas: subquarks and grand unification. My working hypothesis leads to <u>$SU(2)_L \times U(1)_R \times SU(3)_c \times U(1)_{B-L}$</u> for quarks and leptons. The idea proposed by Harari and Shupe is naturally involved in the construction of the spinor of SO(10) by using the *subkomas*. It should be noted that the $SU(3)_c$ symmetry is ascribed to chroms but not to quarks and leptons. All (gauge) bosons are expressed as composites of four *subkomas* and a new selection rule concerning the quasi-spin conservation can exist for the interactions mediated by those bosons. The particles I have constructed are listed in Table 7.

# of *subkomas*	particles
1	*subkomas*
2	quasi-spin waves, wakems
3	chroms
4	(gauge) bosons
5	quarks and leptons

Table 7 Classification of composites

There of course exist lots of problems to be investigated. One of the important issues is to gauge global symmetries or to produce massless vector bosons.[19] I do not know any consisten mechanism which would remove the exotic fields or yields heavy masses characterized by their inverse size, for instance. A proper understanding of the fermion generation mixing should be developed although the gross feature can be given by the mixing between states containing the same constituent.

In conclusion, I am saying, "Subkoma is in coma in the quark-lepton world. Do you think when it comes arround?"

I came across two papers during the workshop which treated the similar topic, "A Dynamical Theory for the Rishion Model" (WIS-80/37/Sept-ph) by H. Harari and N. Seiberg and "Constituent Models for Basic Fields of Grand Unification" (YTP80-25) by F. Mansouri. I thank Dr. X. Lee at VPI for informing me of these two papers. The first paper suggests the possible existence of five-rishion states, of "technicolor" mechanisms and of $SO(4) \times SU(3)_c \times U(1)_{B-L}$ as an effective symmetry which I have discussed in this context but $SU(3)_c$ is originally introduced as a local symmetry of the rishions. The second paper repeats my discussion already presented in Ref. 2) but explicitly introduces five additive quantum numbers of the five subkomas. From my view point, these five quantum numbers are associated with five independent $SO(2)$ rotations in $SO(10)$ and then coupled to the (gauge) bosons specified by $(\uparrow\uparrow\downarrow\downarrow)^{aaaa}$ ($a = 1\sim5$) (Table 8).

Subkomas	$T_L^{(3)}$	$T_R^{(3)}$	$T_c^{(3)}$	$T_c^{(8)}$	B-L	Q^Y
$S^{(1\uparrow)}$	1/4	1/4	0	0	0	1/2
$S^{(2\uparrow)}$	-1/4	1/4	0	0	0	0
$S^{(3\uparrow)}$	0	0	1/4	1/6	-1/3	-1/6
$S^{(4\uparrow)}$	0	0	-1/4	1/6	-1/3	-1/6
$S^{(5\uparrow)}$	0	0	0	-1/3	-1/3	-1/6

Table 8 Five additive quantum numbers of subkomas

These are given by $T_L^{(3)} = (\Sigma^{(1)} - \Sigma^{(2)})/2$, $T_R^{(3)} = (\Sigma^{(1)} + \Sigma^{(2)})/2$, $T_c^{(3)} = (\Sigma^{(3)} - \Sigma^{(4)})/2$, $T_c^{(8)} = (\Sigma^{(3)} + \Sigma^{(4)} - 2\Sigma^{(5)})/3$, $B-L = -2(\Sigma^{(3)} + \Sigma^{(4)} + \Sigma^{(5)})/3$ and $Q^Y \equiv T_L^{(3)} + T_R^{(3)} + \frac{B-L}{2}$ where $\Sigma^{(a)}{}_S{}^{(b\dagger)} = \frac{1}{2}\delta^{ab}{}_S{}^{(b\dagger)}$. In the ordinary SO(10) algebra, these correspond to $\Sigma^{(1)} = S_{12}$, $\Sigma^{(2)} = S_{34}$, $\Sigma^{(3)} = S_{56}$, $\Sigma^{(4)} = S_{78}$ and $\Sigma^{(5)} = S_{910}$ where S_{ij} is the 5 diagonal generators of SO(10) and (i,j) = (1,2)(3,4) for SO(4) and (i,j) = (5,6)(7,8)(9,10) for SO(6).

Acknowledgements

The author would like to express his sincere thanks to Prof. R. E. Marshak for the warm hospitality extended to me during the workshop at VPI. He also thanks Prof. Y. Nambu for giving me an opportunity of presenting this talk at EFI, University of Chicago.

References

1) H. Terazawa, Phys. Rev. D$\underline{22}$ (1980) 184 for the partial references.

2) M. Yasuè, Phys. Lett. $\underline{91B}$ (1980) 85.

3) M. Yasuè, INS-preprint INS-Rep.-394 (Nov., 1980). Similar analysis without a subcolor has been made by R. Casalbuoni, G. Domokos and S. Kovesi-Domokos, SLAC preprint SLAC-PUB-2585 (Aug., 1980) within the rishion model.

4) H. Fritzsch and P. Minkowski, Ann. of Phys. $\underline{93}$ (1975) 193; M. S. Chanowitz, J. Ellis and M. K. Gaillard, Nucl. Phys. $\underline{B128}$ (1977) 506; H. Georgi and D. V. Nanopoulos, Nucl. Phys. $\underline{B155}$ (1979) 52.

5) H. Harari, Phys. Lett. $\underline{86B}$ (1979) 83; M. A. Shupe, ibid. $\underline{86B}$ (1979) 87.

6) G. 't Hooft, Cargèse Summer Institute Lecture Note (1979); H. Terazawa, INS-preprint INS-Rep.-382 (June, 1980) to appear in Prog. Theor. Phys. $\underline{64}$, No.5 (1980); R. Casalbuoni and R. Gatto, Phys. Lett. $\underline{93B}$ (1980) 47; O. W. Greenberg and J. Sucher, U. of Maryland preprint (Aug., 1980). For the rishion model with a subcolor, C. A. Nelson, Phys. Lett. $\underline{93B}$ (1980) 143; E. J. Squires, ibid. $\underline{94B}$ (1980) 54. In relation to forming the light composites, S. Dimopoulos, S. Rabi and L. Susskind, Nucl. Phys. $\underline{B173}$ (1980) 208; S. Weinberg, Talk given at this workshop. For O(n) preons, R. Barbieri, L. Maiani and R. Petronzio, Phys. Lett. $\underline{96B}$ (1980) 63. Within the composite SU(6) scheme, K. Fujikawa, Prog. Theor. Phys. $\underline{58}$ (1977) 978.

7) H. Terazawa, Prog. Theor. Phys. <u>58</u> (1977) 1276.

8) K. Akama and H. Terazawa, INS-preprint, INS-Rep.-257 (1976). The spinor subquark model, (w C h), has been proposed by H. Terazawa and his collaborators, H. Terazawa, Y. Chikashige and K. Akama, Phys. Rev. <u>D15</u> (1977) 480. For the recent works on the horizontal symmetries, T. Maehara and T. Yanagida, Prog. Theor. Phys. <u>60</u>, (1978) 822 and <u>61</u> (1979) 1434; F. Wilczek and A. Zee, Phys. Rev. Lett. <u>42</u> (1979). See also R. Peccei, Talk given at this workshop.

9) Z. Maki, M. Nakagawa, Y. Ohnuki and S. Sakata, Prog. Theor. Phys. <u>23</u> (1960) 1174; Z. Maki, M. Nakagawa and S. Sakata, Prog. Theor. Phys. <u>28</u> (1962) 870.

10) A. Gamba, R. E. Marshak and S. Okubo, Proc. Nat. Acad. Sci. USA <u>45</u> (1959) 881.

11) S. Sakata, Prog. Theor. Phys. Suppl. No.19 (1961) 3.

12) M. Yasuè, Prog. Theor. Phys. <u>61</u> (1979) 269.

13) R. E. Marshak and R. N. Mohapatra, Phys. Lett. <u>91B</u> (1980) 222 and Phys. Rev. Lett <u>44</u> (1980) 1316.

14) O. W. Greenberg and J. Sucher in Ref. 6).

15) R. Casalbuoni and R. Gatto in Ref. 6); K. Kakazu and K. Matumoto, U. of Ryukyus preprint DPUR-2 (Aug., 1980).

16) M. Gell-Mann, Talks given at the Washington Meeting of the American Physical Society, April 1977; T.L. Curtright and P.G.O. Freund, in <u>Supergravity</u> (van Nieuwenhuizen and Freedman, North-Holland, 1979) pp. 197-201; J. Ellis, M.K. Gaillard and B. Zumino, Phys. Lett. <u>94B</u> (1980) 343 which, however, contains spin 5/2 multiplet. More elegant formulation ha

been recently made by K. Shima, Saitama Institute of Technology preprint SIT-80-10, who exploits SO(9) to unify all fundamental interactions and to accomodate at least three generations.

17) M. Gell-Mann, P. Ramond and R. Slansky, Rev. Mod. Phys. $\underline{50}$ (1978) 121.

18) R. Casalbouni and R. Gatto, Phys. Lett. $\underline{88B}$ (1979) 306;
R. N. Mohapatra and B. Sakita, Phys. Rev. $\underline{D21}$ (1980) 1062;
M. Yasùè in Ref. 2).

19) F. Alexander Bais and J.-M. Frère, CERN preprint TH 2911-CERN (July, 1980) and references therein.

A CONSTITUENT MODEL FOR QUARKS AND LEPTONS[*]

Freydoon Mansouri, Dept. of Physics
Yale University, New Haven, CT. 06520

ABSTRACT

A model of quarks and leptons is presented. The symmetries of the constituents with respect to which the composites are non-singlets are all abelian. The number of the constituents and the number of their abelian symmetries is related to the rank of the quark-lepton symmetry group. The low energy gauge symmetry $SU_c(3) \times SU_L(2) \times U(1)$ of quarks and leptons is replaced by the constituent gauge symmetry $SU_{c8}(5) \times [U(1)]^5$. The constituents carry not B- but (B-L) number.

The construction of the constituent models for quarks and leptons can proceed in two ways: (a) One can make use of the same methods and principles which have proved successful at other levels in the past. For constituent models, this means the quark models and variations on them (1). The main drawbacks of this approach are that the symmetry group of the constituents turns out to be even larger than those of leptons and quarks and that quarks and leptons would have to be assigned to the third or higher rank tensor representations of the symmetry group. Most of the attempts so far fall into this category. (b) One can make use of more novel methods hoping that one of them will turn out to be relevant (2,3,4). I would like to describe here a model of latter type (5). Some of the prominent features of this model are the following:

(i) The constituents do not form representations of the same symmetry group as that of quarks and leptons. In this respect it is similar to the Harari-Shupe model.

(ii) The number of constituents is not picked randomly. It is related to the "rank" of the quark-lepton symmetry group. Given a

---

[*] Research (Yale Report YTP80-35) supported in part by The U.S. Department of Energy under Contract No. EY-76-C-02-3075.

ISSN:0094-243X/81/720594-08$1.50 Copyright 1981 American Institute of Physic grand unifying group, it takes some hindsight to find out exactly how the number of constituents is related to this rank. For each SO(4N+2) group, the number of constituents is equal to the rank. The extent to which this approach reduces the number of fundamental fields can be made clear by comparing, e.g., the rank, 11, of the group SO(22) with the dimension, 1024, of its lowest dimensional spinor representation.

(iii) The constituents carry only abelian symmetries. The number of these symmetries is also related to the "rank" of the quark-lepton symmetry group. These symmetries which are characteristically different from those of quarks and leptons are ultimately responsible for the symmetries that the composite objects exhibit.

(iv) Leptons and quarks share their constituents democratically. That is, in each quark or lepton of a given generation, each one of the constituents or their antiparticles but not both are represented.

(v) The choice of quantum numbers for the constituents is not arbitrary. Since there are more quarks and leptons than are constituents, it is remarkable that a (unique) choice can be made. In particular, no assignment of baryon number can be made compatible with the known assignments of the composite fermions. Only (B-L) assignment is possible.

(vi) The existence of generations as well as the absence of low-lying spin-3/2 states can be attributed to a possible "shell structure".

The above features can be incorporated into constituent models which account for the spectrum of leptons and quarks in any grand unified theory. But the most direct incorporation is realized when the grand unifying group is one of the groups SO(4N+2). This is because for these groups the number of constituents turns out to be equal to the rank of the corresponding group. So let me begin with the one-generation group SO(10). It is a rank 5 group, so that in this scheme there will be five constituents $b_1, \ldots, b_5$ and their antiparticles. They are distinguished from each other by up to five abelian symmetries $[U(1)]^5$, i.e., they are one-dimensional representations of these abelian symmetries.

Looking at 5-body states

$$b_1 b_2 b_3 b_4 b_5, \ b_1 b_2 b_3 b_4 \bar{b}_5, \ b_1 b_2 b_3 \bar{b}_4 b_5, \ldots,$$

$$b_1 b_2 b_3 \bar{b}_4 \bar{b}_5, \ b_1 b_2 \bar{b}_3 \bar{b}_4 b_5, \ldots, \ \bar{b}_1 \bar{b}_2 \bar{b}_3 \bar{b}_4 \bar{b}_5$$

we see that they are 32 in number. Each one of these states is again a one-dimensional representations of $[U(1)]^5$ but with quantum numbers which are sums of the corresponding quantum numbers of its constituents. Before these states can be identified with a generation of quarks and leptons one must show that there is a choice of quantum numbers for $b_1,\ldots, b_5$, which is compatible with those of the composite particles. As mentioned above, such a unique choice does exist and is given in Table 1. Moreover, from the quantum numbers given in this table, it is clear that, as far as the symmetries of quarks and leptons are concerned, if we had left the choice of a symmetry group for the constituents open up to now, we would have been forced to take them to be abelian at this stage. The abelian nature of these symmetries is also responsible for limiting the number of 5-body composites to just 32 fermions. Of course, it is possible that the constituents have additional constituent color symmetries, with respect to which quarks and leptons are singlets.

Once a generation of composite fermions with all of their quantum numbers is constructed in this way, then their one-to-one identification with quarks and leptons and their assignment to SO(10) can be made. These are given in Table 2.

Consider next the description of other generations. If one takes the saturation property (shell structure) of the constituents at 5 seriously, then generations can be accounted for in at least two ways: One obvious way is to take other generations as higher excitation of the first (ground state) generation described above. The other way is to allow for the possibility that there are more constituents than just five, but the number as well as the "type" of constituents required for saturation remains the same as for a single generation. Now a glance at the content of Table 1, makes it clear that the color properties of the composite particles are determined by the 1st three constituents, so that the simplest way of assuring the same color properties from one generation to the next is to retain b_1, b_2, b_3 in all generations. This means that the additional constituents must carry the same quantum numbers as those of b_4 or b_5 with respect to the first five U(1)'s. Since there is no á priori reason to have an excess of b_4-type or b_5-type constituents, then the additional constituents must come in pairs. In other words, the number of constituents will always be odd.

For example, if there are seven constituents $b_1,\ldots,b_7$, there will be four generations each consisting of b_1, b_2, b_3, and one of the pairs (b_4, b_5), (b_4, b_7), (b_6, b_5), (b_6, b_7). The generation structure for a few other sets of constituents is given in Table 3 below: Since the number of the

# of constituents	# of generations
5	1
7	4
9	9
11	16
13	25
15	36

Table 3. Number of generation for a given set of constituents.

constituents is not known a priori, the predictions of this scheme in regard to generations is not absolute. But it does predict a pattern of generations.

Finally, let us briefly consider the following questions: (a) At what scale does one expect such a description in terms of constituents to take over, and (b) in what context these abelian symmetries could be understood. A quick answer to question (a) might be that the description in terms of constituents is relevant only above grand unification mass, i.e., above 10^{15} GeV. But it is well to remember that the current trends to extrapolate the behavior of a theory from energies below one TeV to energies of the order 10^{15} GeV involves an implicit assumption: One <u>assumes</u> that the description in terms of quarks and leptons <u>up</u> to those energies remains adequate. If it turns out that somewhere below 10^{15} GeV, e.g., around 10^6 GeV, leptons and quarks begin to show significant structure, then <u>physically</u> such a description will not be applicable beyond this scale even when extrapolation to 10^{15} GeV might be a mathematical possibility.

As for question (b), it appears at first sight that if this picture is correct, it precludes the possibility of a unified theory with a single coupling constant. It must be remembered, however, that description in terms of constituents is relevant only at very high energies where unification with gravity is all the more important. Noting that in the above model there is room for elementary photons, it is quite natural to envisage a unified theory as a generalized Kaluza-Klein theory in a higher dimensional space. Recall that Kaluza-Klein theory involves the unification of Einstein-Maxwell theory in a 5-dimensional world. It is possible to generalize Kaluza-Klein theory to an arbitrary internal symmetry group[6], in particular to $[U(1)]^{\ell}$. So one version of such a unified theory will be a geometric theory of Yang-Mills type in a $(4+\ell)$ dimensional space for which the unbroken group is $SO(4+\ell)$. In such a theory there is no á <u>priori</u> distinction between external momenta and internal charges. But at some stage in the early development of such a $(4+\ell)$ dimensional universe, the ℓ of the $3+\ell$ spacial dimensions become compactified, i.e., the motion along them becomes periodic. Thus the isotropy of the initial universe is broken, and there appear discrete charges associated with internal symmetry, such as our ℓ $U(1)$'s. Therefore, if this picture of constituents turns out to be relevant, a formalism for a genuine unification with gravity with a single dimensionless coupling constant can be constructed along these lines. Preliminary indications are that with $\ell=7$ such a theory will be locally supersymmetric.

It is, of course, possible to augment the above unification scheme to include "constituent color" symmetry. In the present context one such local symmetry is a local $SU(5)$. Thus the low energy local symmetry $SU_c(3) \times SU_L(2) \times U(1)$ or $SU_c(3) \times SU_L(2) \times SU_R(2) \times U(1)$ is replaced by the high energy local gauge symmetry $SU_{cc}(5) \times [U(1)]^5$. Then, one can proceed to unify this symmetry with gauge symmetries of space-time.

REFERENCES AND FOOTNOTES

1. See, e.g., J.C. Pati and A. Salam, Phys. Rev. D10, 275 (1975). J.C. Pati, A. Salam, J. Strathdee Phys. Lett. S.L. Glashaw Harvard Preprint HUTP-77/A005, unpublished. Y. Ne'eman, Phys. Lett. 82B, 69 (1979). G. t'Hooft, Cargese Summer Institute Lectures (1979) unpublished. R. Casalbuoni and R. Gatto Phys. Lett. 88B, 306 (1979); 90B, 81 (1980). G.L. Kane, Private communication. O.W. Greenberg, this Proceedings.

2. See, e.g., H. Harari, Phys. Lett. 86B, 83 (1979); M.A. Shupe, Phys. Lett. 86B, 87 (1979).

3. For a critical analysis of these models and alternative suggestions, see R. Casalbuoni, G. Domokos, S. Kovesi-Domokos, SLAC-PUB 2585, 1980.

4. During the course of this conference, I was informed by Professors J.C. Pati and M. Yasue that they also have models which fall in this category. See J.C. Pati, to be published; M. Yasue, Phys. Lett. 91B, (1980), and Tokyo preprint INS-Rep.-394, Nov.-1980.

5. Based on F. Mansouri, Phys. Lett. B, in Press, available as Yale Preprint YTP80-25. Also, F. Mansouri, in preparation.

6. See, e.g., L.N. Chang, K. Macrae, F. Mansouri, Phys. Rev. 13D, 235 (1976).

Constituent	T_3^w	Y^w	Q	T_3^c	Y^c	B-L
b_1	0	1/3	1/6	$-\frac{1}{4}$	-1/6	1/3
b_2	0	1/3	1/6	$\frac{1}{4}$	-1/6	1/3
b_3	0	1/3	1/6	0	1/3	1/3
b_4	$\frac{1}{4}$	$-\frac{1}{2}$	0	0	0	0
b_5	$-\frac{1}{4}$	$-\frac{1}{2}$	-3/6	0	0	0

Table 1. The additive quantum numbers of the five constituents for SO(10) model. The antiparticles carry quantum numbers opposite to these. The superscripts w and c refer, respectively, to weak and color.

Constituent Structure	T_3^W	Y^W	Q	T_3^c	Y^c	Quark Lepton
$\begin{bmatrix} \bar{b}_1 b_2 b_3 \\ b_1 \bar{b}_2 b_3 \\ b_1 b_2 \bar{b}_3 \end{bmatrix} \cdot \bar{b}_4 b_5$	$-\frac{1}{2}$	1/3	-1/3	$-\frac{1}{2}$ $-\frac{1}{2}$ 0	1/3 1/3 -2/3	d_L
$\begin{bmatrix} \bar{b}_1 b_2 b_3 \\ b_1 \bar{b}_2 b_3 \\ b_1 b_2 \bar{b}_3 \end{bmatrix} \cdot b_4 \bar{b}_5$	$\frac{1}{2}$	1/3	2/3	$\frac{1}{2}$ $-\frac{1}{2}$ 0	1/3 1/3 -2/3	u_L
$\left(\bar{b}_1 \bar{b}_2 \bar{b}_3 \right) b_4 \bar{b}_5$	$\frac{1}{2}$	-1	0	0	0	ν_L
$\left(\bar{b}_1 \bar{b}_2 \bar{b}_3 \right) \bar{b}_4 b_5$	$-\frac{1}{2}$	-1	-1	0	0	e_L^-
$\begin{bmatrix} \bar{b}_1 b_2 b_3 \\ b_1 \bar{b}_2 b_3 \\ b_1 b_2 \bar{b}_3 \end{bmatrix} b_4 b_5$	0	-2/3	-1/3	$\frac{1}{2}$ $-\frac{1}{2}$ 0	1/3 1/3 -2/3	d_R
$\begin{bmatrix} \bar{b}_1 b_2 b_3 \\ b_1 \bar{b}_2 b_3 \\ b_1 b_2 \bar{b}_3 \end{bmatrix} \bar{b}_4 \bar{b}_5$	0	4/3	2/3	$\frac{1}{2}$ $-\frac{1}{2}$ 0	1/3 1/3 -2/3	u_R
$\left(\bar{b}_1 \bar{b}_2 \bar{b}_3 \right) \bar{b}_4 \bar{b}_5$	0	0	0	0	0	ν_R
$\left(\bar{b}_1 \bar{b}_2 \bar{b}_3 \right) b_4 b_5$	0	-2	-1	0	0	e_R^-

Table 2. The structure of left-handed and right-handed quarks and leptons in terms of their constituents. ν_L, e_L^- and the charge conjugate $\hat{d}_R$ of d_R are identified with $\bar{5}$ of SU(5); similarly, $\hat{u}_R$, u_L, d_L, $\hat{e}_R$ are identified with 10 of SU(5). These 15 states together with ν_R constitute 16 of SO(10) group.

COMMENTS AND SPECULATIONS

Presentations

REVIEW OF NEW EXPERIMENTAL UPPER LIMITS ON

FORBIDDEN DECAY MODES OF THE TAU LEPTON [*]

Kenneth G. Hayes and Martin L. Perl
Stanford Linear Accelerator Center, Stanford University
Stanford, CA 94305

ABSTRACT

This paper presents a review of experimental upper limits on the branching fractions for various forbidden decay modes of the tau lepton. These are modes which cannot occur in the conventional model in which the tau and its associated neutrino have a unique, conserved lepton number. The limits are based on data acquired by the Mark II Detector Collaboration at SPEAR.

---

[*] Work supported by the Department of Energy under contract DE-AC03-76SF00515.

I. INTRODUCTION

When we presented this paper at the Workshop on Weak Interactions as Probes of Unification, we reviewed some of the current and much of the proposed research on the tau lepton and its associated neutrino. However most of that material has been published elsewhere.[1-4] Therefore we will restrict this written version to a review of new experimental upper limits[5] on the branching fractions of various tau decay modes which are forbidden by the sequential lepton model[1] of the tau. In that model the tau and its associated neutrino have a unique, conserved lepton number. Hence, in that model, a decay mode of a τ^- or a τ^+ must contain a ν_τ or a $\bar{\nu}_\tau$ respectively. The new results reviewed here are from data collected by the Mark II Detector Collaboration at the SPEAR electron-positron facility.

II. GENERAL SEARCH METHOD

If the τ^- and ν_τ have a unique, conserved lepton number, there are many forbidden decay modes. Some of the simpler forbidden decay modes which do not contain hadrons are:

$$\begin{aligned}
\tau^- &\to e^- + \gamma \\
\tau^- &\to \mu^- + \gamma \\
\tau^- &\to e^- + e^+ + e^- \\
\tau^- &\to e^- + e^+ + \mu^- \\
\tau^- &\to e^- + \mu^+ + \mu^- \\
\tau^- &\to \mu^- + \mu^+ + \mu^- ;
\end{aligned} \quad (1)$$

and some simple, forbidden, hadron-containing, decay modes are:

$$\tau^- \to e^- + \pi^0$$
$$\tau^- \to \mu^- + \pi^0$$
$$\tau^- \to e^- + K^0$$
$$\tau^- \to \mu^- + K^0 \qquad (2)$$
$$\tau^- \to e^- + \rho^0$$
$$\tau^- \to \mu^- + \rho^0$$

Reference 5 contains a discussion of the possible theoretical implications of the existence of some of these decay modes.

We searched for the decay modes listed in Eqs. (1) and (2) using data acquired by the Mark II Detector Collaboration at SPEAR. This data was in the center of mass energy range of 3.9 GeV to 6.7 GeV; and contained 48,108 produced $\tau^+\tau^-$ pair. A produced $\tau^+\tau^-$ pair is a pair that would have been detected if the apparatus had 100% detection efficiency for all decay modes of the τ.

We selected events from this data sample which had the following properties:

(a) 2, 3, or 4 charged tracks;

(b) total charge = 0 for 2 or 4 track events;

(c) total charge = ±1 for 3 track events;

(d) any number of photons; and

(e) the 2 charged track events have an acoplanarity angle[6] greater than 5°.

Such a sample includes the basic signature

$$e^+ + e^- \rightarrow \tau^+ + \tau^-$$

$\ \downarrow\ \ \downarrow$

1 charged track   1 or 3 charged track

allowed decay modes   forbidden decay modes

such as   such as

$$e^+ + \nu_e + \bar{\nu}_\tau \qquad\qquad e^- + \gamma$$

or   or

$$\mu^+ + \nu_\mu + \bar{\nu}_\tau \qquad\qquad \mu^- + \gamma \qquad (3)$$

or   or

$$\pi^+ + \bar{\nu}_\tau \qquad\qquad e^- + e^+ + e^-$$

or

$$e^- + \rho^0$$
$$\hookrightarrow \pi^+ + \pi^-$$

The charge reversed reactions were also used of course. The sample containing a total of three charged tracks allowed us to use events in which the three charged particles in a forbidden decay such as $e^- + e^+ + e^-$ were detected, but the track from the allowed decay was not detected.

III. EXAMPLE: SEARCH FOR $\tau^- \rightarrow \mu^- + \gamma$

The search method for a specific forbidden decay mode depended on the mode. We shall use as an example the search method for the mode

$$\tau^- \rightarrow \mu^- + \gamma \qquad (4)$$

From the sample defined by requirements a.-e., we selected events of the type

$$e^+ + e^- \to x^\mp + \mu^\pm + 1 \text{ or more photons} \qquad (5)$$

Here x is an e, µ, or charged hadron. We suppose that a µ-γ combination comes from the reaction and decay sequence

$$\begin{aligned} e^+ + e^- &\to \ell^+ + \ell^- \\ \ell^- &\to \mu^- + \gamma \end{aligned} \quad ; \qquad (6)$$

where the $\ell^\pm$ have an unknown mass m_ℓ. If this was true

$$E_\mu + E_\gamma = E_b \qquad (7)$$

where E_μ, E_γ, and E_b are the µ energy, γ energy, and beam energy respectively. The mass m_ℓ is given by

$$m_\ell = \left[m_\mu^2 + 2E_\gamma(E_\mu - p_\mu \cos\theta_{\mu\gamma}) \right]^{1/2} \qquad (8)$$

where p_μ is the muon momentum, m_μ is the muon mass, and $\theta_{\mu\gamma}$ is the angle between the momenta of the µ and γ.

In Eq. (8) the quantity with by far the largest measuring uncertainty is E_γ. Therefore we calculated

$$E_{\gamma,\text{ predicted}} = E_b - E_\mu \qquad (9)$$

and

$$D = | E_{\gamma,\text{ predicted}} - E_{\gamma,\text{ measured}} | / E_{\gamma,\text{ predicted}} \qquad (10)$$

The quantity D is normally distributed with a sigma of $0.13/\sqrt{E_{\gamma,\text{ predicted}}}$ (GeV). Therefore we defined

$$Z = D\sqrt{E_{\gamma,\text{ predicted}}} \quad ; \qquad (11)$$

and selected for further study all events of the form of

Eq. (5) with

$$Z < 0.20 \qquad (12)$$

We also required

$$E_{\gamma,\text{ predicted}} > 0.23 E_b \qquad (13)$$

to avoid problems associated with poorly measured or poorly identified low energy photons. The criteria of Eqs. (12) and (13) reduced the event sample to the m_ℓ mass spectrum[8] of Fig. 1. Many of the events come from the electromagnetic reactions

$$e^+ + e^- \to \mu^+ + \mu^- + \gamma \qquad (14a)$$
$$e^+ + e^- \to \mu^+ + \mu^- + \gamma + \gamma \qquad (14b)$$

We eliminated many of these events by a set of electromagnetic background cuts[5]. For example, the μ^+, μ^-, γ in Eq. (14a) are coplanar, hence we required some noncoplanarity in the event. These cuts reduced the mass spectrum to that shown in Fig. 2.

The fundamental question is, of course, how many of these events have a mass m_ℓ compatible with the τ mass of 1782 MeV/c^2? Fig. 3, an expanded scale section of Fig. 2, answers this question. The smooth curve shows the calculated τ mass resolution function. There is no accumulation of events at the τ mass and only one event falls within the resolution function.

It is now easy to calculate an upper limit on the decay mode $\tau^- \to \mu^- + \gamma$. The detector acceptance for this mode, including the signature of Eq. (3) and the selection criteria, is 7.3%. Since one event was found inside the mass resolution function, the 90% confidence upper limit on the branching ratio is given by

$$B(\tau^- \to \mu^- + \gamma) \leq \frac{3.9}{96{,}216 \times .073} = 5.5 \times 10^{-4}$$

Here 96,216 is the number of produced τ's.

IV. RESULTS

Using methods analogous to the example described in the last section we searched for the decay modes listed in Eqs. (1) and (2). We did not find a statistically significant accumulation of events at the τ mass for any of the decay modes. Table I lists the 90% confidence upper limits on the branching fractions. We have also given the detection efficiency for each decay mode including the event selection criteria. Note that the larger upper limits are mostly due to smaller acceptances.

Hence we have not found any decay modes of the τ which violate the concept that the τ and its associated neutrino have a unique, conserved lepton number. This agrees, as do all other published results, with the τ being a sequential lepton.

The upper limits in Table 1 are factors of 10 or more smaller than previously measured upper limits[7]. This improvement comes chiefly from the large data sample. Since it is difficult to increase the detector acceptances by more than a factor of 2 or 3, any substantial reduction in these upper limits will require the acquisition of a yet larger data sample.

A detailed paper on how these limits were obtained is in preparation, authored by the Mark II Detector Collaboration members. This work was supported by the U. S. Department of Energy under Contract Nos. DE-AC03-76SF00515 and W-7405-ENG-48.

Table 1. For the decay modes of the τ lepton listed in column 1, this table gives the upper limits on the branching fractions with 90% confidence, column 2. The detector acceptance for each mode is given in column 3; and the number of events found within the τ mass resolution is given in column 4.

Decay Mode $\tau^- \to$	Upper Limit on Branching Fraction	Efficiency(%)	Number Events Found at τ Mass
$\mu^- + \gamma$	5.5×10^{-4}	7.3	1
$e^- + \gamma$	6.4×10^{-4}	6.3	1
$\mu^- + \mu^+ + \mu^-$	4.9×10^{-4}	4.9	0
$e^- + \mu^+ + \mu^-$	3.3×10^{-4}	7.3	0
$\mu^- + e^+ + e^-$	4.4×10^{-4}	9.3	1
$e^- + e^+ + e^-$	4.0×10^{-4}	10.1	1
$\mu^- + \pi^0$	8.2×10^{-4}	2.9	0
$e^- + \pi^0$	$21. \times 10^{-4}$	3.5	3
$\mu^- + K^0$	$10. \times 10^{-4}$	2.4	0
$e^- + K^0$	$13. \times 10^{-4}$	3.1	1
$\mu^- + \rho^0$	4.4×10^{-4}	5.5	0
$e^- + \rho^0$	3.7×10^{-4}	6.5	0

REFERENCES

1. M. L. Perl, Ann. Rev. Nucl. Part Sci. $\underline{30}$, 299 (1980).
2. G. Flugge, Z. Phys. $\underline{C1}$, 121 (1979).
3. G. Wolf, DESY Preprint DESY 80/13 (1980).
4. S. C. C. Ting, "Test of Quantum Electrodynamics and Study of Heavy Leptons," Proc. of 1980 International School of Subnuclear Physics (Erice, 1980), Ed. by A. Zichichi, to be published.
5. K. G. Hayes, Ph.D.Thesis, Stanford Univ. (1981), unpublished.
6. The acoplanarity angle is defined by
$$\arccos\left[(\underline{u}_1 \times \underline{u}_b) \cdot (\underline{u}_2 \times \underline{u}_b)\right] \Big/ \left[|\underline{u}_1 \times \underline{u}_b| \; |\underline{u}_2 \times \underline{u}_b|\right]$$
where $\underline{u}_1$, $\underline{u}_2$, $\underline{u}_b$, are unit vectors in the direction of each of the final particles and of the beam respectively. This eliminates a very large background from $e^+e^- \to e^+e^-$ and $e^+e^- \to \mu^+\mu^-$.
7. M. L. Perl, Proc. of 1977 Int. Symp. on Lepton and Photon Interactions at High Energies (DESY, Hamburg, 1977) p. 145.
8. The quantity m_ℓ is calculated using $E_{\gamma, \text{predicted}}$ in Eq. (8).

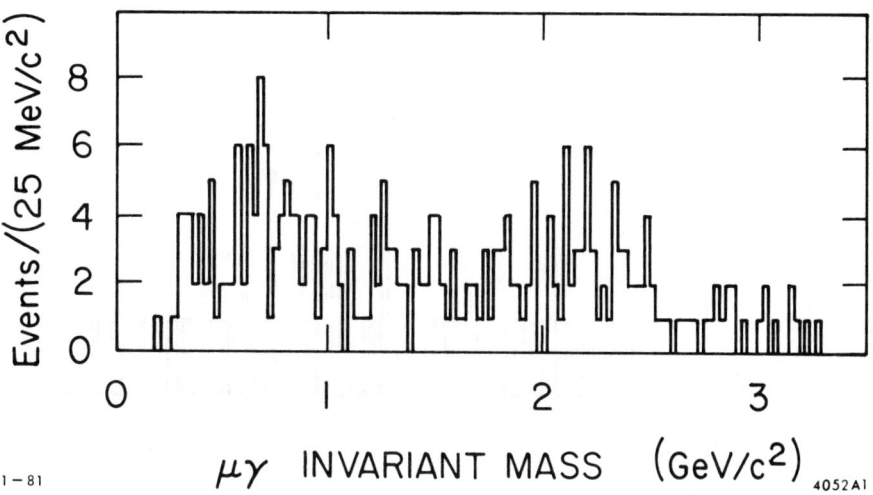

Fig. 1. The μγ invariant mass spectrum with the 5° acoplanarity angle cut and the photon energy cuts of Eqs. (12) and (13).

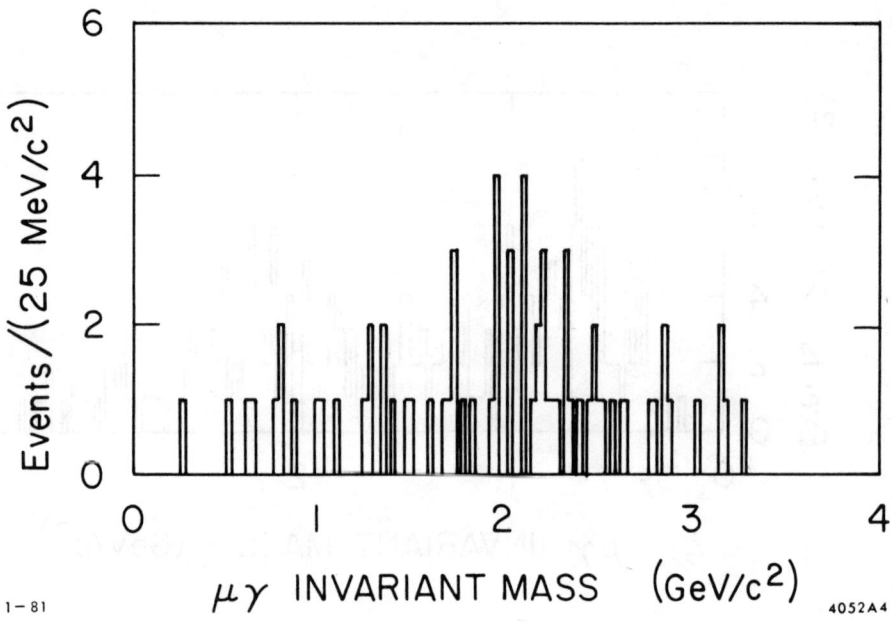

Fig. 2. The μγ invariant mass spectrum with all cuts.

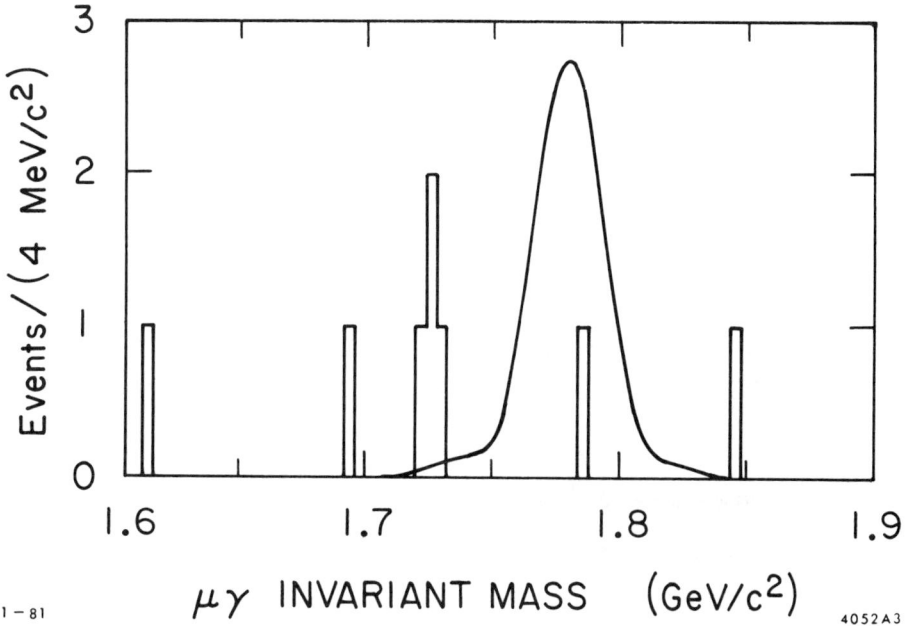

Fig. 3. The μγ invariant mass spectrum in the neighborhood of the τ mass with all cuts. The smooth curve shows the expected τ mass resolution.

Near-Term Prospects for Information on b Decay

Edward H. Thorndike
University of Rochester

The weak decay of the b quark is one of the more hopeful ways of learning about the relations among the three families. Most of the experimental information on B decay in the next few years will come from the large magnetic detector CLEO at the Cornell Electron Storage Ring. Before I give you my estimate of what we will (and will not) learn, let me remind you of what we already know. As Tony Loomis told you yesterday, the b quark decays (it could have been stable). The results are consistent with the standard model. Some specific other models can be ruled out, but at present a wide range of non-standard models are allowed.

Table 1

Results from CLEO (A. Loomis)

1. The b quark decays.
2. Results are consistent with the standard model, but also with other models.

Yield of $\mu + e$ ∼1/4 per B decay

Yield of Kaons ∼1½ per B decay

No information yet on yield of τ, ν, $p/\bar{p}$

Flavor changing neutral current limit too loose to rule out any model.

Now, there's a theorem in particle physics which says that "anything that can happen will happen." It has a corollary concerning the invention of models by theorists, which says "any model that can be written down will be written down."--Indeed, theorists can invent models much faster than experimentalists can understand them, let alone prove them wrong.

Therefore, I won't try to understand any of the models. Rather, starting from a few reasonable assumptions, I'll divide all models into broad classes, and then look for experimental procedures for ruling out entire classes of models.

My assumptions are shown in Table 2. With these assumptions, the possible decay reactions for the **b quark** are also shown.

Table 2

Assumptions

1. The essential features of B decay are revealed by considering b decay (and conversely).
2. b decays exclusively into quarks and/or leptons.
3. b decay conserves color.

Possible Decay Reactions

$b \longrightarrow q\,q\,\bar{q}$ (1)
$ q\,\ell_i\,\bar{\ell}_i \quad i = e,\, \mu,\, \tau$ (2) $\Big]$ mediated by $W^{\pm}$, Z^0

$ q\,\ell_i\,\bar{\ell}_j \quad j \neq i$ (3a)
$ q\,\ell\,\ell$ (3b)
$ q\,\bar{\ell}\,\bar{\ell}$ (3c)   "exotic decays"
$ \bar{q}\,\bar{q}\,\ell$ (4a)
$ \bar{q}\,\bar{q}\,\bar{\ell}$ (4b)

5, 7, 9 quarks and/or leptons (5)

(ℓ = charged lepton or neutrino)

Reactions (1) and (2) can be mediated by $W^{\pm}$ and Z^0, while reactions (3) and (4) cannot. I'll use the term "exotic decays" to refer to reactions (3) and (4). Having listed reaction (5), I won't consider it further today.

My first thought was that all models would split into two classes: those models with b decay mediated by $W^{\pm}$ and/or Z^o, and proceeding exclusively by reactions (1) and (2); and those models in which b decay via $W^{\pm}$ and Z^o is forbidden, with b decay mediated by something new, proceeding exclusively by exotic decays. Of course there is a third class of models, with b decay mediated <u>both</u> by $W^{\pm}/Z^o$ <u>and</u> by something new, proceeding by any combination of the four reactions. The situation is summarized in Table 3.

Table 3
Classes of Theories of b Decay

decay mediated by $W^{\pm}$, Z^o	decay mediated both by $W^{\pm}$, Z^o and by something new	decay via $W^{\pm}$, Z^o forbidden
Examples		
Kobayashi, Maskawa	Kane: horizontal SU(3)	Georgi, Glashow: U_6
Barger, Pakvasa		Georgi, Glashow: SU(7)
Tye, Peskin		Derman: horizontal S(3)
Distinguishing features of Class		
Decays via reaction (1) & (2)	Decays via any reaction	Exotic decays only

---

One near-term experimental goal is to place limits on the fraction of B decays that proceed by reactions (3) and (4). If one could show that these decays constitute less than 100% of B decays, then one of the three classes of models would be ruled out.

How can one rule out 100% exotic decays? One way that I personally believe is doomed to failure is searching for an example of a non-exotic decay. The problem is illustrated in Table 4. Reconstructing a complete B$\bar{\text{B}}$ event, reconstructing a single B decay, reconstructing a D from B decay, all of these I believe are hopeless. The problem is the high multiplicity of the events, which hurts two ways: acceptance and combinatorics. If the acceptance for charged particles is 90% and that for photons is 80% (optimistic numbers), then the acceptance for a single B, summed over all decay channels is about 3%. For B$\bar{\text{B}}$ event reconstruction this number gets squared.

Table 4

Particle Reconstruction

Multiplicity of B$\bar{\text{B}}$ events high ($\sim$18)

∴ Reconstruction difficult because of:
  (a) Acceptance for charged particles, γ's
  (b) ν, K_L^0
  (c) combinatorics

Example: Acceptance for $\pi^{\pm}$ = 90% ⎫
         "   " γ = 80% ⎭ (both optimistic values)

⟹ $\sum$ (Acceptance) × (Branching ratio) $\approx$ 3%

All B decay modes

Acceptance for B$\bar{\text{B}}$ event = $(3\%)^2 = 10^{-3}$

For B$\bar{\text{B}}$ events, B decays, D's from B decay, huge number of wrong combinations.

E.g., consider D → 2, 3, or 4 charged particles.

Acceptance × Branching ratio $\approx$ 5%,

∴ 1/10 correct combination per B$\bar{\text{B}}$ event

  $\sim$350 incorrect   "   "   "   "

  4 background events per B$\bar{\text{B}}$ event

∴ signal/noise = 1/14,000

While the number of reconstructable B's, D's, B$\bar{\text{B}}$'s is low, the number of incorrect ways of combining particles in an attempt to reconstruct B's, etc. is huge. As an easily worked out example, I have considered reconstructing D's. With 1 D every 14 thousand attempts, there is no way the signal can be dug out.

An unexpectedly high branching ratio into some low multiplicity channel which has some characteristic feature making it easy to identify (e.g., $B \to \psi K\pi$) might circumvent my gloomy prediction. However, one can certainly not depend on being able to reconstruct B's, or D's from B decay.

The existence of a large fraction of exotic decays can be spotted by a combination of three measurements, indicated on Table 5. Since B's decay into μ' or e's only 1/4 of the time, the large number of leptons called for in reactions (3) and (4) would have to be neutrinos or taus (half of whose energy goes into neutrinos). Reaction (3), and to a lesser extent reaction (4), will have "missing energy." Reaction (4) also produces one baryon per decay--probably a 50-50 mixture of n and p. Thus by measuring $p/\bar{p}$ yield and missing energy, one can place a limit on the fraction of exotic decays. Within the year I believe CLEO will have ruled out 100% exotic decays (or will have ruled out the standard model.)

Table 5

How to Rule Out Exotic Decays

Use

    (1) Muon and electron yield   (Done, $\sim$22%)
    (2) Proton/antiproton yield   (soon)
    (3) Visible energy   (within a year)

Invisible energy

    Standard model $\lesssim$ 15%
    Reaction (3) $\gtrsim$ 35%
    Reaction (4), with $\gtrsim$ 30%
      n/p 50%/50%

Assume nature is unimaginative, and have the B decay by $W^{\pm}/Z^{o}$. We used to think that this meant b would decay by mixing with s and d, and we could learn whether or not tops existed by looking for flavor changing neutral currents. Now Peskin and Tye's silly model has ruined all that. Nonetheless, looking for flavor changing neutral currents is still a worthwhile enterprise. As Tony Loomis told you yesterday, CLEO's limit on the ratio of branching ratios is ~ 0.2, not yet interesting. Within a year that limit will be improved a factor of 2 (or flavor changing neutral currents will have been seen). In either case one model dies, but in either case, Tye-Peskin survives.

Table 6

If B Decays via $W^{\pm}/Z^{o}$---

Standard Model

$$\begin{pmatrix} u \\ d' \end{pmatrix}_L \begin{pmatrix} c \\ s' \end{pmatrix}_L \begin{pmatrix} t \\ b' \end{pmatrix}_L$$

Top exists; GIM operates.

$$\frac{BR(B \to X\mu^+\mu^-)}{BR(B \to X\mu\nu)} = 0$$

Singlet b Model

$$\begin{pmatrix} u \\ d' \end{pmatrix}_L \begin{pmatrix} c \\ s' \end{pmatrix}_L \quad b'_L$$

No top; Flavor changing neutral currents.

$$\frac{BR(B \to X\mu^+\mu^-)}{BR(B \to X\mu\nu)} \gtrsim 0.13$$

Tye-Peskin Model

$$\begin{pmatrix} u \\ d' \end{pmatrix}_L \begin{pmatrix} c \\ s' \end{pmatrix}_L \begin{pmatrix} c \\ b \end{pmatrix}_R \quad (b'_L)$$

Top not required, but possible.

Mixing of b_L with s, d not needed, but possible

$$0 \lesssim \frac{BR(B \to X\mu^+\mu^-)}{BR(B \to X\mu\nu)} \lesssim 0.13$$

Finally assume nature chooses the ultimate in dullness, the standard model. There still remains the matter of determining the ratio for $(b \to uW^-)/(b \to cW^-)$. As Loomis mentioned yesterday, CLEO observes lots of kaons from B decay, which demonstrates that $b \to cW^-$ is appreciable. Measuring kaon yield, however, will not allow one to determine the u to c ratio to any precision. In particular, one could not distinguish between 10% $b \to u$ and no $b \to u$. At present, I have no sure-fire idea of how to do this. One possibility is to make a precise measurement of the charged lepton spectrum in the decay $B \to X\ell\nu$. The decay $b \to uW^-$ has a higher end point than does $b \to cW^-$. A second possibility is to measure the branching ratio for $B^\pm \to \tau^\pm \nu$. That decay goes only via $b \to uW^-$, by an annihilation diagram. Both possibilities mentioned constitute very difficult experiments, requiring large amounts of data.

Summarizing, within a year or two, certain gross features of B decay will have been determined. In particular, one will be able to choose between the standard model and models with 100% exotic decays, and to choose between the standard model and a topless model with b decaying by mixing.

More fine-grained distinctions are harder to come by. In particular, I don't know how to make a precise determination of the $(b \to uW^-)/(b \to cW^-)$ ratio, nor do I see a way to rule out the Tye-Peskin model.

AXIAL U(1) ANOMALY AND CHIRAL SYMMETRY-BREAKING IN QCD

K. C. Chou*
Virginia Polytechnic Institute, Blacksburg, VA

It is well known that the absence of the ABJ anomaly is necessary for the corresponding gauge field theory to be renormalizable[1]. This condition places severe restrictions on the choice of the possible gauge group, and the representation for fermions, as we have just heard from A. Zee. I would like to report, on the other hand, some consequences of the presence of ABJ anomaly in certain global current conservation equations. This is a work done in collaboration with L. N. Chang.

One important question in QCD is the origin of chiral symmetry breaking. This problem is related to the structure of the theory for large distances, where perturbation theory cannot be used. It has long been conjectured that topologically non-trivial gauge field configurations play some significant role in explaining both confinement and chiral symmetry breaking[2].

Recently, Coleman and Witten[3], and Veneziano[4], have analyzed the question within the context of $1/N_c$ expansion. In particular, Coleman and Witten argue that the axial anomaly and confinement already imply chiral symmetry breaking. Their argument makes no essential use of the non-trivial topological configurations, but relies instead on the absence of analyticity structure in the axial vector vertex, brought about by the anomaly. Veneziano reaches the same conclusion, using the $1/N_c$ expansion and by

---

*On leave from Institute for Theoretical Physics, Peking.

consideration of the fluctuation of topological charge density in the pure Y-M field sector.

In this talk I want to point out that their conclusion on the necessity of chiral symmetry breaking can be obtained without recourse to the $1/N_c$ expansion, if proper attention is paid to the quantum fluctuations in the topological charge density.

We start by recalling that in the presence of non-trivial topological configurations, one could incorporate the θ-vacuum caused by the resultant phenomenon of tunneling by augmenting the conventional Lagrangian with an additional term

$$\mathcal{L} = -(\bar{q}\gamma^\mu D_\mu q) - \frac{1}{4} F^i_{\mu\nu} F^{i\mu\nu} - \theta\nu(x) - J_a(x) O_a(x)$$

$$\nu(x) = \frac{g^2}{32\pi^2} F^i_{\mu\nu} *F^{i\mu\nu}, \quad D_\mu = \partial_\mu - g\hat{A}_\mu \qquad (1)$$

$$*F_{\mu\nu} = \frac{1}{2} \varepsilon_{\mu\nu\sigma\rho} F^{\sigma\rho} .$$

In equation (1), θ is the parameter characterizing the topological structure of the vacuum, while $J_a(x)$ are external sources coupled to various combinations of quark currents $O_a(x)$. Since we are interested in chiral symmetry breaking, our attention will be focused on the densities $\bar{q}(1\pm\gamma_5)q \equiv O_\pm$ and $\bar{q}\gamma^\mu(1\pm\gamma_5)q \equiv O_\pm^\mu$. In the following θ will be considered as a function of x in some intermediate steps of calculation.

The Lagrangian of (1) has an apparent $U(N_f) \times U(N_f)$ chiral symmetry, with N_f flavours, if we set $J_a(x) = 0$. However, due to the ABJ anomaly in the axial current, θ will change to $\theta + \sqrt{2N_f}\, \xi_5$ under the abelian chiral phase transformation $q(x) \to \exp[i\gamma_5 \xi_5] q(x)$.

The best way to study the chiral symmetry structure of (1)

when $J(x) \neq 0$ is through the effective generating functional, which we shall now define. The generating functional, W, for the connected Green's functions implied by (1) can be expressed as

$$W[J,\theta] = -i \ln Z$$
$$Z = \int Dq D\bar{q} DA_\mu \exp\{i\int d^4x \, [\mathcal{L}(x) + \Delta\mathcal{L}(x)]\} \quad . \tag{2}$$

Here $\Delta\mathcal{L}(x)$ includes the gauge fixing and compensating terms necessary to give meaning to the A_μ integration. The classical fields $U(x)$ can now be defined by direct differentiation

$$U(x) = - \frac{\delta W}{\delta J_\pm(x)} \quad . \tag{3}$$

As a result of the axial anomaly, and the formal invariance properties of (1), the generating functional W has to satisfy a Ward identity of the form

$$\partial_\mu \frac{\delta W}{\delta J_{\mu\pm}(x)} = iJ_\pm \frac{\delta W}{\delta J_\pm(x)} \pm N_f \frac{\delta W}{\delta \theta(x)} \quad . \tag{4}$$

This Ward identity can be satisfied by any functional with the following local invariance property

$$W[J(x), J_{\mu\pm}(x), \theta(x)]$$
$$= W[J_\pm(x) \, e^{\pm i\sqrt{\frac{2}{N_f}} \xi_5(x)}, \, J_{\mu\pm}(x) \pm \partial_\mu \left(\sqrt{\frac{2}{N_f}} \xi_5(x)\right)$$
$$\cdot \theta(x) - \sqrt{2N_f} \, \xi_5(x)] \tag{5}$$

where $\xi_5(x)$ is an arbitrary function of x.

We define now the generating functional Γ by using a Legendre transformation on the sources of the scalar currents $0_\pm$

$$\Gamma[U_\pm(x), J_{\mu\pm}(x), \theta(x)]$$
$$= W[J_\pm(x), J_{\mu\pm}(x), \theta(x)]$$
$$+ \int d^4x \left(J_+(x)U_+(x) + J_-(x) U_-(x)\right) \quad (6)$$

Then the Ward identity for the axial U(1) symmetry implies that Γ is invariant under the local transformation

$$U_\pm \to U_\pm e^{\pm i\sqrt{\frac{2}{N_f}} \xi_5(x)}$$

$$J_{\mu\pm} \to J_{\mu\pm}(x) + \partial_\mu \sqrt{\frac{2}{N_f}} \xi_5(x) \quad (7)$$

$$\theta(x) \to \theta(x) - \sqrt{2N_f}\, \xi_5(x)$$

Or in other words, Γ is a functional of the form

$$\Gamma[U_\pm e^{\pm i \frac{1}{N_f} \theta}, J_{\mu\pm}(x) \pm \partial_\mu (\frac{1}{N_f} \theta)] \quad (8)$$

Note that the classical fields $U_\pm(x)$, which are the vacuum expectation values of the corresponding quark bilinear fields, can be determined through the relation

$$\frac{\delta\Gamma}{\delta U_\pm(x)} = J_\pm(x) \quad (9)$$

Any nontrivial solution to (9) when $J_\pm = 0$, $J_{\mu\pm} = 0$ and $\theta(x) = $ constant, would signal the existence of spontaneous chiral symmetry breaking.

Since $U_+ = U_-^*$ we can write them in the form

$$U_\pm + U e^{\pm i \sqrt{\frac{2}{N_f}} \frac{\eta'}{f_\pi}} \quad (10)$$

where η' can be interpreted as the vacuum expectation value of the η'-meson field which corresponds to the axial U(1) pseudo Goldstone field in chiral dynamics.

Notice that Γ is an even function of $\theta - \frac{\sqrt{2N_f}}{f_\pi}\eta'$ as a consequence of (8), and the symmetry under space reflection. It is easily proved that the CP conserving solution of eq. (9) at $J_\pm = 0$, $J_{\pm\mu} = 0$, $\theta(x) = \theta$ is

$$\eta' = \frac{f_\pi}{\sqrt{2N_f}} \theta . \tag{11}$$

Thus all the physical quantity evaluated at this point will be θ independent and CP conserving.

Nevertheless, the nth order derivative of Γ with respect to θ when η' is kept fixed is the Green's function of $n\nu(x)$'s where diagrams with one particle lines of η and U are removed.

$$\frac{\delta^n \Gamma}{\delta\theta^n}\bigg|_{\overline{\eta',u}} = (-i)^n \int d^4x_1 \ldots d^4x_n <0|T(\nu(x_1)\ldots\nu(x_n))|0>_{\text{I.P.I.}} \tag{12}$$

It has been shown by one of us[5] that $\frac{\delta^2\Gamma}{\delta\theta^2}\big|_{\mu'u}$ is proportional to the η' meson mass. This result is a generalization of the result given first by Witten[6] in the leading $1/N_c$ expansion approximation.

The point we wish to emphasize is that (12) can be non-vanishing only if $U \neq 0$, so that if any of the moments defined in (12) were to be nonzero, chiral symmetry would be spontaneously broken. This is the main conclusion of my talk. Now the right hand side of (12) represents a quantum correlation function of the topological charge density, and there is no general reason for (12) to vanish for n even. The case for n odd can be excluded, of course, in the chiral limit when CP is a good symmetry. We therefore argue that QCD will, in general, induce the spontaneous symmetry breaking of flavor chiral symmetry.

The picture we are presenting may therefore be summarized as follows: Owing to the presence of instantons, the QCD vacuum acquires an additional parameter θ. In the absence of any external spontaneous chiral symmetry breaking, like those induced by Higgs couplings, the chiral phases of the quarks will automatically refer themselves to θ. This is the direct consequence of the axial anomaly. However, large scale quantum fluctuations of the topological charge density requires such phases to be defined globally, which can only occur if the chiral symmetry is spontaneously broken. Hence quantum corrections to topologically non-trivial gauge configurations induce spontaneous chiral symmetry breaking.

REFERENCES

1. D. J. Gross and R. Jackiw, Phys. Rev. D6, (1972) 477. C. P. Korthals Altes and M. Perrottet, Phys. Lett. 39B, (1972) 546.
2. G. 't Hooft, Phys. Rev. D14 (1976) 3432. D. G. Caldi, Phys. Rev. Lett. 39 (1977) 121. C. G. Callan, R. F. Dashen and D. J. Gross, Phys. Rev. D17 (1978) 2717.
3. S. Coleman and E. Witten, Phys. Rev. Lett. 45 (1980) 100.
4. G. Veneziano, CERN TH-2872 (1980).
5. K. C. Chou, ASITP-80-005 (1980).
6. E. Witten, Nucl. Phys. B156 (1979) 269.

Mass Spectrum above 40 GeV

M. Veltman[*])
University of Utrecht

I will make an extremely simple comment. Imagine you want to know what particles exist above 100 GeV. It happens that we have at this moment some theories, but we know experimentally only particles up to a mass of 40 GeV. So what can we have with masses above this? Statement: The only window from a phenomenological point of view on the masses of particles over 40 GeV is provided by the ρ parameter. It is the only window at this time we have on the high mass spectrum. I'll tell you precisely what I mean by that. If you work in the standard model, you use the simple Higgs system and you ignore radiative corrections then it turns out that the ρ parameter is equal to 1. That is the ratio of the neutral and charged vector boson masses. Now the radiative corrections to this ρ parameter, and that is the important point, are sensitive to particles with masses in the range anywhere from 100 GeV to infinity. So particles with masses in the SU(5) heavy sector may influence the radiative corrections to the ρ parameter. Technicolor may influence the ρ parameter. For some reason this seems to be neglected. So radiative corrections of anything, any particle, any lepton that you have coupled to the usual vector boson will influence the ρ parameter. Experimentally there is a rather strong constraint on the deviation from one for the ρ parameter, namely about 1.5% (from the latest computation of Langacker, et al.).

It seems that this ρ parameter is compatible with one, so whatever theory you are going to make you must be careful that you do not violate

---

[*]) Co-authors M. Einhorn and T. Jones

ISSN:0094-243X/81/720627-06$1.50 Copyright 1981 American Institute of Physics

this restriction. Let me explain what is going on. We go to the standard model and for simplicity ignore mixing ($\sin^2\theta = 0$). Then the simple Higgs system is the linear σ-model which conserves isospin. So as a consequence of conservation of isospin, the neutral and the charged vector bosons, being members of an isospin triplet, have all equal mass. Thus conservation of normal isospin, that's an isospin extended to include doublets and vector bosons, just tells you (if you ignore mixing) that the charged and neutral vector bosons are equally heavy. So the question is: is isospin conserved? The answer to this is that the isospin is evidently not conserved. If isospin is conserved then all doublets of fermions and leptons must have equal masses. So the electron should have the same mass as the ν_e. The tau should have the same mass as its neutrino, the top and bottom quark mass should be equally large, etc. From this you can say that apparently this isospin is not (at least at this level) taken as a symmetry by nature. And yet this ρ parameter comes very close to one, so you get really mystified. There is another source of this breaking of this very same isospin. This is due to the U(1) electromagnetic interactions. I do not want to enter into this since I have taken the limit of zero mixing. But in any case isospin breaking in this situation is clearly in the theory because the neutrino and electron masses and the quark masses are different. That's an evident isospin breaking.

Now to get an idea of what goes on, let us go to some quantitative effects. I want to take some models to get feeling as to what may happen to this ρ parameter. If you take, for instance, the influence of the tau and ν_τ mass difference, then through radiative corrections (you have basically isospin breaking in the theory), we will see that isospin breaking will appear anywhere you want. In particular, we will see it in the ρ parameter. So the isospin breaking made through the

$\tau - \nu_\tau$ mass difference will give rise to a deviation of the ρ parameter from 1: $\rho = 1 + f(m_\tau, 0)G/2\pi^2$. In general:

$$f(m_1, m_2) = \frac{1}{4}(m_1^2 + m_2^2) + \frac{1}{2}\frac{m_1^2 m_2^2}{m_1^2 - m_2^2} \ln \frac{m_2^2}{m_1^2}$$

so that $f \geq 0$. But in actual fact just to give you an idea, if the tau were as heavy as 300 GeV, the deviation would be bigger than the experimental error. 300 GeV is still rather large but it gives you the idea.

The tau is well known of course, but in addition it follows that the top quark mass shall not be heavier than about 200-300 GeV. As another example I want to take some very heavy particles, maybe 10^{15} GeV, you name it. Take a heavy doublet and a heavy singlet, yet couple them so that they do not need the Higgs system to get their mass. It can be done if you make both left and right handed doublets and a singlet. And then we give some Higgs interaction to them that will produce a mixing between the singlet and the doublet. So here are these particles. The twin members of the doublet have equal mass, and there is the singlet, in terms of isospin, with very high mass. Then the Higgs interacts with them, mixes the singlet with the doublet. You can even imagine that this singlet here is actually the singlet of any of the down quarks or anything of that kind. Generally you expect something like that to happen. If that happens, there will come out some slight mass differences, and again you find some influence on this ρ parameter. Now this gets more complicated in this case because there are a lot more parameters and we are not going to discuss that here. But the point is that we find again that the sign of this radiative correction is positive and the magnitude is sizeable, roughly equal to what we get with the particles at low energies. So again, if there are particles of 10^{15} GeV and we have this sort of little Higgs couplings that produce small variations of the order of a couple of hundred GeV onto these

10^{15} GeV particles, it's excluded. In all examples that Einhorn and Jones and I have been trying to cook up we have found that the fermions' contribution to this thing is always positive. They all add up. So that is now a very important example, because the fact that ρ remains so close to one, I believe, is telling us something. It is this little window that we have on the mass spectrum; when you look out through it what you see is a desert, up to now. So we have been trying to find some construction where maybe the contributions will go the other way, so that you could believe in a cancellation. And we have found some examples with scalar particles that produce, in effect, the other sign, so there is probably no general theorem on this method.

One other point: if you compute radiative corrections to masses of bosonic particles you usually compute the mass squared rather than the mass, and due to this rather trivial thing it happens that bosonic particles have, in general, when they are heavy, no influence whatsoever on this ρ parameter. So for instance, one thing that we investigated is the following. Consider the X and Y bosons of SU(5). These very heavy bosons are actually coupled to the Higgs particle. So this is somewhat like the model that I discussed for the fermions before. So at first instance, one might perhaps expect that we could rule out SU(5). Next we worked out what these heavy bosons X and Y in SU(5) would contribute to the equation for ρ and we discovered that it would give a negligible contribution. So SU(5) by this mass and mass-squared miracle survives this particular test. So we are now trying to push on, to try to rule out as much as possible but I would say that if the experiment increases in accuracy, and remains having the ρ equal to one, this is going to be an extremely strong constraint on whatever people make.

Now, what can we deduce from this? Can we do something with that knowledge? Well there is the other piece of the theory that we are discussing very often these days. And that piece, the important thing, is naturalness. It is argued, as a basis of hypercolor, technicolor, you name it, it's argued that the Higgs boson is not natural. The meaning of this statement is that Higgs boson mass gets a quadrically divergent contribution, and that suggests according to an old argument of Wilson (which actually dated from 1970 and pre-gauge theory) that the Higgs masses are not natural because this quadratic divergence would put up the Higgs mass to some high value. Now this, however, ignores the fact that the quadratic divergence has a coefficient, π or two or something, that will be a function of the parameters in the model. So I have set out to compute this coefficient of this divergence in the standard model, relying thereby on the dimensional regularization scheme. So we compute coefficients of the quadratic divergence in the dimensional regulation scheme. This coefficient turns out to be $3\left(\frac{1}{2} + \frac{1}{4e^2}\right) + \frac{3m_H^2}{2M^2} - \sum_f \frac{m_f^2}{M^2}$. So if by some chance this coefficient turns out to be zero, then it means that the Higgs could be perfectly natural. So the question is whether it could actually be zero. In this case all you need is that the top quark gets a little bit heavier, and another possibility is that we have one more generation to come up, starting with a heavy lepton of 40 GeV or so. If that is the case, then this formula could very well be true. Then it means that the quadratic divergence of the scalar Higgs is zero due to the coefficient and the basis of **technicolor** and **hypercolor goes down the drain.**

So then, given that this goes down the drain, we may ask just what happens in the theory. So you then have a big desert? Well could be. We ought to consider, in this context, once more that in SU(5) there are two scalars, a light one and a heavy one. Imagine that the coefficient of the quadratic divergence of the light one cancels to zero by some symmetry, probably super**symmetric** relations. Then with SU(5) we discover that the other Higgs, the heavy one, is not coupled to the usual fermions, and if I assume what I suspect from the ρ parameter, namely that there are not many fermions any more, then I would say that maybe this heavy Higgs is not coupled to fermions at all. If that is the case, then there can be no such cancellations for this heavy Higgs. If there is no such cancellation for the heavy Higgs, then it has a quadratic divergence so its mass will go up. But that is fine because it is actually of the order of 10^{15} GeV in the theory. So if this scenario is true, (including the remark of Yao yesterday, that for the non-quadratic divergence, the theory is perfectly natural) then I would say that it is quite conceivable that SU(5) is a completely natural theory. Thus, initially all particles are around 100 GeV, and then through radiative corrections, one of the Higgs is pulled up to 10^{15} GeV and with it the vector bosons, and that could be a possibility. From here on it gets too speculative. so at this point I think I can stop, being the first speaker to stop before time.

Strings and Vortices

Y. Nambu

Univ. of Chicago, Chicago, IL

I will discuss briefly some speculations concerning the possible existence of heretofore unknown fields. While this topic may not be relevant to the current conference, perhaps if we interpret the word weak in a generic sense it will be relevant. There are a couple of motives. One is the general expectation which has impressed many people, like Einstein and Dirac, that a theoretically appealing and elegant principle usually governs natural phenomena. The other one is the observation that we have learned over the last 20 years or so that the vacuum in space time looks more and more like the ordinary medium of condensed matter, so whatever condensed matter physics offers to us in terms of various excitations may find a counterpart in the vacuum.

Okay, so while I go on I will use only one figure so that you can see what is going on. We have learned in the more recent developments in field theory that there are extended objects, solitons, monopoles or string-like objects, instantons and so forth. If we take field theory seriously, we also have to take these objects seriously. Now a string, of course, has found its relevance in the quark confinement problem, but apart from that, I will discuss the possibility of the existence of such objects in a different context.

Going down from the top of Figure 1, I will compare in classical mechanics the mass point with the string. If you take the mass point, you have a trajectory which is a world line. In the case of strings, you have a world sheet. The world line has a tangent vector at each point which essentially corresponds to momentum. In the case of the string, you have a tangent plane at each point. From there on you go to dynamics. You set up a Hamilton-Jacobi kind of mechanics for the motion of strings or mass points, free or otherwise. Here, I will talk only about free mass points or strings. The momentum or tangent vector is given as the gradient of the scalar function. A Hamiltonian function, correspondingly, in the case of the tangent plane, is given as the curl of a vector field, namely as a gauge field like the Maxwell field, although it is subject to different equations than the Maxwell equations.

Next we ask the question: What is the natural field of force that can be associated with a mass point or a string? For a mass point, it is Maxwell's electromagnetic field satisfying the gauge principle, as everybody knows. For a string, it is the Kalb-Ramond field described by an antisymmetric tensor potential. Physically, it can be interpreted as the velocity field in a relativistic hydrodynamics of vortices. Maxwell discovered electromagnetic waves by amending Faraday's Law. In the same way, the relativistic formulation of hydrodynamics naturally leads to the existence of massless excitations or acoustic waves, which travel with the speed of light. This is an analogy I'm drawing between the two. Of course, you can carry on this sort of analogical reasoning even further. You can introduce two-dimensional objects instead of the one-dimensional string and so on and so forth. And then, correspondingly you can introduce higher order tensors. For example, more recently, Freund and Curtwright observed that a three-index tensor of mixed nature also can be associated with a certain natural gauge field and Curtwright tells me that one can interpret it as a dislocation field in crystals.

So my point is the following: perhaps, in nature, there are as yet unknown long-range fields. In the case of strings, that might be interpreted as some kind of vortices in the vacuum which give rise to a velocity field. But the question is how to find them, and the other question is what are the sources? Like vortices, do they correspond to a local sort of elementary particles or excitations which give rise to vortices and the small quanta of vortices? Or could they be non-local and microscopic in nature? I haven't found any local field for sources of the closed vortices, but it might be possible. I don't know. Now the scale of such vortices **or** vortex lines might be very small and elementary on an atomic scale, or it might be on a human scale, or it might be of an astronomical or galactic scale.

I have no idea whatsoever. Anyway, I am proposing as a possibility that there might still be unknown long-range fields. Closed strings do not have long range interactions of the Coulomb type; their interactions are more of the dipole-dipole type. Nevertheless, there might be such fields and there might be associated quanta of massless particles.

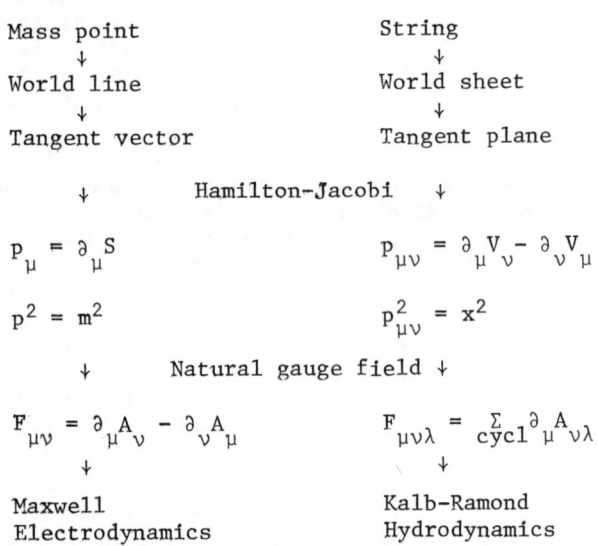

Figure 1.

Exceptional Groups for GUTs and Remarks on Octonions

Feza Gürsey
Yale University, New Haven, CT 06520

I. Introduction

In this talk, I would like to review some aspects of work done on the use of exceptional groups of the E series as models for grand unified theories. Since all exceptional groups are endowed with octonionic structure, I will also add speculative remarks on a possibly more fundamental role octonions might play in Physics.

Let me first sketch briefly a few of the features that make exceptional groups interesting and relevant to the particle physicist.

1. The exceptional groups are related to unique algebraic and geometrical structures. The (27) representation of E_6 is the only realization of an exceptional Jordan algebra. The coset spaces $E_6/SO(10) \times SO(2)$ and $E_7/E_6 \times SO(2)$ are the only exceptional hermitian symmetric spaces (positivity domains), providing definite models for new finite Hilbert spaces to describe internal symmetries. This aspect of exceptional structures provided the initial motivation[1] to introduce them in connection with the nature of fundamental multiplets, assuming that a principle of uniqueness would be operative in Nature's choice of fundamental symmetries.[2]

2. The exceptional groups are natural generalizations of the groups SU(3) and SU(3)×SU(3) which so mysteriously emerge in Particle Physics. Thus, the color group $SU(3)^c$ is naturally embedded in the octonionic structure of the exceptional groups, while another SU(3) flavor group is associated with the triality property of the octonionic algebra. The

latter is the SU(3) group the Cartan subalgebra of which yields naturally the electric charge and the B-L quantum number.

3. The E series of exceptional groups consist of E_6, E_7 and E_8. By chopping off their Dynkin diagrams one can define E groups of lower rank which, however are insomorphic with classical groups. There are two different extrapolations to rank four, namely E_4 and E'_4. We have the isomorphisms.

$$E'_4 \sim SO(8), \quad E_4 \sim SU(5), \quad E_5 \sim SO(10).$$

These are all gauge groups arising in grand unification: E'_4 unifying color and electric charge in supergravity[3] E_4 and E_5 giving the oldest successful GUT models.[4,5]

4. The exceptional series associated with octonions (in the same way that orthogonal groups, unitary groups and symplectic groups are respectively associated with real numbers, complex numbers and quaternions) provide the most symmetrical Lie groups, in the sense that they maximize the ratio D/R where D is the dimension and R is the rank of the group.

5. All exceptional groups (including E_6 which is complex like unitary groups) are free of Adler-Jackiw-Bell anomalies, leading to renormalizable theories with fermions in one representation.

6. Should the top quark continue to elude observation exceptional groups provide the most natural topless option[6] for the grouping of fundamental fermions.

II. The E_6 Models (standard and topless versions)

All known quarks and leptons can be accommodated either in two 27-plets (topless version) or three 27-plets (standard version) of E_6.

The topless version was proposed first[6] and more extensively studied recently[7]. Each 27-plet contains 3 quarks (u,d,b) and (c,s,h) with charge structure (2/3, -1/3, -1/3) and, besides neutral leptons, two charged leptons (e^-, τ^-) and (μ^-, M^-). The heavy quarks b and h with charge -1/3 decay through their interactions with 3 new neutral, weak bosons associated with a new[8] weak group SU(2) arising from the decomposition E_6 SU(6) x SU(2) where SU(6) contains the "GUT" SU(5) which, in turn, contains the usual electro weak group SU(2) x U(1). Hence the electroweak group of the topless version[7] of E_6 is $SU(2)^W$ x U(1) x SU(2) . This model is still in agreement with experiment[9] although it requires another undiscovered charged lepton M^-.

The standard version[10] of E_6 is a straight forward generalization of the SO(10) model reexamined recently by several authors.[10]

In the SO(10) model, basic fermions fit in the 16-dimensional spinor representation of SO(10) consisting of u, d, e, ν_L and a possibly superheavy ν_R, with two other similar families associated with the muon an tau lepton. The basic Higgs scalar fields required are in the 10, 16 and 45 dimensional representations of SO(10). The gauge bosons belong to the adjoint (45) representation. Through two-loop diagrams involving gauge bosons and fundamental Higgsons, an effective Higgs field belonging to the 126 dim. representation is generated.[11]. This effective field is responsible for small masses for the left handed neutrinos.[12]

The standard E_6 model proceeds similarly. There are 3 families (27-plets) of quarks and leptons. Since with respect to SO(10) we have the decomposition

$$27 = 16 + 10 + 1 \qquad (2.1)$$

the model also makes use of the vector and singlet representations of SO(10), which in this version are associated with much heavier fermions.

The gauge bosons belong to the adjoint (78) representation which, besides the 45 bosons of the SO(10) theory, also includes a singlet and two spinor representations 16 and 16'.

Two basic Higgs fields are required. One (78) that has no Yukawa coupling to fermions can give superheavy masses to leptoquarks that mediate the decay of the proton. The other Higgson in a (27) representation can have Yukawa coupling to basic fermions and split their masses. If the (27) also has some superheavy vacuum expectations values (v.e.v.s) some of the quarks and leptons could take superheavy masses.

Again we have the possibility of a two-loop Witten diagram. Provided that the gauge hierarchy works[13] this diagram will generate an effective Higgs field belonging to the 351' representation of E_6 that contains the 126 of SO(10) and gives right handed neutrinos superheavy masses, and, upon diagonalization of the mass matrix, provides left handed neutrinos with small masses. This property of the E_6 model was noted recently[14] and found independently by Ramond and Bowick.[14]

III. The B-L Generator in E_6 as a Flavor Hypercharge

I shall now consider the standard assignment with 3 families (the third containing the top quark). Let $J_{27}^{(e)}$ represent the electron family containing the u and d quarks.

Its decomposition under $SU(3) \times SU(3)' \times SU(3)^c$ is as follows

$$L = \begin{pmatrix} \hat{N}_R & \hat{E}_R & \hat{e}_R \\ E_L^- & N_L & \hat{\nu}_R^e \\ e_L^- & \nu_L^e & N_L' \end{pmatrix}, \quad Q_L^i = \begin{pmatrix} u_L^i \\ d_L^i \\ B_L^i \end{pmatrix}, \quad \hat{Q}_R^i = \begin{pmatrix} \hat{u}_R^i \\ \hat{d}_R^i \\ \hat{B}_R^i \end{pmatrix}, \quad (3.1)$$

where i = 1,2,3 is the color index. We have used the notations

$$\psi_L = \tfrac{1}{2}(1+\gamma_5)\psi \;,\quad \hat{\psi}_R = i\sigma_2 \psi_L^* = \tfrac{1}{2}(1+\gamma_5)\psi^c \tag{3.2}$$

for 2-component parts of the Dirac spinor ψ and its charge conjugate ψ^c. If A and B denote 3x3 unitary matrices associated respectively with the first and the second flavor SU(3) groups we have the transformation law

$$L' = BLA^\dagger \;,\quad Q_L'^i = A Q_L^i \;,\quad \hat{Q}_R'^i = B^* \hat{Q}_R^i \;. \tag{3.3}$$

$SU_W(2)$ is contained in A, while the neutral $SU'(2)$ is contained in B. Accordingly, the $SU_W(2)$ doublets are

$$\begin{pmatrix}\nu_L^e\\e_L^-\end{pmatrix},\begin{pmatrix}N_L\\E_L^-\end{pmatrix},\begin{pmatrix}\hat{E}_R\\\hat{N}_R\end{pmatrix},\begin{pmatrix}u_L^i\\d_L^i\end{pmatrix}, \tag{3.4}$$

while the $SU'(2)$ doublets are

$$\begin{pmatrix}E_L^-\\e_L^-\end{pmatrix},\begin{pmatrix}N_L\\\nu_L^e\end{pmatrix},\begin{pmatrix}\hat{\nu}_R^e\\N_L'\end{pmatrix},\begin{pmatrix}\hat{d}_R'^i\\\hat{B}_R^i\end{pmatrix}. \tag{3.5}$$

Here E^- is an additional charged lepton, N_L, N'_L, N_R additional neutral leptons and B^i an additional quark of charge $-1/3$. The additional fermions could in principle have heavy or superheavy masses depending on the V.E.V's of the Higgsons. Standard phenomenology will be reproduced if the neutral vector bosons associated with $SU'(2)$ are an order of magnitude more massive than the weak bosons associated with $SU_W(2)$.

Let us now consider the SU(3) subgroup of the flavor SU(3) x SU'(3) obtained by taking A = B. Its Cartan subalgebra consists of the electric charge q and the hypercharge y associated with the 3x3 matrices

$$q = \tfrac{1}{2}(\lambda_8 + \sqrt{3}\,\lambda_3)\;,\quad y = \tfrac{1}{\sqrt{3}}\,\lambda_8\;. \tag{3.6}$$

The transformation laws induced by these generators is obtained by putting

$$q:\; A = \exp(iq\alpha)\;,\quad B = \exp(iq\alpha)\;, \tag{3.7}$$

$$y:\; A = \exp(iy\phi)\;,\quad B = \exp(iy\phi)\;. \tag{3.8}$$

(3.7) is nothing but the electric charge U(1) subgroup of the electroweak $SU_w(2) \times U_w(1)$. Hence, we identify $U(1)_w$ with the $(I_8 + I'_8 + \sqrt{3}\,I'_3)$ generator of $SU(3)$, the charge operator being given by

$$q = (I_3 + \tfrac{1}{\sqrt{3}} I_8) + (I'_3 + \tfrac{1}{\sqrt{3}} I'_8)\;. \tag{3.9}$$

(3.8) corresponds to the generator

$$y = \tfrac{2}{\sqrt{3}}\,(I_8 + I'_8) \tag{3.10}$$

which induces the transformations

$$(e^-_L, \nu^e_L) \to (e^-_L, \nu^e_L)\exp(-i\phi)\;,\quad \begin{pmatrix}\hat{e}_R\\ \hat{\nu}^e_R\end{pmatrix} \to \exp(i\phi)\begin{pmatrix}\hat{e}_R\\ \hat{\nu}^e_R\end{pmatrix} \tag{3.11}$$

$$\begin{pmatrix}u^i_L\\ d^i_L\end{pmatrix} \to \exp(\tfrac{i}{3}\phi)\begin{pmatrix}u^i_L\\ d^i_L\end{pmatrix}\;,\quad \begin{pmatrix}\hat{u}^i_R\\ \hat{d}^i_R\end{pmatrix} \to \exp(-\tfrac{i}{3}\phi)\begin{pmatrix}\hat{u}^i_R\\ \hat{d}^i_R\end{pmatrix}\;. \tag{3.12}$$

Thus we identify y with the baryon number minus the lepton number:

$$y = B - L\;. \tag{3.13}$$

The additional leptons E, N, H' have zero hypercharge (B-L = 0) while the extra quark B has hypercharge or B-L quantum number -2/3. Hence, in the standard E_6 scheme the usual leptons and quarks are doublets under the isospin group $\vec{I} + \vec{I}'$. Singlets and triplets are either heavy or superheavy.

IV. The Induced (351') Yukawa Term in the Effective Lagrangian

Because of the Clebsch-Gordan series

$$(27 \times 27)_{sym.} = \overline{27} + 351' \qquad (4.1)$$

we can define two invariant tensors d_{abc} and h_{ab}^{mn} that project out respectively the $\overline{27}$ and the 351' parts of the symmetric direct product of two 27-plets H^a and H^b. Let H^a denote a $\overline{27}$-Higgs and G_{mn} an effective 351'-Higgs fields. Then we can have Yukawa couplings of the fundamental fermion fields ψ_L^a that transform like a 27-plet with H^a and G_{mn} in the form

$$\mathcal{L}_{\psi\psi H} = g_H (\psi_L^a)^T i\sigma_2 \psi^b d_{abc} H^c, \qquad (4.2)$$

$$\mathcal{L}_{\psi\psi G} = g_G (\psi_L^a)^T i\sigma_2 \psi^b h_{ab}^{mn} G_{mn}. \qquad (4.3)$$

In addition we also have the vector coupling of ψ to the gauge bosons $V_{c\mu}^a$ which reads

$$\mathcal{L}_{\bar{\psi}\psi V} = g (\psi_L^a)^\dagger \bar{\sigma}_\mu \psi_L^b k_{bm}^{an} V_n^{m\mu}, \qquad (4.4)$$

g being the universal gauge coupling constant and k_{bm}^{an} the invariant tensor that projects out the (78) part out of the Clebsch-Gordan series

$$27 \times \overline{27} = 1 + 78 + 650. \tag{4.5}$$

The gauge coupling of H^a is similarly

$$\mathcal{L}_{H\bar{H}VV} = g^2 \left(k^{an}_{bm} V_n^{m\mu} H^b \right) \left[k^{b'n'}_{am'} V_{n'\mu}^{m'} (H^b)^\dagger \right]. \tag{4.6}$$

Finally we need the following cubic self coupling of the Higgson H^a:

$$\mathcal{L}_{HHH} + \mathcal{L}_{\bar{H}\bar{H}\bar{H}} = \mu \left[H^a H^b H^c d_{abc} + (H^a)^\dagger (H^b)^\dagger (H^c)^\dagger d^{abc} \right], \tag{4.7}$$

where μ has the dimension of a mass.

The two loop Witten diagram arises through Wick contractions in the integrand

$$\mathcal{L}_{\psi\psi H} \mathcal{L}_{\bar{\psi}\psi V} \mathcal{L}_{\bar{\psi}\psi V} \mathcal{L}_{H\bar{H}VV} \mathcal{L}_{\bar{H}\bar{H}\bar{H}}, \tag{4.8}$$

giving an effective interaction $\mathcal{L}_{\psi\psi\bar{H}\bar{H}}$ of the form (4.3) with

$$G^{eff.}_{mn} = (H^m)^\dagger (H^n)^\dagger \tag{4.9}$$

and

$$g^{eff}_G \sim g_H g'^4 \mu K, \tag{4.10}$$

the factor K coming from the propagators.

Now, if the v.e.v's of H are denoted by v this means fermion masses m of the order of $g_H v$ from $\mathcal{L}_{\psi\psi H}$ and induced mass contributions δm of the order of $g^{eff}_G v^2$ from $\mathcal{L}^{eff}_{\psi\psi\bar{H}\bar{H}}$ for fermions that are left massless in $\mathcal{L}_{\psi\psi H}$. We need an effective 351' to give mass to neutrinos. As shown by Witten[11] the neutrinos will acquire masses of the order $m_q m_w \alpha_s^{-2} M$

from the two loop diagram, where M is the grand unification mass, the quark mass and m_w a weak boson mass.

V. E_8 as a Possible Family Group and Cabibbo Mixing

Recently Bars and Günaydin[15] have shown that the adjoint (248) representation of E_8, which is also its smallest fundamental representation can accommodate four 16-dimensional SO(10) families together with their conjugate families. Out of these 8 families two are superheavy, three can be identified with the observed e, μ and τ families and three similar conjugate families must lie higher in mass. Here, I would like to point out that, owing to the decomposition

$$248 = (78,1) + (1,8) + (27,3) + (\overline{27},\overline{3}), \tag{5.1}$$

of E_8 under E_6 x SU(3), there are three 27-plets and three $\overline{27}$-plets contained in (248), SU(3) acting as a "horizontal" family group and E_6 as a "vertical" gauge group[16]. One (78) and eight E_6 singlets also arise as additional heavy fermions.

Under E_7 x SU(2) the same multiplet has the decomposition

$$248 = (133,1) + (1,3) + (56,2). \tag{5.2}$$

When the family group SU(3) is broken down to SU(2) x U(1) and E_7 down to E_6 x SU(2) it can be verified that 56 contains one 27 and one $\overline{27}$. Thus the term (56,2) accommodates two families and their conjugates while (133,1) contains one 27 and its conjugate. It is tempting to associate (56,2) with the e and μ families and (133,1) with the τ family. Then, as long as U(1) is good, the e and μ families will undergo Cabibbo mixing while the 27 in 133 will remain unmixed due to its different E_7 and U(1) quantum numbers. Thus, we are led to consider the chain

$$E_8 \supset E_6 \times SU(3) \supset E_6 \times U(1) \supset SU(3) \times SU(3)' \times SU(3)^c \qquad (5.3)$$

with further breaking of SU(3) x SU(3) considered above. A detailed realistic symmetry breaking mechanism for E_8 is not yet available.

VI. Concluding Remarks on Octonions

If the exceptional groups, which are all connected with octonions, play an important role in Particle Theory, could the octonions enter in a fundamental way in the formulation of basic laws?

A - The action of E_6 for instance can be formulated in a Hilbert space associated with an exceptional Jordan algebra.[1] The color singlet subspace then becomes an ordinary projective space corresponding to the standard complex Hilbert space. However, the separation of colored states from the color singlets does not seem necessary at a fundamental level since confinement can apparently be understood within the standard Quantum mechanical framework, using the usual complex Hilbert space. Thus for the moment there is no physical motivation to develop an octonionic Quantum Mechanics. However, the construction of multiparticle states within octonionic Q.M. is not the main difficulty as this problem can be solved by choosing a direction, (say the seventh) for the vacuum (no particle state), thereby reducing the overall unobservable phase group to the $SU(3)^c$ subgroup of the automorphism group G_2 of the octonion algebra.

B - Another fundamental application of octonions lies in the possibility of constructing an exceptional Grassmann algebra out of the direct product of octonions[17] in perfect analogy to the Jordan-Wigner construction of ordinary Grassmann numbers out of direct products of quaternions (Clifford numbers).

Such octonionic Grassmann numbers can be used to define non-associative Fermi fields with color associated with the phase group $SU(3)$. The color singlet subspace is associative.

Another application is to the possible construction of new algebras in a superspace with coordinates that are partly bosonic, partly colored (exceptional) fermionic. If we consider products of such nonassociative coordinates with nonassociative generators of a new algebra, such that the products lie in an associative subspace of the superspace, then we are led to the construction of new kinds of supergroups with non-associative generators. Such supergroups have not yet been classified. Do some of them contain exceptional groups as bosonic subgroups? If the answer is yes, exceptional groups would have a role to play in supergravity.

Adknowledgements

I would like to thank Professor R. Marshak for giving me the opportunity to air these remarks and Professor P. Ramond and Professor M. Serdaroğlu for valuable discussions.

References

[1] M. Günaydin and F. Gürsey, Phys. Rev. $\underline{D9}$, 3387 (1974), F. Gürsey, P. Ramond and P. Sikivie, Phys. Rev. $\underline{D12}$, 2166 (1975).

[2] F. Gürsey, in to Fulfil a Vision, ed. Y. Neeman, p. 22 (Addison-Wesley, New York, 1980).

[3] E. Cremmer, B. Julia and J. Scherk, Phys. Lett. $\underline{76B}$, 409 (1978).

[4] H. Georgi and S. L. Glashow, Phys. Rev. Lett. $\underline{32}$, 438 (1974).

[5] H. Fritzsch and P. Minkowski, Ann. Phys. $\underline{93}$, 193 (1975).

[6] F. Gürsey, P. Ramond and P. Sikivie, Phys. Lett. $\underline{60B}$, 177 (1976).

[7] Y. Achiman and Stech, Phys. Lett. $\underline{77B}$, 389 (1978). P. Ramond "The Family Group in Grand Unified Theories," Cal. Tech. Preprint 68-709 (1979), F. Gursey, in 2nd Workshop on Current Problems in High Energy Particle Theory, ed. G. Domokos and S. Kövesi-Domokos, p. 3 (Johns Hopkins Univ. 1978).

[8] H. Georgi and S. L. Glashow, Harvard preprint HUTP-79/A073 (1980) F. Gürsey, in First Workshop on Grand Unification, ed. P. H. Frampton, S. L. Glashow and A. Yildiz, p. 39 (Mat. Sci. Press, Brookline, Mass. 1980).

[9] H. Georgi and A. Yildiz, Harvard preprint (1980).

[10] F. Gürsey in Ref. 8, R. Barbieri and D. V. Nanopoulos, Phys. Lett. $\underline{95B}$, 43 (1980). F. Gürsey and M. Serdaroğlu, to be published, P. Ramond, "The Family Group in Grand Unified Theories," Cal. Tech. preprint 68-709 (1979), P. Ramond in these proceedings.

[11] P. Ramond, in First Workshop on Grand Unification, ed. P. H. Frampton, S. L. Glashow and A. Yildiz, p. 105, (1980).

[12] E. Witten, Phys. Lett. $\underline{91B}$, 81 (1980).

[13] See M. Veltman, these proceedings.

[14] F. Gürsey, in Ref. 8, F. Gürsey and M. Serdaroğlu in Ref. 10, P. Ramond and M. Bowick, to be published and P. Ramond in these proceedings.

[15] I. Bars and M. Günaydin, Phys. Rev. Lett. $\underline{45}$, 859 (1980).

[16] F. Wilczek and A. Zee, Phys. Rev. Lett. $\underline{42}$, 421 (1979), A. Davidson, K. C. Wali and P. D. Mannheim, Phys. Rev. Lett. $\underline{45}$, 1135 (1980).

[17] See F. Gürsey in Ref. 7.

LEFT-RIGHT SYMMETRY, COMPOSITE MODELS AND BARYON NUMBER OF THE UNIVERSE

by

Rabindra N. Mohapatra[*]

Max-Planck-Institut für Physik und Astrophysik,
München, Fed. Rep. of Germany

und

Physik Department, Ludwig-Maximilians-Universität,
München, Fed. Rep. of Germany

Abstract: The present theoretical and phenomenological outlook for left-right symmetric theories of weak interaction is discussed. The identification of the U(1) generator of these theories with the B-L quantum number leads to baryon non-conserving processes obeying $\Delta B = 2$ selection rules, such as N-$\bar{N}$ oscillations. We also summarize the recent theoretical investigations into composite models that lead to $SU(2)_L \times SU(2)_R \times U(1)_{B-L}$ electroweak symmetry at low energies and make a plausible estimate of $\sin^2\theta_W = 1/4$ in these models in agreement with recent neutral current data. The scale of right-handed interactions is estimated to be around 10^6 GeV based on composite models as well as from the considerations of cosmological baryon production. We outline the scenario for baryon generation in these models.

---

[*] Alexander von Humboldt-Fellow. On sabbatical leave from the City College of New York, New York. Work supported in part by U.S. National Science Foundation and CUNY-PSC-BHE Award No. RF 13406.

1. INTRODUCTION

It is the purpose of this talk to summarize the present theoretical and experimental outlook for the left-right symmetric theories of weak and electromagnetic interactions. This is a particularly appropriate moment to embark on such a task for the following reasons: It is by now generally and widely accepted that low energy electroweak dynamics is generated by an effective local $SU(2)_L \times U(1)$ symmetry[1]. This idea has received striking confirmation by the low energy neutral and charged current interactions. The next obvious question to ask is, is $SU(2)_L \times U(1)$ part of a bigger symmetry? The most appealing bigger symmetry is the left-right symmetric gauge symmetry [2] $SU(2)_L \times SU(2)_R \times U(1)$, which restores parity to the status of a short distance symmetry of weak interactions and accepts the observed parity asymmetry in low energy weak interaction as the result of the vacuum being parity asymmetric. The mass threshold associated with parity breakdown could be (though need not be!!) as low as 300 GeV (i.e. $m_{W_R} \gtrsim 300$ GeV).

Secondly, a new generation of high energy ep colliding machines (such as HERA, CHEER/NEVIS and TRISTAN) are being planned, which are expected to explore energy regions, where signals of left-right symmetry are likely to appear if $m_{W_R} \approx 300$ GeV and could thus reveal exciting new physics. Also, a number of interesting low energy experiments such as the ones to detect $N-\bar{N}$ oscillations [3] $(\beta\beta)^\circ$-decay [4] and better measurements of μ-decay parameters [5] are either already under way or being planned, which could also test the hypothesis of left-right symmetry.

Finally, on the theoretical side, investigations into the possible subcomponent picture of quarks and leptons seem to indicate that if low energy electroweak dynamics is assumed to result as a residual effect of some kind of confining dynamics [6], then the associated electroweak symmetry corresponds to [7] $SU(2)_L \times SU(2)_R \times U(1)_{B-L}$.

In this talk, after a brief summary of the basic theoretical framework, we focus our attention on two recent theoretical developments; the first deals with the subcomponent picture of quarks and leptons and its implications for low energy electroweak symmetry. In the second part, we discuss the problem of baryon generation of the universe in partial unification models where the scale of baryon non-conservation is of the order of 10^5 or 10^6 GeV.

2. REVIEW OF THE BASIC IDEAS IN LEFT-RIGHT SYMMETRIC MODELS

In this section, I would like to review the basic ideas in left-right symmetric models discussed in Ref. (2). We work within the framework of the gauge group $SU(2)_L \times SU(2)_R \times U(1)$, with parity invariance making the $SU(2)_L$ and $SU(2)_R$ gauge couplings equal, i.e. $g_L = g_R = g$. The quarks and leptons are assigned to left-right symmetric doublets (denoted collectively by $\psi_{L,R}$) as follows:

$$\begin{pmatrix} u_L \\ d_L \end{pmatrix} (\tfrac{1}{2}, 0, \tfrac{1}{3}) \quad ; \quad \begin{pmatrix} u_R \\ d_R \end{pmatrix} (0, \tfrac{1}{2}, \tfrac{1}{3}) \tag{2.1}$$

$$\begin{pmatrix} \nu_L \\ e_L^- \end{pmatrix} (\tfrac{1}{2}, 0, -1) \quad ; \quad \begin{pmatrix} \nu_R \\ e_R^- \end{pmatrix} (0, \tfrac{1}{2}, -1) \tag{2.2}$$

It is thus clear that the Lagrangian conserves parity prior to spontaneous breakdown of the gauge symmetry. An extremely important point to note at this point is that the U(1) generator in this model can be identified with the B-L quantum number [7,8], thereby leading to several interesting predictions [7]. To see this, we write down the formula for electric charge Q in this model:

$$Q = I_{3L} + I_{3R} + \frac{B-L}{2} \tag{2.3}$$

Since $SU(2)_L \times U(1)$ symmetry is expected to be a good symmetry above $E \gtrsim 10^2$ GeV, it then follows from Eq.(2.3) that,

$$\Delta I_{3R} \simeq -\tfrac{1}{2} \Delta(B-L) \tag{2.4}$$

This equation is extremely interesting as it relates the scale of parity breakdown to the scale of baryon and lepton number breakdown. In fact, in a model in which the spontaneous breakdown of symmetry is achieved only via fermion composites involving the basic fermions in the theory (such as $\bar{\psi}_L \psi_R$, $\psi_L^T C^{-1} \psi_L$ and $\psi_R^T C^{-1} \psi_R$), it is clear that $\Delta I_{3R} = 1$. Therefore, Eq.(2.4) leads to the following two selection rules in baryon and lepton non-conservation. [9].

(i) $\Delta B = 0$, $\Delta L = 2$ leading to [10] Majorana neutrinos, $(\beta\beta)_0$ decay, etc.

(ii) $\Delta L = 0$, $\Delta B = 2$ leading to [7] $N \leftrightarrow \bar{N}$ oscillations, $N+P \to \pi$'s, etc.

To study the implications connected with baryon and lepton non-conservation, we introduce the effective Higgs fields having the same quantum numbers as the deleptons and diquarks: We denote them by [11]:

$$\Delta_{\ell\ell,L} : (1,0,-2) \quad ; \quad \Delta_{\ell\ell,R} : (0,1,-2)$$

and
$$\Delta_{qq,L} : (1,0,\tfrac{2}{3}) \quad ; \quad \Delta_{qq,R} : (0,1,\tfrac{2}{3}) \tag{2.5}$$

The $\bar\Psi_L \Psi_R$ composite Higgs fields is denoted by

$$\phi_1 \equiv \phi : (\tfrac{1}{2},\tfrac{1}{2},0) \quad ; \quad \phi_2 = \tau_2 \phi^* \tau_2$$

The Yukawa couplings of the fermion multiplets Ψ_q, Ψ_ℓ can be written down in a straightforward manner:

$$\mathcal{L}_Y = \sum_i \left[h'_{2i} \bar\Psi_{qL} \phi_i \Psi_{qR} + h_{2i} \bar\Psi_{\ell L} \phi_i \Psi_{\ell R} + h.c \right]$$

$$+ h_q \left(\Psi_{qL}^T C^{-1} \tau_2 \vec\tau \cdot \vec\Delta^\dagger_{qq,L} \Psi_{qL} + L \leftrightarrow R \right) \tag{2.6}$$

$$+ h_\ell \left(\Psi_{\ell L}^T C^{-1} \tau_2 \vec\tau \cdot \vec\Delta^\dagger_{\ell\ell,L} \Psi_{\ell L} + L \leftrightarrow R \right) + h.c.$$

To study the detailed spectrum of fermion and gauge boson masses in this theor we allow the spontaneous breakdown of symmetry to occur via the following [12]:

$$\langle \phi_1 \rangle = \begin{pmatrix} K & 0 \\ 0 & K' \end{pmatrix} \quad ; \quad \langle \vec\tau \cdot \vec\Delta^\dagger_{\ell\ell} \rangle = \begin{pmatrix} 0 & v_L e^{-i\theta_L} \\ 0 & 0 \end{pmatrix}$$

$$\tag{2.7}$$

and
$$\langle \vec\tau \cdot \vec\Delta^\dagger_{\ell\ell,R} \rangle = \begin{pmatrix} 0 & v_R e^{-i\theta_R} \\ 0 & 0 \end{pmatrix}$$

A detailed investigation of the (allowed) minima of the Hamiltonian consistent with Eq.(2.7) leads to two natural possibilities:

(i) Parity-conserving solution:

$$V_L = V_R \quad ; \quad \langle \phi \rangle \neq 0 \qquad (2.8)$$

(ii) Parity-violating solution in which case the following patern emerges naturally:

$$V_L \ll \kappa, \kappa' \ll V_R \qquad (2.9)$$

with $V_L \simeq \gamma \frac{\kappa^2}{V_R}$, where γ is an arbitrary parameter related to the detailed Higgs interaction and is expected to be of order 10^{-3} to 10^{-1}.

It is clear that the first solution is unphysical whereas solution (ii) has all the desirable features, i.e.

(a) low energy weak interactions are given by the standard $SU(2)_L \times U(1)$ model with corrections usually of order $\frac{\kappa^2}{V_R^2}$ in the amplitude.

(b) Weak $\Delta I_W = \frac{1}{2}$ rule which is satisfied by the low-energy neutral current data to an accuracy of $\frac{\kappa^2}{V_R^2}$. Thus, in principle, any deviations from this rule would constitute evidence in favour of left-right symmetric models.

What is, however, not clear about the above pattern of symmetry-breaking is the magnitude of V_R. In subsequent chapters, we argue in favour of $V_R \approx 10^5$ GeV based on theoretical (composite models) and phenomenological considerations.

For the time being, we accept the above symmetry-breaking and study its implications for the neutrino masses.

2a. <u>Relation Between Small Left-Handed Neutrino Mass and Suppression of V + A Currents</u>

The neutrino mass matrix now consists of a combination of Dirac and Majorana masses as follows (in a two-component basis [13]):

$$\begin{pmatrix} & \nu & N \\ \nu & h_\ell r \frac{k^2}{v_R} & hk \\ N & hk & h_\ell v_R \end{pmatrix} \qquad (2.10)$$

We note that the eigenvalues of this matrix are [10]:

and
$$m_{N_e} \simeq h_\ell v_R$$
$$m_{\nu_e} \simeq r' k^2/v_R \qquad (2.11)$$

where:

$$\nu_e = \nu \cos\zeta + N \sin\zeta$$
$$N_e = -\nu \sin\zeta + N \cos\zeta \qquad (2.12)$$

with $\tan\zeta \simeq \frac{hk}{h_\ell v_R} \ll 1$

An important implication [10] of this formula is that, for $V_R \to \infty$ (i.e. $m_{W_R} = gv_R/\sqrt{2} \to \infty$), $m_{\nu_e} \to 0$: Thus, there is a smooth analytic connection between a small non-vanishing neutrino mass and the observed maximality of parity violation in low energy weak interaction in left-right symmetric models. The consequence of spontaneous parity violation is, the result that ν_e is a light, two-component Majorana particle with its right-handed counterpart N_e being rather heavy (i.e. $m_{N_e} \simeq \frac{h_\ell}{g} m_{W_R}$). It is then obvious that, in low energy μ and β decay experiments, only V-A currents would participate.

In the domain of neutral currents, however, deviations from $SU(2)_L \times U(1)$ model predictions would be expected. A detailed investigation [14] of low-energy neutral current data implies that:

$$m_{W_R} \gtrsim 180 \text{ GeV}$$

and $m_{Z_2} \gtrsim 250$ GeV

If, indeed, the scale of parity breakdown is so low, one would expect to see its effects in the proposed ep-colliding machines such as HERA, CHEER and TRISTAN. For a detailed estimate of the magnitude of these effects, we refer the reader to Ref.(15). The important signatures to be looked for are (a) the polarization asymmetry in ep scattering i.e.

$$A \equiv \frac{\sigma(e_L^-) - \sigma(e_R^-)}{\sigma(e_L^-) + \sigma(e_R^-)}$$

and (b) production of N_R, i.e.,

$$\sigma(e_R^- p \to N_R X)/\sigma(e_L^- p \to \nu_L X)$$

at very high E and Q^2.

In the domain of leptonic physics, there exist other signatures of this idea; namely it predicts $\tau_{\beta\beta}$, $B(\mu \to e\gamma)$ within experimentally accessible ranges if the scale of W_R mass is indeed low enough. Typical expectations are $\tau_{\beta\beta} \gtrsim 10^{23}$ yrs for ^{82}Se decay and $B(\mu \to e\gamma) \simeq 10^{-11}$ to 10^{-13}.

2b. Baryon- Non-Conservation and N-N̄ Oscillation

This subject has been extensively discussed in the literature by Marshak and this author [7]. I only wish to repeat the salient points of this discussion for completeness.

The main point is that invariance under the local symmetry $SU(2)_L \times SU(2)_R \times U(1)_{B-L}$ allows couplings of the form:

$$\mathcal{L} = \lambda \, \epsilon_{i_1 j_1 k_1} \epsilon_{i_2 j_2 k_2} \Delta^a_{\ell\ell} \Delta^a_{q_{i_1} q_{i_2}} \Delta^b_{q_{j_1} q_{j_2}} \Delta^b_{q_{k_1} q_{k_2}} + h.c. \quad (2.13)$$

(we have suppressed the L,R indices for simplicity: they can be supplied without much effort by the reader). Now it follows that since $\langle \Delta_{\ell\ell,R} \rangle \neq 0$, the coupling in Eq.(2.13) introduces baryon violation into the theory and the selection rule that follows is $\Delta B = 2$, leading to N-N̄ oscillation or $N + P \to \pi$'s decay mediated by Δ_{qq} (see Fig.1 for a typical contribution).

The existence of such B-violating forces raises the possibility that in conjunction with CP-violating interactions present in the theory, it may provide an understanding of the origin of baryon number in the universe. This will be discussed in Section 4.

3. DYNAMICAL INTERPRETATION OF LEFT-RIGHT SYMMETRY

In recent days a great deal of theoretical discussion has focussed on the possiblity that quarks and leptons may be composites of more "elementary" constituents. Apart from the obvious motivation to discover the simplest picture that can explain the continuing proliferation of quarks and leptons, it is the hope of many physicists that this may provide a dynamical basis for the low energy electroweak symmetry and a theoretical insight into the various quantum numbers associated with it, such as weak isospin, B-L, etc.. One result of several of these investigations has been that the "low" energy electroweak symmetry is $SU(2)_L \times SU(2)_R \times U(1)_{B-L}$. In order to summarize these findings we first lay down the basic working hypotheses of these models. They are the following:

(i) Quarks and leptons are composites of all spin 1/2 fermions, f (such as for instance fff) or a fermion and scalar boson multiplets (such as fff...s). The f and s will be collectively called subcomponents [17].

(ii) The subcomponents are bound by a subcomponent analog of color force, to be called hyper-color [18,19,20] (and the associated dynamics QHCD), which confines all HC-non-singlets with all low mass (massless) particles (i.e. observed leptons and quarks) being singlets under the group.

(iii) The weak-electromagnetic forces are identified with the residual Van der Waal type forces resulting from QHCD in the same way that nuclear foreces are residual effects of QCD. This analogy then implies that the range of QFD forces must be characterized by Λ_{QHCD}.

The way to identify the unbroken electroweak (gauge) symmetry in such models is to look at the global symmetry of the QHCD singlet composite fields. It is then immediate that, in models where we work with a doublet of subcomponent fields binding with other objects which represent the color degree of freedom, in the limit of these composite fields being massless, the surviving symmetry [18,20] is $SU(2)_L \times SU(2)_R \times U(1)$.

In the work of Greenbeg and Sucher [18], this symmetry was identified with $SU(2)_L \times SU(2)_R \times U(1)_L \times U(1)_R$ but it turns out that the axial $U(1)_A$ generator is not a symmetry due to hyper-color anomalies leaving only the $U(1)_{B-L}$ part. We can use their model to illustrate these ideas:

All quarks and leptons can be assumed to be composites of two sets of fundamental fermions $\begin{pmatrix} f_u \\ f_d \end{pmatrix} \equiv f$ and $C_i, i = 1,2,3,4$, where the last index is for leptonic color. We further assume f and C to carry an extra hyper-color index denoting the fact that they transform as vector representation under $SU(N)_H$ as well as a color index for ordinary QCD. It will be assumed that the hyper-color degree of freedom will be confined with light composite fermions being hyper-color singlets. If we denote by Λ_H the scale of hyper-color interaction (i.e. the value of Q at which $\alpha_H(\Lambda_H) \simeq 1$), any composite of f and C_i which is a hyper-color non-singlet will have mass of order Λ_H. These ideas are based on intuition gained from the QCD theory of strong interaction.

It is then clear that if we write the Lagrangian for $f^a_{u,d}$ and C^a_i (a = 1, ... N for $SU(N)_H$), then this has the global symmetry generated by $\bar{f}_L \gamma_\mu \tau_i f_L$, $\bar{f}_R \gamma_\mu \tau_i f_R$ and $\bar{f} \gamma_\mu f$. The axial U(1) symmetry associated with the current $\bar{f} \gamma_\mu \gamma_5 f$ is not a symmetry due to the presence of hyper-color anomalies. Thus, if we assume that the electroweak dynamics of the composite fermions is generated by the maximal local symmetry a la hypothesis (iii), the low energy electroweak symmetry must be $SU(2)_L \times SU(2)_R \times U(1)_{B-L}$.

2c. Composite Dynamics and $\sin^2\theta_W$

Below we present a possible way to evaluate the electroweak mixing angle $\sin^2\theta_W$ in the framework of composite models. Before we enter into this discussion, we like to present an ansatz in the L-R models that leads to [21] $\sin^2\theta_W = \frac{1}{4}$. In left-right models there exist two independent gauge couplings: $g_L = g_R \equiv g$ the SU(2) couplings and g' the U(1) coupling. We can write the electric charge e, as:

$$\frac{1}{e^2} = \frac{2}{g^2} + \frac{1}{g'^2} \tag{3.1}$$

If we now make our ansatz that the effective vector gauge couplings in this theory are equal, this implies that:

$$g' = g/\sqrt{2} \tag{3.2}$$

This immediately implies that $\sin^2\theta_W = \frac{e^2}{g^2} = \frac{1}{4}$. It appears that this ansatz may have quite a plausible basis in the framework of composite models. Note that g' is the analog of the ω-coupling g_ω to the isosinglet current and to that of the ρ-coupling, i.e.

$$\langle 0|\bar{u}\gamma_\mu u + \bar{d}\gamma_\mu d|\omega\rangle = m^2/g_\omega$$

and

$$\langle 0|\bar{u}\gamma_\mu u - \bar{d}\gamma_\mu d|\rho\rangle = m^2/g_\rho \tag{3.3}$$

In the framework of QCD, it is quite plausible that $g_\omega \approx g_\rho$ as can be seen by evaluating the graphs in Fig.2. In a similar manner, QHCD should imply Eq.(3.2)

There are some further remarks that we wish to make in regard to such composite models. The first remark is that one would, of course, like to replace the scalar C particles by some further subcomponents. In such a picture the subcomponent dynamics may be governed by $SU(N)_H \times SU(3)_C$ with the preons transforming as (N, 3) under this gauge symmetry. The quarks and leptons will then be N-particle composites. Obviously, if the quarks have to be color-triplets, then one would require N = 3p + 1, where p is an integer. To construct leptonic states, we may introduce another fermion, χ_α, which transforms as

(N, 1) under $SU(N)_H \times SU(3)_C$.

Without discussing this model any further we can make the following qualitative comment about it. Since the subcomponent dynamics is governed by $SU(N)_H$ symmetry and since N can be rather large (it is at least 4), one can apply the ideas based on 1/N expansion of QCD as discussed by 't Hooft [23] and Witten [24]. Since the gauge bosons in this picture are bound states, it is immediately obvious that the gauge coupling of the effective low energy gauge group [25] g goes like $1/\sqrt{N}$, thereby providing a rationale for smallness of g.

The second point worth some speculation is that in the composite picture the electroweak interactions could arise out of a picture of vector dominance of the symmetry currents without any reference to unbroken local symmetries. The estimates for various parameters of the theory could then be made using the spectral sum rules involving the currents involving subcomponent fields.

Finally, we note that in the composite picture the scale of the right-handed W bosons can be assumed to be Λ_H, the scale at which compositeness is supposed to reveal itself. The reason for this is that, at $E \sim \Lambda_H$, composite quarks and leptons are supposed to form, and the residual interactions from QHCD can form dilepton and diquark bound states (the Δ-type Higgs bosons) that can break the left-right symmetry. Since the current upper limits on the sizes of quarks and leptons, derived from $\mu \to e\gamma$ and g-2 of muon, are put around 10^{-17} to 10^{-18} cm $(r_\ell \sim \Lambda_H^{-1})$, we would expect on the basis of the above reasoning $m_{W_R} \approx \Lambda_H \approx 10^5$ to 10^6 GeV. We, however, point out that the phenomenological lower limits on m_{W_R} are much lower, being around 200 GeV as discussed earlier.

4. $\Delta B = 2$ TRANSITION AT INTERMEDIATE MASS SCALE AND COSMOLOGICAL BARYON PRODUCTION

One of the most interesting applications of the grand unification hypothesis is the possibility that combined with CP violation it may provide an understanding of the observed matter-antimatter asymmetry of the universe starting with arbitrary initial conditions in the standard big bang model of the universe. According to the conventional picture of baryon generation in grand unified theories, the characteristic temperature at which baryon production starts is $T \approx m_U \approx 10^{15}$ GeV where m_U is the scale of baryon non-conserving interaction. A crucial ingredient in this discussion is that at least one of the particles involved in producing the baryon asymmetry must be out of thermal equilibrium [26]. It has been argued by Weinberg [27] that if there exist baryon-violating forces at intermediate mass scales, i.e. $m_{\Delta B \neq 0} \ll 10^{15}$ GeV, then, for a large range of temperatures 10^{15} GeV $> T > m_{\Delta B \neq 0}$, these forces will be in equilibrium and wipe out the baryon excess generated earlier unless some quantum number B + a L (a $\neq$ 0, -1) is conserved by these new interactions. Since ΔB = 2 interactions are supposed to be operative at $m_{\Delta B = 2} \simeq 10^5$ -10^6 GeV and, of course, they do not conserve any linear combination such as B+a L, therefore, the existence of ΔB = 2 interactions would wipe out any pre-existing baryon excess. However, Masiero and myself [28] have argued recently that fresh baryon production can start at $T \simeq m_{\Delta B = 2}$, when all required conditions are met and can produce enough baryons. This not only makes ΔB = 2 interactions consistent with the observed matter-antimatter asymmetry but also provides a new picture of baryon-generation at lower temperatures. We will also observe that this would put a lower bound on the mass of W_R of around 10^6 GeV.

Let us now outline this picture starting from $T \gg m_{W_R}$. At these temperatures, the left-right symmetry is exact and there are B-violating forces which, however, conserve only B - L quantum number. According to the argument of Weinberg [27], these interactions would wipe out any pre-existing baryon excess. As the universe cools below $T \simeq m_{W_R}$, $SU(2)_L \times SU(2)_R \times U(1)_{B-L}$ symmetry breaks down to $SU(2)_L \times U(1)$. In order to continue our discussion further, we need to display the mass spectra of the particles.

The particles relevant for our discussion are Higgs bosons $\Delta_{qq, L\ or\ R}$, $\Delta_{\ell\ell, L\ or\ R}$ and the Majorana lepton N. Their masses respectively are assumed to be:

$$m_{\Delta_{\ell\ell}} \approx 10^5 \text{ GeV} \approx m_{W_R}/10$$
$$m_N \approx 10^4 \text{ GeV} \qquad (4.1)$$
$$m_{\Delta_{qq}} \gtrsim 10^3 \text{ GeV}.$$

We take these as typical values for the masses and they could be smaller or larger than these values by factors of 2 to 3. As far as the Higgs field is concerned, since the temperature is much higher than $\langle \phi \rangle$, there are two possible points of view:

(i) If we adopt the philosophy that Higgs fields are composites made of technicolor fermions [29], above the phase transition temperature, the Higgs field, decompose into their component fermion fields, which are heavy ($m_{F_T} \approx$ several tens of TeV's).

(ii) Alternatively, if ϕ's are treated as elementary Higgs fields for $T \gg \langle \phi \rangle$, we expect these particles to have mass $m_\phi \approx T\sqrt{\lambda_\phi}$, where λ_ϕ is the self-ϕ^4 coupling.

Next, we discuss the decay modes of $\Delta_{\ell\ell}$ and N which involve both $\Delta B \neq 0$ as well as $\Delta B = 0$ transitions.

(a) $\Delta_{\ell\ell}$ Decays: Its various decay modes are $\Delta_{\ell\ell} \to 3\Delta_{qq}$, $\ell\ell \Delta_{qq}\bar{\Delta}_{qq}$, etc., of which the dominant one is $\Delta_{\ell\ell} \to \Delta_{qq}\bar{\Delta}_{qq}$ which conserves baryon number and has a decay width

$$\Gamma_{\Delta_{\ell\ell} \to \Delta_{qq}\bar{\Delta}_{qq}} \approx \frac{\lambda_{q\ell}^2 V_R^2 m_{\Delta_{\ell\ell}}^3}{m_{\Delta_{qq}}^2} \approx 10^3 \text{ to } 10^7 \text{ Gev}$$

depending on whether $\lambda_{q\ell} \approx 10^{-4}$ to 10^{-2}. The $\Delta B \neq 0$ transition $\Delta_{\ell\ell} \to 3\Delta_{qq}$ has a decay width $\approx \lambda^2 m_{\Delta_{\ell\ell}} \approx 10^{-1}$ GeV for $\lambda \approx 10^{-3}$. We note that above $T \approx m_{W_R} \approx V_R$, only the first two decay modes survive which vary with temperature as $\approx \lambda^2 m_{\Delta_{\ell\ell}}^2 / \sqrt{T^2 + m_{\Delta_{\ell\ell}}^2}$

(b) N Decays: A priori there are the following possibilities: $N \to \ell\phi$, $\ell 3(\Delta_{qq})$. However, if we adopt either of the above two pictures for the Higgs field, then for $T \gtrsim m_N$, $N \to \ell\phi$ is forbidden. The only decay mode of N is therefore the one of $\Delta B = 2$ type given by Fig.3. It is obvious that since Δ_{qq} have a net baryon number 2/3, producing Δ_{qq} is equivalent to producing net baryon number. We further note that, for $T < m_{\Delta_{\ell\ell}}$ this decay rate goes like $(\lambda h)^2 T^5 / m_{\Delta_{\ell\ell}}^4$.

Next, let us discuss the collision and annihilation rates for $\Delta_{\ell\ell}$ and N. The collision rates involving $\Delta_{\ell\ell}$ are given by $\alpha^2 T$ (or $h^4 T$ if it is electrically neutral), whereas for $T < m_\Delta$ is given by:

$$\approx \frac{\alpha^2}{m_\Delta^2} (m_\Delta T)^{3/2} e^{-m_\Delta/T}$$

Comparison with the expansion rate of the universe leads one to the conclusion that $T \lesssim m_{\Delta_{\ell\ell}}/10$; $\Delta_{\ell\ell}$ decouple after which they are in free expansion and decay immediately via the $\Delta_{qq}\bar{\Delta}_{qq}$ decay mode.

The main processes operative after the disappearance of $\Delta_{\ell\ell}$ are those involving the heavy right-handed Majorana particle N_R. There are mainly two types of processes responsible for pair annihilation of N's: $N_R N_R \to e_R \bar{e}_R$ caused via W_R exchange and $N_R N_R \to e_L \bar{e}_L$ caused via ϕ exchange. The former interaction rate is given by $\alpha^2 T/m_{W_R}^4$ whereas the latter is given by $\alpha^2 (\frac{m_e}{m_W})^4 T$. The second process is negligible whereas the first annihilation process becomes slow compared to the expansion rate T^2/M_P for $T \approx m_N$, $(m_N/m_{W_R}) \lesssim 1/500$. For $m_{W_R} \approx 5 \times 10^6$ GeV, this implies $M_N \approx 10^4$ GeV. Since by $T \approx m_N$, all annihilation processes are stopped, the thermal distribution of N's are out of equilibrium (30), leading to baryon generation via the B-violating decay mode $N \to \ell (3 \Delta_{qq})$ and $\bar{\ell} (3 \bar{\Delta}_{qq})$ in the presence of CP violation. The detailed discussion of the one-loop inteference graph and CP violation that leads to an estimate of the baryon to entropy ratio is discussed in Ref.(25). It is possible to get $N_B/N_\gamma \approx 10^{-10}$ to 10^{-11} in this manner.

The important point is that m_{W_R} must be $\approx 10^6$ GeV, a value which is of the same order as the one derived from the considerations based on the composite model.

ACKNOWLEDGEMENT

I wish to thank Y. Chikashige, R.E. Marshak, A. Masiero, R.D. Peccei, G. Senjanović and M. Turner for many useful discussions and comments.

REFERENCES

(1) S.L. Glashow: Nucl. Phys. 22, 579 (1961); S. Weinberg: Phys. Rev. Lett. 19, 1264 (1967); A. Salam: In Elementary Particle Theory: Relativity Groups and Analyticity, ed. by N. Swartholm (Wiley, New York) 1968.

(2) J.C. Pati and A. Salam: Phys. Rev. D10, 275 (/1974); R.N. Mohapatra and J.C. Pati: Phys. Rev. D11, 566 2558 (1975); G. Senjanović and R.N. Mohapatra: Phys. Rev. D12, 1502 (1975).

(3) R. Wilson: Proc. of the First Workshop on Grand Unification, New Hampshire (1980); M. Baldo-Ceolin: Padova Preprint (1980); P. Ratti: Report to this Conference; G. Young: Report to this Conference; T. Kamae: Report to this Conference; H. Anderson: Report to this Conference.

(4) F. Avignone: Report to this Conference.

(5) M. Strovnik et al.: Berkeley Proposal (1980).

(6) H. Harari and N. Seiberg: Weizmann Inst. Preprint (1980); O.W. Greenberg and J. Sucher: Univ. of Maryland Preprint (1980); K. Matsumoto et al.: Preprint(1980).

(7) The identification of the U(1)-generator of the left-right symmetric group with the B-L quantum number and its implications were first studied in R.E. Marshak and R.N. Mohapatra: Phys. Lett. 91B, 222 (1980), and R.N. Mohapatra and R.E. Marshak: Phys. Rev. Lett. 44, 1316 (1980).

(8) See also A. Davidson: Phys. Rev. D20, 776 (1979).

(9) We wish to point out, however, that in all discussions of left-right symmetry prior to 1980, the choice of Higgs multiplets introduced two extra U(1) symmetries, i.e. $\psi_{lepton} \to e^{i\alpha_l} \psi_{lepton}$, $\psi_{quark} \to e^{i\alpha_q} \psi_{quark}$ into the theory, which persisted even after the gauge symmetry broke down. These two extra symmetries could be identified in the broken theory as the baryon and lepton number, thus leading to exact baryon and lepton conservation in the model despite Eq.(2.4). In such models, the neutrino is a Dirac particle with low mass and new effects associated with the V + A currents in μ and β decay would be expected. For a summary of such effects, see M.A.B. Bég, R. Budny, R.N. Mohapatra and A. Sirlin: Phys. Rev. Lett. 38, 1252 (1977). Experiments are currently being planned to measure the electron polarization parameter ξ, which can test these kinds of models; see: M. Strovnik et al.: Berkeley Proposal (1980), and J. Bowman et al. In this version, however, the smallness of m_ν remains a mystery; see: G. Branco and G. Senjanović: Phys. Rev. D18, 1621 (1978) and G. Costa, M. d'Anna and A. Masiero: Nuovo Cim. 53A, 455 (1979).

(10) R.N. Mohapatra and G. Senjanović: Phys. Rev. Lett. 44, 912 (1980), and Phys. Rev. D (to be published).

(11) We wish to note here that there is no need to go beyond $SU(2)_L \times SU(2)_R \times U(1)_{B-L} \times SU(3)_c$ to have $\Delta B = 2$ transitions.

(12) An intriguing possibility in these models is to introduce CP violation into the Majorana condensates Δ_R and Δ_L so that the strong CP problem can be solved by having real quark mass matrices in the manner described in the papers: M.A.B. Bég and H. Tsao: Phys. Rev. Lett. 41, 278 (11978), and R.N. Mohapatra and G. Senjanović: Phys. Lett. 79B, 283 (1978) (i.e. by having $M = M^+$). To get adequate weak CP violation then, it becomes necessary to introduce at least two sets of Majorana condensates $\Delta^i_{L,R}$, $i = 1,2$. The detailed nature of CP violation and generation mixing in this model is the subject of a forthcoming publication by A. Masiero, R.N. Mohapatra and R.D. Peccei, in preparation.

(13) See K.M. Case: Phys. Rev. 107, 307 (1957) for a review.

(14) G. Ecker: University of Vienna preprint (1979);
J. Bajaj and G. Rajasekaran: Phys. Lett. 93B, 461 (1980)

(15) T. Rizzo: Phys. Rev. D (1980);
T. Rizzo and D.P. Sidhu: Phys. Rev. (1980);
D.P. Sidhu: Phys. Rev. D (1980);
L.M. Sehgal: Aachen Preprint.

(16) See R.N. Mohapatra and R.E. Marshak: Ref.(7) and Phys. Lett. 94B, 183 (1980), and
R.N. Mohapatra: Proc. of the First Workshop on Grand Unification, ed. by P. Frampton, S. Glashow and A. Yildiz (New Hampshire) (1980).

(17) K. Matumoto: Prof. Theor. Phys. 52, 1973 (1974);
J.C. Pati, A. Salam and J. Strathdee: Phys. Lett. 59B, 265 (1975);
J.D. Bjorken, C.H. Woo and W. Krolikowski: unpublished;
E. Nowak, J. Sucher and C.H. Woo: Phys. Rev. D16, 2874 (1977);
J. Terazawa, K. Akama and Y. Chikashige: Phys. Rev. D15, 480 (1977);
H. Harari: Phys. Lett. 86B, 83 (1978);
M. Schupe: Phys. Lett. 86B, 87 (1979);
J.C. Pati: Bhubaneswar preprint (1980);
H. Terazawa: INS-Rep.-382 (1980).

(18) O.W. Greenberg and J. Sucher: Univ. of Maryland preprint (1980).

(19) H. Harari and N. Seiberg: Weizmann Inst. preprints WIS-80/37-Sept.-Ph. and WIS-80/48-Nov.-Ph.

(20) K. Matumoto et al.: Preprint (1980).

(21) R.E. Marshak and R.N. Mohapatra: Festschrift for Maurice Goldhaber, Proc. of the N.Y. Academy of Sciences, ed. by G. Feinberg and A. Sunyar (New York) (1980).

(22) In Harai's model this relation is obtained in a somewhat different way.

(23) G. 't Hooft: Nucl. Phys. B72, 461 (1974).

(24) W. Witten: Nucl. Phys. B160, 57 (1979).

(25) It has been brought to our attention by Y. Chikashige that the fact that $g \sim 1/\sqrt{n}$ was also noted by S. Nussinov, Maryland preprint (1980).

(26) J. Fry, K. Olive and M. Turner: Enrico Fermi Inst. preprint (1980), and references therein.

(27) S. Weinberg: Phys. Rev. D (to appear).

(28) A. Masiero and R.N. Mohapatra: Max-Planck-Inst. preprint (1980).

(29) S. Weinberg: Phys. Rev. D19, 1277 (1979);
L. Susskind: Phys. Rev. D20, 2619 (1979).

(30) I wish to thank M. Turner for useful discussions and clarifications of this point.

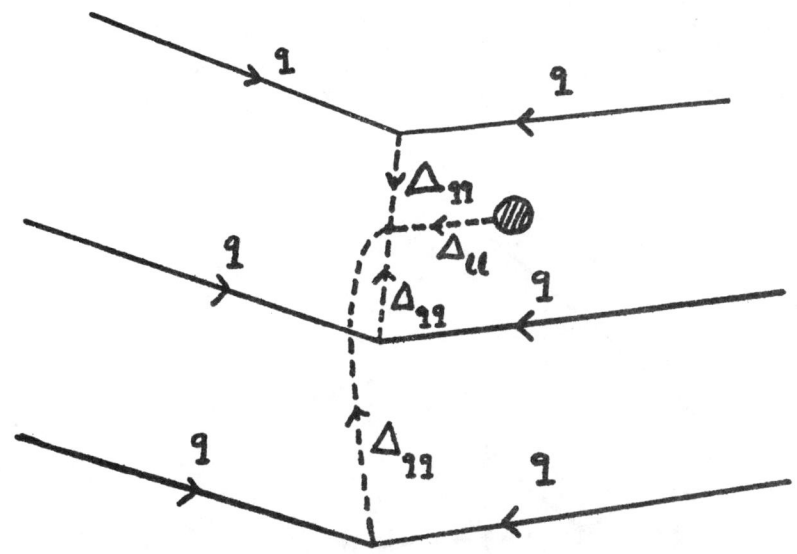

Fig. 1

The Feynman diagram responsible for $\Delta B = 2$ transitions.

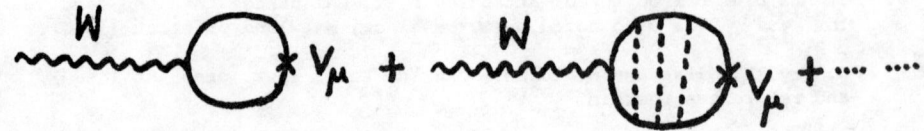

Fig. 2

Vector Dominance type graphs that contribute to $<0|V_\mu|W>$ type matrix element in the subcomponent model. The cross stands for the current, solid line for subcomponent fermions, dotted line for hyper-color and wavy line for the vector meson composite.

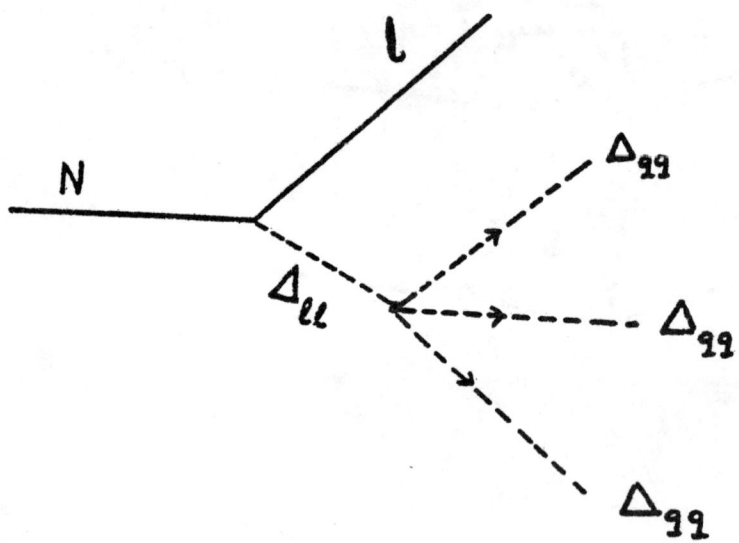

Fig. 3.

Baryon number non-conserving decay of the heavy Majorana lepton.

γ_5 Invariance, B-L Symmetry and Naturalness

R. E. Marshak
Virginia Polytechnic Institute, Blacksburg, VA

I had not planned to participate in this session on "Comments and Speculations" because I had conceived of it as a forum for a representative sampling of views by young "futurists". T. D. persuaded me to enter the fray despite my hoary age. I shall offer three comments which draw heavily upon the past but, hopefully, are future-directed. The first comment has to do with γ_5 invariance. The mathematical properties of the γ_5 operator were understood fairly soon after the Dirac equation was invented but it was not until 1957 (after parity violation had been observed) that the γ_5 invariance concept was related to the massless neutrino with a definite helicity[1] (Weyl equation). The brash generalization of γ_5 invariance to weak charged baryon and lepton currents[2] led to the purely left-handed charged weak current theory. It was understood, and so stated, at the time that the neutral vector electromagnetic current - a combination of left-handed and right-handed currents - shared the property of γ_5 invariance with the charged weak current. What was not understood in the early days - and may soon be understood - was why the same global γ_5 invariance principle was operative for the manifest gauge invariant electromagnetic interaction (with its massless photon) and the apparent non-gauge invariant weak current (with its massive intermediate vector boson). Within the framework of spontaneous breaking of the gauged electroweak group, I suggest that the clue to the explanation lies with the fact that before the symmetry is broken, all quarks and leptons are massless - with strict application of γ_5 invariance possible - and after the symmetry is spontaneously broken, the deviations must be small because of the small values of the Higgs-generated quark and lepton masses compared to the Higgs-generated W and Z^0 masses. Whether the Higgs bosons are composite or not may be relevant to this question.

There is another aspect of γ_5 invariance which has a past, and I believe an interesting future. This bears on Heisenberg's attempt to derive hadron masses from massless Weyl fields.[3] Heisenberg's famous equation contained a four-fermion interaction which was invariant[4] under a SU(2) x U(1) group, in which the first group was identified with isospin and the second with baryon number. The generalization of Heisenberg's approach to N (massless) Weyl fields[5] led to the observation that the chiral group, SU(N) x SU(N), was a higher hidden symmetry group which was broken down to the SU(N) group by a finite common mass of the N fermions. There is much work to do to understand the connection between the relatively small masses of the quarks and leptons after the gauge symmetry is broken and the exact chiral invariance which holds before the symmetry is spontaneously broken. It is conceivable that the correct theory will be able to relate the much smaller

ISSN:0094-243X/81/720665-04$1.50 Copyright 1981 American Institute of Physics

(but possibly finite) masses of the left-handed neutrinos – compared to the charged leptons and the quarks – to their Majorana character. I should note that the γ_5 invariance argument still leads to a left-handed weak charged current independently of whether a finite mass neutrino is Dirac or Majorana. Of course, if the Higgs structure gives rise to right-handed W and Z^0 bosons, deviations from the left-handed charged weak current will set in when the energies become comparable to the masses of the right-handed W and Z^0 bosons.

The second comment that I should like to make concerns the role of global and local B-L symmetry in unification models. The so-called "Kiev symmetry" proposed[6] in 1959 comprised three key elements: (1) a baryon-lepton symmetry principle for weak interactions couched in the form: $\nu \to p$, $e \to n$, $\mu \to \Lambda$; (2) a "weak" Gell-Mann-Nishijima relation between the third component of weak isospin (I_{3w}) and weak hypercharge (Y_w) hypothesized for baryons and leptons; and (3) the statement that $Y_w = B-L + T$ with $T = 0$ for weak isodoublets and nonzero for weak isosinglets. There were no quarks, no SU(3), no Cabibbo angle, only one neutrino, and no gauge theory of weak interactions. But there was the highly promising left-handed charged weak current theory, apparent electron-muon universality, and the intriguing symmetry between the two charged lepton currents, ($\bar{\nu}e$) and ($\bar{\nu}\mu$), and the $\Delta Y = 0$ and $\Delta Y = 1$ weak hadron currents, ($\bar{p}n$) and ($\bar{p}\Lambda$), respectively (here Y is the "strong" hypercharge). In the light of recent developments, it is clear that baryon-lepton symmetry is replaced by quark-lepton symmetry, the six quarks and six leptons are assigned to three families, and the values of I_{3w} and Y_w in the "weak" Gell-Mann-Nishijima relation depend on the helicity representations of the weak isospin.

If we now accept the standard gauge group $SU_L(2) \times U^{Y_w}(1)$ and use the expression $Y_w = B-L + T$, then evidently $T = 0$ for all left-handed isodoublet representations (of quarks and leptons) and $T = +1$ (-1) for all right-handed isosinglet up (down) representations. By insisting on $Y_w = B-L$ for left and right-hand representations, we are immediately led to the left-right symmetric electroweak group[7]: $SU_L(2) \times SU_R(2) \times U_{B-L}(1)$, where B-L is converted from a global into a spontaneously broken local symmetry and parity is restored at energies above m_{W_L}. The "weak" Gell-Mann-Nishijima relation becomes $Q = I_{3L} + I_{3R} + \frac{B-L}{2}$, which reduces to the standard model relation by the simple device of setting $I_{3R} = \frac{T}{2}$, thus explaining the absence of triangle anomalies. The physics of the left-right symmetric model is, however, quite different from the standard electroweak model. At energies above m_{W_L}, the left-right symmetric model tells us that $-\Delta I_{3R} = \Delta(B-L)/2$, thereby relating spontaneous parity breakdown to the spontaneous breakdown of B-L local symmetry. Some of the consequences of the left-right model [right-handed W and Z^0 bosons, two Majorana neutrinos (one light and one heavy) per generation, neutron oscillations and novel cosmological effects] have been discussed at this workshop by Riazuddin and Mohapatra. I should conclude this comment by pointing out that the B-L is globally conserved in SU(5) but is not a generator of this grand unified group. On the other hand, if one wishes to treat B-L as a spontaneously broken local symmetry, one can proceed through a partial unification group like $SU_L(2) \times SU_R(2) \times SU(4')$ (where B-L is the fourth "color") or deal with it at the GUT level in groups like SO(10) or E(6).

The third and final comment relates to the criterion of "naturalness" recently proposed by 't Hooft[8]. Roughly speaking, this criterion states that anomalously small physical quantities are "unnatural" and should have an explanation in some symmetry principle. I should like to mention an interesting example supporting this viewpoint. I refer to early attempts to understand the low values of $K_L \to \mu\bar\mu$ and $\Delta m(K_L-K_S)$. In the absence of neutral currents, these processes were calculated as higher order charged weak current effects. Ioffe and Shabalin[9] showed that the observed upper limit on the $K_L \to \mu\bar\mu$ branching ratio required a weak interaction cutoff substantially below the unitarity limit. Mohapatra and I[10] used the observed $\Delta m(K_L-K_S)$ to argue that the weak interaction cutoff is ~3-4 GeV, a very small value compared to the mass of W. With the neutral weak current in the picture, the problem became even more glaring and required for its explanation a new symmetry principle which could operate through a new quark inducing cancellations in the transition amplitude (i.e. the G.I.M. mechanism[11]). The low values of $K_L \to \mu\bar\mu$ and $\Delta m(K_L-K_S)$ become "natural" when the charmed quark takes its place as the "up" component in the second generation of quarks.

References

1. L. Landau, Nucl. Phys. **3**, 127 (1957); T. D. Lee and C. N. Yang, Phys. Rev. **105**, 1671 (1957); A. Salam, Nuovo Cim. **5**, 299 (1957).

2. E. C. G. Sudarshan and R. E. Marshak, Proc. of Padua-Venice Conf. on "Mesons and Newly Discovered Particles" (1957), reprinted in P. K. Kabir, "Development of Weak Interaction Theory", Gordon and Breach (1963). R. P. Feynman and M. Gell-Mann, [Phys. Rev. **109**, (1958)] and J. J. Sakurai [Nuovo Cimento **7**, (1958)] later derived the (V-A) interaction in ways that were equivalent to γ_5 invariance.

3. H. P. Dürr, W. Heisenberg, H. Mitter, S. Schlieder and K. Yamazaki, Zeits. f. Naturfor, **14a**, 441 (1959); cf. also Y. Nambu and G. Jona-Lasinio, Phys. Rev. **122**, 345 (1960).

4. Heisenberg et al (Ref. 3) showed how their equation was invariant under the transformations: $\psi \to \alpha\psi + \beta\gamma_5 C^{-1}\psi$ ($|\alpha|^2 + |\beta|^2 = 1$) and $\psi \to e^{i\alpha\gamma_5}\psi$, giving rising to the SU(2) isospin and U(1) baryon number groups respectively. Note the role of γ_5 in both transformations.

5. Cf. R. E. Marshak and S. Okubo, Nuovo Cim. <u>19</u>, 1226 (1961). M. Gell-Mann [Phys. Rev. <u>125</u>, 1067 (1962)] was led to the chiral SU(3) x SU(3) (corresponding to N = 3) by looking for the group generated, under equal-time commutation, by the space integrals of the vector and axial vector weak hadron (quark) currents.

6. A. Gamba, R. E. Marshak and S. Okubo, Proc. Nat. Acad. of Sci. <u>45</u>, 881 (1959). Cf. also Proc. of Kiev Conf. (1959), Vol. <u>2</u>, p. 297.

7. R. E. Marshak and R. N. Mohapatra, Phys. Lett. <u>91B</u>, 222 (1980); R. N. Mohapatra and G. Senjanovic, Phys. Rev. Lett. <u>44</u>, 912 (1980); R. N. Mohapatra and R. E. Marshak, Phys. Rev. Lett. <u>44</u>, 1316 (1980); R. E. Marshak, Riazuddin and R. N. Mohapatra, Prof. of TRIUMF Workshop on Muon Physics (Vancouver, 1981).

8. G. 't Hooft, Univ. of Utrecht preprint (1980).

9. B. L. Ioffe and E. P. Shabalin, J. Nucl. Phys. (USSR) <u>6</u>, 328 (1967).

10. R. N. Mohapatra, J. Subba Rao and R. E. Marshak, Phys. Rev. Lett. <u>20</u>, 19 (1968).

11. S. L. Glashow, J. Iliopoulos and L. Maiani, Phys. Rev. <u>D2</u>, 1285 (1970).

PANEL DISCUSSION

LEE

The first Rochester Conference was held thirty years ago. Since then, every decade there have been several important discoveries made in particle physics; some of these are experimental and some theoretical. I would like to ask each panelist to look into his crystal ball and to forecast: During the coming decade, what will be the three most important discoveries that will be made in particle physics? I shall ask the panelists to respond in reverse order to their earlier comments.

MARSHAK

I think and I hope, that one important discovery will be to answer Pauli's question, "Why is God a left-hander?" I would speculate that during the next ten years someone will come along with a good argument to explain why nature has chosen the left-handed way as the dominant if not the exclusive way for the basic weak interaction. I believe that the starting point of the explanation will lie in the masslessness of the quarks and leptons before the gauge symmetry is broken and their relatively small mass after the symmetry is spontaneously broken.

The second speculation concerns (B-L) which surfaced a long time ago, as far as I am concerned, as a sort of global quantum number but acquired a much deeper meaning when one started gauging the theories. I think something spectacular might happen here. Certainly, if neutron oscillations and heavy Majorana neutrinos are seen, the standard electroweak model will have to be modified in the direction of a model that restores parity to an intermediate energy symmetry. This would also favor SO(10) over SU(5) at the GUT level. I think that the physical interpretation of weak hypercharge as B-L is very attractive. Many people quote Einstein as saying that "symmetry dictates dynamics" but, of course, it has to be the right symmetry. Somehow, I have a feeling that the B-L symmetry is so elegant that it more than compensates for the fact that it implies finite-mass neutrinos. I am also impressed by the fact that B-L gives a "natural" explanation for freedom from triangle anomalies at the electroweak level.

MOHAPATRA

I think Prof. Marshak just gave two of the predictions that I would have given, so I only have to give one more. I would like to see the electric dipole moment of the neutron found in the range of $\sim 10^{-26} - 10^{-29}$, because in the area of CP violation there are then two choices. One is you go the way of hard CP violation of Kobayashi and Maskawa which basically predicts an extremely low electric dipole moment of 10^{-33} or so. The other route is the direction of soft CP violation. If you assume that Nature is a soft CP violator, then why would you like to think that is the case? There have been several arguments. One is that the KM model probably doesn't solve the θ problem, and unfortunately the axion probably is not going to appear. So that leaves us with the route of soft CP in a theory with left-right symmetry like Beg, Tsou and Senjanovic and myself have discussed. Also there are a lot of interesting dynamical symmetry-breaking models in which also you would expect CP to be a softly broken symmetry. Within this framework you can try to make some predictions, and they are generally at a much higher level than the expectation from the KM model, about 10^{-26} to 10^{-29}. So it would be nice if the e.d.m. is discovered in that range.

GURSEY

I take the Fifth Ammendment unless the chairman answers his own question at the end of the discussion. (Lee agrees.)

I don't know what will be important, but sometimes big advances are due to the discovery of new calculational techniques, like covariant conservation theory and Feynman diagrams. Maybe we will have such an advance in theoretical physics, and we will learn a new way of calculating Feynman path integrals from those of you who use partition functions in statistical mechanics. That would free us from either extremely strong-coupling calculations or extremely weak-coupling calculations. That would certainly be welcome.

Other great advances are made by revolutionary ideas. We have not had a true revolution in physics since general relativity and quantum mechanics. And according to Mao that's a very bad thing. So maybe the time is ripe for a new revolution. The revolution should replace quantum mechanics by something else. Now what could this be? No one knows. But quantum mechanics is based on a flat Hilbert space, which is very similar to the flat space-time of pre-Einsteinian physics. It is possible that the Hilbert space is an infinite-dimensional space with durvature. So we may have a curved Hilbert space. What would happen then? Then the superposition principle in that space would not be satisfied in cases where you have energies of curvature. The experimentalists may well find the failure of the superposition principle, which would prompt the physicists to make a Neo-Einsteinian theory of curved Hilbert space and derive the curvature equation. So that would be, I think, truly revolutionary. I doubt it will happen. But in purely experimental thinking, the nice thing about experiments is that they show surprises.

One thing that I would wish to happen would be the discovery of a truly stable array of heavy elementary particles, so that we could store the anti-particles and particles separately, and bring them together to solve our energy problem. Otherwise high energy won't be funded.

NAMBU

I interpreted Professor Lee's question to mean not only important but also totally unexpected or contradictory types of surprises. That reminded me of a kind of joke I used to make about the quarks: when and if the theorists agree on confinement, the experimentalists will discover quarks, real quarks. I think we theorists are approaching that state now, so I expect to hear any day now of their discovery. Perhaps these quarks will be really superheavy.

The other thing is that in order to understand the true validity of the concept of spontaneous breakdown and other things in physics, I think it will be desirable to see, and there might turn out to be, a domain structure in the universe. One part of the universe might have a different phase from another. In this case we can compare the two phases because at the junction we have two phases separated by a gap, and then you really appreciate the meaning of the order parameter. So this is what I would like to see.

The third one is also experimental. I have been wondering for some time about the e^-e^+ colliding beam experiments measuring the ratio R. Of course, people in the future will see huge resonant peaks due to Z^0 and so forth. I am wondering, what is this R rises to a certain level and suddenly starts coming down again, like a giant resonance in nuclear physics.

VELTMAN

Well you gave me your fair warning yesterday evening. I started thinking about it and of course I came into a panic. Let me be nostalgic for a while, I discovered a new means of getting out of the situation. I watched the great debate of Reagan and Carter and I know that Carter gets his main political insight from his daughter, Amy. So I took the phone and I called my daughter. And I asked her, "Can you tell me how the universe is made?" And she said to me, "Why don't you jump out of the window." I said, "What kind of answer is that?" She said, "Well, think about it," so I starting thinking about it. What happens if I jump out of the window? Well, I get hurt because there is gravity. So I started imagining that I was living in a world in which there was no gravity, say in interstellar space. So imagine there are all these people that all this time are living in interstellar space and they are doing physics like we do except there is no gravity.

What will they do? All the things we are doing. Beta decay, you name it. They will have their own Einstein and so everyone will have discovered General Relativity, neutron stars, etc. After that we all have some, well you name it, some Bohr, some Heisenberg, everybody. Perhaps even Veltman, who knows? Anyway then he will have another such student like 't Hooft and so on. There will be Weinberg and Salam and you will all work for me though you will all hate one another. Then the physics comes up and let's say they have come to the stage of 1971 when everyone starts believing in gauge theories and then somebody, starts constructing a local supergauge theory. Then someday one of these guys stands up and says, now I have a terrific prediction. In the supersymmetry there is a strange spin 2 field that must be there that follows precisely from special relativity and quantum mechanics. If this is true and he works and works on Einstein's equations and finally he comes up with the prediction that light should be deviated by a star. And by God, they do the experiment and they note that the ray of light is deviated by a star. So now experiment has, then, for those people absolutely made certain that supergravity exists. So when I have seen that my daughter had put me on the right track I realized that supergravity must necessarily be the ultimate theory of this earth.

If that is true, then the sort of scenario you can think of is to go all the way up to SU(5), 10^{15} GeV; we have nothing then but top-heavy quarks, SU(5) is it, and somehow the nature of things, perfectly true to the prediction of Ramond, coming in the wake of Ellis, Zumino and Gaillard, is from supergravity and that's maybe the world. If that is true, then I have instead of a prediction no prediction. Well there are some predictions, maybe they will find the top quark, and maybe they will find another generation but they will not find technicolor. They will not find anything else. They will find the Higgs, which could be relatively light and that would be a thing to be found. All of it within the next 10 years. They would or would not, it is hard to say, discover the decay of the proton. And theoretically I would expect somebody to finally take the theory of Ellis, Zumino and Gaillard and turn it into a decent theory, namely by removing the usual Veltman theory. So if that could be done then that is the kind of thing to look for.

WEINBERG

Gee, I don't have anything amusing to say. I think the important discoveries in at least the next 5 years, if not the next 10 years, will be experimental ones. It seems to me that theorists have been looking at the same problems now for about 5 years, so much so that like comedians at a banquet, they only have to give the number of the joke and everyone laughs. We just have to say the strong CP problem or the gauge hierarchy problem and everyone groans. I think most of us feel that we have rigorously proven that we ourselves cannot solve these problems and it may well be that some fellow that is a graduate student now will solve them. But I rather suspect that we are going to be shaken out of this feeling of desperation by experimentalists. Then it is not so hard to make predictions because I don't have to do the experiments. Other people do and that's fine; I'm happy to talk about that.

I think one thing that will be clarified in the next year or so perhaps, is the thing Mohapatra mentioned, the CP problem, especially through the experiment at Grenoble on the neutron electric dipole moment and I agree with his description of it. I think there are some kinds of CP violation, Kobayashi-Maskawa or superweak, which tend to give very small neutron electron dipole moments. There are others, like Higgs boson exchange, which tend to give rather large neutron electron dipole moments of a magnitude that will be discovered this year, if they exist. And on both sides of that distinction, we can have hard or soft, and merely having a soft CP violation doesn't solve the strong CP problem anyway. But that is a technicality, I generally agree with what Mohapatra said about that.

Another discovery which seems to me quite likely in the quite near future is proton decay. One thing is absolutely clear: Proton decay is discovered in the foreseeable future, then the lifetime of the proton is between $10^{30} - 10^{33}$ yrs, because if it is any longer than that it ain't going to be discovered and if it is shorter than that, it would have been already. So that's a safe prediction. I also think that if it is discovered, it will be conventional proton decay, which sounds like a contradiction in terms, but by which I mean proton decay into anti-lepton plus meson, say $e^+ + \pi^0$ rather than any of the other exotic modes. I would guess that that is what it will be. If that's what it is, then I think one would be justified in reaching a reasonable confident conclusion that it is caused by the exchange of particles with a mass scale of the order of 10^{15} GeV. Simply because if it is conventional proton decay, it's got to be a dimension six operator, then the dimensional analysis tells you it has got to be 10^{15} GeV in order to give the lifetime range which it has. However, I don't think we will know much more than that for a long time. The 10^{15} GeV is a prediction which was made in an old paper by Georgi, Quinn and me and I think clearly this discovery will then tend to validate the general assumptions made in that paper. But those assumptions are very broad and apply to a very large range of theories. Not only to theories like SU(5) and SO(10) but also even to a variant of the $SU(4)^4$ model. Pati and Salam have discussed $SU(4)^4$ and they tend to discuss it in **versions** which obscure its similarity to the other theories. But, in fact, if you assume the symmetry breaks in one step to the familiar gauge group of weak, electromagnetic and strong interactions, then the Georgi-Quinn-Weinberg analysis applies to it as well as to the other, and it again predicts a mass of 10^{15} GeV for the scale that is producing the proton decay. That of course is a completely different theory. It is not related group theoretically at all to SU(5). Neither one is a subgroup of the other and it will be a long time before one can distinguish between them. I think if proton decay is discovered one prediction that can be made with absolute certainty is that there will be by the end of the decade of the order of 1000 proton decays. Because people will not rest with just one or two and with that many proton decays it might be possible to begin to distinguish different models. For example, by measuring the polarization of the muon and $p \to \mu^+ + \pi^o$.

The third prediction which we can make with some real confidence is that the W and Z will be discovered, possibly next Winter by Carlo Rubbia at CERN. I think beyond that there is a whole world of exciting physics in the neighborhood of 100 GeV to 1 TeV and about that I am reasonably confident because our model seems to divide into two classes. One class of models has dynamical symmetry breaking in which, as Beg said yesterday, there has to be pseudo-Goldstone bosons in that mass range. And the other class has good old symmetry breaking with elementary scalars in which case there has to be Higgs bosons in that range. So I have no doubt that in the energy range from 100 GeV to 1 TeV, one will find a rich variety of phenomena, in fact it will be very hard at first to tell what it is we are seeing. And that is going to be a preoccupation for a long time. Eventually, out of that will come an answer to the old question of what is the actual mechanism by which symmetries are spontaneously broken.

THORNDIKE

Well, my suggestions as an experimentalist focus on experimental results. I tried to think of the most striking experimental results that I could and they turned out to be null results. In fact it's sort of surprising. The theoretical framework in which we find ourselves is that we are making theoretical predictions of experimental results which some time ago would have seemed preposterous. Yet we hold to them with great certainty. We are predicting that the matter of which we are all made is unstable, we're predicting the existence of particles of a mass an order of magnitude larger than any we have ever seen, and the most shocking effect of experiments on our theoretical thinking would be if these experiments turn in null results. I would predict that the W searches will, over the next 10 years, reveal no W, the Z^o searches over the next 10 years will reveal no Z, and the proton decay searches over the next 10 years will find no proton decay. And then the ball will be back in the theorists' court.

PERL

When Professor Lee proposed the question of what major discoveries will be made in the next ten years in elementary particle physics, I began to think of how major discoveries might be made. This

started me thinking about the new experimental techniques that
may be needed to make the major discoveries. And so I will talk
about three areas in high energy experimentation where the major
advances are being made, or where major advances need to be made.

One area where new techniques and inventions are needed, is the
colliding beams accelerator. The need is most apparent in electron-
positron colliding beams storage rings using conventional radio-
frequency power systems. Such accelerators are facing several
problems. Construction and operating costs increase rapidly with
energy. This is because the ring diameter is proportional to the
square of the beam energy; and for a fixed diameter, the power
used is proportional to the fourth power of that energy. The LEP
electron-positron storage ring, with an ultimate maximum total
energy of about 150 to 200 GeV, is probably the highest energy
e^--e^+ ring that the elementary particle community can afford.
Another problem of such accelerators is that we don't know how
to get the 10^{33} cm^{-2} sec^{-1} luminosity required in the several
hundred GeV energy range. Indeed we are finding that even at
lower energies we don't understand the luminosity limitations
produced by the multiple pass, beam-beam interaction. And so
existing machines are not operating at their design luminosities.

There are two new directions in accelerator physics and technology
which may allow us to overcome these problems. At SLAC we are
actively pursuing the development of colliding linear accelerators.
In such a device an electron beam from one linear accelerator
collides just once with a positron beam from a second linear
accelerator. The construction cost and operating cost are proportional to the energy; hence higher energies, compared to storage
rings, appear to be feasible. Of course there is a great deal of
research and development to be done on this new technique. The
other new direction in accelerator technology is the development
of superconducting radio frequency power systems. This concept
being actively pursued at several laboratories, particularly at
Cornell, can lead to reduced operating costs. However it does not
reduce construction costs.

Antiproton-proton colliding beams storage rings are much less
sensitive to the problems I have mentioned. Therefore we can
expect that the techniques now being developed at CERN and Fermilab for these machines have a rich future. However these machines
are not yet operating, and their full utilization may also require
new techniques and inventions in accelerator physics.

Another experimental area where new techniques are being developed
is the use of low energy experiments to study high energy physics.
The most notable class of such experiments are the current searches
for proton decays. Another interesting class are the high precision
studies of the various muon decay modes. These classes of experiments may be opening up new directions in elementary particle
experimental physics. But, as an old experimenter, I want to add

a word of caution. I'm a little worried about whether we will be able to find proton decays or if we will be able to find minute deviations from conventional theory in muon decay. It is very hard when one works almost at the limit of sensitivity to find a result and to make sure it is right. If you look through the history of experimental physics, you will find that new discoveries are usually made by turning to new techniques rather than by pushing old ones to their limits. Therefore these low energy experiments may eventually require substantially new methods in order to realize their potential.

The third experimental area where invention is required has to do with the size and complexity of modern particle detectors. Look at the detectors which are being built for use at the CERN and Fermilab antiproton-proton colliding beams storage rings; or look at the detectors being designed for LEP. I think that we will all agree that these detectors are at the upper limits of size, complexity, and cost. For example, we don't know how to maintain systems with more than 10^5 separate electronic information channels; and we can't expect to spend substantially more than 50 million dollars for a detector.

And so we need a new way to collect data, a new kind of comprehensive particle detector. That seems to be the most difficult thing to do and I have not the least idea how one is going to do it. I want to point out, that what we are doing now would be understood by Wilson, who invented the cloud chamber, and by Geiger. We are really not doing anything different than those people did, except that we have modern electronics. We clearly need a revolution in detector technique. I hope that will occur in the 1980's.

CHOU

Since I got the warning, I really didn't know what the important thing was, so I decided to try a Chinese method. That is, I go to the masses to ask them what they think is the really important thing. Because time was limited, I only chose some American-Chinese theorists here, about 5-6 people, and I get the following impressions, which I now report to you. The first impression is that every prediction made by the standard model will be confirmed in the next few years. This is an objective statement. The subjective statement is that we hope that one or some of them will be proved wrong; the feeling is that that could help solve the job problem in this country. The second impression is that we want the desert to bloom. The third impression is that we are now at the stage of coming to the next layer of structure of matter. And as we go from atom to nucleus, from nucleus to hadrons, and from hadrons to quarks and so on, the particles' behavior changes from individual to collective. So we would like to see some experimental results which show more explicitly the collective behavior of the particles. For example, if the particle has some kind of collective structure, then we could find some other internal quantum numbers like the intrinsic angular momentum in the spheroid of the nucleus, or of some of the particles are solitons, their mass might be finite but the moment of inertia might be infinite. Under such circumstances, there is no angular momentum conservation in scattering, but it does not contradict the isotropy of space. Some direct evidence of this kind will be very welcome.

LEE

1. The discovery of the Regge trajectories of leptons and quarks.

Suppose that neither the lepton nor the quark is elementary. They are not point particles, but instead they have a size. This size cannot be too large. Consider, for example, the electron; its size ℓ has to be less than $\sim 10^{-16}$ cm due to tests of QED. Now $\ell \neq 0$ implies that there should be a nonzero moment of inertia, and through that the electron can acquire an angular momentum in a collision, resulting in a high-spin $j = 3/2, 5/2, \ldots$ heavy electron. The slope of its mass vs. j should be

$$\sim \frac{1}{\ell} \gtrsim 100 \text{ GeV}.$$

On the basis of known interactions, we can predict the minimal phenomena that such a heavy electron, called E, would exhibit. We expect among its dominant decay modes

$$E \to e + \gamma,$$
$$\to e + \text{lepton pairs}.$$

Thus, if E can be produced we will see high-energy jets of pure γ and lepton showers, which will make the observation quite spectacular.

2. The possiblity of doing "vacuum engineering".

Based on either the spontaneous symmetry-breaking mechanism or the quark confinement phenomena, we believe our physical vacuum to be quite complicated. Like solid, it can carry some long-range-order parameters. Since the vacuum is Lorentz-invariant, the simplest representation of the long-range order is to introduce a phenomenological Higgs-type scalar field ϕ. Such a field may have quantum numbers, but with a nonzero vacuum expectation value

$$< \phi > = \phi_{vac} \neq 0$$

Depending on ϕ_{vac}, different particles can acquire different masses and different interactions. By introducing either a high-matter density or high-energy density over a large volume, we may alter the expectation value $< \phi >$. Then, within this volume the masses and interaction strengths of particles can be quite different from the familiar values. A change in $< \phi >$ may be viewed as a change in the vacuum; this feat will be called vacuum engineering. To explore this possibility, we have to study high energy coherence phenomena, for example by using ultra relativistic heavy ions. It is conceivable that some of our high-energy machines, such as ISR, may be converted into heavy-ion accelerators in the future. If we are able to alter the vacuum, then we may encounter some totally unexpected phenomena.

3. Deviation from conventional quantum mechanics.

In quantum mechanics, the system that is being observed satisfies the time-dependent Schroedinger equation

$$H \psi_{system} = i \dot{\psi}_{system} \qquad (1)$$

in which the observer is not included as part of its dynamical variables. The effect of the observation on the system is described by the well-known rules of probability amplitude decomposition of ψ_{system}. However, the converse effect of the system on the observer is not at all well described by our usual formulation of quantum mechanics. To be sure, such an effect must be there, as is obvious if we consider the example of a physicist who inadvertently drinks a cup of the liquid helium superfluid that he is observing. However, while the effect of the observer on the system is quantum-mechanical, the converse effect is invariably expressed only in classical terms. Yet, by reciprocity such an asymmetry must be viewed only as an approximation.

We may wonder how to express the effect on the observer in terms of quantum-mechanical probability amplitude and not classical probability. At present, there even seems to be a lack of vocabulary adequate to express such a phenomenon. A way out of this dilemma is to consider a state vector in a larger Hilbert space

$$\psi = \psi_{observer} \otimes \psi_{system} ,$$

which satisfies the usual quantum-mechanical law of motion. By taking into account the quantum-mechanical effect of ψ_{system} on $\psi_{observer}$ and then reiterating the effect of $\psi_{observer}$ back onto ψ_{system}, we may derive an improved equation for ψ_{system} which carries a correction to (1). The result of this corrected equation can then be expressed in terms of the usual language which ignores the non-classical effect of the system on the observer. By carrying out successive approximations in this direction, hopefully during the next decade we may be led to a more fundamental equation, which in turn may bring us new phenomena and deeper understanding of quantum mechanics.

AUDIENCE DISCUSSION

ABASHIAN: I'd just like to make a few remarks as an experimentalist regarding how we might look into the crystal ball and see what might be on the horizon. I've found in just listening these last couple of days a whole treasury of experiments that would be very, very exciting to do. One of them, for example, is the one Bob has been pushing, the oscillation of neutrons into antineutrons. Another is the proton decay. A third one is the neutrino oscillations. I guess to some extent my preferences are dictated by my personal history of being interested in that type of process, oscillations, decays, etc. Most of these have something to do with the breaking of some conservation law. This has a long history, of course.

Some of you here in this room, T.D. and Frank, had the guts to actually come out and say, well maybe parity was violated. But I think most of us at that time tended to think that was a far-out idea and it just couldn't be. Well, after parity was found to be violated, we patched it all up and said CP is conserved and T is conserved. We went on for a few years until that was found to be broken. And on and on the story goes. I do believe that as time goes on we are going-to find more and more of our sacred principles dropping by the wayside. But the one that I think will definitely be false is that the neutrino has a zero mass.

KANE: There is one thing I expected to hear mentioned which I didn't. One of our oldest problems involves why there are so many generations. And no one addressed that. Every theoretical idea that has had anything to do with that subject, and there are certainly no ideas that many people find compelling, always leads to large numbers of transitions among generations. So if I were to express what I would expect conventionally, I would say that one should expect to discover in the near future transitions among generations, flavor-changing neutral currents or that sort of thing. They should be much larger than the existing limits in the light sector and maybe they will be larger in the heavier sectors. Maybe there is some reason, or one can make up reasons, why they are suppressed in the light sector. What would be even more surprising and extraordinary in its impact if there were no transitions among generations discovered.

LANDE: I just wanted to reinforce Marty Perl's comments that many of the experiments, many of the things being tested are now at the very level of sensitivity of the apparatus. I think in the past it has always been clear that you need an order of magnitude or two of safety if you are going to proceed. We are probably rapidly using up the safety factors that we have in our accelerators, in our techniques and our technology, and certainly one thing that we have to worry about in the next decade is how to have a radical change in technology to be able to explore the various questions in nuclear decay and neutrino oscillations, etc. that are now being discussed.

PRASAD: I would like to ask the panel what type of equation of motion one should look into that will predict the number of generations? Is there any idea one can get from some equation of motion that will predict the number of generations?

MARSHAK: I'm not going to try to answer that directly, but Prof. Kane's question raises a point that I would like to direct to our chairman. One of your predictions has to do with higher spin for quarks and leptons. In some quantum mechanical systems, at a certain point, there is confusion as to whether you have rotational or other degrees of freedom. Perhaps you will meet such confusion if and when you reach the 4th or 5th generation of quarks and leptons. Is it clear that higher spin states can be separated from the generation problem?

LEE: I don't think it is clear, it just opens up the game in a different ballpark. If you have a Regge trajectory and several daughters and granddaughters, generation problems, of course, will come up too, but in a different light. You can regard this as evidence for compositeness.

VELTMAN: I wish to comment just for a minute, that all these things that we speculate on would actually influence this parameter ρ and you cannot do very much. You mention spin 3/2. If that structure affects ρ, I don't think it is possible.

LEE: Presumably, that is where the theorists can be of some help.

RAMOND: One thing that I would like to mention is that first of all, one big theoretical problem that exists is the absence of a cosmological term in our theories, which has not been mentioned. And also the fact that quantum mechanics, gravity and the world of elementary particles do not exactly mesh. That to me seems to be a very important problem. Now whether or not supersymmetry is an answer, I do not know. But then let me point out that supersymmetry as of next month will be 10 years old. Yang-Mills theory was 16 years old before 't Hooft disrobed it. Maybe the discovery of the decade will be the discovery of supersymmetry as an important symmetry, not at our scale obviously but at a different scale. And comment number two is maybe this decade will see the discovery of gravitational waves.

MESHKOV: One thing that would be nice to have happen would be the discovery of new forms of matter, for example, glueballs, that is something which is non-quarkful, not a lepton but made just of gluons. And it would be nice if we could present it with other forms that we haven't thought of. I mean, in general, as has been said, the things that are most spectacular are the ones that you can't sit here and think of.

WALI: The idea of compositeness of quarks and leptons had the embarrassment that it is hard at present energy levels to get evidence that quarks and leptons are composite. The suggestion I am going to make is really more of a wish than a prediction. The suggestion I would make is in analogy with previous experience, that the calculation by Bohr of levels of the hydrogen atom gave evidence for that model, and the Gell-Mann-Okubo mass formula gave evidence for SU(3), so a calculation of the masses of the quarks and leptons in the different generations on the basis of compositeness might give compelling evidence that composite structure really exists before we have seen more direct evidence. There will be associated with compositeness new levels of confinement and I would suggest that these new levels of confinement will operate perhaps at mass scales less than we presently expect. And the confinement mechanism will provide the new form, the new method for symmetry breaking and would replace the elementary Higgs. Finally, I would suggest that the true role of supersymmetry would be at the level of subquarks and subleptons. That is where the supersymmetry would play its role. All of this, of course, is in the nature of wishful thinking rather than prediction.

GREENBERG: Just as the derivation of the Balmer formula for the energy levels of the hydrogen atom by Bohr and the Gell-Mann-Okubo formula for the baryon octet gave strong support to Bohr's model and SU(3), respectively, I speculate that the derivation of a mass formula for quarks and leptons, including, hopefully, the correct prediction of quarks and leptons as yet unseen, on the basis of a composite model will give strong support to the idea of compositeness. The compositeness will be associated with new levels of permanent confinement, perhaps at lower energies than we now expect. A very detailed spectrum of excited quarks and leptons will appear in the region above 100 GeV; in particular the lepton spectrum will be of comparable complexity to the present hadron spectrum. Finally, I speculate that supersymmetry will play an important role at the composite level.

AVIGNONE: This is a question to Professor Gürsey or Lee. In your comments about finding some new theoretical breakdown, are you giving a new respectability to investigations of foundations of quantum theory, a la Aharanov-Bohm or that sort of very fundamental thing?

LEE: Not me. No. In wave mechanics we have the Schroedinger equation. But the observer obeys classical mechanics and has no phase relations; now that can't be right. So in this sense it is in the opposite direction from Bohm. It is simply that it seems to me the observer must also obey a similar kind of mechanics. Right now the symmetry is an approximation, is a projection because we don't have the right kind of language to say what kind of quantum mechanical effect the liquid helium can have on me, beyond life and death, the example I just gave. So we must describe it by a coherence. Now this coherence is very difficult to measure because it is somehow embedded into our consciousness. And the one way to bypass us is to project it back onto the system itself. Then we can measure it. Now you say, once you take two rounds, then you make a classical approximation, you can see the fact. So this is, or at least as I visualize it, in the opposite direction from Bohm and I believe it should be there. There will be a deviation which has to be a macro-quantum system. How to extract that information in terms of precise theoretical prediction and so it can be measured accurately is a job, but we have ten years to work on that.

SIRLIN: My comment is very prosaic in comparison with the profound statements that have been made about proton decay, quark confinement, etc. But I recall that the last time we came across a renormalizable field theory, like QED, people like Schwinger, Feynman etc. proceeded to do some difficult calculations and QED is still, I believe, one of the glories of theoretical physics. Now again we are looking at a renormalizable theory, so at a more prosaic level than you gentlemen have been discussing, one of the interesting things could be to try to use the renormalizability of that theory to do something with it beyond using it as a criterion for model building and try to see whether we can make some predictions, etc. There are now a number of groups working in that direction and there are a number of predictions. So it might be interesting to see whether the experiments agree with this kind of effort.

TYE: I just want to make a comment on a point which none of the panelists seems to have mentioned. It is, I think, very clear that machines can reach probably energies of only TeV and beyond that we have to search elsewhere. One way to extend the energy range is to see the interplay of particle physics and cosmology. At this moment the relationship or the application of particle physics to cosmology is very poorly understood and mostly it is more like an art form than a science. However, a lot of the uncertainities can be settled theoretically and it is quite likely that in the next 10 years we will be able to understand a great deal more about this relationship and, we hope, change art into science.

MILLER: I would like to see in the next 10 years a neutron beam reactor with a liquid deterium moderator and large access ports. It is unfortunate that the reactor in Grenoble has the neutron reflecting guide tube for every port. So it becomes difficult to do an experiment such as neutron oscillations. But I hope that some place in the world will see fit to build a really high flux, research reactor with a liquid deterium moderator at the site of the flux trap.

WEINBERG: I wanted to comment on Abashian's remark about the neutrino mass. I agree with it, but with a qualification. I think the reason that we believe in baryon non-conservation is basically, as shown by an old argument of Lee and Yang, that baryon conservation is not a gauge symmetry, and all the other non-gauge symmetries in physics like isotopic spin and strangeness and so on, have now been explained dynamically within the context of deeper principles, particuarly within quantum chromodynamics. Baryon non-conservation is almost the only remaining non-gauge symmetry in physics and therefore there is no reason for it and we don't believe it really to be true at a fundamental level. We understand in terms of existing theories why the protons live so long. But there is no particular reason to believe that it lives forever.

And exactly the same remarks apply to lepton non-conservation. In fact, even the Lee and Yang argument for baryon conservation also applies to lepton conservation. So I think it is entirely plausible that lepton number is not conserved, and therefore that the neutrino gets a Majorana mass. You can make an order of magnitude estimate of the neutrino mass and say, well, if the proton lifetime is in an observable range due to a superheavy gauge boson of 10^{15} GeV, what neutrino mass does that give? It gives a neutrino mass of about 10^{-4} volts, which is large enough to be important for the solar neutrino problem but probably unobservable in any terresterial experiment. So I think proton decay is a more sensitive test of extremely high mass scales than the neutrino mass, although I am well aware there are many theories in which you can get a larger neutrino mass. But I tend to think if the experiments are borne out, that's fine; but if they are not borne out, then I would tend to think the neutrino mass is nonzero but unobservable by terrestrial means.

People also ask about quantum gravity and I would like to offer a guess there. We are probably much further from understanding the physics of the energy scale at the Planck mass than we now think we are. We think that general relativity somehow or other has to be patched up and that it will then work at 10^{19} GeV. I don't think that is true at all. I don't think that's true because we can understand general relativity so well as a low energy approximation, an effective Lagrangian in the sense that Yao was using it the other day, which describes the long wave length limit of almost any conceivable theory that you have at very short distances. I mean, if you take almost any theory and assume that the real scale is 10^{19} GeV, and ask what kind of physics you can observe at ordinary energies, a few hundred GeV, the answer is the only things you will be able to see will be conventional renormalizable interactions like the gauge theories, plus a few unconventional things that you can observe because they violate symmetries like baryon or lepton conservation, plus long-range forces which can be observed because microscopic bodies can produce them in a coherent buildup over large scales. And the only kind of force of that type that can exist is gravity. So essentially we are seeing all that we can hope to see, and we are seeing what we would expect to see, whatever the physics is at really small scales. So I suspect that we really don't know anything about the physics of 10^{19} GeV.

VELTMAN: I would like to make a short comment on the question of all these symmetries going down the drain. I wish to point out that the inverse process has happened as well. We now have $U(1) \times SU(2) \times SU(3)$ symmetries that are supposed to be very good symmetries and to say at higher energies you have less symmetry is definitely not true. I want to emphasize that when you go to higher and higher energies and you go to more and more substructures, you can have increasing symmetry but not decreasing symmetry. So the idea that you have a random series somewhere, which is an idea brought forward by Nielsen, in particular, is I think not tenable. We have recognized some symmetries; they are going to stay and if anything we are going to find some more symmetries.

PERL: I just want to say, as one of the participants in this conference, I really want to thank those who organized it, and those who chaired the sessions, for what I have found to be a tremendously stimulating and exciting meeting.

PARTICIPANTS

Alexander Abashian
Virginia Polytechnic Institute

Stephen Adler
Institute for Advanced Study

Herbert L. Anderson
Los Alamos Scientific Laboratory

Richard A. Arndt
Virginia Polytechnic Institute

Julius Ashkin
Carnegie-Mellon University

Frank T. Avignone, III
University of South Carolina

Vernon Barger
University of Wisconsin

Mirza A. B. Beg
Rockefeller University

David Berley
National Science Foundation

Lawrence C. Biedenharn, Jr.
Duke University

Marvin Blecher
Virginia Polytechnic Institute

J. David Bowman
Los Alamos Scientific Laboratory

Douglas A. Bryman
University of Victoria

William M. Bugg
Oak Ridge National Laboratory

Nicola Cabibbo
New York University (University of Rome)

Frank Calaprice
Princeton University

Roger Carlini
Los Alamos Scientific Laboratory

Lay Nam Chang
Virginia Polytechnic Institute

Ngee-Pong Chang
City College of New York

Ta-Pei Cheng
University of Missouri

Norman Christ
Columbia University

Kuang-chao Chou
Virginia Polytechnic Institute
 (Institute for Theoretical
 Physics, Peking)

Hans Cohn
Oak Ridge National Laboratory

George B. Collins
Virginia Polytechnic Institute

Tom Curtright
University of Florida

Yuan-ben Dai
Institute of Theoretical Physics
 (Beijing)

Ashok Das
University of Maryland

Raymond Davis
Brookhaven National Laboratory

Brian Dennison
Virginia Polytechnic Institute

Tom Dombeck
Los Alamos Scientific Laboratory

John R. Ficenec
Virginia Polytechnic Institute

Paul M. Fishbane
University of Virginia

H. Fritzsch
University of Munich

Mo-Lin Ge
State University of New York
  at Stony Brook

Terry Goldman
Los Alamos Scientific Laboratory

Eugene Golowich
University of Massachusetts

Matthew Goodman
Harvard University

Kazuo Gotow
Virginia Polytechnic Institute

Oscar W. Greenberg
University of Maryland

Piet C. Gugelot
University of Virginia

Feza Gursey
Yale University

Moo-Young Han
Duke University

Martin Herzlinger
Cornell University

Bo-yu Hou
Yale University

Kerson Huang
Massachusetts Institute of
  Technology

Tau Huang
Stanford University

Vernon Hughes
Yale University

David Jackson
University of California

David Jenkins
Virginia Polytechnic Institute

Prabahan K. Kabir
University of Virginia

T. Kamae
Lawrence Berkeley Laboratory

Gordon L. Kane
University of Michigan

David Kaplan
Virginia Polytechnic Institute

C. S. Lam
McGill University

Kenneth Lande
University of Pennsylvania

Paul Langacker
University of Pennsylvania

Tsung Dao Lee
Columbia University

Ling-Fong Li
Carnegie-Mellon University

Xiaoyuan Li
Virginia Polytechnic Institute

Dale D. Long
Virginia Polytechnic Institute

W. Anthony Loomis
Harvard University

Chin-Yuan Ma
Columbia University

Freydoon Mansouri
Yale University

William Marciano
Northwestern University

Robert E. Marshak
Virginia Polytechnic Institute

Sidney Meshkov
National Bureau of Standards

Philip D. Miller
Oak Ridge National Laboratory

Richard Mischke
Los Alamos Scientific Laboratory

Luke W. Mo
Virginia Polytechnic Institute

Rabindra N. Mohapatra
Max Planck Institute for Physics
  & Astronomy (Munich)

Darragh E. Nagle
Los Alamos Scientific Laboratory

Yoichiro Nambu
University of Chicago

Yee Jack Ng
University of North Carolina

Hwa-Tung Nieh
State University of New York
  at Stony Brook

Susumu Okubo
University of Rochester

Jogesh C. Pati
University of Maryland

Harry Payne
Virginia Polytechnic Institute

Roberto Peccei
Max Planck Institute for Physics
  & Astronomy (Munich)

Martin Perl
Stanford University

Vinod Prasad
Virginia Polytechnic Institute

Pierre Ramond
University of Florida

Sergio Ratti
University di Pavia

Michael Reiner
Virginia Polytechnic Institute

Riazuddin
Virginia Polytechnic Institute
  (Quaid-i-Azam University,
   Islamabad)

Peter Rosen
Purdue University

Minic Rushton
University of California

William Sauder
Virginia Military Institute

Goran Senjanovic
Brookhaven National Laboratory

Robert Shrock
State University of New York
  at Stony Brook

Joseph Slawny
Virginia Polytechnic Institute

Alberto Sirlin
New York University

Jack H. Smith
State University of New York
  at Stony Brook

Pierre Sokolsky
Columbia University

Mark Srednicki
Princeton University

Mark Strovink
University of California

E. C. G. Sudarshan
University of Texas

M. K. Sundaresan
Carleton University (Ottawa)

Vigdor L. Teplitz
U. S. Arms Control and
   Disarmament Agency

Edward Thorndike
University of Rochester

Peter Trower
Virginia Polytechnic Institute

T. Laurence Trueman
Brookhaven National Laboratory

Y. S. Tsai
Stanford University

Hung-Sheng Tsao
Rockefeller University

San Fu Tuan
University of Hawaii

Wu-Ki Tung
Illinois Institute of Technology

Michael Turner
University of Chicago

Henry Tye
Cornell University

Hendrik Van Dam
University of North Carolina

Martinus Veltman
University of Michigan
   (University of Utrecht)

Kamishwar Wali
Syracuse University

Ling-Lie Wang
Brookhaven National Laboratory

Steven Weinberg
University of Texas

P. K. Williams
Department of Energy

Lincoln Wolfenstein
Carnegie-Mellon University

Chien-Shiung Wu
Columbia University

Dan-di Wu
Harvard University

Ding-chang Xian
State University of New York
   at Stony Brook

M. Yan
State University of New York
   at Stony Brook

Chen-Ning Yang
State University of New York
   at Stony Brook

York-Peng Yao
University of Michigan

Masaki Yasue
University of Tokyo

Glenn R. Young
Oak Ridge National Laboratory

Anthony Zee
University of Washington

Bao-heng Zhao
State University of New York
   at Stony Brook

Xue Zhou
City College of New York

Runsheng Zhu
Columbia University

Royce Zia
Virginia Polytechnic Institute

Paul F. Zweifel
Virginia Polytechnic Institute

AIP Conference Proceedings

		L.C. Number	ISBN
No.1	Feedback and Dynamic Control of Plasmas	70-141596	0-88318-100-2
No.2	Particles and Fields - 1971 (Rochester)	71-184662	0-88318-101-0
No.3	Thermal Expansion - 1971 (Corning)	72-76970	0-88318-102-9
No.4	Superconductivity in d-and f-Band Metals (Rochester, 1971)	74-18879	0-88318-103-7
No.5	Magnetism and Magnetic Materials - 1971 (2 parts) (Chicago)	59-2468	0-88318-104-5
No.6	Particle Physics (Irvine, 1971)	72-81239	0-88318-105-3
No.7	Exploring the History of Nuclear Physics	72-81883	0-88318-106-1
No.8	Experimental Meson Spectroscopy - 1972	72-88226	0-88318-107-X
No.9	Cyclotrons - 1972 (Vancouver)	72-92798	0-88318-108-8
No.10	Magnetism and Magnetic Materials - 1972	72-623469	0-88318-109-6
No.11	Transport Phenomena - 1973 (Brown University Conference)	73-80682	0-88318-110-X
No.12	Experiments on High Energy Particle Collisions - 1973 (Vanderbilt Conference)	73-81705	0-88318-111-8
No.13	π-π Scattering - 1973 (Tallahassee Conference)	73-81704	0-88318-112-6
No.14	Particles and Fields - 1973 (APS/DPF Berkeley)	73-91923	0-88318-113-4
No.15	High Energy Collisions - 1973 (Stony Brook)	73-92324	0-88318-114-2
No.16	Causality and Physical Theories (Wayne State University, 1973)	73-93420	0-88318-115-0
No.17	Thermal Expansion - 1973 (lake of the Ozarks)	73-94415	0-88318-116-9
No.18	Magnetism and Magnetic Materials - 1973 (2 parts) (Boston)	59-2468	0-88318-117-7
No.19	Physics and the Energy Problem - 1974 (APS Chicago)	73-94416	0-88318-118-5
No.20	Tetrahedrally Bonded Amorphous Semiconductors (Yorktown Heights, 1974)	74-80145	0-88318-119-3
No.21	Experimental Meson Spectroscopy - 1974 (Boston)	74-82628	0-88318-120-7
No.22	Neutrinos - 1974 (Philadelphia)	74-82413	0-88318-121-5
No.23	Particles and Fields - 1974 (APS/DPF Williamsburg)	74-27575	0-88318-122-3
No.24	Magnetism and Magnetic Materials - 1974 (20th Annual Conference, San Francisco)	75-2647	0-88318-123-1
No.25	Efficient Use of Energy (The APS Studies on the Technical Aspects of the More Efficient Use of Energy)	75-18227	0-88318-124-X

No.26	High-Energy Physics and Nuclear Structure - 1975 (Santa Fe and Los Alamos)	75-26411	0-88318-125-8
No.27	Topics in Statistical Mechanics and Biophysics: A Memorial to Julius L. Jackson (Wayne State University, 1975)	75-36309	0-88318-126-6
No.28	Physics and Our World: A Symposium in Honor of Victor F. Weisskopf (M.I.T., 1974)	76-7207	0-88318-127-4
No.29	Magnetism and Magnetic Materials - 1975 (21st Annual Conference, Philadelphia)	76-10931	0-88318-128-2
No.30	Particle Searches and Discoveries - 1976 (Vanderbilt Conference)	76-19949	0-88318-129-0
No.31	Structure and Excitations of Amorphous Solids (Williamsburg, VA., 1976)	76-22279	0-88318-130-4
No.32	Materials Technology - 1975 (APS New York Meeting)	76-27967	0-88318-131-2
No.33	Meson-Nuclear Physics - 1976 (Carnegie-Mellon Conference)	76-26811	0-88318-132-0
No.34	Magnetism and Magnetic Materials - 1976 (Joint MMM-Intermag Conference, Pittsburgh)	76-47106	0-88318-133-9
No.35	High Energy Physics with Polarized Beams and Targets (Argonne, 1976)	76-50181	0-88318-134-7
No.36	Momentum Wave Functions - 1976 (Indiana University)	77-82145	0-88318-135-5
No.37	Weak Interaction Physics - 1977 (Indiana University)	77-83344	0-88318-136-3
No.38	Workshop on New Directions in Mossbauer Spectroscopy (Argonne, 1977)	77-90635	0-88318-137-1
No.39	Physics Careers, Employment and Education (Penn State, 1977)	77-94053	0-88318-138-X
No.40	Electrical Transport and Optical Properties of Inhomogeneous Media (Ohio State University, 1977)	78-54319	0-88318-139-8
No.41	Nucleon-Nucleon Interactions - 1977 (Vancouver)	78-54249	0-88318-140-1
No.42	Higher Energy Polarized Proton Beams (Ann Arbor, 1977)	78-55682	0-88318-141-X
No.43	Particles and Fields - 1977 (APS/DPF, Argonne)	78-55683	0-88318-142-8
No.44	Future Trends in Superconductive Electronics (Charlottesville, 1978)	77-9240	0-88318-143-6
No.45	New Results in High Energy Physics - 1978 (Vanderbilt Conference)	78-67196	0-88318-144-4
No.46	Topics in Nonlinear Dynamics (La Jolla Institute)	78-057870	0-88318-145-2
No.47	Clustering Aspects of Nuclear Structure and Nuclear Reactions (Winnepeg, 1978)	78-64942	0-88318-146-0
No.48	Current Trends in the Theory of Fields (Tallahassee, 1978)	78-72948	0-88318-147-9
No.49	Cosmic Rays and Particle Physics - 1978 (Bartol Conference)	79-50489	0-88318-148-7

No. 50	Laser-Solid Interactions and Laser Processing - 1978 (Boston)	79-51564	0-88318-149-5
No. 51	High Energy Physics with Polarized Beams and Polarized Targets (Argonne, 1978)	79-64565	0-88318-150-9
No. 52	Long-Distance Neutrino Detection - 1978 (C.L. Cowan Memorial Symposium)	79-52078	0-88318-151-7
No. 53	Modulated Structures - 1979 (Kailua Kona, Hawaii)	79-53846	0-88318-152-5
No. 54	Meson-Nuclear Physics - 1979 (Houston)	79-53978	0-88318-153-3
No. 55	Quantum Chromodynamics (La Jolla, 1978)	79-54969	0-88318-154-1
No. 56	Particle Acceleration Mechanisms in Astrophysics (La Jolla, 1979)	79-55844	0-88318-155-X
No. 57	Nonlinear Dynamics and the Beam-Beam Interaction (Brookhaven, 1979)	79-57341	0-88318-156-8
No. 58	Inhomogeneous Superconductors - 1979 (Berkeley Springs, W.V.)	79-57620	0-88318-157-6
No. 59	Particles and Fields - 1979 (APS/DPF Montreal)	80-66631	0-88318-158-4
No. 60	History of the ZGS (Argonne, 1979)	80-67694	0-88318-159-2
No. 61	Aspects of the Kinetics and Dynamics of Surface Reactions (La Jolla Institute, 1979)	80-68004	0-88318-160-6
No. 62	High Energy e^+e^- Interactions (Vanderbilt, 1980)	80-53377	0-88318-161-4
No. 63	Supernovae Spectra (La Jolla, 1980)	80-70019	0-88318-162-2
No. 64	Laboratory EXAFS Facilities - 1980 (Univ. of Washington)	80-70579	0-88318-163-0
No. 65	Optics in Four Dimensions - 1980 (ICO, Ensenada)	80-70771	0-88318-164-9
No. 66	Physics in the Automotive Industry - 1980 (APS/AAPT Topical Conference)	80-70987	0-88318-165-7
No. 67	Experimental Meson Spectroscopy - 1980 (Sixth International Conference, Brookhaven)	80-71123	0-88318-166-5
No. 68	High Energy Physics - 1980 (XX International Conference, Madison)	81-65032	0-88318-167-3
No. 69	Polarization Phenomena in Nuclear Physics - 1980 (Fifth International Symposium, Santa Fe)	81-65107	0-88318-168-1
No. 70	Chemistry and Physics of Coal Utilization - 1980 (APS, Morgantown)	81-65106	0-88318-169-X
No. 71	Group Theory and its Applications in Physics - 1980 (Latin American School of Physics, Mexico City)	81-66132	0-88318-170-3
No. 72	Weak Interactions as a Probe of Unification (Virginia Polytechnic Institute - 1980)	81-67184	0-88318-171-1
No. 73	Tetrahedrally Bonded Amorphous Semiconductors (Carefree, Arizona, 1981)		0-88318-172-X